Facade Construction Manual

HERZOG
KRIPPNER
LANG

BIRKHÄUSER – PUBLISHERS FOR ARCHITECTURE
BASEL · BOSTON · BERLIN

EDITION DETAIL
MUNICH

This book was produced at the Institute for Design and Building Technology, Faculty of Architecture, Chair of Building Technology, Munich Technical University
www.gt.ar.tum.de

Author

Thomas Herzog
Prof., Dr. (Rome University), Dipl.-Ing. Architect
Chair of Building Technology, Munich Technical University

Co-authors:

Roland Krippner
Dr.-Ing. Architect
(Dimensional coordination; Concrete; Solar energy)

Werner Lang
Dr.-Ing., M.Arch. (UCLA) Architect
(Glass; Plastics; The glass double facade)

Scientific assistants:
Peter Bonfig, Dipl.-Ing. Architect (The structural principles of surfaces)
Jan Cremers, Dipl.-Ing. Architect (Internal and external conditions; Metal)
András Reith, M.Sc.Arch. (Budapest University), guest scientist (Stone; Clay)
Annegret Rieger, M.Arch. (Harvard University) Architect (organisational coordination; Timber)
Daniel Westenberger, Dipl.-Ing. Architect (Edges, openings; Manipulators)

Student assistants:
Tina Baierl, Sebastian Fiedler, Elisabeth Walch, Xaver Wankerl

Editors

Editorial services:
Steffi Lenzen, Dipl.-Ing. Architect

Editorial assistants:
Christine Fritzenwallner, Dipl.-Ing.

Susanne Bender-Grotzeck, Dipl.-Ing.; Carola Jacob-Ritz, M.A.; Christina Reinhard, Dipl.-Ing.; Friedemann Zeitler, Dipl.-Ing.; Christos Chantzaras, Manuel Zoller

Drawings:
Marion Griese, Dipl.-Ing.
Elisabeth Krammer, Dipl.-Ing.

Drawing assistants:
Bettina Brecht, Dipl.-Ing.; Norbert Graeser, Dipl.-Ing.; Christiane Haslberger, Dipl.-Ing.; Oliver Klein, Dipl.-Ing.; Emese Köszegi, Dipl.-Ing.; Andrea Saiko, Dipl.-Ing.; Beate Stingl, Dipl.-Ing.; Claudia Toepsch, Dipl.-Ing.

Translation into English:
Gerd H. Söffker and Philip Thrift, Hannover

DTP & production:
Peter Gensmantel, Cornelia Kohn, Andrea Linke, Roswitha Siegler

Reproduction:
Karl Dörfel Reproduktions-GmbH, Munich

Printing and binding:
Kösel GmbH & Co. KG, Altusried-Krugzell

Specialist articles:

Winfried Heusler, Dr.-Ing. (Planning advice for the performance of the facade)
Director, Objekt-Engineering International, Bielefeld

Michael Volz, Prof. Dipl.-Ing. Architect (Timber)
Frankfurt am Main Polytechnic

Consultants:
Gerhard Hausladen, O. Prof. Dr.-Ing. (Edges, openings)
Institute for Design and Building Technology, Chair of Building Climate and Building Services, Munich Technical University

Stefan Heeß, Dipl.-Ing. (Concrete)
Dyckerhoff Weiss, Wiesbaden

Reiner Letsch, Dr.-Ing. M.Sc. (Plastics)
Chair of Building Materials and Materials Testing, Munich Technical University

Volker Wittwer, Dr. (Solar energy)
Fraunhofer Institute for Solar Energy Systems (ISE), Freiburg

A CIP catalogue record for this book is available from the Library of Congress, Washington, D.C., USA

Bibliographic information published by Die Deutsche Bibliothek
Die Deutsche Bibliothek lists this publication in the Deutsche Nationalbibliographie; detailed bibliographic data is available on the Internet at http://dnb.ddb.de.

This book is also available in a German language edition (ISBN 3-7643-7031-9).

Editor:
Institut für Internationale Architektur-Dokumentation GmbH & Co. KG, Munich
http://www.detail.de

Birkhäuser – Publisher for Architecture, P.O. Box 133, CH-4010 Basel, Switzerland
Part of Springer Science+Business Media

Printed on acid-free paper produced from chlorine-free pulp. TCF ∞

Printed in Germany

ISBN 3-7643-7109-9

9 8 7 6 5 4 3 2 1

http://www.birkhauser.ch

Contents

Preface

This construction manual dealing with facades comes some 30 years after the publication of the first construction manual.

For centuries architects focussed their artistic skills on the creation of a successful external appearance for their structures. This was often the subject of intense controversy surrounding the style to be chosen, but was also exploited as a medium for conveying new artistic stances.

The fact that facades are again becoming a subject for discussion is no doubt due to the growing importance of the building envelope with respect to energy consumption and options for exploiting natural forms of energy. In addition – and usually contrasting with this – there is also the search for the self-expression and "location marketing" of those clients whose "packaging" of their often trivial interiors has long since become a substitute for good, quality architecture. The booming cities of Asia demonstrate this only too clearly.

The layout of this book and the sequence of the various chapters are based on a discerning approach to designing and developing a facade construction. Those aspects generally relevant to the building envelope, i.e., the requirements they have to satisfy, their principal functions or their structural form, are treated separately from the special circumstances of individual cases. Accordingly, this is not just a compendium of different buildings in terms of location and context, genre and technology. Instead, features are sorted according to the different materials used for the envelope or its cladding.

The first part of this book deals with the requirements placed on a facade as a result of internal needs, depending on the use of the building. Inevitably, these have to face up to very diverse local climatic conditions, which vary from region to region. This juxtaposition gives rise to the functional requirements placed on the respective envelopes. This collection of requirements then becomes the architect's brief and is initially open to a number of solutions. Construction details are for this reason not legitimate in this part of the book. The criteria are presented in pictorial form by way of diagrams and schematic graphics showing the morphology of surfaces and openings. Furthermore, there is a direct interaction between the building envelope and the other subsystems: loadbearing structure, interior layout and building services. The interrelationships present or to be defined must be geometrically coordinated in three dimensions in every construction system. The dimensional and modular conditions and the proportions must be clarified so that the building can evolve as a unified, overall concept. Once we combine the aforementioned aspects, we create the boundary conditions for a practical realisation based on the materials and methods of construction available to us.

If the materials and the technologies required for their production now become the criteria for the elaboration of further details, then the parameters in terms of physics, materials, erection requirements and aesthetics have to be harmonised with each other.

The layout of the second part of the book is based on these interrelationships. Again, there are separate chapters covering general aspects preceding the examples. Each of these chapters begins with a brief history of humanity's use of the respective materials, and their properties. The use of the materials is initially not limited to buildings for the simple reason that as civilisation evolved, so technologies developed in different ways through the interplay with the materials, and the first applications were often in totally different fields. The significance of stone, ceramics and metal, for instance, was so great that they gave their names to entire cultural epochs. Even today, a large share of technical innovation in building, and particularly in modern facade design, is due to technology transfer from completely unrelated sectors. This is true for many areas, such as forming technology, surface treatments, robots. A selection of built examples related to the respective material follows each introductory chapter. The idea behind this is to provide an insight into the scope of possibilities and to serve as inspiration for further ideas. We have employed drawings of the principal facade details ex-plained by way of legends because this is the medium normally used by architects to convey information.

We have selected new projects that exhibit interesting facade configurations and also "classics" whose architectural qualities still set standards. Furthermore – also in conjunction with projects within existing, older buildings – some of the details may well be of great practical value to architects and engineers.

Each example concentrates on the facade and not the building as a whole. This is why, apart from the architects, other collaborators on the projects are mentioned only rarely, and specialist engineers only when they made a particular contribution to the facade construction.

When looking at the construction details the reader will sometimes notice approaches differing from those normally employed in Germany, or from the German codes and standards. The authors consider this to be justified in a book containing international examples.

For those readers wishing to find out more about a particular project, we have included references to other publications, as indicated by the symbol 🕮.

It is certainly possible to regard buildings – even when they are major technical accomplishments – not as perplexing, possibly even hardly manageable systems comprising many different components, but rather, tersely and simply, as equal measures of powerful and sensitive design. However, the developments of recent decades, with the enormous increase in the requirements placed on the building envelope, have resulted in multiple layer constructions in which every single layer has to perform specific functions. In the meantime, this has become a common feature of modern building for most materials. This led us to include special topics of facade design beyond those related just to the materials.

The chapter entitled "Manipulators" takes a new look at the centuries-old principle of varying and individually regulating the "permeability" of openings in the facade – whether for reasons of energy economy, the interior climate, lighting or security – based on new requirements and new ideas. In our opinion the widespread use of multiple layers and double-leaf facades over the last ten years warrants separate treatment because there is still great uncertainty about their use among architects and engineers, who often feel the need to follow the fashion instead of allowing the advantages of these concepts to be fully and properly utilised. This frequently results in fundamental errors, because the relationships in terms of construction and energy economy as well as the individual variations available for this type of design are not fully appreciated.

The integration of direct and indirect solar systems into the building envelope is still new territory for many in the construction industry. The happy co-existence of serviceability, technical and physical functions, and mastery of architectural and engineering requirements is still the exception, even though the first pioneering applications were built many decades ago.

We would like to thank all those persons, institutions, architects, photographers and companies whose competent assistance has helped us to produce this book.

Munich, spring 2004
Thomas Herzog

Envelope, wall, facade – an essay

This book about the construction of facades concentrates mainly on the functional and technical aspects. Nevertheless, before we start, it is worth taking a moment to look beyond those aspects and consider this very complex, culture-related subject in other contexts, at least briefly. For this subject also concerns our direct perception of architecture.

The protective envelope

The building envelope, as it provides protection against the weather and against enemies, and for storage provisions, represents the primary and most important reason for building. In contrast to structures such as bridges, towers, dams or cranes, buildings contain rooms whose creation and utilisation must be regarded as intrinsic elements of human civilisation, closely linked with the necessities forced upon us by climate.

This is clear when we consider that building work is reduced to a minimum in those regions where the external climatic conditions essentially match the ambient conditions we humans find comfortable. However, the greater the difference between exterior climate and interior conditions, the greater is the technical undertaking required to create the conditions necessary for occupying the interior.

For much of history, therefore, people have searched for suitable, ready-made natural shelters for themselves and their animals. These included, for example, holes in the ground, caves in the rock or very dense vegetation. In other words, sheltered places where the conditions were conducive to survival (fig. 2).

As humans changed from a nomadic to a sedentary lifestyle, people built their own shelters using the materials at hand in conjunction with appropriate building processes. Roofs and external walls were born. The outside surfaces of these human-made shelters became important because they had to fulfil numerous functions, the most important of which was protection from the weather (fig. 3).

The mass of stone or earth surrounding natural holes and caves was reduced to a relatively thin skin in these human-made structures. The building now had an inside and an outside.

The components of the term "external wall" designate both the location, i.e., "external", and the character of this construction "subsystem", i.e., the "wall". However, throughout the history of building – at any rate into the 20th century – the vast majority of walls constituted not only an enclosure but also a significant part of the loadbearing structure. (They transfer the imposed loads, their self-weight, the dead loads of the floors they support and the wind loads – via the stiffening effect of their massive construction – down to the foundations.) We therefore associate the term "wall", especially the external wall, with stability, robustness, mostly heavyweight construction, indeed even unfriendliness, the demarcation between private and public. In this way, the external wall is a primary factor determining the nature of the building with respect to its surroundings.

The outer surface now provided an additional communication medium, complementing the inner surfaces, which had long been used as the main medium of communication (e.g., cave paintings). From now on the outer surface would serve as a "hoarding" for secular and sacred social structures, and for conveying hierarchies of values and claims to power.

2

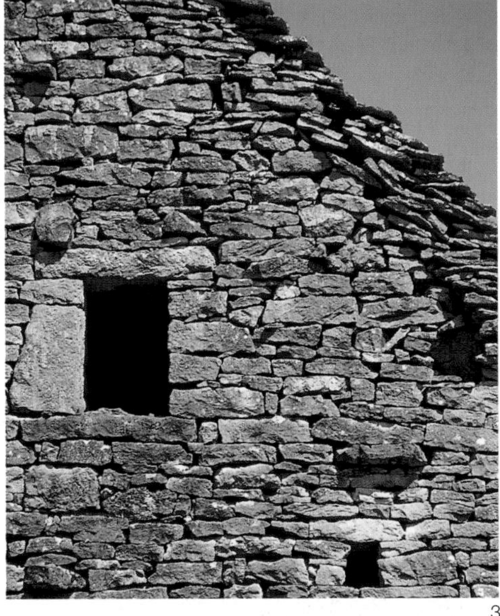

3

2 Cave dwelling
3 External wall made from local natural stone, Auvergne (F)

1 Moldavian monastery, Sucevita (RO), 16th century

4

5

4 Farmhouse Museum, Amerang (D)
5 Majolica frieze on the “Ospedale del Ceppo”, Pistoia (I)
6 Alhambra, Granada (E)
7 Cathedral, Lucca (I), (12th–15th) century
8 Casa Batlló, Barcelona (E), 1906, Antoni Gaudí

Materials and construction

The space created within the external walls now had to fulfil demands and functions concerning usage and comfort. In order to accomplish this, the local conditions and the users' requirements had to be established in more detail, manipulated and then fulfilled by means of a suitable construction.

The technical solution grows out of the context of materials, construction, jointing, sequence of production, and also the demands due to gravity plus other internal and external physical influences and circumstances. The outsides of buildings therefore reflect the technological progress of a region and hence a substantial part of the respective local culture.

The decision in favour of a certain material, for example, is based not purely on the loads and stresses resulting from internal or external actions, but is made rather with regard for the rules re-lated to the production process of the respective building envelope. Here, it is not just the individual utilisation requirements that determine the form of the facade. Instead, these requirements must always be considered in conjunction with the questions of jointing, the method of construction, and hence the technical realisation within an overall system of construction, the legitimacy of the materials and the geometrical order (fig. 4). Most importantly in this field, we have to see the professional competence of the architects in their role as “master builder”. They alone are aware of all the relationships and the multiple correlations within and between the composition of the architecture and the logic of the construction.

Form

External walls are also customarily referred to as “facades”. Now, in contrast to the aforementioned fundamental functions of protection from weather and control of the interior climate, another aspect takes centre stage – that of the perception of the building by way of its “face”, derived in a roundabout way from the Latin *facies* via the French *façade*. What this means is something constructed, something that “looks onto” its surroundings, or rather is perceived from its surroundings as the prime and governing semantic message [1].

Surfaces formed by humans have always served to convey information. This information has portrayed the things that governed social life, determined transcendental and religious visions, conveyed objectives and reports: praise to God, hunting or rituals, battles, weddings, wealth and death – all long before writing was available, as an abstract form of communication (fig. 5).

The characteristics of external surfaces in terms of their pictorial effect should be regarded in a similar light to the aforementioned internal surfaces with respect to graphic features, textures, colouring, engraving and relief, combinations of information conveyed by means of materials, writing and pictures. Over the course of history the entire spectrum has been rendered visible – "the thrill of creation and the hideousness of death" [2].

Only humans create buildings with a differentiated "personal" form: Three-dimensional objects that can be perceived from the outside as a whole or as different components and which, in comparison to true wall surfaces, exhibit other features. For example, this is done by way of proportions in space, volumes, and both in relation to the existing environment.

As walls became more differentiated thanks to the increasing refinement of the construction, so a similar effect was noticeable around the openings. Here too, function prevailed at first, with the engineering of spanning over the aperture in the wall with a lintel or an arch using the same or a different material. Requirements concerning maximum admittance of light through a minimal aperture by means of angled reveals (tapering inwards and outwards), refraction of light, privacy and controlled ventilation by means of elements placed across or in the openings became key components in the overall architectural effect due to their form and artistic complexity (fig. 6).

Like the walls, the fittings to the openings, with fixed or movable parts, frequently made use of local materials. This resulted in true gems, whose sides and surfaces were painstakingly and elaborately decorated.
The composition of multi-layered main facades led to a grand interplay between wall and opening, as in the achievements at the cathedrals in Lucca and Ferrara, where this resulted in three-dimensional facades with all details taking on a sculpted form (fig. 7).

In the course of these developments the facade gave rise to additional effects due to superimposition or penetration, to the changing exposure of objects. This led to diverse or alternating brightnesses, light-and-shade effects over the entire volume and parts thereof. The realm of stereometric order was abandoned in favour of unfettered sculptural developments. There ensued an alternation of surfaces in single and double curvature, in juxtaposition with flat areas that are horizontal, vertical, sloping, folded, or broken up in some other way (fig. 8).

6

7

8

9

10

11

The socio-cultural environment

Local circumstances, the type of society found in a certain region, the history and ethnography, their *Weltanschauung*, the local climate (which can differ considerably over short distances), or the availability of local resources have all played a key role in the design of the building envelope.

Such relationships influence the essence of regional or local cultures, which characterises societies, gives them stability and orientation, and forms the foundation of civil conventions. Coexistence demands cultural covenants. And the appearance of buildings presents these as permanent time capsules [3].

Against such a background, the external elevations of buildings take on a special significance which goes well beyond the effect of the individual building. We have only to think of the effect of street frontages, of plazas and squares, of whole districts. In these cases the cumulative effect of the external walls regulates the public spaces.

The characteristics of the facades in terms of the effects of the materials, colours, proportions, volumes and pictorial information indicate their functions or the importance attached to them.

However, there is also the danger of arbitrary additions or adulterations lending buildings new semantic meanings, which could lead to them being estranged from their real nature and therefore losing their "dignity". And this could be due to an exaggerated tolerance with respect to an overwhelming self-portrayal, or the result of the wrong objective.

This does not rule out purely fashionable trimmings within temporary settings. For example, we could cite forms of art in which a chronological progression or a chronological confinement is a feature, like a play, opera, ballet or film. But if such aspects are allowed to dominate architecture, this easily leads to a destabilisation of the aesthetic identity, even a loss of orientation with respect to this cultural witness. Nevertheless, the visual effect may not be evaluated within a limited set of rules.

For that would mean that art in essence only prevails when it is frozen, is no longer evolving. It is therefore a feature of cultural processes that we deal creatively with the built environment that we inherit (fig. 12).

The awareness surrounding the significance of the external elevation of a building in terms of its effect within the public space should instead be seen as a vital aspect founded on communication within a community. He who erects a structure informs the outside world of his intentions and hence announces his own identity, but also determines the extent of the desired assignment or classification in an existing spatial and building context. Generally, the architect is also involved in the further development of this "announcement" [4].

Numerous examples (e.g., fig. 10) illustrate to what extent people during the Renaissance, as humanism flourished and hence the importance of the spiritually independent individual grew, emphasised the effect of the external walls as an "exhibit". This effect became more pronounced during the Baroque era. As a rule, these facades facing a street or open public space, in contrast to the other elevations, were decorated at great expense with precious materials and expressive artistic means, and became – almost detached as a whole from the building itself – exacting, giant backdrops.

Much more than being just a technical or utilitarian aspect, the facade played a key role as a vehicle for architectural impact. The external wall made use of relief, sculpture, painting, mosaics and writing to become a visual medium where all primary functional parts became objects for maximum decorative moulding (fig. 9).

The multimedia facades of the present day, possible worldwide through the integration of new forms of design and communications technology, and which demonstrate new kinds of graphic and colour effects in transparent and translucent glass and membrane surfaces, continue this tradition of the building envelope as a "hoarding". Just how much this change can lead to contrasts, indeed to denaturalisation, is shown in the example from London (fig. 11), where two originally similar blocks face each other across the street. As soon as the brightness of the competing daylight diminishes sufficiently and artificial light can dominate, electronically controlled LEDs and videos have long since taken over as the prevailing aesthetic factor on the outside of buildings – conveying information and architectural import (fig. 13).

When the historical predecessors employed materials and their graphical or sculptural form to determine wholly the effect of the facade, they intensified the perception with respect to the building itself. Its own, original components were the cause of this. It is different when the non-physical semantic message is conveyed via a non-self-determining neutral medium such as a computer program and projection techniques. The variable software permits complete independence with respect to the contents displayed and, to a large extent, also with respect to the form of its presentation.

Facades with extremely intensive effects relying on constant change are the main reason behind the appeal of such urban spaces. This type of constantly fluctuating facade, relying on the integration of ever newer technologies, can be seen in Times Square in New York – one of countless examples. This results in a completely new intensive cultural reference which employs other media for its effects. Here, the aesthetic significance of the building facade itself retreats into the background (fig. 13).

12

9 Painted street facade, Trento (I)
10 San Giorgio Maggiore, Venice (I), 1610, Andrea Palladio
11 Piccadilly Circus, London (GB)
12 Old – new, detail of transition
13 Times Square, New York (USA)

The facade and building services

In the history of European building, building services have been integrated into the external walls as functionally important elements in many ways. Examples are Wells in southern England, where the masonry external walls continue upwards to form chimneys, and Europe's first development of terrace houses from a more recent period, which become characteristic elements in the streetscape (fig. 14).

The positioning of radiators or convectors beneath windows on the inside or – in hot climates – decentralised air-conditioning units on the outside of the building is commonplace. The fact that the support brackets for such technical equipment can also be elegantly

13

integrated into prefabricated facades is shown by the example of the semiconductor assembly plant in Wasserburg am Inn (see p. 168).

Large air ducts are also positioned on the facade, primarily to keep interior spaces clear of all obstacles, as is the case with manufacturing and exhibition buildings. This technical motif was turned into a form of expression on a large scale at the Centre Georges Pompidou in Paris (Renzo Piano, Richard Rogers, 1977), becoming the governing architectural element (fig. 15). In a similar way, the air-conditioning systems at the Sainsbury Centre for Visual Arts (see p. 172) are fitted to the periphery of the building; but here some parts are only visible from the outside through the glass, which protects them from the weather.

The use of such elements, essentially originating from the realm of mechanical engineering, as a primary construction subsystem and indeed as a programmatic element in the "exposed sides" of buildings, represents a key change [5]. Their importance in the course of generating an artificial interior climate with the increasing use of energy – and their dependence thereon – must be investigated, particularly from the current viewpoint. However, these (large-scale) technical installations could still be advisable if they can be justified from the viewpoint of saving resources, e.g. by using renewable energy sources more and more.

For reasons of easy accessibility, maintenance and renewability alone, the de-integration of loadbearing structure and protective envelope is certainly expedient.
If we refrain from providing service voids in ceilings and floors in order to be able to activate the thermal storage capacity of the loadbearing components, but at the same time – as is the norm in office buildings – partitions must remain demountable, then the construction of the external wall must include suitable facilities for distributing and providing access to high- and low-voltage networks plus cooling, heating and ventilating systems. More recently, we have seen an increase in the number of small, decentralised ventilation units being fitted to facades. These are designed as counter-flow systems to minimise ventilation heat losses and thus guarantee efficient heat recovery during the heating period.

This has been achieved in the new ZVK offices in Wiesbaden (see p. 282–83) by means of a duct in the spandrel panel, services cabinets, integral involute lighting units, small convectors on all office grid lines and controllable vents with baffle plates behind.

Totally different effects relying on natural, organic systems, like those influencing the microclimate at the facade, can be achieved with the specific, functional use of vegetation (fig. 16). In terms of controlling dust, balancing humidity and providing shade, plants are very useful, occasionally – particularly during hot periods and in more southerly regions – providing a considerable natural cooling effect. This is definitely an opportunity for an impressive combination of functionality and aesthetic intent [6].

Ageing

If we assume that, a certain time after being completed, a structure becomes part of the history of building, the issue of ageing behaviour is intrinsic. This applies especially to the external appearance, the building envelope, which is exposed to the weather.

Over the course of time the envelope is subjected to many strains and stresses. These result not only in technical and functional changes, but ultimately in a change in its appearance.

There are facades that rot, degrade, become "shabby" – in other words age badly due to their form of construction and choice of materials. And there are others that show hardly any signs of age at all, although the same, i.e., technical, criteria apply. For instance, glass installed many centuries ago may have suffered superficial damage, but its substance and aesthetic characteristics will have changed little.

And there are also materials that undergo severe changes within a very short period of time but age acceptably and possibly even become more beautiful thereby. An example of this is patination (fig. 17). Their serviceability is not impaired, nor is their technical appropriateness (because, for example, parts have rotted or sections have been weakened by corrosion).

The architectural and engineering conception and construction of a facade also means guaranteeing that ageing does not reduce its quality, its value. Society's general readiness to accept such aesthetic changes can be seen

14

15

when materials are well-known beyond their natural environment, where a high value may be placed on the preservation of historical fabrics and precious details.

This applies to stone, copper and bronze, for example. But the most characteristic example of this is probably wood. Where wood is a local resource, people grow up with its countless variations. And as we know, wood is a material whose appearance never reaches a final condition – convincingly demonstrated by Peter Zumthor's extension to a timber house in Versam (Grisons, CH, 1994) (fig. 18).

16

17

18

Notes

[1] The fact that this is not always regarded as a positive effect is demonstrated by the saying "It's just a facade", which means that the real qualities of a person or thing do not correspond with his or its outward appearance.

[2] Jochen Wagner, Tutzing Protestant Academy, in a television programme, Feb 2004.

[3] This provides psychological stability for both the individual and society. The built environment forms an important "prospect" for the awareness of belonging, an attachment to our homes and an understanding of our own identity.

[4] In his essay *Zukunft bauen* (Building the Future) Manfred Sack writes: "... every facade, indeed much more: every structure is a public affair – and to hell with the architect, who thought it would be easy. The facade belongs to all of us; only the bit behind belongs to those who have to use it. And therefore it is also clear that the facade should not be just a cosmetic matter. For a city to be perceived as attractive is – which some do not expect – a social, a communal, a political task." Translated after: Sack, Manfred: *Verlockungen der Architektur*, Lucerne, 2003.

[5] The first person to have thoroughly investigated this scientifically is Rudi Baumann, who as part of his dissertation showed the huge potential for regulating the climate in temperate zones through the proper use of vegetation in the form of climbing plants. For further details see *Begrünte Architektur*, Munich, 1983.

[6] Even though "key change" has almost become a buzzword in recent years, in this case we can see that the παραδειγμα originally designated an architectural model specially produced for a competition.

14 Vicar's Close, Wells (GB)
15 Centre Georges Pompidou, Paris (F), 1977, Renzo Piano/Richard Rogers
16 Facade with climbing plants
17 Patinated bronze oriel window, Boston (USA)
18 Weathered wooden facade, Grisons (CH), 1994, Peter Zumthor

Part A Fundamentals

Outside	Facade	Inside
Conditions specific to the location		**Requirements**
solar radiation, temperature, humidity, precipitation, wind — severe fluctuations in external climate		minimal fluctuations internally — comfortable temperature/humidity range, amount and quality of light (lighting environment), air exchange rate/fresh air supply at comfortable air velocity, comfortable sound level
sources of noise in the surroundings amount of gas and dust mechanical loads electromagnetic radiation		visual relationship with external surroundings demarcation between private and public zones protection against mechanical damage fire protection if necessary limitation of toxic loads
urban/formative surroundings local resources socio-cultural context		
	Protective functions by way of permanent and also variable conditions (increasing or reducing the effect)	
	insulation/attenuation seals/barriers filters storage redirection physical barriers	
	Regulatory functions	
	controlling/regulating responding/changing	
Supplementary measures with a direct effect		**Supplementary measures with a direct effect**
thermal insulation sunshading (e.g. shutters, blinds, brise-soleil, lamellas)		antiglare protection privacy provisions (e.g. curtains) redirection of daylight
measures influencing the microclimate (e.g. vegetation, bodies of water)		utilisation of internal component (floors, walls, ceilings) for storing and later release of energy for heating/cooling
Supplementary building services	**Facades with integral services**	**Supplementary building services**
external collectors photovoltaics heat pipes, heat sondes, etc.	integral air/water collectors solar walls media transport/distribution heat recovery	convectors/radiators artificial lighting air conditioning (centralised/decentralised)

A 1.1

A 1 Internal and external conditions

The facade is the separating and filtering layer between outside and inside, between nature and interior spaces occupied by people. In historical terms, the primary reason for creating an effective barrier between interior and exterior is the desire for protection against a hostile outside world and inclement weather. Diverse other requirements have been added to these protective functions: light in the interior, an adequate air change rate, a visual relationship with the surroundings but, at the same time, a boundary between the private sphere and public areas. Special measures make it possible to regulate such openings. So this leads to control and regulatory functions being added to the protective functions.

All of these requirements can be divided into two groups, depending on how the facade is being considered: the external conditions specific to the location and the requirements governing the internal conditions. The general aspects can also be subdivided into numerous individual requirements. A comprehensive understanding of these fundamentals and their interdependence form the foundation for taking decisions during the design and construction of a facade.

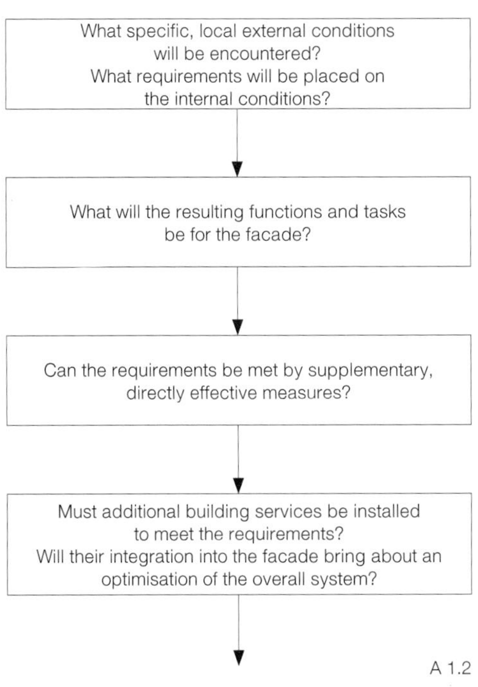

A 1.2

A 1.1 Requirements placed on the facade from inside and outside; protective, regulatory and communication functions; supplementary passive measures and building services
A 1.2 Key issues/Procedures for determining boundary conditions and requirements

The demands placed on the facade from outside and inside

As a rule, the external conditions cannot be influenced by the design. They therefore represent a primary criterion even at the stage of choosing a plot of land. Specific, unique external conditions prevail at every location. These require careful analysis because their nature and intensity varies according to district, region, country and continent. In addition, the direct surroundings and microclimate play a distinct role. Besides the climate specific to the location, with defined, statistically calculated amounts and distribution of precipitation (rain, snow and hail), other factors (e.g., a neighbouring industrial estate with higher levels of noise and odour-laden air) call for special measures in the design of the facade.

On the other hand, the requirements placed on the internal conditions are not defined right at the start but are instead determined during the design phase by way of a catalogue of requirements which is drawn up bearing in mind the intended use of the interior. Accurate knowledge of these targets is crucial to the success of the design because they directly influence the construction. They determine the amounts of energy and materials required for construction and operation in the long term. The requirements placed on the interior climate, which are essentially determined by the term "comfort" (see p. 22, fig. 1.12), are in certain circumstances supplemented by far-reaching measures derived from various other quality demands, such as the desire for high artistic merit, or special security against intruders. These conditions and requirements, illustrated graphically in fig. A 1.1, mean that the facade has to provide protective and regulatory functions. The former is essentially protection against the intensity of external influences, primarily those of the weather. The latter regulates as necessary to achieve an acceptable interior climate as required, aiming for "thermal comfort" (see p. 22).

If we see the facade as the human body's "third skin" (after that of the body itself and our clothing), the analogy of the design objective becomes clear: the fluctuations of the external climatic conditions on our bodies have to be reduced by each of these functional layers in turn in order to guarantee a constant body temperature of approximately 37°C. However, the climatic conditions also give rise to requirements that cannot be exclusively allocated to either side; rather, they are due to the difference between inside and outside. They lead to mechanical loads on the materials of the facade and the construction details, and are primarily the result of temperature, moisture and pressure differentials. Such loads must be accommodated by suitable means such as expansion joints and flexible connections.

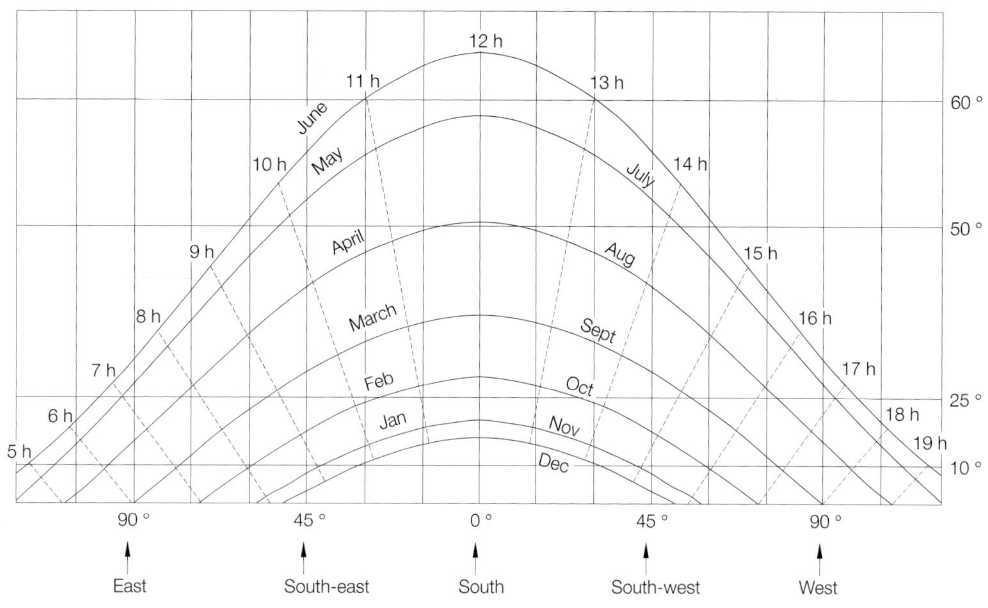

A 1.3

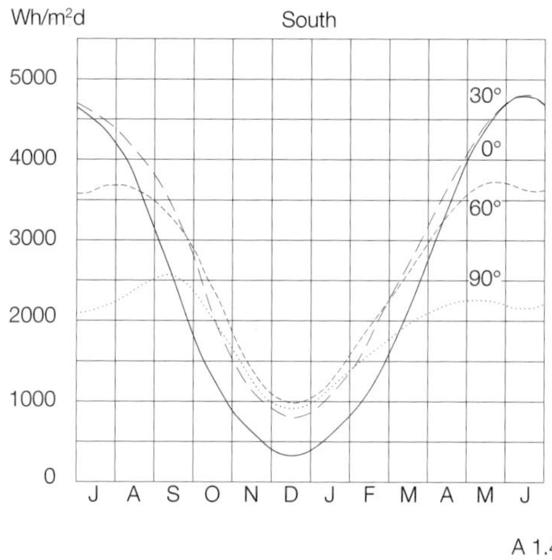

A 1.4

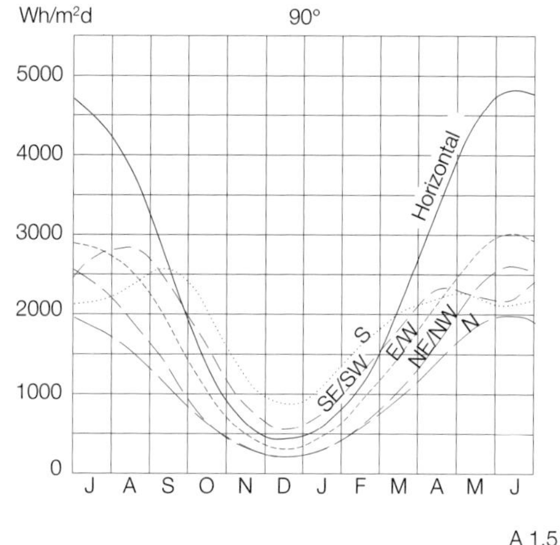

A 1.5

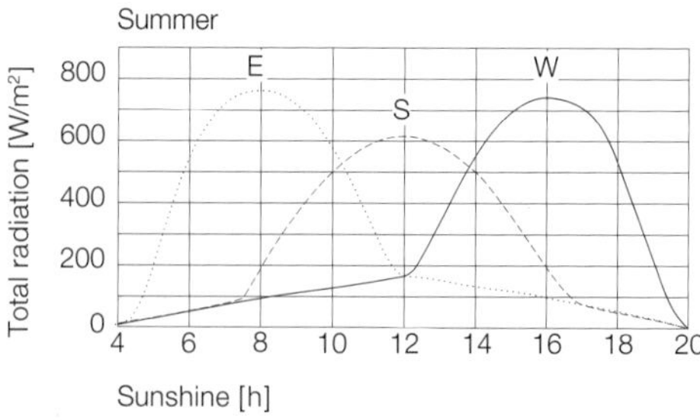

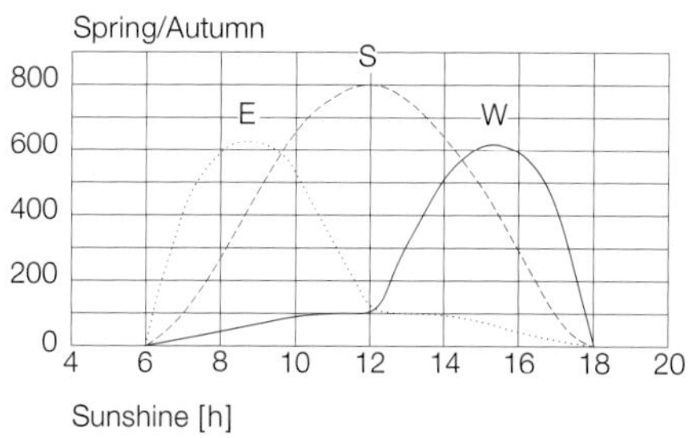

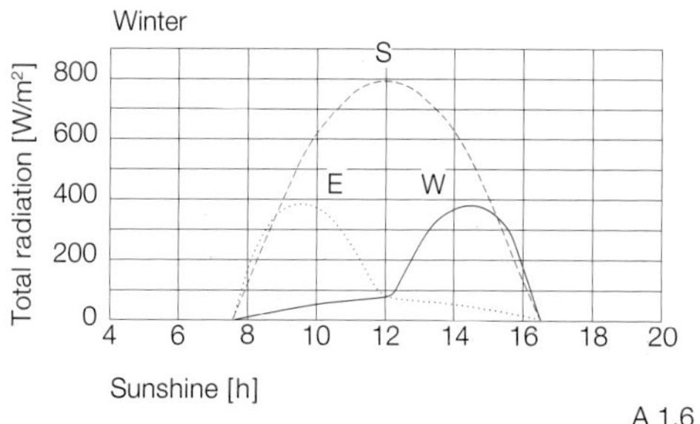

A 1.6

A 1.3 Solar trajectory diagram (50°N)
A 1.4 Incident radiation on south-facing surfaces with different inclinations
A 1.5 Incident radiation on vertical surfaces facing in different compass directions
A 1.6 Total radiation on to wall surfaces facing in various directions on sunny days at various times of the year

The efficiency of a facade

A facade should be able to handle climate-related tasks as comprehensively as possible because this allows additional measures such as air conditioning, to be minimised and even omitted altogether. To achieve this design objective, knowledge of the relevant physical principles is indispensable.

Supplementary measures with a direct effect can assist this task on both sides of the facade. For instance, it is possible to "activate" other components inside the building; e.g., by interim storage of energy in walls and floors.

Open bodies of water for cooling (due to evaporation) or dehumidifying (with a sufficient water–interior air temperature differential) can be employed externally or in intermediate zones. Suitable measures allow the energy peaks to be cushioned to be used otherwise. The heat radiation against which the building has to be protected can, for example, be converted into electricity or absorbed by collectors and used for hot water provision. The same is true for the use of higher external temperatures, wind and precipitation.

The other requirements that cannot be adequately handled by construction measures must be dealt with by installing the necessary building services. These regulate temperature, lighting levels, air quality, air change rates and humidification/dehumidification. However, such supplementary technical measures always require additional energy and call for expensive maintenance, elaborate means to convey media, and so on.

We can incorporate technical facilities for such requirements directly into the facade (see p. 13). When the equipment required is not installed in central plant rooms but rather in the facade directly where it is required, it is referred to as "decentralised facade services".

Apart from the above influencing factors, we also have to employ a similar approach to conditions resulting from the context of the building as a whole. These include dimensional coordination, constructional interdependencies, tolerances and erection sequences. These aspects will be dealt with in later chapters.

External conditions: solar radiation

In terms of specific local external conditions, the Sun plays the key role. It is the most important direct and indirect source of energy and the foundation for all life on our planet. The quantity of solar energy incident on the Earth is equivalent to approx. 10,000 times the current global energy demand (an average energy flux of 1353 W is incident on every square metre of the outer limits of the Earth's atmosphere), and on a human scale is inexhaustible, free and environmentally friendly. In order to be able to utilise this source of energy, it is vital to consider the intensity and duration of radiation in conjunction with the orientation and inclination of the facade.

Facade design also calls for a thorough analysis of the following aspects and relationships:

- solar altitude angle in relation to location, time of day, and season
- quantity of radiation, depending on orientation and inclination of surface, location, time of day, and season
- the various types of radiation (diffuse, direct and various wavelengths) and their magnitudes in relation to weather conditions, orientation, location, time of day, and season
- the interaction with surfaces and materials
- the anticipated quantities of incident energy in relation to weather conditions, orientation, location, time of day, and season
- the relationship of the radiation with the heat requirement resulting from the intended utilisation.

Figs A 1.3–1.11 illustrate a selection of the important relationships.
For example, in Germany, we can take the following values for the available solar radiation as our starting point:

1400–2000	sunshine hours/year
700–800	sunshine hours/heating period

The proportion of diffuse radiation related to the total available solar radiation in a year is approximately:

South facade	30%
East and West facades	60%
North facade	90%
(100% = direct radiation)	

However, the available solar radiation also embodies certain risks for humans (overheating, premature aging of the skin, skin cancer) which we have to protect ourselves against in suitable ways.

Thermal comfort

The internal climatic conditions have to meet certain requirements, which can be summed up by the term "thermal comfort".
The relevant influencing factors related to the construction of the facade are (fig. A 1.12):

- temperature of interior air (a)
- relative humidity of interior air (b)
- surface temperature of building components enclosing the room (c)
- airflow across the body (d).

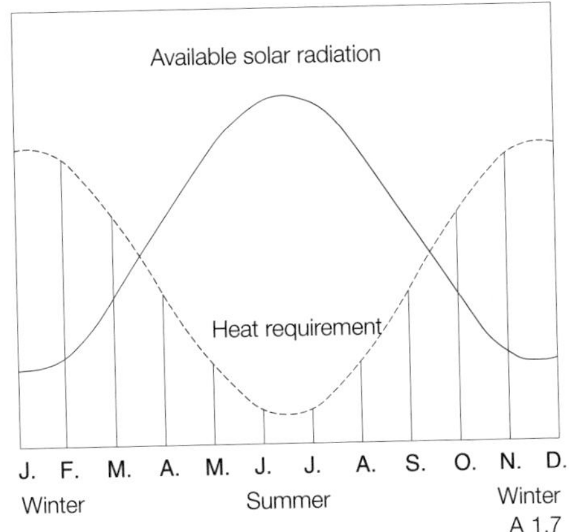

A 1.7

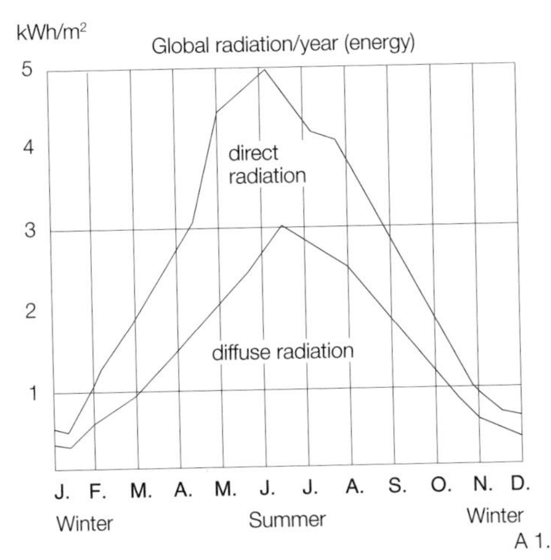

A 1.8

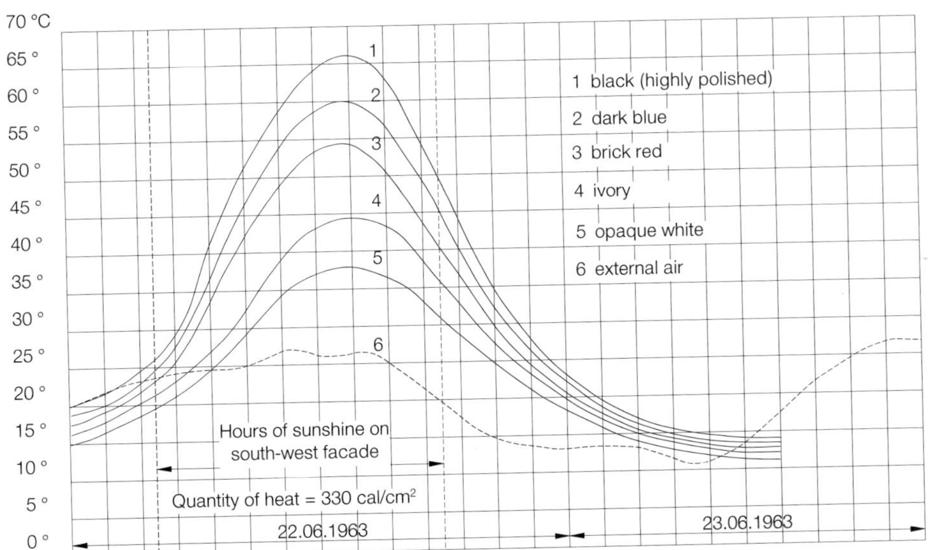

A 1.9

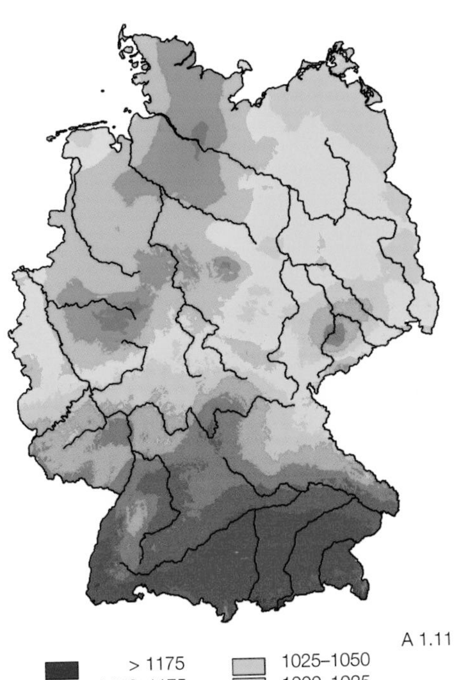

A 1.11

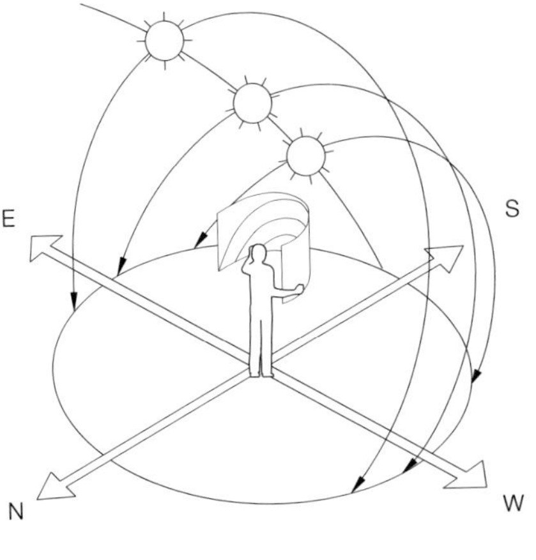

A 1.10

A 1.7 Heat requirement/Hours of sunshine (schematic)
A 1.8 Intensity of radiation (daily average) for central Germany (50°N)
A 1.9 Measured surface temperatures on a sunny day for south-facing facade surfaces of various colours
A 1.10 Principle of projection diagram for solar trajectories
A 1.11 Local distribution of annual global radiation in Germany [kWh/m²]

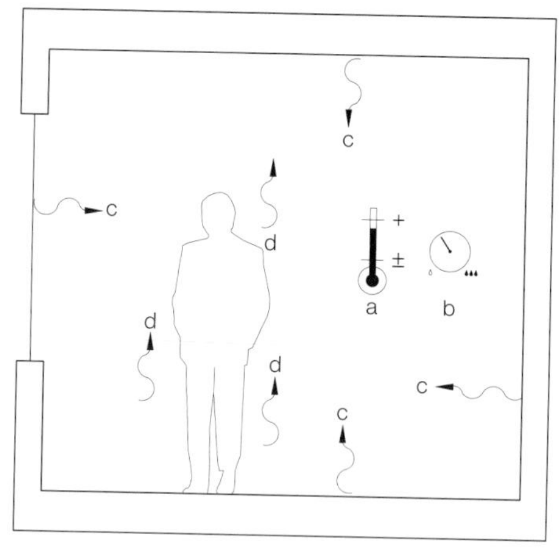

A 1.12

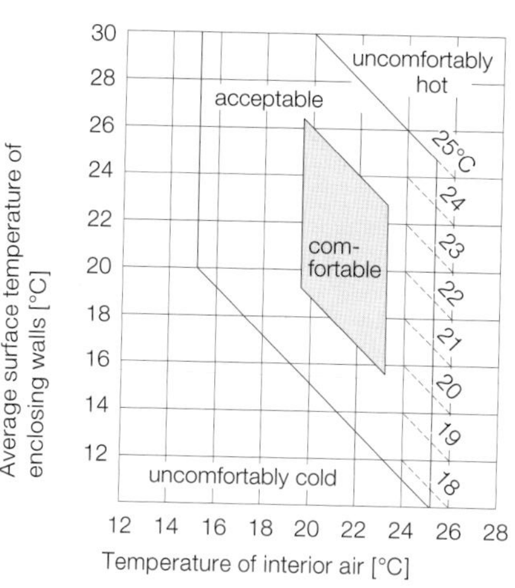

A 1.13

A 1.12 Factors influencing thermal comfort:
a temperature of interior air
b relative humidity of interior air
c surface temperature of building components
d airflow across the body

A 1.13 Comfort zone in relation to air temperature and average (only slightly differing) surface temperature of enclosing walls (after Reiher and Frank)

A 1.14 Comfort zone in relation to air temperature and relative humidity (after Leusden and Freymark)

A 1.15 Room temperature/Air movement
Comfort zone in relation to air temperature and air movement (after Rietschel-Raiß)

Applicable ranges for figs. A 1.13–1.15:
- relative humidity from 30 to 70%
- air movement from 0 to 20 cm/s
- essentially equal temperatures of 19.5 to 23°C for all surfaces enclosing the room

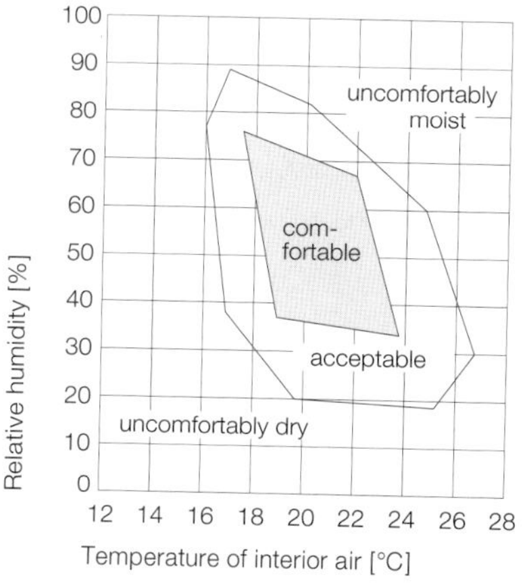

A 1.14

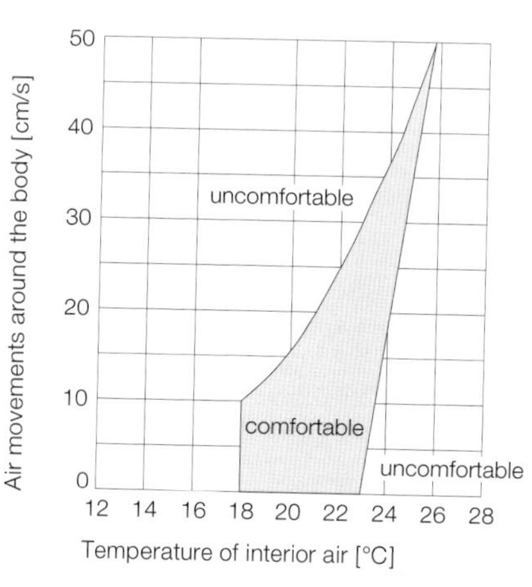

A 1.15

Heat flow
– K° + K°

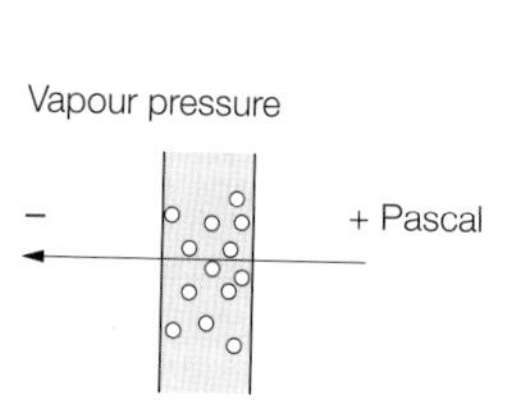

Vapour pressure
– + Pascal

A 1.16

Heat transmission
radiation
conduction
convection

A 1.17

Water, granite, concrete, aluminium, calcium-silicate brick, screed, loose gravel, dry sand, aerated concrete, lightweight clay brick
[kWh/m³K]
Bulk density [kg/m³]

A 1.18

These quantifiable variables – in conjunction with region, habits, clothing, activities and individual sensitivity – determine the level of thermal comfort. The ranges in which the values of the individual influencing factors should fall are called "comfort zones" (figs. A 1.13-1.15). There are no fixed target values for any of these variables; they are all mutually dependent on each other. The room air temperature and the average radiation temperature of the surfaces enclosing the room are both roughly equally responsible for the "perceived room temperature". The term "comfort" is increasingly taking on an interpretation which goes beyond purely climatic requirements:

- "lighting environment" and "visual comfort" – quantity and quality of light and luminance contrast (antiglare protection)
- hygiene comfort (low level of hazardous substances and odours)
- acoustic comfort (noise level)
- electromagnetic compatibility.

Even psychological factors (e.g. materials, colours) and cultural aspects play a role here and should be taken into account.

Underlying physical principles

To understand the functions of the facade, we must look at the scientific principles of the construction, e.g. heat flow, water vapour pressure, radiation transfer (fig. A 1.16).

Heat transfer
Thermal energy always flows from the hotter (higher-energy) side to the colder side. Three basic principles govern the transfer of heat energy (fig. A 1.17):
- conduction
- radiation
- convection.

The thermal transmittance – U-value (W/m²K) – can be calculated for planar components.

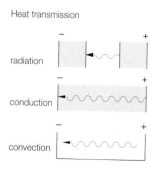

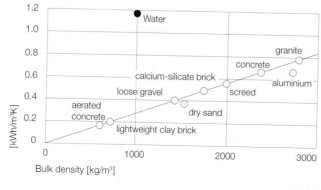

A 1.16 A selection of the basic scientific principles affecting the building
A 1.17 The principles of heat transmission
A 1.18 Volume-related heat capacity of selected materials

A 1.19 Hot air is lighter and rises
A 1.20 Positive and negative pressure effects as air flows around a building
A 1.21 Wind speeds increase with height above ground level
A 1.22 A radiation-permeable leaf in a facade allows the air in the cavity to heat up so that it then rises ("stack effect")
A 1.23 Geometric solutions to improve the air flow
A 1.24 Wind: regional frequencies and directions taken as an annual average for the city of Munich.
Wind speed:
a up to 3 m/s
b more than 3 m/s

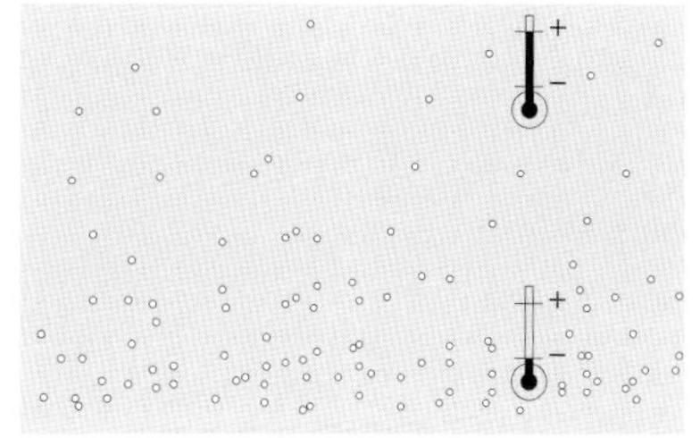

A 1.19

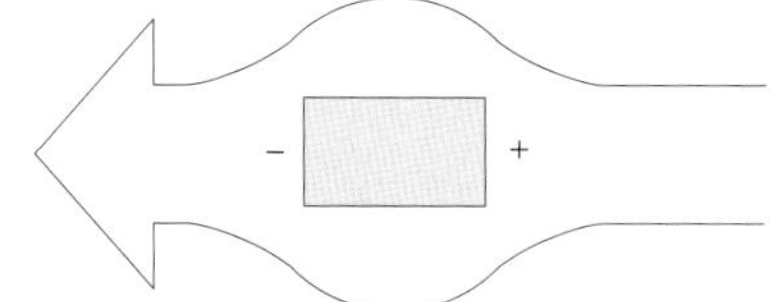

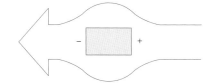

A 1.20

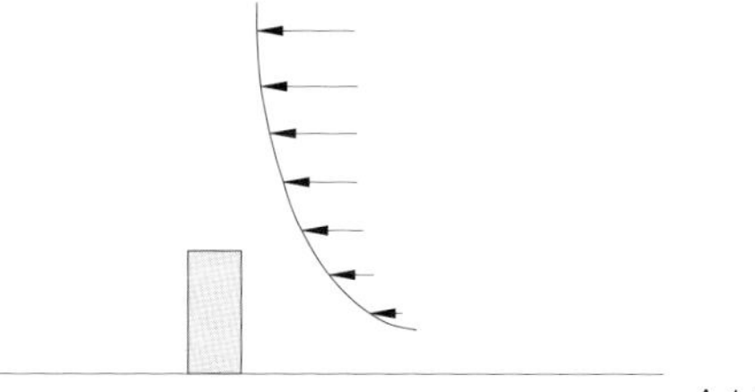

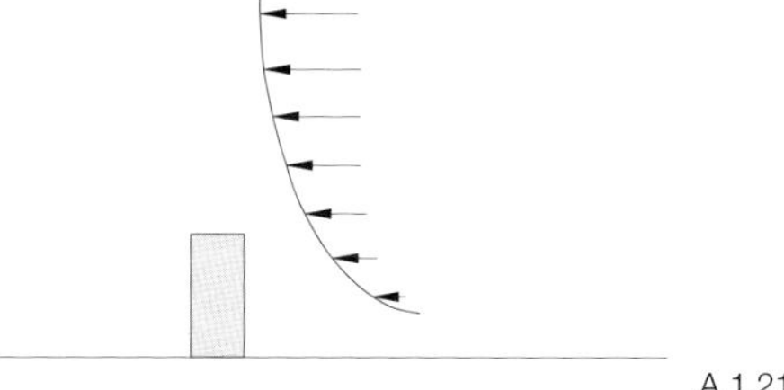

A 1.21

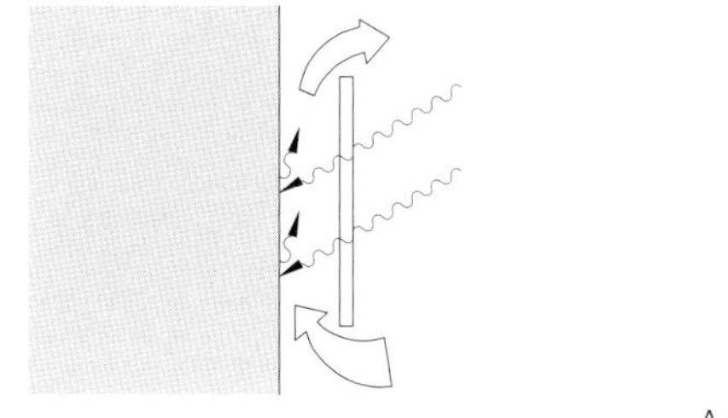

A 1.22

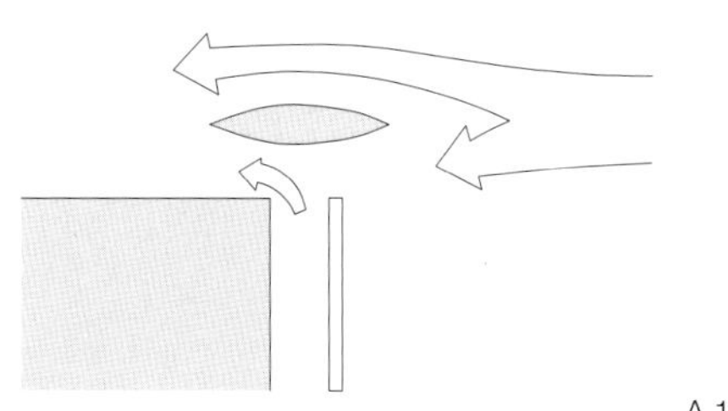

A 1.23

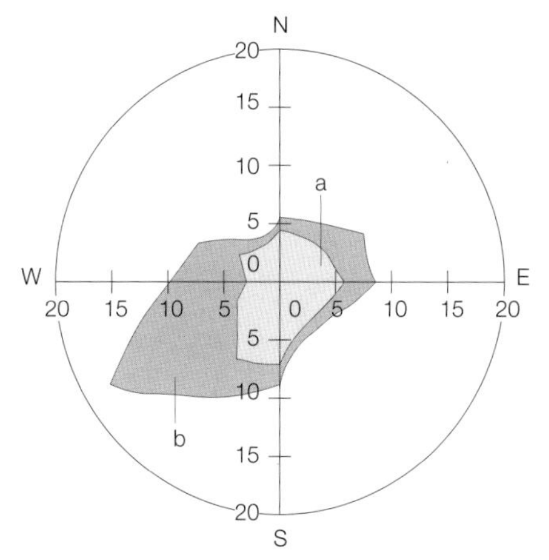

A 1.24

Thermal conductivity and heat capacity
Both of these factors depend on the properties of the material and generally increase with bulk density. However, the heat capacity of water represents a clear exception in comparison with common building materials (fig. A 1.18).

Relative humidity
Air can absorb water vapour until it reaches its saturation point, which depends on the temperature. We therefore speak of the "relative humidity" (of the air). Moist air is fractionally lighter than dry air at the same temperature.

Water vapour pressure
Water vapour flows from the side with the higher vapour pressure (partial pressure) to the side with the lower pressure. If there is a simultaneous severe temperature gradient, the temperature drops below the dew point and the water condenses out of the air (and hence leads to the risk of condensation collecting on surfaces and mould growth).

Radiation transfer
Radiation incident on a component is reflected, absorbed or transmitted (fig. A 1.16, p. 22). The heat radiation behaviour essentially depends on the properties of the surface of a material, especially its colour (fig. A 1.9, p. 21).

The underlying principles governing wind, thermal currents and natural ventilation

Currents of air in the atmosphere (wind), the interaction via openings linking inside with outside, and thermal effects in the boundary air layers are phenomena that affect every building; they represent another external condition. Since the specific local weather and wind conditions vary considerably in terms of severity and direction (fig. A 1.24), only statistical values can be incorporated in the design.
The currents of air that ensue due to the geometric properties of bodies in specific wind situations can only be investigated in wind tunnel tests and by means of dynamic, highly complex flow simulations. However, basic principles of facade design also play a role, as these are based on fundamental thermal principles (figs. A 1.19–1.23).
The aim of the facade design should be to enable natural ventilation of the building wherever possible. This approach can minimise the risks associated with the well-publicised "sick building syndrome" [1]. In this respect, the following problems associated with natural ventilation should be avoided as far as possible:

- increased heat requirement
- too high interior air temperatures in summer
- draughts in the building
- too low interior humidity in winter
- inadequate ventilation in calm air

The more the air heats up (i.e. absorbs energy), the faster the gas molecules move (fig. A 1.19). The air pressure rises, the air becomes less dense and hence lighter per volumetric unit, and it rises. In an enclosed room, this therefore results in different air temperatures, with a layer of warmer air at the top and cooler air below. Objects create an obstacle to the flow of air, splitting up the current of air as it flows past (fig. A 1.20). Besides creating turbulence, this also leads to higher air pressure on the windward side of the building and lower relative pressure (suction) on the leeward side. We should remember that the direction of the wind fluctuates considerably (fig. A 1.24) and that such effects can change very quickly.
Near the ground, wind speeds are generally lower due to the interaction with the surface (roughness) and obstacles (fig. A 1.21). The wind speed – and hence wind pressure and suction – increases with the height of the building.
If radiation energy passes through a transparent or translucent layer before striking a building component separated from this outer leaf by an air layer, the component heats up due to the absorption process (fig. A 1.22). It transfers part of this heat energy to the air in the cavity, which heats up and rises (similar to fig. A 1.19); the air starts to circulate. This effect is reinforced when the air can escape at the top of the cavity and is replaced by air flowing in at the bottom. Objects with suitable

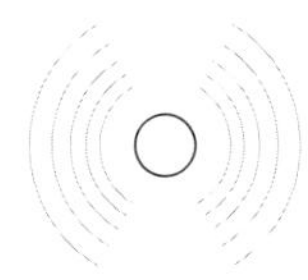

A 1.25

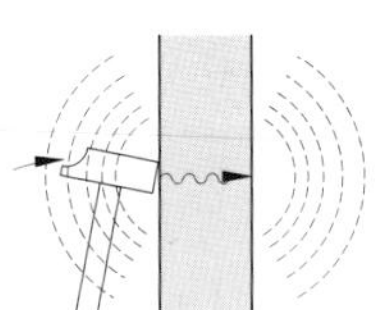

A 1.26

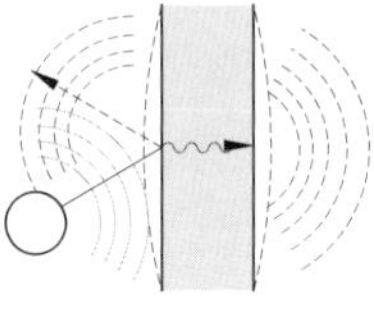

A 1.27

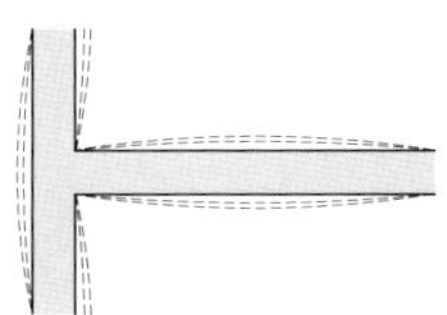

A 1.28

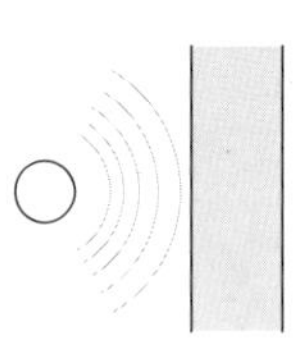

A 1.29

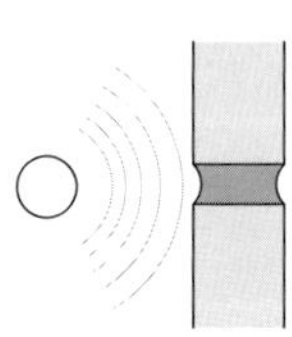

A 1.30

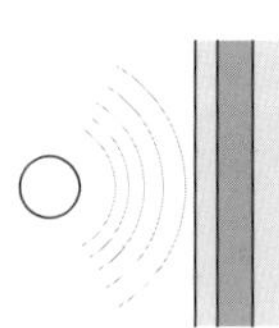

A 1.31

A 1.25 Sound source
A 1.26 Excitation of a mass by mechanical action
A 1.27 Excitation of a mass by airborne sound and transmission in the material (structure-borne sound)
A 1.28 Transmission of sound over long distances within building components ("flanking transmissions")
A 1.29 Strategy 1 to combat transmission of airborne sound: mass
A 1.30 Strategy 2: efficient joint sealing
A 1.31 Strategy 3: mass–spring–mass principle

shapes can exploit the available incident air flows around a building to generate an additional negative pressure (fig. A 1.23) and therefore reinforce the stack effect, or even to accelerate the dissipation of hot air from rooms at a lower level.

The underlying principles of sound transmission

At the facade, sound is both an external condition and an internal requirement (acoustic insulation) because sources of sound can be on either or both sides of the facade.
Sound insulation calls for especially careful design and workmanship because the transmission of sound can take place even across minimal acoustic bridges. Sound waves propagate in roughly spherical form through the air from the sound source and into the interior (airborne sound, fig. A 1.25). They are reflected to a greater or lesser extent by all the surfaces enclosing the room and by all the objects in the room. The smoother and harder the surface, the less distorted and more complete is the reflection.

If a solid material is caused to vibrate, e.g. by mechanical influences (footsteps on the floor), sound waves propagate through the building components (structure-borne sound, fig. A 1.26). When a solid body is excited by airborne sound, structure-borne sound propagates within the body (fig. A 1.27). This can in turn excite the layer of air on the other side, which then transmits the waves again in the form of airborne sound.

Sound waves can travel great distances in the form of structure-borne sound transmissions (fig. A 1.28). If the "solid" components of a building are joined together, sound may in some circumstances propagate throughout the entire building by this means. We speak then of "flanking transmissions".

One possible strategy to combat airborne sound transmission is to increase the mass of the components (fig. A 1.29). The object is made as heavy as possible and so given a high inertia, i.e., it consists of a material with a high density and therefore airborne sound waves cause it to vibrate only to a limited extent. Another way of dealing with airborne sound transmission is to provide an efficient seal (fig. A 1.30), which prevents the airborne sound propagating directly through "leakage points" such as joints, gaps and seams. There is also the possibility of attenuating airborne sound transmission by using a double-leaf construction with an insulated cavity (fig. A 1.31). This approach is particularly efficient when the two leaves have different thicknesses and weights, and hence exhibit unequal natural frequencies. However, the success of such measures should not be put at risk by including rigid fixings between the two leaves (principle: mass–spring–mass). Other aspects of sound insulation are dealt with in chapter A 3 "Planning advice for the performance of the facade".

Practical realisation

The aforementioned internal and external conditions, the resulting functional requirements and the underlying thermal, acoustic and other scientific principles give rise to a directly related interaction between the building components chosen for the facade construction.

Due to, for example, transmission, a radiation-permeable component leads to an energy gain inside the building (fig. A 1.32). When the radiation strikes surfaces in the interior, part of the energy is transferred to the material by way of absorption and from there continues to be transmitted by way of conduction (fig. A 1.33). Another part is "stored" in accordance with the heat capacity of the material. The energy is returned to the room by way of radiation after a time-lag, the length of which depends on, among other parameters, the thermal conductivity specific to the material (fig. A 1.34). Careful choice of materials and component sizes can help to exploit this effect to even out temperature peaks without needing to introduce further energy in the form of heating or cooling.

Convection processes enable energy to be transferred between inside and outside by regulating or controlling the ventilation (fig. A 1.35). This can take place in both directions. Skilful use of thermal effects can render mechanical ventilation totally unnecessary (e.g. figs. A 1.19, 1.22, 1.23).

Greenhouse effect
When high-energy short-wave solar radiation strikes surfaces in the interior, a major part of the energy is released into the interior in the form of diffuse, long-wave radiation in the infrared range (fig. A 1.36, p. 25), where it contributes to heating up the interior air and other surfaces. The very low radiation permeability of the building envelope in the long-wave range (in simple greenhouses, for example, glass, but primarily layers of insulation or thermally insulating double or triple glazing, possibly with additional coatings to enhance this effect) prevents the radiation from escaping. It is therefore "captured" in the room and we speak of the "greenhouse effect".

If this effect is desirable, it is possible to influence the degree of efficiency by aligning the radiation-permeable surface with the source of the radiation (i.e. usually the orientation with respect to the Sun) and hence changing the associated angle of incidence of the radiation (fig. A 1.37).

The shallower the angle of incidence, the greater the proportion of reflected radiation, which does not enter the interior (fig. A 1.36). With an angle of incidence of 90°, only a minimal proportion is reflected from the surface. The exact magnitude of this reflected component – like the absorption component – is a variable specific to the material, which can be modified by means of additional measures, e.g. coatings (see B 1.6 "Glass", p. 186).

Opening and angle of incidence

The quantity of radiation entering through openings (of identical size and identical orientation) varies considerably depending on the angle of incidence (fig. A 1.37). This effect, in conjunction with seasonal variations of the solar altitude angle, plays a decisive role in the design of openings and sunshading systems (see A 2.2 "Edge treatments, openings", pp. 40–42).

Consequences for the plan layout/zoning

The arrangement of the rooms according to the principle of "thermal onion" can influence the requirements to be met by the facade even during the early planning of the interior layout. Rooms requiring a higher temperature are surrounded by areas with less stringent requirements (fig. A 1.38). These "buffer zones" generally allow the heating and/or cooling loads to be reduced effectively.

One consequence of the trajectory of the Sun is that it can be useful to gain solar energy by exploiting the greenhouse effect and "capturing" the solar radiation in a projecting zone (according to fig. A 1.39) with interior surfaces designed to store heat. In Central Europe, the north elevation can hardly be used for solar gains and this should therefore be insulated accordingly. However, this concept readily leads to overheating, primarily in summer, and therefore calls for appropriate shading and ventilation provisions.

A 1.32

A 1.33

A 1.34

A 1.35

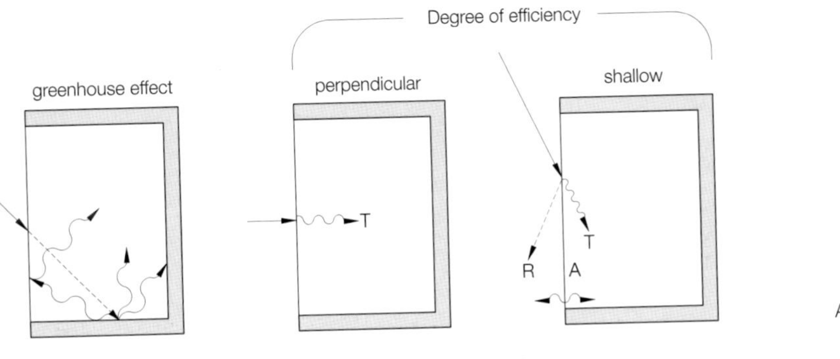

A 1.36

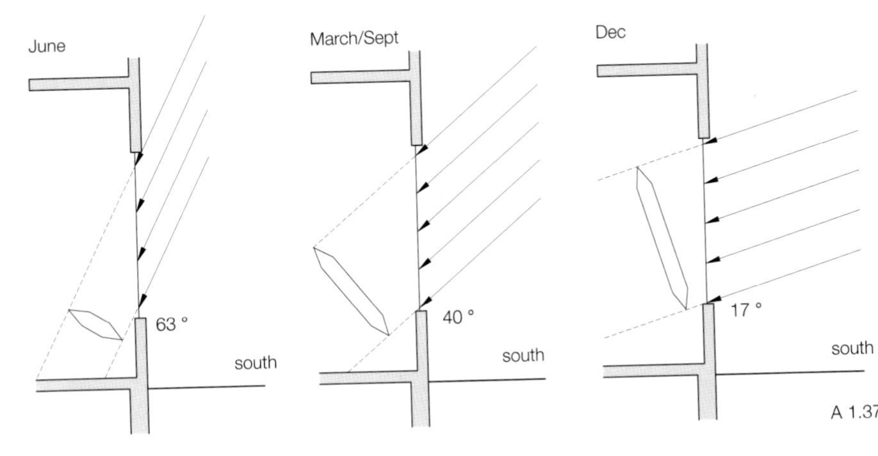

A 1.37

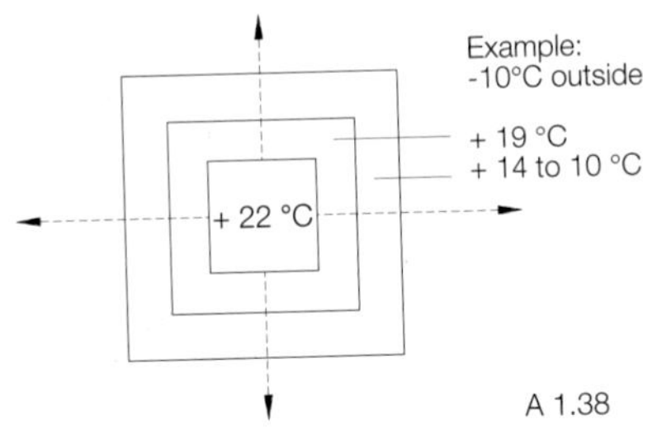

A 1.38

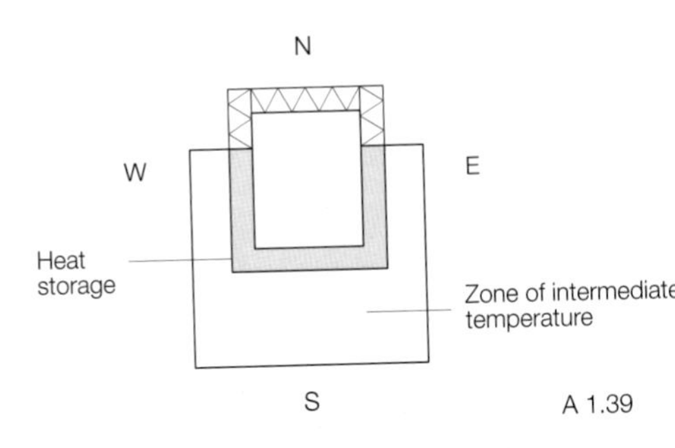

A 1.39

Notes

[1] For further information about "sick building syndrome" see: Dompke, Mario, et al. (ed.): Sick Building Syndrome, 2nd Proceedings from the 1996 workshop in Holzkirchen, published by Fraunhofer Institute for Building Physics and Bundesindustrieverband Heizungs-, Klima-, Sanitärtechnik e.V., Bonn, 1996.

A 1.32 Transmission
A 1.33 Heating up – conduction
A 1.34 Thermal storage masses – radiation
A 1.35 Convection – distribution – regulation
A 1.36 Exploiting the greenhouse effect
A 1.37 Angle of incidence of solar radiation through an opening
A 1.38 "Thermal onion – temperature-related zoning of plan layout
A 1.39 Orientation of building – heat storage – thermal insulation

A 2.1 The structural principles of surfaces

Facades are primarily vertical and planar (i.e. two-dimensional) structures positioned between the external and internal environments.

Regardless of what materials are employed, there are various generally applicable features and engineering design principles that are valid for facades, and these are described below. A knowledge of these is helpful for the design process. A design principle indicates a fundamental solution for a defined construction task in accordance with predetermined functions [1]. Here, physical, chemical and geometrical "effects" are utilised and their interactions linked together in a suitable structure [2]. The structure of the facade is considered

- in the plane of the facade, and
- perpendicular to the plane of the facade.

Facades can be assigned certain performance profiles, which can vary across the surface, according to the catalogue of functions and requirements. The engineering principles and materials employed may require several functional and constructional "levels" perpendicular to the plane of the facade. Additional constructional elements, which are not themselves part of the building envelope (e.g. horizontal sunshading systems, daylight redirection systems, maintenance walkways) may prove to be advisable.

The objective should be to create a structure that is an efficient assembly of individual components.

Classification of design principles [3]

Functional criteria
Definitions of performance profiles as objectives for the surfaces of the facade involve more than just the general protective functions such as insulation and waterproofing; permeability with respect to air and light – or rather, radiation – and are particularly important. The degree of permeability determines the character of the building envelope, and the usefulness and the quality of the interior. It also has a major influence on the energy balance of a building. Just how much a facade surface can react to changing conditions, whether the surface is variable and, if necessary, can even regulate itself automatically, are important criteria for making decisions.

Permeability with respect to air
Natural ventilation strategies call for variable and controllable air permeability. In addition, the dissipation of excessive heat, water vapour and, during a fire, hot and toxic gases may require a certain permeability.

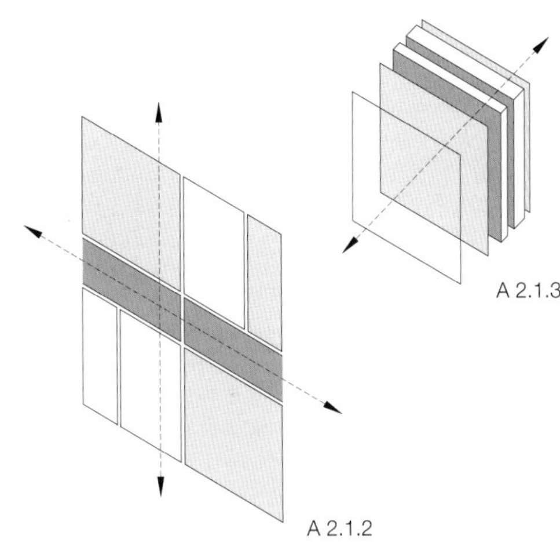

A 2.1.2 Aspects to be considered in the plane of the facade:
- type of surface
- allocation of performance profiles
- loadbearing ability
- design principle
- jointing

A 2.1.3 Aspects to be considered perpendicular to the plane of the facade:
- realisation of performance profiles
- construction in terms of layers and leaves
- joining of layers and leaves

A 2.1.4 Classification according to functional criteria
A 2.1.5 Classification according to constructional criteria

Air permeability	closed
	partly permeable open
Light permeability	opaque
	translucent semi-transparent transparent open
Energy gains	none
	heat electricity
Variability	none
	mechanical physical (struct. change) chemical (material change)
Control	manual, direct/indirect "self-regulating" with control circuit

A 2.1.4

Part of loadbearing structure	non-loadbearing
	loadbearing
Construction in layers	single layer
	multiple layers
Construction in leaves	single leaf
	multiple leaves
Ventilation	no ventilated air cavity
	ventilated air cavity
Prefabrication	low
	high

A 2.1.5

A 2.1.1 Studio house (D), 1993, Thomas Herzog

Permeability with respect to light
The nature and degree of light – radiation – permeability will control the level of daylighting and the general character of a room, create visual relationships between inside and outside and govern how much heat energy enters and leaves the building. In the case of perforated, semi-transparent surfaces, certain perception-specific phenomena, such as sunshading and antiglare elements, can be beneficial. Even surfaces with a very small proportion of perforations are "transparent" (our brains fill in the gaps) to the ob-server looking in the direction of the brighter lighting environment when the perforations are small and closely spaced. Looking in the direction of the darker lighting environment, however, such a surface appears opaque because the eye cannot adjust to the lower level of light seen through the perforations.

Energy gains
Surfaces permeable to solar radiation allow direct energy gains to be made due to the heating-up of components such as floors and walls within the building. Special technical facilities (e.g., photovoltaic systems, absorber walls with transparent thermal insulation) allow heat or electricity for the operation of the building to be gained directly in the structure of the facade itself.

Variability
The surface of the facade can react to changing external conditions by modifying the position or the properties of components, for example:

- by moving parts of the facade mechanically (position of louvres, degree of opening of shutters, etc.).
- by triggering reversible changes to material properties using electrical, thermo-sensitive or photo-sensitive processes, which have an effect on the light permeability, for instance; these changes are either of a physical nature (e.g., changing the state of the element, changing the orientation of crystalline structures) or a chemical nature (changing the chemical bonds) [4].

Control
Variability calls for control. The following means can be used to track changing conditions:

- manual or mechanical actuation, direct or indirect, e.g. "at the touch of a button"
- "self-regulating", e.g., the light permeability of thermotropic glass is altered by thermo-sensitive processes
- the use of sensors and microprocessor-controlled stepping motors working with control circuits.

Fundamental design criteria
Important, fundamental decisions regarding the design must be clarified to prepare the way for the realisation in terms of structure and materials.

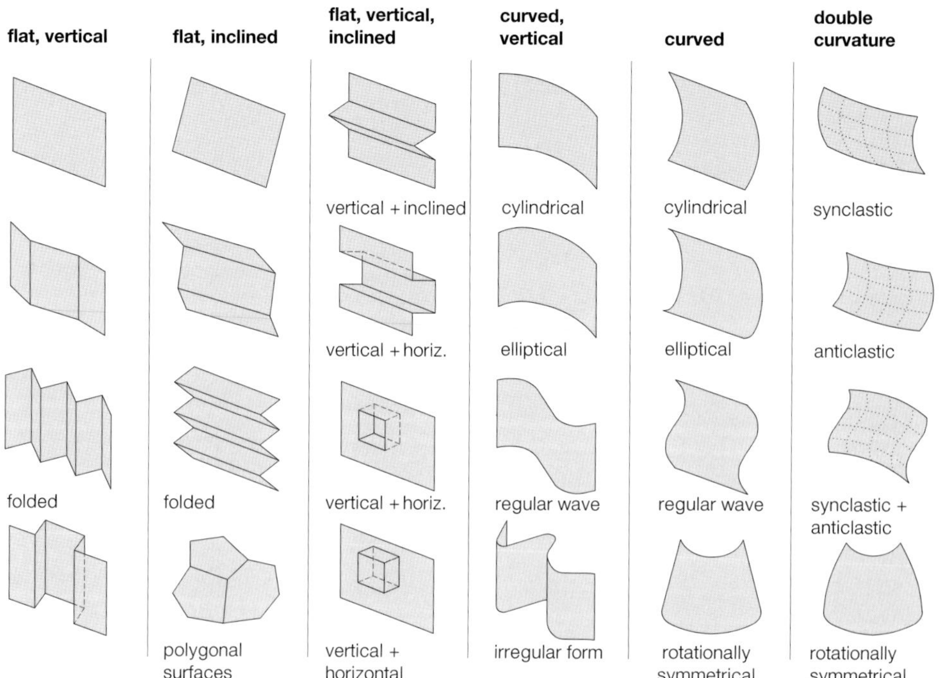

A 2.1.6

Relationship to loadbearing structure
"Non-loadbearing" facades do not carry any loads nor assume any loadbearing tasks involving the stability of the building.

Structure in terms of "layers" and "leaves"
Different materials with different thicknesses and structures can be optimised to perform certain tasks or parts thereof and then added together according to constructional and other performance principles to form a functional unit – the facade. Numerous different combinations with matching performance profiles can be achieved. The design thicknesses of the individual functional "levels" can vary between fractions of a millimetre (e.g., low-emissivity coating on double glazing) and several metres (e.g., air cavity in a multi-leaf glass facade). The sequence of the levels must be correct in order for the facade to function efficiently, and to avoid damage.
Functional levels that are unimportant, or at best secondary, in terms of load-carrying ability are classified as "layers", but those with a structural function and self-supporting are "leaves" (see page 36) [5].

Ventilation
Facade structures with integral ventilation have one or more layers of air which dissipate condensation and/or heat efficiently by means of thermal currents ("stack effect"). Such systems are always multi-leaf arrangements.

Prefabrication
The desired degree of prefabrication has a major effect on the design principles, the nature of the elements, the actual size of the individual components and the conditions governing the erection of the facade and, possibly, its later dismantling.

Structure in the plane of the facade

Types of surfaces
When establishing the external three-dimensional geometry of the building, it is worth remembering that the enclosing envelope is governed by its own laws. Every facade is made up of several flat or curved surface units which intersect or make contact with each other and with the roof surfaces along certain lines (edges or arrises). The shape of the surfaces and their positions in space, whether they are vertical or inclined or nearly horizontal, has a decisive influence on the architectural and engineering detailing of the facade. Intersecting edges and, in particular, "corners", at which three surfaces coincide, require special attention. Diverse factors determine the three-dimensional conception of the arrangement of surfaces. These factors seldom apply singly, but rather mostly in combinations with differing weightings, for example:

- plan and elevational geometry of the building
- utilisation aspects (e.g., creation of recesses for concealed open spaces)
- structural concept of the envelope itself (e.g., folded plate)
- thermal performance aspects (e.g., minimising the envelope area/volume ratio A/V)
- constructional aspects (e.g., water run-off)
- materials aspects
- architectural intentions.

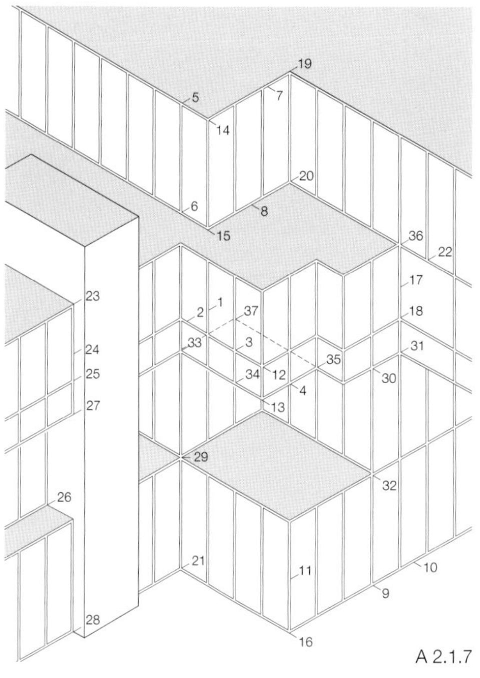

A 2.1.7

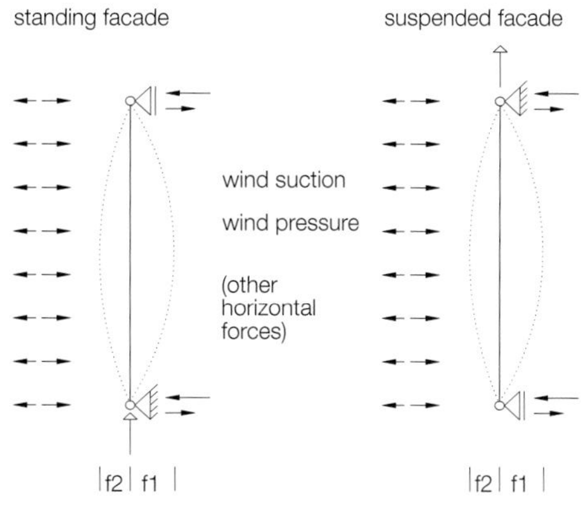

A 2.1.8

A 2.1.6 Schematic views of typical types of surface, which can be combined to form numerous variations.
A 2.1.7 Examples of different junction details based on a vertical, orthogonal facade system
A 2.1.8 Schematic diagrams of standing/suspended facades

Appraisal of different types of surfaces

Vertical surfaces

Water run-off is not hampered, folds and projections increase the external surface area, and internal edges require proper constructional and geometric detailing. Acute angles between adjoining surfaces may cause problems for construction and utilisation. The detailing of vertical edges benefits from the fact that they run in the flow direction of facade run-off water. Folded surfaces can be designed as structurally effective folded plates.
Fig. A 2.1.7 shows 37 different geometric cases in which facade surfaces intersect with each other or with floor or roof surfaces at edges and corners. A unique detail has to be designed and built for each one of these positions. Points at which more than three different surfaces coincide (e.g., No. 29) are almost impossible to detail properly in terms of engineering and architecture. And if different angles or even curvatures are involved, the number of geometrical and hence constructional cases multiplies considerably.

Inclined surfaces

Additional problems occur with every non-vertical surface, especially at projections and returns within significantly angled surfaces. Water run-off is hampered, accumulations of snow and the formation of ice lead to additional loads, larger horizontal surfaces have to be treated like roofs with appropriate drainage, the surface area increases, waterproofing and insulation layers are "cranked", thus causing weak spots in the construction where these layers change direction.

Every window reveal, every oriel, every loggia, etc., signifies an offset in the surface both vertically and horizontally. Furthermore, this results in internal and external edges and corners.

Curved surfaces

When they are vertical, the water run-off is not hampered. In many cases it is not possible to create a continuous curved surface. Instead, the surface is polygonal due to the initial geometry of the materials and semi-finished products used.

Surfaces in double curvature

Such surfaces are not necessarily only encountered with membrane structures. Such geometries are often created as translational surfaces which can be built as polygons with individual flat facets.

Structural principles

Actions

The facade must safely withstand the forces to which it is subjected and transmit these to the (primary) loadbearing structure. Every facade design, even a non-loadbearing one, must be conceived and designed as a secondary loadbearing structure to carry the following loads:

- Vertical loads:
  dead loads, special loads (e.g., sunshading, plants, temporary scaffolds), imposed loads (e.g., persons), snow and ice loads (e.g., to be calculated in every case for facade planting)
- Horizontal loads:
  wind load (pressure and suction generally occur in the ratio 8:5, but near the edges the suction loads can be considerably greater), imposed loads (e.g., impacts)
- Restraint forces, caused by thermal or moisture-related volume changes.

Normally, the loads acting on the facade surfaces are transferred into the floors, walls and columns of the loadbearing structure. Vertical and horizontal loads can be carried and transferred separately into different components of the loadbearing structure.

Standing and suspended facades

One fundamental distinction regarding the loadbearing behaviour is whether the facade is supported from below (standing) or from above (suspended), i.e., whether the planar or linear components need to be designed for tension and bending, or compression and bending and hence also buckling (stability problems).

The suspended form in which the dead loads are transferred into the loadbearing structure at the top of the facade components (e.g., into the floor slabs) has become established worldwide owing to the advantages of this principle:

- The component assumes a stable position immediately after being erected (hung) (in contrast to the imperfect stability of the standing component), which is very important in terms of site safety, especially during the erection of high-rise buildings.
- The self-weight acts as a tensile force along the longitudinal axis of the component. The ensuing pretension has a stabilising effect (= lower buckling load). This avoids the unfavourable superimposition of buckling due to compression and bending.

In the case of long spans in particular, the suspended solution has many advantages over the standing solution. However, deformations perpendicular to the plane of the facade are not reduced by any significant amount.

Anchor points, sliding supports

The facade and the loadbearing structure – at least where these are separate systems – are subjected to different temperature fluctuations and loads and the resulting geometrical changes. This fact calls for an unrestrained coupling between the two, with anchor points and sliding supports. Relative movements in both directions must be accommodated (positive and negative tolerances). The interfaces between the two subsystems usually represent the meeting points between different trades, methods of construction and tolerance ranges. This is why the fixings require adequate adjustment options in all directions. Likewise, connections between facade components exhibiting different longitudinal expansion (due to mechanical, thermal and moisture-

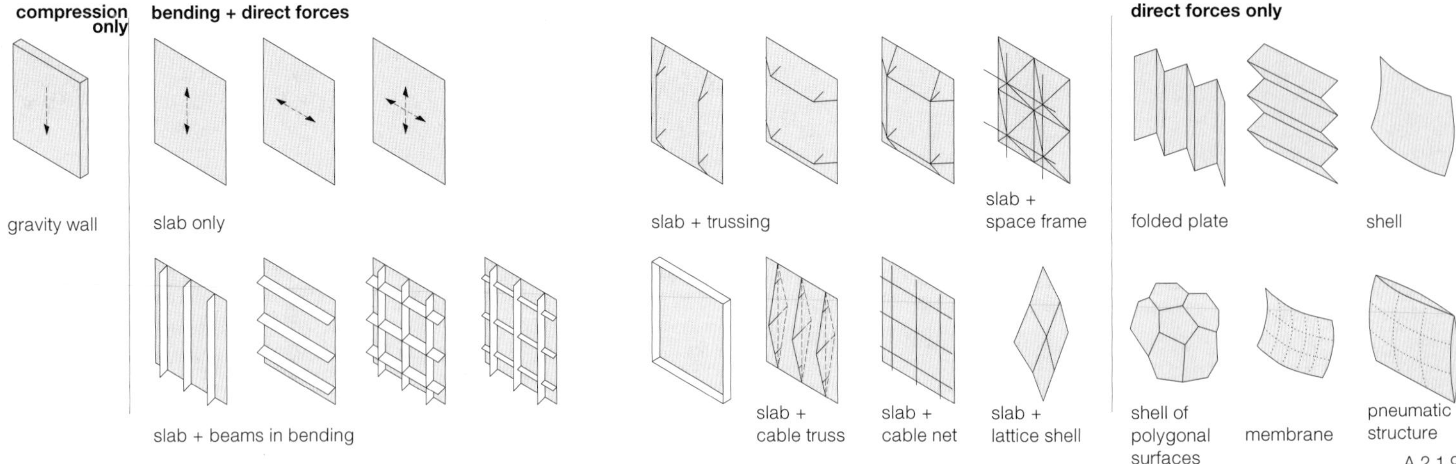

A 2.1.9

related actions) must be designed to eliminate any restraint in order to prevent damage.

Loadbearing structures

It is the nature of all enclosing envelopes to include planar components as a key element in the facade. Depending on the type of loadbearing structure (figs. A 2.1.9 and 2.1.10), these are subjected to direct forces (tension and/or compression) in the plane of the components, or, in addition, bending perpendicular to the plane of the components. These planar elements can be provided with linear loadbearing elements such as trussing systems, beams in bending, etc., either in the same or higher up the structural hierarchy. Combinations render possible hierarchical systems with primary and secondary loadbearing members.

Planar and linear elements act either as a constructional unit (e.g., T-beams, trussed slabs), or separately, which allows planar components to be more readily detached and replaced.

The logic of such structures is not only based on the efficient use of materials to carry the loads in the finished condition; prefabrication and erection issues also play a part. Transportation and erection procedures can lead to different loading cases which must be accommodated. It is often not the permissible bending stresses but rather the limits placed on deflection – particularly in the case of glass – that govern the design.

Gravity walls

In walls whose joints cannot transfer any tensile forces, the resultant vertical and horizontal forces must lie in the kern of the plan area of the wall in order to guarantee its stability and prevent the occurrence of cracks (at the joints). The horizontal forces are cancelled out by the vertical loads.
It can be advantageous in such cases to add further vertical dead loads from floors and roofs to the self-weight of the wall itself, i.e., the wall becomes part of the primary loadbearing structure (= a loadbearing facade). Masonry external walls are usually built according to this principle.

Slabs

A slab is a planar loadbearing element that carries horizontal forces by way of bending stresses (in one or two directions) perpendicular to its plane. The bending strength and stability (with superimposed compressive forces) are essentially defined by the "structural depth" of the component (i.e., the effective thickness of the component perpendicular to the plane of the facade). The cross-sectional shape, with material concentrated at the boundaries of the component, should be chosen to match the actions. Continuous-span effects are useful for reducing the bending moments. The simultaneous application of horizontal and vertical loads always results in bending moments and direct forces being added together. Vertical forces can also be carried by bending in the plane of the planar component, transferring the forces horizontally to the sides.

Slab + trussing

Trussing arrangements, forming a constructional unit with the slab by way of structural fixings between the two, enable the structural depth to be increased in a way which is economical in terms of materials. Trusses may be fixed to one or both sides. The slab is additionally subjected to compressive forces in its plane. The connections of the tension and compression members lead to point loads which have to be analysed for their punching effects. Trussing arrangements do not result in any additional support reactions that have to be carried by the primary loadbearing structure.

Slab + beams in bending

Linear loadbearing members subjected to bending, and possibly compression too, reduce the span of a planar component. The beams collect the point and/or line loads from the planar components (slabs) and – via bending – transfer these as single loads to components higher up the structural hierarchy. When only wind loads are involved, the bending loads are carried in one direction, but this can be reversed (wind pressure and wind suction). The superimposition of compression and bending exacerbates the stability problem, with a risk of buckling in the direction of the weaker axis of the particular section. Pinned connections enable individual bending members to be coupled together to form loadbearing structures, also curved or polygonal, covering large areas (e.g., post-and-beam facades). Manufacture, transportation and erection limit the size of frame constructions. However, they can be combined with identical or other elements to create prefabricated facades.

Slab + linear members, subjected only to direct forces

The linear loadbearing structures include:

- Space trusses: three-dimensional structures composed of tension and compression linear members, suitable for large spans
- Cable trusses, cable nets: pretensioned structures subjected only to tension which are advisable when the high tensile forces required for the pretensioning do not require elaborate additional measures in order to be carried by the loadbearing structure of the building; such delicate-looking structures are ideal when a high level of transparency is required
- Lattice shells.

Folded plates, shells, membranes

Stressed skin structures subjected to tension and/or compression in their plane are highly suited to carrying loads uniformly distributed over a wide area. Alternating distributed and/or point loads place additional bending stresses on these systems. Appropriate pretensioning guarantees that the deformation of membranes subjected exclusively to tension is kept to a minimum, also for alternating loading cases.

Stresses in components	Structural principle
Primarily direct forces only	
compression only	gravity principle
compression + tension	folded plate
compression + pos. tension	shell
tension only	pneumatic structure membrane structure
Bending and direct forces	
bending + compression	standing slab
bending + tension	suspended slab

A 2.1.10

The form of planar facade components

Planar components can be divided into basic forms which can be achieved with various materials and can frequently be combined. The variations shown in fig. A 2.1.11 do not indicate the entire facade construction but rather just the methods of construction used for layers or leaves. A slab of solid material subjected to bending can be used as the complete system for a storey-height, single-leaf and single-layer construction, or might be just a small-format segment of the cladding to an external wall. The governing criteria for choosing a suitable principle are:

- load-carrying capacity with respect to the structural requirements (fig. A 2.1.10)
- constructional relationships – size of component, workability, fixing options, joints, deformations, changes in length, degree of prefabrication, resistance to moisture and frost, etc.
- performance issues – specific weight, thermal conductivity, heat capacity, vapour diffusion resistance, light permeability, etc.
- appearance.

"Continuous" internal structure

This is made up of solid sections with isotropic or anisotropic behaviour. The planar components are prefabricated in a factory or constructed on site, often in formwork with joints between the individual production steps. The size and shape of the components depends on the materials and production techniques. Such components can be given specific loadbearing abilities by designing them as composite sections with reinforcement (metal bars, glass fibres, natural fibres, synthetic fibres, etc.) to resist tension and/or compression. This principle is applied in the same way to slabs of solid material subjected to bending for example, and to membranes made from a composite material and subjected only to tension .

"Continuous" internal structure (solid material)
a basic material
b mixture of materials, composite material
c reinforced/fibre-reinforced composite material

Internal structure with high air content
d porous, foamed
e spherical structure
f three-dimensional grid/network

Internal structure with voids
g voids, cells (discrete, linear)
h offset voids
i multiple-web twin-wall sheets

Layered internal structure, friction bond and/or interlocking connections
j irregular units, friction bond
k regular units, friction bond and interlocking connections
l multiple-web twin-wall sheets

Layered internal structure, material bond
m linear units
n planar units
o linear and planar units

Sandwich construction
p with closed-cell core
q with open, cell-like core (honeycomb, webs, etc.)
r with profiled core

Ribs/frames and slabs
s ribs with planks on both sides to form a constructional unit
t frame with planks on both sides to form a constructional unit
u frame and isolated infill panel

Profiled elements
v separate sections
w trapezoidal profile
x ribbed profile

A 2.1.9 Loadbearing structures for facades
A 2.1.10 Stresses in the planar facade components due to vertical and horizontal loads
A 2.1.11 Overview of constructional assemblies for planar facade components

a b c d e f g h i j k l m n o p q r s t u v w x

A 2.1.11

Internal structure with high proportion of air or cavities/voids
Various production methods can increase the air content in components to achieve the following objectives:

- to reduce weight and use of material
- to lower the thermal conductivity (= increase the thermal insulation effect)
- to create voids for building services.

If it is possible to concentrate the material at the boundaries of the element, we can expect only a minor reduction in the bending strength compared to a solid section. A large reduction in material resolves the section into boundary zones in tension or compression and webs in shear.

Layered internal structure with friction bond and/ or interlocking connections
The layering of small-format, irregular units without a binding agent is a traditional method of building which is still used for facing leaves. Enclosing these units in metal cages section by section (gabion walls) increases the stability considerably.
Dimensionally coordinated units of regular size and shape can be added together using a friction bond and/or interlocking connections to form larger components. Adaptability is improved when the modular increments are small.

Layered internal structure with material bond
Linear, planar or three-dimensional arrangements (e.g. honeycomb, mesh) can be joined together by way of a material bond (e.g., an adhesive) to form larger, slab-type units. Sandwich construction is a special form of this approach.

Sandwich construction
Bonding thin-walled tension- and compression-resistant facings to a shear-resistant core (mostly with a highly open or porous structure) results in a constructional unit with high bending strength and economical use of materials. A core with good thermal insulation properties creates a unit generally suitable for use as a lightweight, opaque facade panel.

Planks with ribs or frames
Ribs/frames and planking or infill panels stabilise each other and result in planar components with a good load-carrying ability and economical use of materials. Voids/cavities can be filled with thermal insulation materials.

Profiled elements
This principle results in a high stiffness using minimal material. Simple U- and Z-shapes can be classed as profiled elements and added together to form larger planar elements. Profiled elements can be fabricated from a large number of tension- and simultaneously compression-resistant materials, e.g., by forming, extruding or casting techniques.

Joining facade components
Almost every facade consists of an assembly of individual components and therefore contains numerous joints. These represent interruptions in the layers and leaves (e.g., weatherproof leaf) and in many case potential weak spots which must be sealed in the best possible way. In other cases joints are of the open drained type:

- to relieve vapour pressure
- to allow air to enter or escape (for ventilation)
- to allow drainage of facade run-off water or condensation water
- to allow relative movements
- to allow the passage of light.

Although the seams of the building may be very different, we must pay special attention to them, because these are points where many aspects relevant to the design are concentrated (fig. A 2.1.12). Besides the functional and engineering aspects, joints have an overall effect on a facade (inside and outside), demarcating the individual components and lending order to the geometry and the construction.

Joints in external facade surfaces are exposed to the full force of the weather. The wind load increases with the height of the building. The currents concentrate at the edges of the building and hence lead to even higher wind speeds, and during rainfall to a concentration of facade run-off water, accumulating as it drains down the building. The position of joints with respect to the direction of precipitation and run-off water, which is determined by gravity and wind, is an important factor in joint design. Joints nearly parallel to the direction of flow of run-off water (vertical joints) are generally less at risk than those primarily transverse to the flow of water. Changes in length or volume of adjoining components due to loading, temperature fluctuations and water absorption/ release place extra stresses on any type of joint. This is most obvious with prefabricated facades, but wet-in-wet methods of construction are also not rigid assemblies.

Design principles for sealing joints
It is the task of a joint seal [6] to slow down or stop the air/water mixture (fluid) within the joint. As sealing elements can never be applied to the mating faces of facade components without any gaps whatsoever, the efficiency of the seal is always only relative. Only a material bond can achieve a complete seal. If a joint cannot be sufficiently filled with a sealing element at one position with the cross-section, other strategies are called for. Providing a barrier in several stages and, if necessary, using different jointing elements is a proven method (multi-stage jointing system).
Joint design is based on just a few basic principles, which can be implemented in many ways (fig. A 2.1.13). When choosing a jointing system it is vital to be aware of the calculated or anticipated magnitude and direction of movement of the components.

Moisture	rainwater/facade run-off w. capillary water water vapour/condensation ice formation, snow
Atmospheric/ wind pressure	airtightness reduce wind press./suction air inlets/outlets
Sound	airborne sound structure-borne sound
Light	lighting UV resistance of joint material
Force transfer	element–element supporting construction–element
Compensation for tolerances	manufacturing tolerances erection tolerances movement tolerances
Erection	adjustability, fixing sequence independence from weather
Maintenance	necessity options/access
Dismantling	detachability recycling reuse
Pattern of joints	overlapping highlighted undercut profiled change of material colour

A 2.1.12

Opendrained jointing systems
Building components are intentionally erected with spaces in between and their edges are designed to create turbulence and thus hinder the flows in the joint. This principle permits large relative movements and is ideal as the first stage in a multi-stage jointing system. In a labyrinth joint the line of the joint is additionally joggled to create overlaps.

Butt joints
The simple abutting of two edges without any further sealing (not illustrated in fig. A 2.1.13) is the archetypal joint form. Owing to the unevenness of the mating faces, the gap can only be reduced – even in the case of elastic or plastic materials – not closed completely, even if a force is introduced.

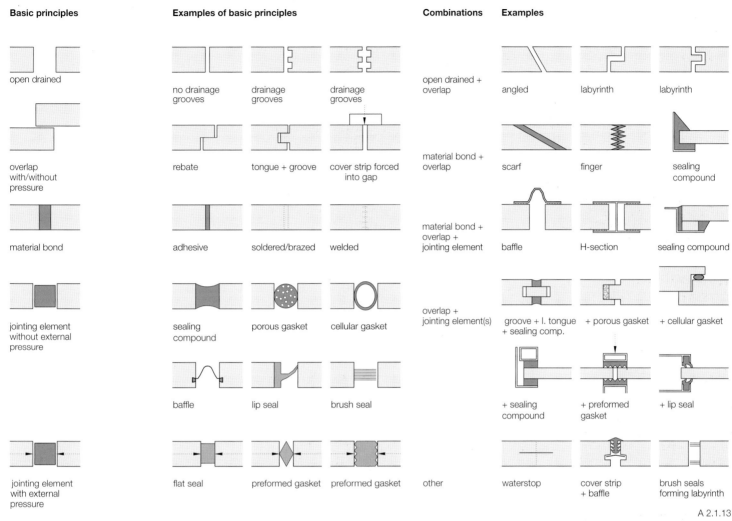

Basic principles	Examples of basic principles			Combinations	Examples		
open drained	no drainage grooves	drainage grooves	drainage grooves	open drained + overlap	angled	labyrinth	labyrinth
overlap with/without pressure	rebate	tongue + groove	cover strip forced into gap	material bond + overlap	scarf	finger	sealing compound
material bond	adhesive	soldered/brazed	welded	material bond + overlap + jointing element	baffle	H-section	sealing compound
jointing element without external pressure	sealing compound	porous gasket	cellular gasket	overlap + jointing element(s)	groove + l. tongue + sealing comp.	+ porous gasket	+ cellular gasket
	baffle	lip seal	brush seal		+ sealing compound	+ preformed gasket	+ lip seal
jointing element with external pressure	flat seal	preformed gasket	preformed gasket	other	waterstop	cover strip + baffle	brush seals forming labyrinth

A 2.1.13

Overlaps

This is probably the simplest, earliest and most effective principle, and is used in many jointing systems. The arrangement of the overlap must reflect the flow direction of the facade run-off water.

Fig. A 2.1.14 shows examples of run-off water safely conveyed over horizontal joints employing the overlapping principle. Some variations allow horizontal movements between the components (e.g., at opening lights).

Material bond

The use of adhesives, welding, soldering/brazing or rolling to join the components can result in a complete seal in some individual cases. However, at best, only limited relative movements are possible.

Sealing compounds

These are particularly suitable for uneven mating faces. The sealing effect is based on the adhesive bond between the sealing compound and the sides of the joint. Rigid sealing compounds can create structural connections if required. Sealing compounds with plastic or elastic deformation behaviour can accommodate small relative movements. Unfortunately, if workmanship is poor, it does not become noticeable until some time later.

Preformed porous and cellular gaskets

These are larger than the maximum volume of the joint prior to being inserted. That means they have to be compressed and are therefore constantly in a state of prestress. They can accommodate small relative movements transverse to the line of the joint, but movements parallel to the joint call for precautions to be taken to prevent the gaskets becoming dislodged.

Preformed cellular gaskets are more suitable for frequently alternating loads and larger movements owing to their having higher internal prestress than porous versions.

Baffles

These can accommodate large relative movements both perpendicular to and parallel to the line of the joint. The connection to the adjoining components can be achieved in many different ways, such as a press fit or adhesive.

Lip seals

An elastically deformable element with one or more sealing lips which are pressed against the sides of the joint by means of internal spring forces. They can accommodate large translational movements parallel to the line of the joint, but, depending on the exact type of seal, only limited movement transverse to the joint.

A 2.1.12 Aspects of joint design
A 2.1.13 Joint sealing principles, schematic
A 2.1.14 Examples of horizontal joints for draining facade run-off water (left side = outside)

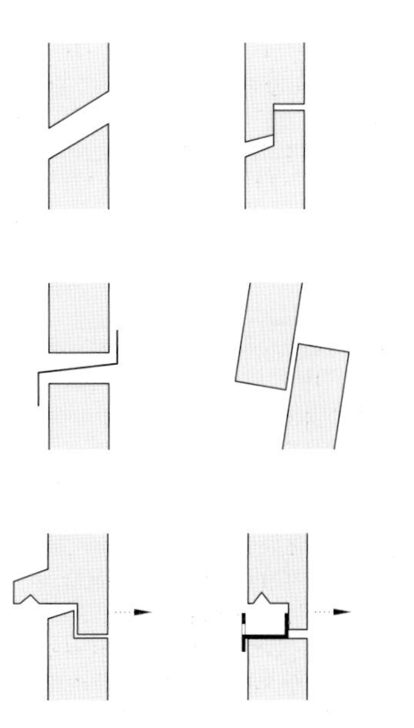

A 2.1.14

Jointing elements relying on external pressure
The application of an external force achieves contact between the jointing element and the sides of the joint over the entire contact area. The pressure acts on small joint–component interfaces in the case of preformed gaskets. Capillary water is stopped in the voids and wind pressure diminished through turbulence. Relative movements are difficult to accommodate. It is important to hold the jointing element in position.

Combinations
The basic principles can be combined to form more complex, highly efficient, usually multi-stage jointing systems. As the barrier effect must always been considered in a relative sense, supplementary measures (e.g., glass rebate ventilation/drainage) should be provided to compensate for failure or partial failure of the jointing system. In a two-stage system the first, outer barrier prevents the ingress of surface water and the second, e.g., with a preformed cellular gasket, the passage of air. Turbulence in the intermediate gap (in labyrinth form if required) reduces the wind pressure even further and any water that has managed to pass the outer barrier can drain away.

Erection sequence
Generally, joints based on the overlapping principle can be divided into two categories in terms of erection and dismantling:

- The individual parts can be erected only in a certain order and dismantled only in the reverse order. The replacement of individual parts in such a chain is only possible to a limited extent and not without damaging parts (e.g., jointing elements or rebates). Special solutions may be required for joining and sealing parts which have been replaced or refitted (e.g. tongue-and-groove and groove + loose tongue joints in fig. A 2.1.13).
- There is no fixed sequence for the erection and dismantling of individual components, and replacement within the system is possible at any time (e.g., open-drained, cover strip, and sealing compound joints in fig. A 2.1.13). This principle is especially recommended when there is a risk of damage (e.g., at the base) and there is a high likelihood of replacement.

Structure perpendicular to the plane of the facade

From "monolithic" to multi-layer/multi-leaf
A homogeneous envelope construction made from primarily just one material (often called "monolithic") is hardly likely to fulfil the stringent thermal performance requirements placed on modern building envelopes.
Differentiated assemblies splitting up the individual functions between different layers and leaves made from certain materials with a certain structure allow the performance profile of the facade to be matched very accurately to the requirements. The variability of layers or leaves enables the properties of the envelope to track changing external conditions. Furthermore, individual layers and leaves can be added later, or replaced by others, which allows the building envelope to be adapted to other requirements profiles during the course of its life. For instance, an external weatherproof leaf applied as a "wearing course" can be renewed after a certain period of exposure without the underlying facade structure having to be changed. This principle is also useful for subsequent refurbishment and optimisation of existing external wall constructions.
The allocation of individual functions to layers and leaves can also have disadvantages, depending on the quality of the materials chosen and the method of construction:

- the creation of many interfaces between different materials and components carries risk of incompatibility between those materials
- an increase in the proportion of joints and hence potential weak spots
- the creation of uncontrolled voids and cavities
- fixing problems: penetration of water run-off or thermal insulation layers, creation of bending moments when anchoring facing leaves
- high production costs
- increased maintenance costs
- several trades and responsibilities within one wall construction, meaning increased coordination work and overlapping liabilities
- problems with separation and hence disposal of different layers.

We have noted the following trends in the current economic climate:

- an increase in the efficiency of functional layers/leaves
- a reduction in the space requirements of layers (e.g., vacuum insulation), right down to miniaturisation of functional structures (e.g., prismatic light redirection systems with a thickness < 0.1 mm)
- surface coatings from the field of nanotechnology
- combining of several functions in one polyvalent layer.

The tasks of layers and leaves
The following functions (singly or in combination) can be realised in their own layers or leaves, for example:

- visual effects, information medium
- physical protection
- protection against driving rain
- airtightness
- preventing/reducing vapour permeability
- redirecting/scattering light
- reflection of light/heat radiation
- absorption of heat radiation
- reflection of electromagnetic radiation
- absorption of sound
- reflection of sound
- heat storage
- reducing the heat transmission
- carrying and transferring of loads
- dissipation of heat
- absorption and release of water vapour
- conversion of solar energy into thermal or electrical energy.

Further layers and leaves arise depending on the requirements of the construction, for example:

- removal of water vapour
- drainage of condensation water or infiltrating surface water
- compensating for unevenness
- layers for bonded joints (adhesive layers)
- measures for stabilising layers (e.g., preventing the "swelling" of thermal insulation)
- supporting constructions for coupling layers and leaves
- separating layers necessary because of incompatible materials
- sliding layers for unrestrained movements.

Typical assemblies and how they work
Fig. A 2.1.15 shows a small selection of assemblies in schematic form. They are classified according to functional and constructional criteria (see "Classification of design principles" section at the start of this chapter on p. 27). The number and thickness of layers and leaves varies quite noticeably. We can divide the assemblies into solid and lightweight forms, suitable for use in temperate climate zones.

Protection against driving rain
When absorbent materials are used, protection against frost is necessary and any moisture entering the material must be able to evaporate completely from time to time. Drainage of the facade run-off water is also possible at several positions within the cross-section. In a design that uses a ventilated weatherproof leaf with open drained joints, some of the run-off water drains down the rear face of this leaf. In this case there is a lower risk of soiling because any dust and dirt collecting on horizontal ledges is washed off regularly.

Airtightness
Air barriers – which must be positioned on the outside of layers of thermal insulation – are particularly effective when the wind pressure is first diminished by means of baffles closer to the outside face. Joints must be designed as overlaps.

non-permeable
non-variable

loadbearing or non-loadbearing
single layer
single leaf
no ventilation cavity

Internal structure of material determines performance, adaptability by way of wall thickness only, infiltrating moisture must be able to evaporate completely from time to time.

non-permeable
non-variable

loadbearing or non-loadbearing
multiple layers
single leaf
no ventilation cavity

Improved thermal performance with layer of insulation, wearing and protective layers/leaves inside and outside, interior benefits from heat storage capacity.

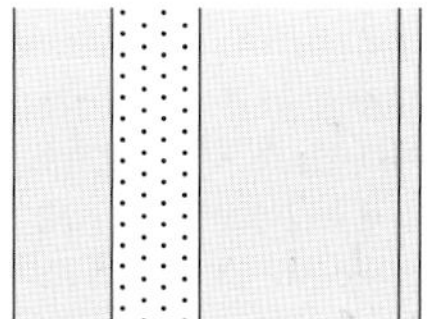

non-permeable
non-variable

loadbearing or non-loadbearing
multiple layers
two leaves
no ventilation cavity

Outer leaf provides robust physical protection to insulation and also protection against driving rain, outer and inner leaves may be partly coupled together, but do not form a single constructional unit.

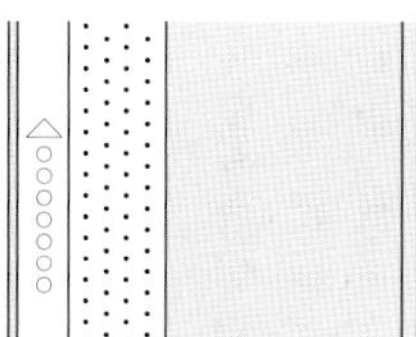

non-permeable
non-variable

loadbearing or non-loadbearing
multiple layers
two leaves
ventilation cavity

Replaceable facing leaf, fixings should not hinder rising air currents, condensation water and infiltrating moisture are reliably removed, air inlet and outlet openings required.

non-permeable
variable for purpose of energy gains

loadbearing or non-loadbearing
multiple layers
three leaves
vent. cavity behind outer leaf

Ventilated leaf of light-redirecting louvres, light-permeable leaf of transparent thermal insulation in front of solid absorber wall, total assembly not light-permeable, energy gains are variable and possibly controlled via control circuit.

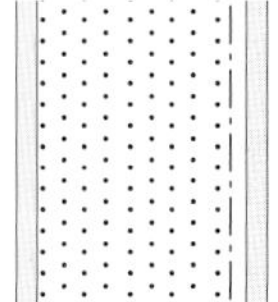

non-permeable
non-variable

loadbearing or non-loadbearing
multiple layers
single leaf
no ventilation cavity

Lightweight construction, inner and outer leaves usually coupled together to form a single constructional unit, barrier on inside prevents vapour trap, in the form of a timber stud wall also part of the loadbearing structure, sandwich construction in special cases.

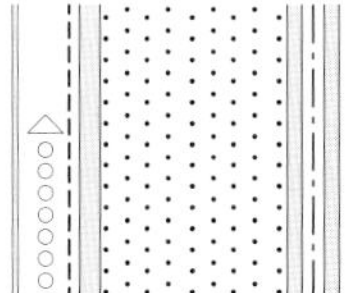

non-permeable
non-variable

loadbearing or non-loadbearing
multiple layers
two leaves
ventilation cavity

Ventilated protective and wearing leaf on the outside, diffusion resistance decreases towards the outside, separate layer for airtightness, interior lining as separate layer.

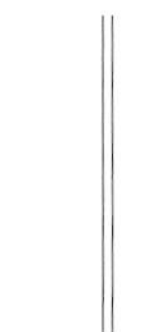

light-permeable
non-variable

non-loadbearing
single leaf

Assembly itself does not provide energy gains even when permeable to solar energy, which is absorbed by components in the interior, no thermal insulation.

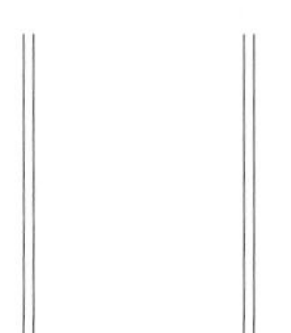

light-permeable
non-variable

non-loadbearing
single layer
two leaves
with/without ventilation cavity

Low thermal insulation value because air circulates in cavity (heat losses due to convection), leaves do not form a single constructional unit, risk of condensation water collecting in cavity.

light-permeable
if required, variable and regulated

non-loadbearing
multiple layers
single leaf

Functional unit comprising several light-permeable or light-redirecting layers, if necessary with radiation-reflecting coatings, light permeability variable if required.

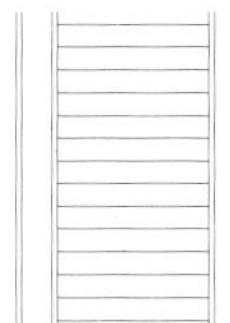

light-permeable
variable if required

non-loadbearing
multiple layers
single leaf

Functional unit comprising several light-permeable layers/leaves, improved thermal performance with transparent thermal insulation, light permeability variable if required and self-regulating, e.g., by use of thermotropic glass.

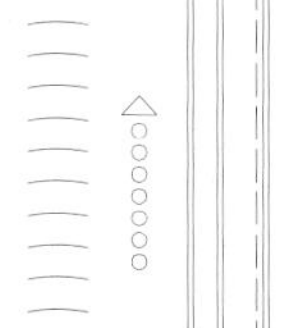

light-permeable
variable

non-loadbearing
multiple layers
two leaves
ventilation cavity

Good thermal insulation thanks to two stationary air/inert-gas layers and, if required, radiation-reflecting (low-emissivity) coatings, adjustable or rigid louvres as facing, ventilated leaf.

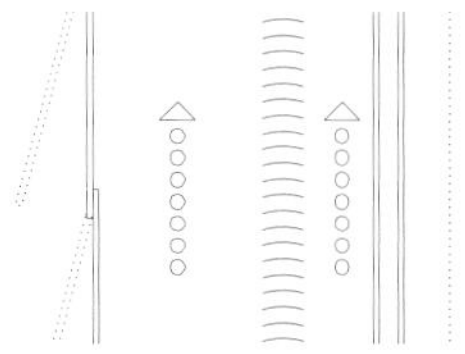

light- and air-permeable
variable

non-loadbearing
multiple layers
four leaves
ventilation cavity

Double-leaf facade, outer and, if required, inner glazing can be opened, air cavity between leaves with controlled ventilation, louvres and antiglare protection on the inside of some leaves to regulate light permeability.

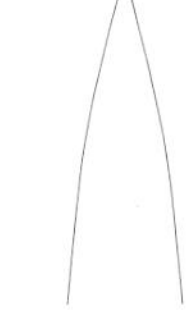

light-permeable
non-variable

non-loadbearing
multiple layers
single leaf

Pneumatic structure with light-permeable layers which, depending on the system, form a single constructional unit, therefore single-leaf construction.

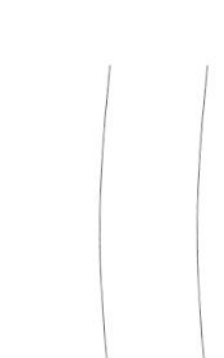

light-permeable
non-variable

non-loadbearing
single or multiple layers
two leaves

Membranes as two independent leaves, air cavity and, if required, controlled ventilation to remove water vapour and heat, but heat losses due to convection.

opaque material structure
light-permeable material structure
opaque thermal insulation
transparent thermal insulation
light-redirecting system

ventilated cavity
airtight barrier
vapour barrier
reflected radiation

scattering of light, antiglare protection

A 2.1.15 Structures/assemblies perpendicular to the plane of the facade (left side = outside)

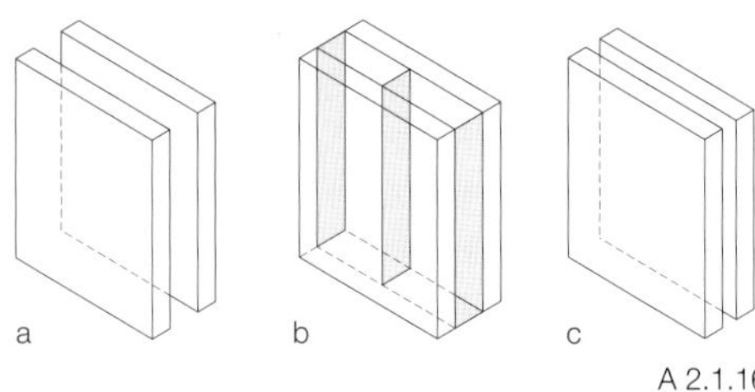

A 2.1.16

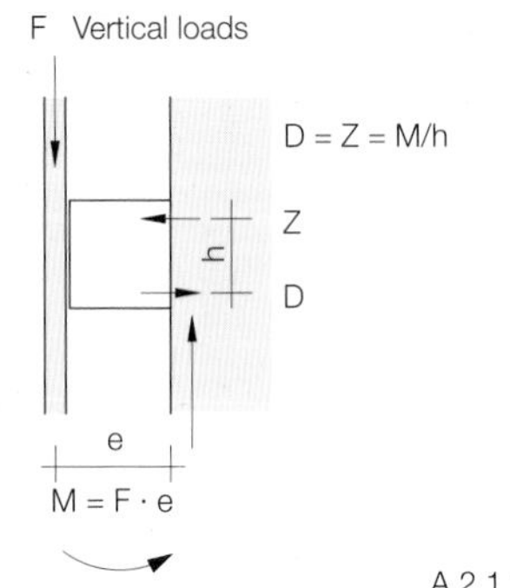

A 2.1.17

Geometry	discrete linear entire surface area
Detachability	detachable
	not detachable
Method	interlocking connections force-transfer connections material bond
Stresses	compression tension bending shear torsion
Movement	non-sliding
	sliding in one direction sliding in two directions
Adjustability	not adjustable
	in one direction in two directions in three directions

A 2.1.18

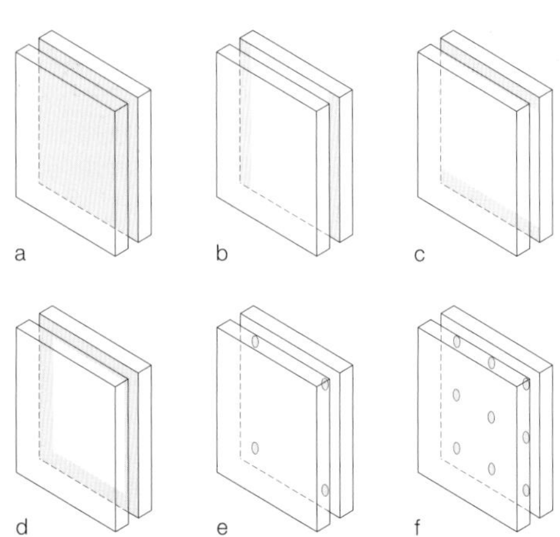

A 2.1.19

Thermal performance
Layers of material with a large proportion of stationary air inclusions guarantee good thermal insulation properties. Open-pore insulating materials that can absorb moisture and water through capillary action and hence ruin their function, need to be protected against moisture.

Water vapour diffusion
The vapour diffusion resistance of the layers should generally decrease from inside to outside in order to combat the formation of condensation water within the component (avoidance of a vapour trap). Condensation water that accumulates within the facade construction during the heating period must be able to evaporate completely during the warmer months.

Ventilation
An effective ventilation cavity behind a facing leaf should be at least 20 mm deep and be provided with adequate air inlets and outlets measuring at least 50 cm^2 per 1 m of facade length [7]. This enables moisture (facade run-off water and/or condensation) and heat (summer thermal performance) to be carried away efficiently. Stationary air layers (no ventilation) act as additional thermal insulation.

Heat storage
Layers on the inside with good heat capacity properties can be activated to regulate the interior climate.

Sunshading
Sunshading devices are most effective when fitted on the outside to reduce the energy gain via radiation-permeable layers. The ventilation cavity behind the devices counteracts the heating-up of the surfaces that would otherwise radiate heat into the interior. The characteristics of such functional items mean that they can be called leaves.

Coupling of layers and leaves
Layers and leaves must be connected together to form a constructional unit – the facade. Functional and performance aspects are more important than constructional aspects for determining the sequence. Different loads are effective depending on the position of the functional layer/leaf within the assembly. Owing to their material properties and/or thicknesses, certain planar components cannot fully or even partly withstand and transfer forces (e.g., thin foils/films, soft fibrous insulating materials, loose fill, air cavities). The load-carrying abilities therefore call for clear hierarchies that determine which planar component is carried by which other component.

The designation of a functional level of a facade assembly as a layer or leaf depends on the degree of structural autonomy.

Layers have, at best only, very limited loadbearing capacity and/or are parts of a constructional unit further up the structural hierarchy. Examples: structurally irrelevant foils, films and coatings, air cavities, insulation, plaster and render, the individual panes of double/triple glazing, the individual membranes of an air-inflated structure.

Leaves are essentially loadbearing, partly or completely three-dimensional and/or structurally autonomous. A leaf can comprise several layers. Examples: the inner and outer leaves of double-leaf facades, components separated by air cavities (e.g., for ventilation) or non-loadbearing layers of insulation.

As a rule, additional constructions link the separate leaves unless each leaf is a stable unit on its own.

A construction (e.g., posts and rails) is either higher up the structural hierarchy and links together several leaves, or it acts as a supporting construction (e.g., brackets) connecting a constructional component further down the structural hierarchy (e.g., facing leaf) to a more important structural element. In the latter case the vertical loads on the lower-hierarchy leaf cause bending moments due to the distance e (= lever arm) which then have to be resisted by the supporting construction or the leaf higher up the structural hierarchy. Fig. A 2.1.17 shows that increasing the distance h between the fixings greatly reduces the tension and compression forces to be transferred. This does not affect the shear load; however, wind suction forces also subject the fixings to tension.

Anchors or fixings for facing leaves often have to penetrate thick layers of insulation, which leads to a large lever-arm effect. Metal fixings with good thermal conductivities represent thermal bridges at which condensation water can collect. These fixings must therefore be made from a rustproof material – even galvanised steel is not permissible [7]. The insulating material must be packed tightly around the fixings in order to avoid aggravating this weak spot in the construction even further. It is advisable to minimise the cross-sectional area via which the heat can flow. Another possible approach is to isolate the fixing itself or the connection by means of a thermal break. The fixings must include suitable drips perpendicular to the construction to guarantee that facade run-off or condensation water is not drawn into the layer of insulation or other layers and leaves by way of adhesion.

The coupling of layers, in contrast to leaves, is less problematic owing to their closer spacing. Wherever possible, fixings should not penetrate or damage functional layers (e.g., weatherproof leaf, waterproofing, airtight barrier, vapour check, thermal insulation), otherwise its performance is reduced, leading to problems and, in

A 2.1.16 The effects of adding functional planes
Position of surfaces in relation to each other:
- a with gap but not coupled
- b with gap and coupled via supporting construction
- c without gap and coupled directly without supporting construction

A 2.1.17 Force relationships for fixing facing leaves
A 2.1.18 Criteria for fixing layers and leaves
A 2.1.19 Fixing of planar components
- a over entire surface
- b linear, vertical
- c linear, horizontal
- d linear, peripheral
- e discrete
- f discrete

A 2.1.20 Examples of principles for supporting facing leaves

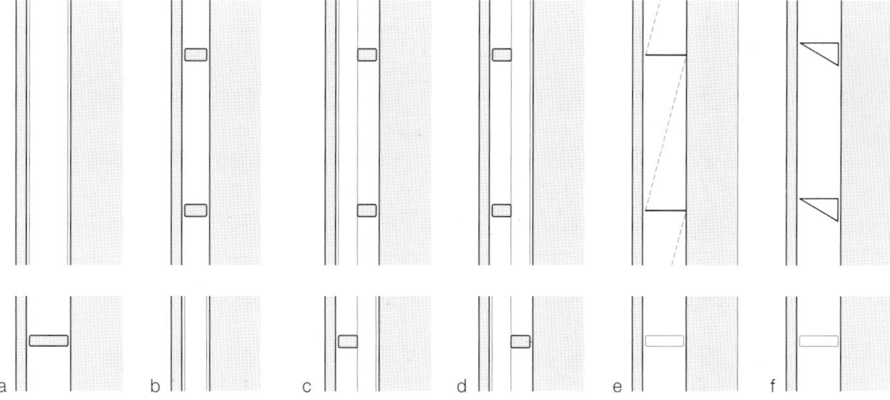

A 2.1.20

the end, damage to the construction. Uncontrolled voids, cavities and continuous joints should always be avoided (e.g., by staggering the joints). Air cavities between leaves should generally be ventilated and, if necessary, drained. Horizontal supporting constructions must not reduce the ventilation cross-sections. Air cavities must be permanently protected against birds, insects and rodents by gratings, perforated plates or nets. Direct contact between functional layers/leaves or with fixings must be avoided when incompatibility between materials is anticipated. This is also important when direct contact is not possible but water could act as a medium promoting incompatibility.

Fixing strategies
There are many ways of fixing of layers together or leaves to supporting constructions (and vice versa). The following points must be considered in particular:

- reliable transfer of all actions
- if required, unrestrained support for components, with anchor and sliding points
- clarification of the erection sequence and subsequent replaceability
- demarcation of interfaces between different trades and/or companies
- adjustability when connecting components by different trades with differing manufacturing tolerances.

Fixing of facing leaves
Facing leaves or facings with a ventilated air cavity behind, are fixed to and clear of (space required for insulation and/or ventilation) planar components higher up the structural hierarchy by means of supporting constructions. A suspended (top-supported) arrangement should always be preferred to a standing (bottom-supported) arrangement. Supporting constructions can be divided into several basic types (see fig. A 2.1.20). Which type is the best solution in any particular case depends on:

- size and weight of individual facing elements
- facing fixing options (e.g., discrete or linear force transfer)
- ventilation requirements
- options for fixing and transferring forces to a leaf higher up the structural hierarchy (e.g., can high tension forces be transferred and resisted?)
- performance aspects (significance and risks of thermal bridges).

Very heavy facing leaves or other elements (balconies, plant trellises, etc.) projecting beyond the thermally isolating envelope should be separate constructions with, if necessary, their own foundations to carry vertical loads. The leaves should then only need to be an-chored to transfer horizontal forces and, if necessary, to prevent buckling.

Principles of supporting construction (fig. A 2.1.20):

a posts
b rails
c and d
vertical and horizontal loadbearing elements; ventilated air cavity and, if required, drainage should not be impaired by horizontal loadbearing members – variation d presents problems in this respect
e supporting construction consisting of tension/compression members and diagonal hangers to carry vertical loads – if necessary in combination with further linear loadbearing members (vertical or horizontal)
f separate brackets cantilevering out from the loadbearing leaf – combinations with further linear loadbearing members (vertical or horizontal) are conceivable

Adjustment of connections
The following basic strategies enable adjustment:

- shims, washers, spacers
- threaded spacers
- elongated holes or rails (e.g., "Halfen" cast-in channels)
- oversized pockets, later filled with grout
- unrestricted, adequate positioning options for attaching fixings to surfaces, e.g. bonded fixings (areas of adhesive, welding plates), accurate positioning of bolts, anchors, etc. during erection

Notes

[1] VDI directive 2221, Düsseldorf, 1993, p. 39
VDI directive 2222, Düsseldorf, 1996, p. 5
[2] VDI directive 2221, Düsseldorf, 1993, p. 39: "Effect: the never-changing, foreseeable physical, chemical or biological effects governed by the laws of nature".
[3] The revised classification is based on typological investigations within the scope of a research project on building envelopes: Herzog, Thomas; Krippner, Roland: Gebäudehülle. Synoptische Darstellung maßgeblicher baulicher Subsysteme der Gebäudehülle mit Schutz- und Steuerungsfunktionen als Voraussetzung für die experimentelle Arbeit an ihrer energetischen und baukonstruktiven Optimierung (final report – unpublished manuscript). Munich Technical University, 2000. Herzog, Thomas; Krippner, Roland: Synoptical Description of Decisive Subsystems of the Building Level. In: Pontenagel, Irm: Building a new Century. 5th Conference on Solar Energy in Architecture and Urban Planning, Proceedings; Eurosolar (ed.), Bonn, 1999, pp. 306–10.
[4] See: Themeninfo I / 02 "Schaltbare und regelbare Verglasungen"; BINE Information Service (ed.), Karlsruhe, 2003.
[5] Leaves have been given different, sometimes contradictory, definitions in the literature. The definition given here seems to be the most plausible one. Confusion reigns when the classification refers to only one type of construction (e.g. single-leaf concrete wall) and not to the complete system of the envelope (e.g. double-leaf construction with concrete wall plus weatherproof leaf of profiled aluminium cladding). See: "Coupling of layers and leaves" in this chapter.
[6] The description and, to a certain extent, the breakdown of the jointing systems is based on the following research report: Scharr, Roland; Sulzer, Peter: Beiträge zum methodischen Vorgehen in der Baukonstruktion. Außenwanddichtungen; VDI (ed.), Düsseldorf, 1981. The analysis of built construction elements and the composition of jointing systems in the external walls of buildings is investigated and demonstrated by means of scientific methods.
[7] See: DIN 18516 part 1; Berlin, 1999; does not apply to "small-format slabs" with an area 0.4 m^2 and a self-weight ≤ 5 kg.

A 2.2 Edges, openings

Edges

Up to this point we have considered the building envelope as a continuous surface, and its structure in terms of its thickness. However, as the surfaces of the building envelope are finite, each surface is defined by its edges. Whenever the constructional, functional and architectural properties within the building envelope change, we can speak of definable, different areas. As a rule, the changes relate to the permeability.

Openings are those parts of the building envelope permeable to flows of energy and materials. This is generally the case for parts that can actually be opened completely, such as windows. However, it seems reasonable to extend the term "opening" to include the relationship with the respective physical process. For example, a rooflight or skylight is an "opening" in the roof surface through which light enters.

The change in the properties (performance) is also linked with a change in the construction. The term "edge" as used in this chapter designates not the edge of a component, which as an individual part is joined together with many identical parts to form a whole (e.g., a clay brick in a masonry wall), but rather the transition from surface to opening.

Reveal

The depth of the reveal is primarily determined by the construction of the facade (fig. A 2.2.5). The depth can be increased by the use of additional elements, but cannot be decreased. The geometrical form of the reveal has a direct influence on the admittance of daylight and the visual relationship between inside and outside. Fig. A 2.2.4 illustrates a number of basic features. The design of the reveal surfaces is related to the (constructional) termination of the components used at the openings (e.g., windows). For example, the reveal can also be used to (re)direct daylight into the interior. Besides the geometry, the properties of the respective surfaces are also important here. There is also always a direct relationship between the depth of the reveal and the size of the opening, and the latter is always related to the surface of the facade itself. The plastic effect of the facade within its immediate surroundings is essentially created by the offsetting of the individual surfaces within the facade and the resulting shadows.
Constructional aspects of reveal design are:

- the transfer of wind loads
- carrying the dead loads of the construction
- sealing against wind, precipitation, etc.

Openings

If the interior of building is to be used and provided with light and air, then openings in the building envelope are unavoidable. The protection and supply functions result in a need for the openings to be provided with variable permeability because there is a desire to create constant internal conditions despite fluctuating external conditions. The openings in the building envelope also have the task of creating a link between inside and outside, i.e., enabling a controllable exchange between interior and exterior climates. The individual parameters such as heat, light, air, sound, moisture, all fall under the expression "regulation of permeability".

We use "opening closure elements" [1] to regulate the permeability. The best-known form is the window, which, by using suitable materials, controls the light permeability even when closed, but enables an exchange of air only when it is open. Of course, the lighting and ventilation functions can also be kept separate. The simplest form of this is a fixed window with a separate (opaque) opening for ventilation [2].

The appearance of the first facades with large expanses of glass (e.g., palm houses) in the 18th century and the erection of structures like the Crystal Palace in London (1851) or the Glaspalast in Munich (1854) mark the consummation of a transition. The window had been transformed from a transparent element in an opaque wall surface to the opening element in a totally transparent facade. Like the windows in a solid wall construction, the opening elements in a (transparent) glass facade are therefore also called windows.

Position and geometry

The arrangement and geometrical form of the opening is always related to the interior layout behind. The position and geometry of the opening have a fundamental effect on the admittance of daylight, the ventilation and the occupants' view of the outside world. There is also a relationship between the use of the interior and the horizontal and vertical positions of the openings. A change of use and hence a change in the plan layout can alter the horizontal relationship to the openings. On the other hand, the relationship to the vertical arrangement of the openings cannot normally be changed because raising or lowering the floor levels is not possible.

Dividing up the area of the facade
The facade to a storey can be divided into three principal areas (fig. A 2.2.2):

- high-level windows
- eye-level windows
- low-level windows

At the top and bottom (junctions) we also have the following additional terms, which originate from perforated (fenestrated) facades:

- lintel area: the area between window/door and underside of floor above

A 2.2.1 Private house, Paderborn (D), 1995, Thomas Herzog

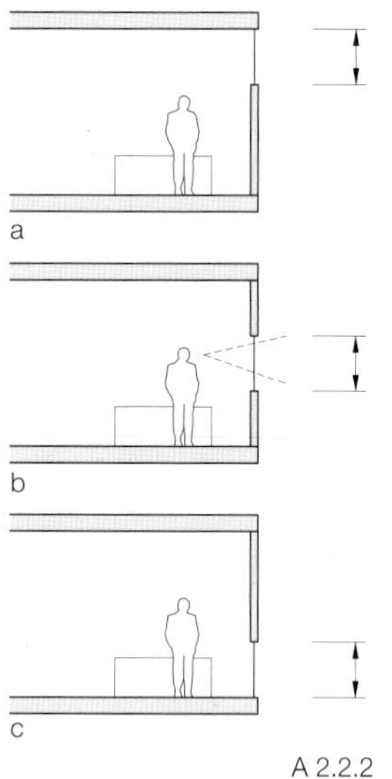

A 2.2.2

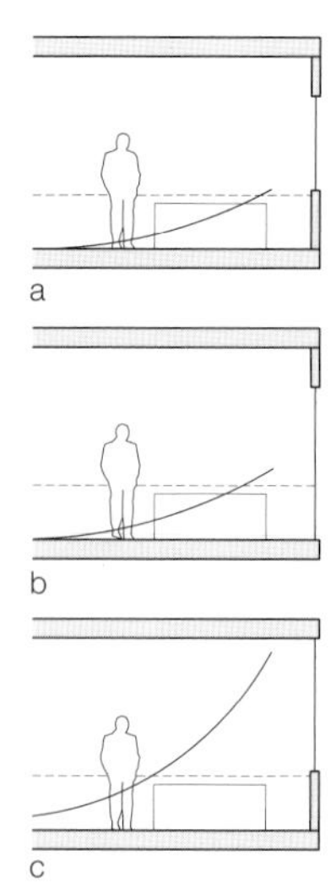

A 2.2.3

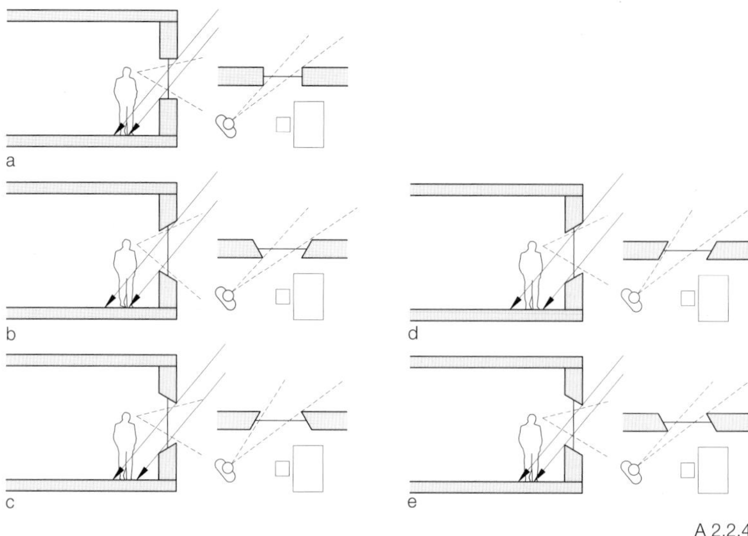

A 2.2.4

- threshold area: the area between (glazed) door and top side of floor below

Visual relationships
The desire for a supply of fresh air is often coupled with the need to stand directly adjacent to the opening (at an open window). Therefore, when designing an opening, contact with the outside world must also be considered in addition to the manual operation of fittings to open the window. The opening must make this possible but, on the other hand, still form a boundary with the outside. In this respect we distinguish between visual and physical links. The average eye heights for different postures are as follows [3]:
- approx. 1750 mm when standing
- approx. 1300 mm when seated and working
- approx. 800 mm when seated
- approx. 700 mm when lying down (300 mm above floor level).

Both the position and the subdivision of the opening must match to the type of utilisation and the position of the occupants.

Lighting
The amount of daylight entering via the facade decreases as we proceed further into the interior (fig. A 2.2.3 [4]). The amount of light is expressed by the daylight factor D as a percentage. This specifies the ratio of interior to exterior illuminance (diffuse light only) under normal conditions [5]. The external influencing variables are:

- orientation in relation to compass direction
- time of day
- local factors influencing the incidence of solar radiation (climatic conditions, local shading due to vegetation and/or buildings).

In the plane of the facade the position and geometry of the openings is crucial. High-level window openings increase the amount of incoming daylight.
The actual level of illumination within the room is essentially determined by the degree of reflectivity of the inner surfaces, which depends heavily on the main colours [6].

Ventilation
Expressed simply, ventilation means the "replacement of interior air by exterior air" [7]. The renewal of the air in the interior fulfils hygiene requirements and also contributes to other performance aspects (e.g., removal of pollutants from the air, removal of moisture). In ventilation we make a fundamental distinction – based on the driving forces – between mechanical ventilation (air movement driven by mechanical forces) and natural, i.e., non-mechanical, ventilation. The latter employs pressure differentials between the interior and exterior to move the air. These pressure differentials are a result of forces caused by natural conditions [8]:

- Wind forces:
  Pressure differentials between the interior and exterior, induced by the wind adjacent to the facade, which bring about an exchange of air.
- Thermal buoyancy:
  Forces caused by different densities due to temperature differentials (thermal stratification); as the wind pressure increases, so this is superimposed on the thermal currents.

Fig. A 2.2.6 [9] illustrates the basic principle of the exchange of air at an opening in a facade due to thermal stratification (ignoring wind influences). There is no movement of air in the region of the theoretical neutral zone N. This zone can be shifted up and down by changing the vertical position of the opening, or by taking wind forces into account.
Besides the shape and arrangement of ventilation openings in the facade, their variability is a crucial factor in the context of the physical properties of the envelope and the mass of a building [10].
Continuous ventilation requires small, easily regulated ventilation openings. The flow of air within the interior is important here because this form of ventilation takes place over a longer period of time:

- Single-sided ventilation:
  Efficient utilisation of thermal buoyancy requires two openings spaced as far apart as possible vertically; easily adjustable regulation prevents undesirable cooling effects and draughts.
- Cross-ventilation:
  In order to exploit thermal buoyancy in this case, the vertical distance between the air inlet and air outlet should be as large as possible; however, this distance is not important in the case of wind-induced pressure differentials.

Surge ventilation requires openings with the largest possible ventilation cross-section:

- Single-sided ventilation:
  Owing to the neutral zone in the middle of the opening, the area can be divided into two parts horizontally.
- Cross-ventilation:
  Owing to the cross-ventilation effect, the air flows in one direction only.

When we consider comfort, the global air change rate is only one factor; the movement of the air is also relevant [11]:

- the air velocity at the inlet
- the maximum air velocities occurring in the room
- the average velocity of the air in the room
- the average velocity of the air at the level of the occupants (1 m above floor level).

An air velocity of 0.2 m/s is regarded as the maximum comfortable level. In offices and buildings with a similar usage, papers will be disturbed when the air velocity is higher than this [12]. A draught is defined as a flow of air

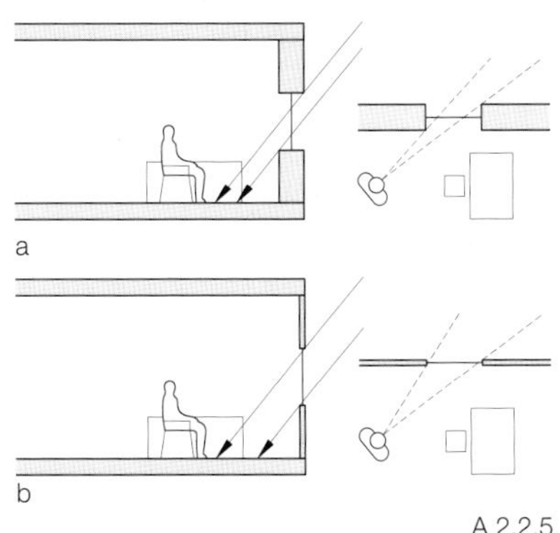

A 2.2.5

A 2.2.2 Areas derived from the utilisation
a high-level window
b eye-level window
c low-level window
A 2.2.3 How position and size of opening affect daylight penetration
a middle
b lower
c higher
A 2.2.4 How the shape of the reveal (same form on all sides) affects daylight penetration and the occupant's view out
a parallel
b tapering inwards
c tapering outwards
d parallelogram, sloping inwards
e parallelogram, sloping outwards
A 2.2.5 How the thickness of the wall affects daylight penetration and the occupant's view out
a thick wall construction
b thin wall construction
A 2.2.6 Principle of air exchange through a facade opening based on thermal stratification but ignoring the effect of wind forces; neutral zone N at 1/2H

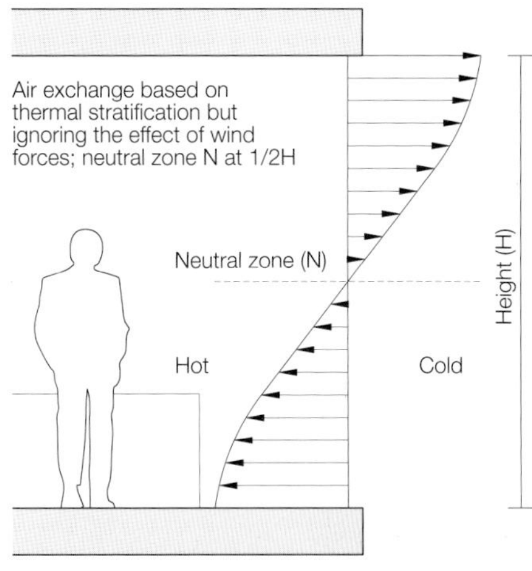

A 2.2.6

that causes an "undesirable local cooling of the human body" [13, 14]. So a draught is not an absolute value. We therefore speak of a draught risk [15]. To avoid draughts, it is best to allow the incoming air to be well distributed around the interior.

One problem with the "comfortable supply of fresh air" is the ingress of warmer exterior air in summer and the draughts which can be expected in winter due to the ingress of colder exterior air (exacerbated by the cold air descending down the facade). Local units to preheat/precool the incoming air positioned near the facade openings can help to combat this problem.

As the operation of mechanical ventilation is more predictable than the fluctuations of external conditions, many approaches and studies consider artificial ventilation as the primary source of ventilation. It is only in recent years that we have seen designers taking into account the fluctuating conditions of natural ventilation in models and measurements. As our understanding of natural ventilation grows and the importance of using natural energy sources increases, so ventilation via the windows will gain in significance. In a similar way to mechanical ventilation, where exact values are available for all the components, windows also require calculation of aerodynamic variables for the air entry points (window gap, shape of profile). Some of the effects familiar in air-conditioning systems can also be transferred to windows.

Displacement ventilation, which is characterised by relatively low air velocities, tries to separate the upward, displacing flow of incoming fresh air and the polluted exhaust air. The fresh air is introduced at a lower temperature near the floor. Internal heat sources then draw up the incoming air from the floor-level layer, due to thermal buoyancy, and force out the exhaust air at ceiling level. As a rule, displacement ventilation is used in conjunction with mechanical ventilation. Displacement ventilation can be used with natural ventilation when the inlet openings in the facade enable a regular entry of fresh air into the interior at floor level. The "Coanda effect" can be exploited if the incoming air can penetrate as deep as possible into the room. When laminar jets of air are blown through slits not immediately adjacent to the ceiling but instead at a certain distance below, then the jet of air follows – "sticks to" – the surface due to the induced turbulence. This effect is occasionally called the turbulent boundary layer effect [16]. This effect, familiar in mechanical ventilation systems, can also be transferred to window-based ventilation in certain circumstances. The flow of external air is guided over smooth surfaces as tangential ventilation. The effectiveness within the depth of the room is guaranteed by reducing turbulence to a minimum. The corresponding surfaces must be positioned in the immediate vicinity of the air inlet point. Furthermore, the position and geometry of the air inlet (window opening) must be taken into account.

The lower the temperature of the incoming air compared to the interior air, the greater the risk of draughts. The external air entering the interior can be preheated by positioning heat sources adjacent to the inlets. The incoming air should be able to heat up on components by means of convection.
If the guidelines regarding comfort are to be maintained, window-based ventilation is possible only above a certain external temperature. The literature on this subject gives minimum external temperatures of 0 to 6°C, depending on the type of window [17].
When the external temperature is close to the thermal comfort threshold, the incoming air should be able to reach the position of the occupants in the room as directly as possible without picking up heat from warmer components. During the hotter months, the incoming air can be cooled down (slightly) by convection as it passes over cooler components.
Thermally effective masses can release absorbed heat energy during night-time ventilation or as components cool down. To maintain the level of comfort, window-based ventilation during daytime hot periods is therefore of limited usefulness.

The arrangement of ventilation openings in the facade and the type of ventilation (single-sided ventilation or cross-ventilation) determines the depth of the room in which natural ventilation via openings in the facade is still effective. This also makes a major contribution to thermal comfort. Without considering more detailed information regarding the arrangement of the opening light, the rule of thumb that generally applies is that rooms with single-sided ventilation options can be ventilated by natural means when their maximum depth is 2.5 times their clear height (H). When cross-ventilation is possible, the maximum depth is 5 times the clear height [18]. Single-sided ventilation and a high-level opening reduces the maximum room depth to 2H. When a low-level opening is added, the effectiveness of the ventilation is increased to 3H [19]. These figures are in no way absolute and can serve only as guidelines, and the type of opening has not been considered.

Small window openings must be accurately positioned and detailed because in an otherwise sealed envelope the effect of the incoming air is like a jet of air being blown from a small nozzle! If it is not possible to regulate the ventilation by means of windows, additional elements (e.g., flaps) can be incorporated in the facade.

The tables in DIN 5034 for calculating the minimum window sizes for residential buildings are based on daylight requirements; the size of the opening for ventilation purposes cannot be derived from these tables.

A 2.2.7 Shading effects due to louvres; effect of compass direction
a south facade, horizontal louvres
b east/west facade, vertical louvres

A 2.2.8 Principles of sunshading; screening/filtering direct sunlight
a overhang: screening
b overhang as light shelf: shading to due to screening plus redirection for daylight utilisation
c louvres: screening
d louvres: screening plus redirection for daylight utilisationg
e covering: screening
f perforations: filtering

A 2.2.9 Classification of types of movement for windows

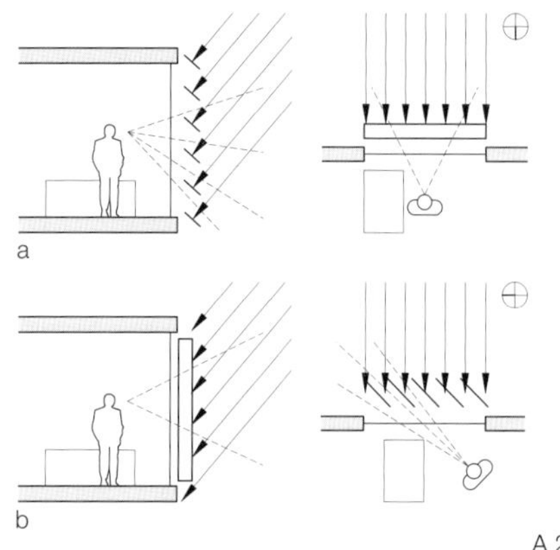

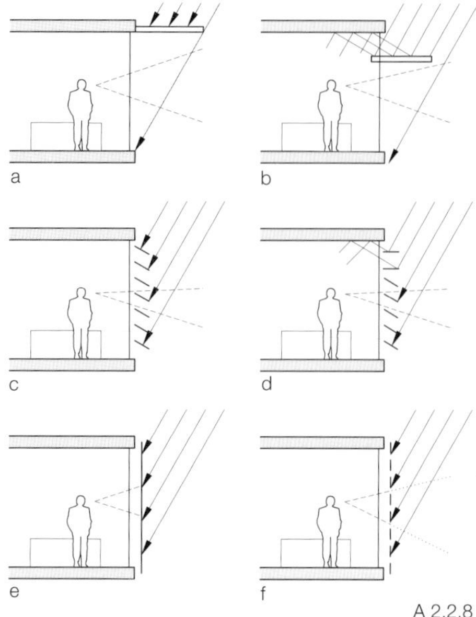

Changing the permeability

The property of permeability can be influenced by the construction. Both rigid and variable elements are used for this.

Rigid elements

As the incidence of solar radiation and hence the climatic conditions change in relation to time of day and season, the effects (shade, reflection, redirection of light) of non-variable elements also change, depending on the respective solar altitude angle.
We can make a distinction between the various shading principles (fig. A 2.2.8):

- total, direct shading of the facade surface
- shading by means of a cantilevering element
- shading by means of the cumulative effect of smaller elements (e.g., louvre or grid structures).

Louvre structures can be broken down into two categories in terms of their arrangement, resulting from their orientation and hence the associated solar altitude angle:

- horizontal louvres prevent solar radiation striking a south facade at a steep angle from penetrating to the interior
- vertical louvres prevent solar radiation striking an east or west facade at a shallow angle from penetrating to the interior.

Despite this shading effect, occupants can still enjoy a view of their surroundings (fig. A 2.2.7).

Variable elements

The chapter on manipulators (B 2.3, p. 258) deals with movable and variable elements at openings, using detailed examples. Here, we shall consider only the variability of windows. The primary function of the window is to provide an option for the partial opening and closing of the building envelope. Of all the various normal features of a window (material, form of opening, construction of frame, fixing to masonry) it is the form of opening (type of light), as the facade opening function, that determines the constructional and architectural properties of a window.
When distinguishing between different types of windows, the form of opening can be categorised by specifying one of four aspects with its respective distinguishing criteria (fig. A 2.2.9) [20]:

- facade surface, distinguished according to variability
- degree of variability
- type of movement
- other distinguishing features.

First aspect: variability of facade surface
In terms of variability, facade surfaces can be divided into fixed and opening areas. The window opening in turn is subdivided according to structural (load-bearing) and constructional (elements for fixed glazing and opening lights) aspects.
The size of the individual light-permeable areas depends on the availability of materials (e.g., panes of glass) and therefore defines the subdivision.

Second aspect: degree of variability
The degree of variability is determined by the degree of freedom, which in turn is determined by the frame, sash construction and type of fittings.

Third aspect: type of movement
The degree of variability can be further subdivided. The respective movement is reflected in the names of the different windows:

- Partial change of position, movement about a vertical axis (rotation):
  - vertical pivot
  - side-hung (left or right)
- Partial change of position, movement about a horizontal axis (rotation):
  - hopper (bottom-hung)
  - top-hung
  - horizontal pivot
- Complete change of position, but without changing the element (translation):
  - sliding
  - push-out
- Complete change of position plus a change to the element (transformation):
  - folding
  - roller
- Combinations of the above.

The folding windows customarily employed are, strictly speaking, side-hung sliding windows because the glazed area itself does not fold but instead is made up of several individual sashes. This is similar to the folding partitions used to subdivide rooms, where – at least in terms of the surface – the entire wall surface is "folded".
As part of the building envelope, one of the fundamental functions of the facade is to form a vertical partition between two areas. The types of movement can therefore be further differentiated with respect to the plane of the facade – generally outside/inside and above/below – for example:

- rotating: opening inwards/outwards
- pivoting: opening inwards/outwards
- sliding: horizontally (to the left/right) or vertically (upwards/downwards).

Other distinguishing features
We can also distinguish according to construction principles and the resulting features. Besides distinguishing according to number of lights, which applies to all movable surfaces, we can also use specific features for the respective type of opening.

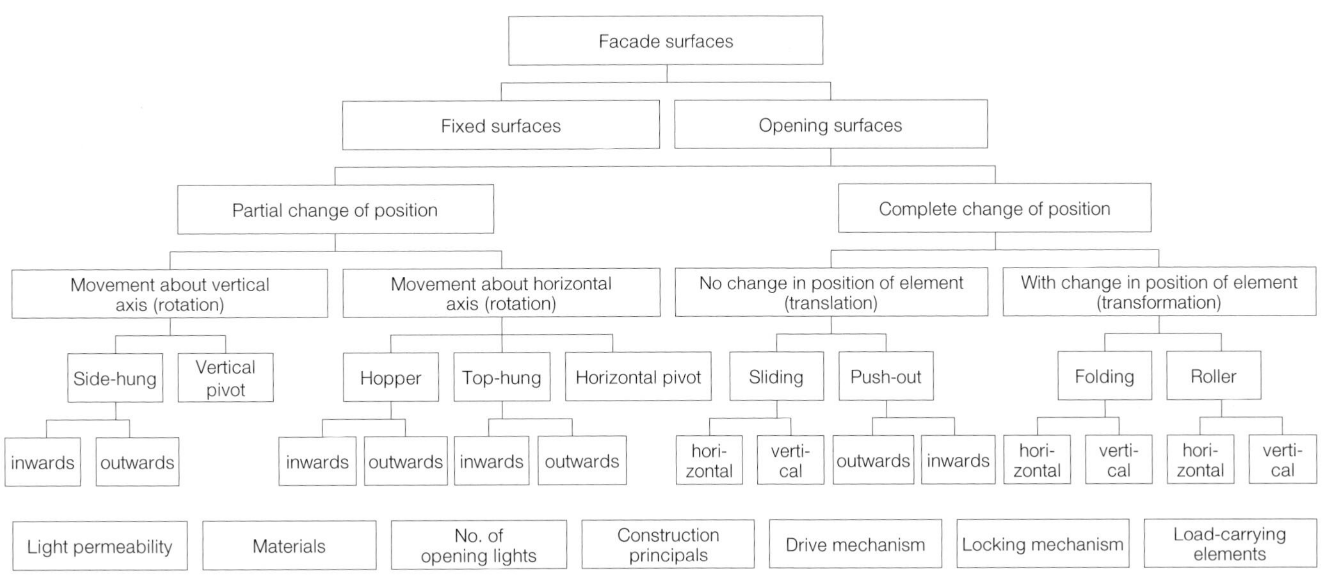

A 2.2.9

The number of lights (opening lights, lights opened only for the purpose of cleaning or maintenance, fixed lights) clarifies the various opening options. One distinguishing criteria that has gained importance in connection with controlled natural ventilation is whether the movement is performed manually or mechanically.

Specific construction principles describe the various types of opening. We can therefore differentiate only within a movement principle. Furthermore, when considering windows, regardless of any categorisation, there are features that concern primarily the construction and the opening mechanism only as a secondary aspect.

Performance spectrum for type of movement
The movement mechanisms exhibit different properties which are of fundamental importance for function, construction and design [21]. The performance spectrum of an opening element in the building envelope is primarily determined by functional properties (fig. A 2.2.10). To be able to use windows efficiently (in terms of energy balance and comfort of occupants) as components in the building envelope, we require detailed information about the types of movement and the associated performance profiles [22].

Combinations
The terms used illustrate the many types of movement which result from the combinations possible:

- side-hung with side-hung/sliding fittings
- side-hung hopper
- top-hung: projecting top-hung
- folding wall (rotation plus sliding)
- folding: folding/sliding
- horizontal pivot; horizontal pivot/sliding
- vertical pivot
- sliding: vertical, lifting/sliding, suspended sliding, drop (window/door), horizontal
- push-out light: bottom-hung/push-out, side-hung/push-out

The types of movement have evolved from a number of steps to a multitude of variations. Most of the variations that were in use in the mid-20th century are no longer produced today. The reason for this development, besides the problem of joints, is the higher demands placed on performance, which calls for heavier panes and hence places greater demands on fittings and frames. The problem of sealing joints has resulted in an adequate air change rate being sacrificed in order to reduce heat loss (partial optimisation) instead of considering the problem in its full context.

The facade system

As the building envelope cannot normally be produced in one piece, it is necessary to break it down into individual parts. When considering this system, the basic scientific terms can be broadened to five steps for the architect. This results in the following sequence (fig. A 2.2.11):

- system
- subsystem
- component
- element
- material.

The choice of scale, or rather step, may result in a shift within this system (e.g.. in urban planning: town = system, building = element).

Erection and assembly sequence
The process of building is linked with the chronological progression of erection and assembly. Besides the final condition, there are various interim states. Depending on the situation, external conditions can influence the progress of the building work. On inner-city sites in particular, the supply and removal of materials may be restricted, especially for larger building projects. In addition, climatic conditions have a direct influence on the progress of the building work. A change in the weather can lead to delays which have an effect on all subsequent work. The erection of a facade as protection against the weather enables the fitting-out of the building to take place more or less independent of the weather conditions.

Prefabricated components made from elements
So that progress on the building site can continue regardless of the weather (as far as possible), individual parts may be prefabricated off the site under controlled factory conditions. This can considerably reduce the actual erection time on site and the associated risks. Prefabrication also results in considerably better accuracy and tighter tolerances.
In fenestrated facades, the windows are fitted into openings in the facade construction. In addition, two different construction principles are employed for non-loadbearing external walls for facades with a high proportion of glazing. The difference lies in the erection procedure.

Prefabricated facade
This term is used for facades that comprise individual, prefabricated elements which are then assembled on site to create a complete facade. This designation is not based on the sequence of terms given above, but instead highlights the prefabrication and the erection procedure. In the case of glass facades the prefabricated parts are generally panes of glass fitted into frames. Prefabricated facades are highly suitable for high-rise office blocks. The prefabricated elements are lifted by crane to the corresponding place; erection takes place without scaffolding.

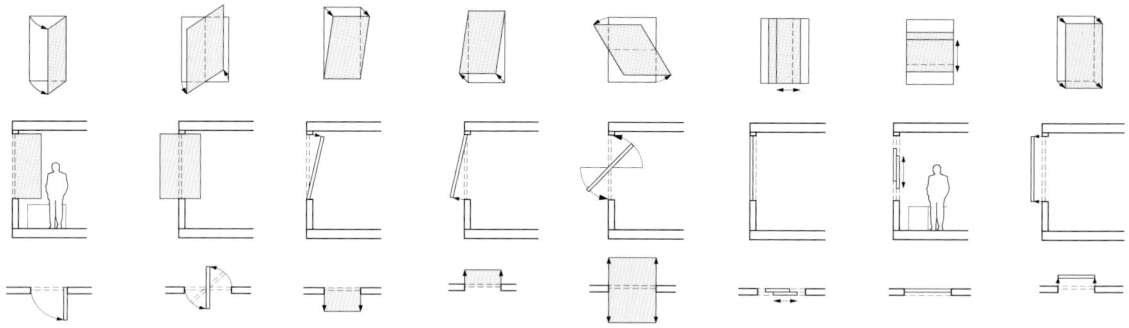

Comparison of types of movement for windows for determining their various performance profiles	Side-hung window opening inwards	Vertical-pivot window	Hopper window	Top-hung window	Horizontal-pivot window	Horizontal sliding window	Vertical sliding window	Push-out window
Impairment of usable floor area related to depth of room	width of opening	1/2 width of opening	minimal	none (when opening outwards)	1/2 width of opening	none	none	none (when opening outwards)
Options for locating in circulation areas	yes (when opening outwards)	no	yes	yes (when opening outwards)	only by limiting opening	yes	yes	opening outwards
View through: maximum unobstructed opening area and subdivisions	100 %	100 % with vertical division	no unobstructed opening	no unobstructed opening	100 % with horizontal division	50 % with vertical division	50 % with horizontal division	no unobstructed opening
Geometric description of minimum/small openings possible	1 gap at side, triangular openings top and bottom	2 gaps at sides, 2 triangular openings top and bottom	2 triangular openings at sides, gap at top	2 triangular openings at sides, gap at bottom	4 triangular openings at sides, gaps top and bottom	2 gaps at sides	top and bottom	peripheral gap
Geometric description of maximum/large openings possible	complete opening area	complete opening area, vertical division	2 triangular openings at sides, gap at top	2 triangular openings at sides, gap at bottom	complete opening area, horizontal division	50 % of opening area, divided vertically	50 % of opening area, divided horizontally	peripheral gap
Suitability for ventilation through gaps	limited	limited	limited	limited	limited	good	good	good
Suitability for surge ventilation	good	good	no	no	good	good	good	no
Adjustability of openings	no (only with additional fittings)	(only with additional fittings)	only for maximum opening position	by means of fitting necessary for opening	no	good	good	good (mechanical drive)
Protection against weather (against precipitation) with minimal opening	no	no	yes	yes	yes	no	top: yes bottom: limited	limited (with add. element over top opening)
Type of movement provides protection against being blown closed by the wind	no	no	no	with additional fittings	no	yes	yes	yes
Possibility of combining with internal manipulators	no	no	limited	yes	no	yes	yes	yes
Possibility of combining with external manipulators	yes	no	yes	no	no	yes	yes	limited
Possible to clean the outside from inside	yes	yes	with detachable fitting	no	yes	no	with additional (detachable) fitting	no
Notes regarding sealing	Also strikes seals outwards (during wind and rain)	Horizontal seals offset	Only limited rebate possible at bottom	Employed in regions with high winds	Vertical seals offset	Bottom rebate possible, contact pressure needs extra feature	Bottom rebate possible, contact pressure needs extra feature	No weather protection even at minimum opening
Notes regarding fittings	Cantilevering light generates moment	Load carried centrally	Secure light against falling	Light must be locked in open position	Light sags when open	Tall, narrow formats can jam	Compensation for self-weight, can jam	Scissors mechanism must transfer wind forces

A 2.2.10

Term	Example
System	Building
Subsystem	Envelope: roof, facade, loadbearing structure, services, interior layout, access
Component	Window light in frame (sash)
Element	Sections, insulating glass, fittings, seals
Material	Sheet metal, glass

A 2.2.11

A 2.2.10 Comparison of types of movement for windows for determining their various performance profiles

A 2.2.11 Basic terms for considering the system from the architect's viewpoint

Post-and-rail facade

In contrast to the prefabricated facade, the post-and-rail facade consists of individual parts: the vertical facade posts and the horizontal facade rails, which are connected together on site. The designation is based on the principle of the construction. Post-and-rail facades are primarily used for low-rise buildings these days.

Notes:

[1] Dietze, Lothar: Freie Lüftung von Industriegebäuden, Berlin, 1987, p. 18.

[2] This distinction was made, for example, by Le Corbusier at the Dominican Friary of La Tourette, 1957.

[3] Pracht, Klaus: Fenster – Planung, Gestaltung und Konstruktion, Stuttgart, 1982, p. 102.

[4] Graphic after: Müller, Helmut; Schuster, Heide: Tageslichtnutzung; in: Schittich, Christian (ed.): Solares Bauen, Munich/Basel, 2003, p. 63.

[5] VDI Guideline 6011, Düsseldorf, 2001.

[6] Miloni, Reto: Tageslicht-ABC; in: Fassade/Façade, 01/2001.

[7] Meyringer, Volker; Trepte, Lutz: Lüftung im Wohnungsbau, published by the Federal Ministry for Research and Technology, Karlsruhe, 1987, p. 11.

[8] The distinction between driving forces is related to the local situation on the building because wind forces are caused by climatic relationships, which are always attributable to incident solar radiation and hence temperature differentials.

[9] Graphic based on: Zürcher, Christoph; Frank, Thomas: Bauphysik. vol. 2: Bau und Energie – Leitfaden für Planung und Praxis, Zürich/Stuttgart, 1998, p. 80.

[10] ibid. [7], pp. 33–36.

[11] Givoni, Baruchi: Passive and Low Energy Cooling of Buildings, Van Nostrand Reinhold, New York/London/Bonn, 1994, p. 42.

[12] Please note that air velocities < 0.15 m/s can be perceived subjectively. Hanel, Bernd: Raumluftströmung, Heidelberg, 1994, p. 6.

[13] Fanger, Ole: Behagliche Innenwelt; in: Uhlig, Günther et al.: Fenster – Architektur und Technologie im Dialog, Braunschweig/Wiesbaden, 1994, p. 217.

[14] When there are a large number of closely-spaced windows the inlets are therefore positioned over radiators to guarantee the minimum air supply.

[15] Apart from noise, draughts are one of the main reasons for dissatisfaction with air-conditioning and ventilation systems. Recknagel, Hermann; Schramek, Ernst-Rudolf (ed.): Taschenbuch für Heizung und Klimatechnik einschließlich Warmwasser- und Kältetechnik, Munich, 2001, p. 59.

[16] Recknagel, Hermann; Sprenger, Eberhard; Schramek, Rudolf (ed.): Taschenbuch für Heizung + Klimatechnik, Munich, 1999, p. 1207.

[17] Zeidler, Olaf: Freie Lüftung in Bürogebäuden; in: HLH, vol. 51, 07/2000.

[18] Daniels, Klaus: Gebäudetechnik – ein Leitfaden für Architekten und Ingenieure, Zürich/Munich, 1996, p. 260.

[19] Baker, Nick; Steemers, Koen: Energy and Environment in Architecture, London, 2000, p. 58.

[20] Westenberger, Daniel: Vertikale Schiebefenster – Zur Typologie der Bewegungsarten von Fenstern als Öffnungselemente in der Fassade; in: Fassade/Façade, 03/2002, pp. 23–28.

[21] Westenberger, Daniel: Vertikal verschoben – Eigenschaften und Leistungsspektrum von vertikalen Schiebemechanismen bei Fensteröffnungen; in: db 09/2003, pp. 86–91
Westenberger, Daniel: Verschiebliche Manipulatoren; in: Fassade/Façade, 03/2002, pp. 10–16.

[22] This chapter contains excerpts from an ongoing dissertation by Daniel Westenberger, Chair for Building Technology, Munich Technical University. The work concerns the use of vertical sliding mechanisms for windows and other facade openings, paying particular attention to the resulting combinations.

A 2.3 Dimensional coordination

Buildings normally consist of a multitude of individual parts (components, elements), most of which are incorporated at different times and produced and erected by different companies. We therefore need overriding geometrical rules to enable the construction of a flawless whole. Such a form of "grammar" covers the overall building technology context of the (building-related) subsystems of loadbearing structure, facade, internal fitting-out and services, and is generally known as dimensional coordination [1].

From column orders to the modular coordination system

The dimensional coordination of "elements of the building structure" is in no way a new idea. Back in the 1st century BC Vitruvius designated a mathematically calculated part as a *modulus* (= module), a basic dimension – based on the column diameter at the base – on which the "symmetria [= harmonious relationship] ..., [as an] interrelationship of the individual parts separately for designing the building as a whole", is based [2]. In the architecture of antiquity as well as during the Renaissance, primary dimensions (column spacing and height, entablature height and overhang) were specified in terms of column diameter (intercolumnation). As the construction and the shape of the columns were based on the form of the human body, there is a close relationship between the "sizes of modules and the human body" [3].
In connection with column orders and proportional systems plus the associated "theory of modulation" there are also square grids drawn across building plans and facades, in which the spacing between the individual lines is also called a module. This invisible module represents an abstract basic unit of a (theoretical) geometrical system of dimensional coordination for the organisation and construction of a building in three dimensions.
Geometrical and modular coordination systems are not found only in European architecture. Japanese living spaces, for instance, are essentially determined by the principle of the *tatami* mat dimensions – unique in the history of building. These firmly compacted, rectangular straw mats with their side ratio of about 1:2 are laid on the floor in living areas and form the basic module for the structure and proportions in three dimensions. However, the *tatami* mat is only one element in a modular system for Japanese timber houses. What evolved out of the desire to create standardised component dimensions resulted in not only one ideal size but, related to two defined column spacings, one module for the town (954 x 1900 mm) and one for the country (909 x 1810 mm). The deviations in the modular system resulting from the *tatami* mat dimensions are also the outcome of manual building work (figure A 2.3.2) [4].
The work of Jean-Nicolas-Louis Durand represents a key change in the (modular) conception

A 2.3.1 Eames house, Pacific Palisades (USA), 1949, Charles and Ray Eames

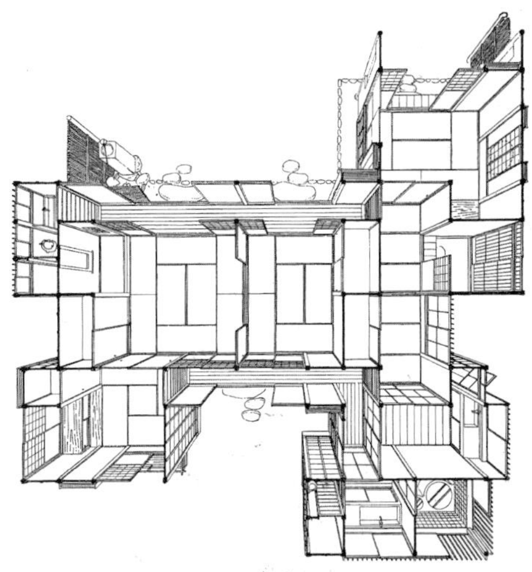

A 2.3.2

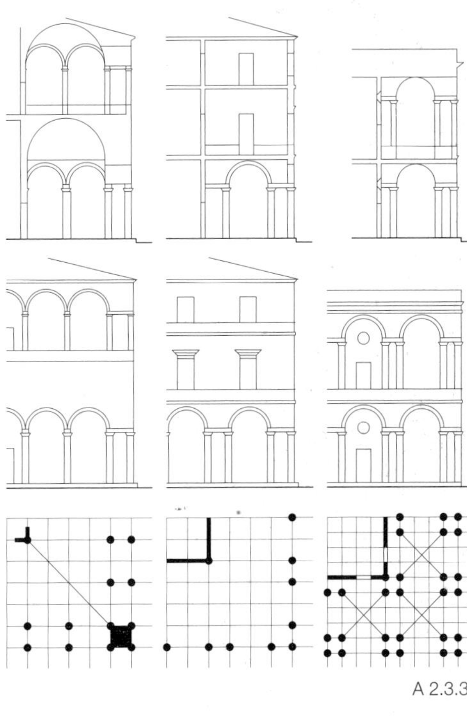

A 2.3.3

A 2.3.2 Perspective plan view of a typical single-storey Japanese house
A 2.3.3 Arcade systems
A 2.3.4 "ARMILLA" set of tools for computer-supported service route planning in high-tech buildings; layout of secondary lines, Fritz Haller

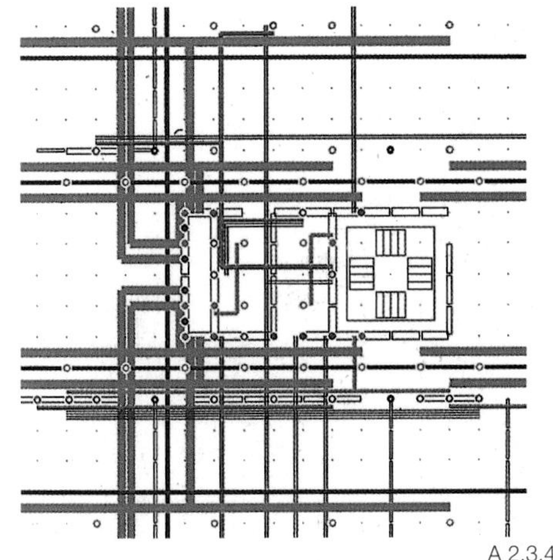

A 2.3.4

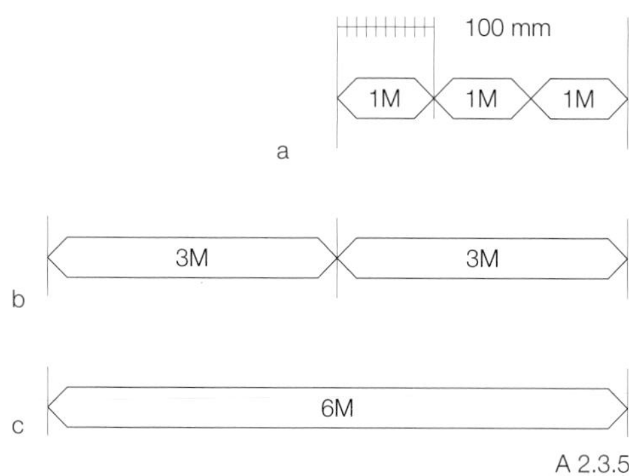

A 2.3.5

A 2.3.5 Dimensional coordination
a Basic module
The basic module is the unit size used as the starting point in a dimensional coordination system. The basic module (M) agreed throughout the EU is 100 mm.
b Multimodule
The multimodule is the standardised multiple of the basic module with a whole-number multiplier, e.g. 3M, 6M, 12M.
c Building module
The building module is a multiple of the multimodule and establishes the coordinating size for the loadbearing structure.

1
2 5 3
4 10 6 15 9
8 20 12 30 18 45 27
16 40 24 60 36 90 54 135 81
32 80 48 120 72 180 108 270 162 405 243

A 2.3.6

A 2.3.6 Preferred increments
Preferred increments are selected multiples of modules. Their numerical values, in conjunction with the modules, result in preferred dimensions in the form of multimodular or modular sizes. For practical reasons, they should be limited to certain multiples of the module. It is preferable to use these for deriving the coordinating sizes. Preferred increments are:
1, 2, 3 to 30 times M
1, 2, 3 to 20 times 3M
1, 2, 3 to 20 times 6M
1, 2, 3... times 12M

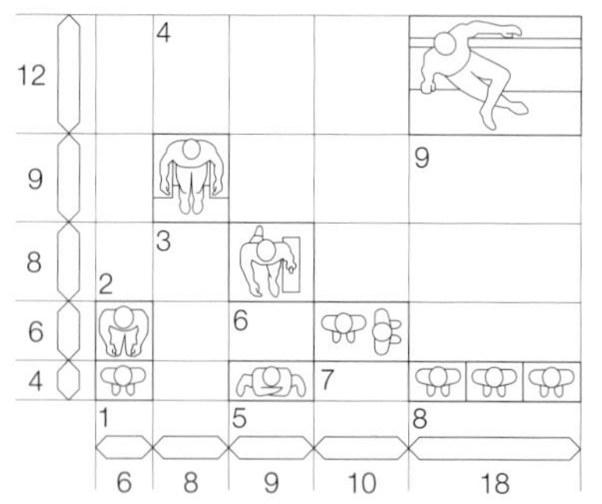

A 2.3.7

A 2.3.7 Numerical values for length and width of frequently used dimensions expressed as modules, based on the size of the human body:
1 person standing
2 person seated
3 person seated in armchair
4 person reclining
5 person standing with legs apart
6 person walking with suitcase
7 two persons standing
8 three persons standing in a row
9 person seated on sofa

of structures. Around 1800 he abandoned anthropometric and hierarchical architectural theories and based all construction tasks, as well as architectural elements, on the same grid as rational modular proportions (figure A 2.3.3). The starting point for this system is the column spacing, which as a "constructional, material-related dimension of the load-carrying beam" also considers aspects of economy and the purposefulness of the design [5]. Durand's work forms an important foundation for the use of the modular system, which later became the basis for the development of industrialised building.
Konrad Wachsmann, more than any other, deals with this subject in detail in his book *The Turning Point in Building*, which looks at the industrialised production and coordination of standardised elements.
Systems of dimensional coordination cover not only square grids or flat surfaces, but can also be used in three dimensions as well as plan layouts and facades. Coordination systems of this kind are the result of detailed theoretical and practical studies of "measured values, methods of measurement, dimensional provisions, sizing of the smallest parts right up to the complete building" [6].
The transition from building work on site determined by manual workers to the (partial) industrialisation of the construction process calls for the possible play in the positional relationships of the individual parts to be defined increasingly accurately. As technology-based manufacturing procedures enable a high degree of dimensional accuracy, the definition and control of tolerances represents a key requirement for geometrical modular coordination.

The "Modulor" of Le Corbusier is totally different from this technological approach and from the – usually – identical modular grids. Although his "Modulor" reference system is based on a sequence of numerical values, they are not derived from a common basic dimension. This is therefore a theory of proportion based on an "isotropic, dynamic structure" [7].

Dimensional coordination and modular systems

The modular system is a form of dimensional coordination based on modules and application rules for the dimensional coordination of technical elements whose arrangement and function in a system must be harmonised. This regulates, "with the help of grids and coordination systems, the position, size and joining of technical elements" by using modules [8]. Dimensional coordination serves to establish rules for the sizes of components on which to base the planning, production and erection. It serves to coordinate the processes and those involved in these processes, and is a prerequisite for the present degree of industrialisation in building work.
Every component can therefore be defined in terms of its position and the dimensions important for connections, and related to other, adjoining or associated components in a system of dimensionally coordinated relationships. The objectives of modular coordination are:

- the overall geometrical and dimensional coordination of the building
- the replaceability of products
- restricting the diversity of products
- prefabrication, with controlled and harmonised erection on the building site.

Definitions and units

Module

Modules are ratios of technical variables. The basic module (M) for dimensional coordination in Europe is defined as 100 mm (fig. A 2.3.5a). To limit the diversity of possible component dimensions and to ensure sensible modular sizes and component functions, preferred increments – multimodules – have been defined, i.e., multiples of M (M = n x M). Multimodules, or planning modules, determine the systematic evolution of the design (figure A 2.3.5b). DIN 18000 "Modular coordination in building" [9] proposes various multimodules (3M, 6M, 12M) derived from the basic module. Multiples of the planning module result in the building module, which determines the composition and coordination of the construction (figure A 2.3.5c). We distinguish between a number of customary building modules (e.g., 36M, 54M, 72M) depending on the type of utilisation.
The addition or subtraction of building modules results in parts or multiples, designated as preferred increments in DIN 18000. For practical reasons, preferred increments should be limited to a certain number of multiples. Highly practical and highly useful preferred increments are characterised by their multiple divisibility (figure A 2.3.6).
Based on these preferred increments, or rather multimodules, we can define functional modular sizes for various human activities such as standing, sitting, lying down, walking (figure A 2.3.7) [10].

Reference systems

In order to determine the position and the general dimensions of the modular component, as well as its relationship with neighbouring components, we require reference planes, reference lines or reference points.

Grids

A grid is a three-dimensional geometrical coordination system with a regular sequence of equally spaced reference lines, the grid dimensions. As the selected planning dimensions, these determine the spacing and position of the grid lines. The grid dimensions are based on a module or a multiple thereof. In most cases the basic form of the grid is a rectangle or a square. The position of every component and its relationship to other components can be coordinated with the help of the grid. We also speak of axial controlling lines, which – based on the building modules – determine the spacing of the grid lines of the construction and form the coordination system.

Types of reference

The types of reference are rules for allocating modular and non-modular parts to a coordination system. Basically, there are two ways of relating components to a modular grid:

- axial controlling lines (structural grid, centre-to-centre measurements)
- face controlling lines (planning grid, nominal measurements).

Axial controlling lines or centre-to-centre measurements create a relationship between the component and the reference system by ensuring that the axis of the component coincides with a reference line, i.e., the component is centred on the reference line. This establishes only the position of the component and designates the centre-to-centre spacing of the components, but defines neither its cross-sectional shape nor its size. The dimensions of adjoining components cannot be derived in this case (figure A 2.3.8a). Face controlling lines or nominal measurements are where the component is defined by at least two reference lines of the reference system. This means that its position and also its general size (in two dimensions) are established (fig. A 2.3.8b).
The combination of centre-to-centre and nominal measurements defines a component both in terms of its position in one dimension and also, in a second dimension, in terms of its size (figure A 2.3.8c).

Components are three-dimensional and can be uniquely defined in all three dimensions with the help of types of reference within the coordination system. Here, the choice of the respective types of reference and their combination depends on the individual case. DIN 30798 part 3 specifies the following rules of thumb for classifying technical elements:

- face controlling lines in all three dimensions (cubic parts/rooms)
- face controlling lines in two dimensions, axial controlling lines in one dimension (planar parts/wall elements)
- face controlling lines in one dimension, axial controlling lines in two dimensions (linear members/columns)
- axial controlling lines in all three dimensions (discrete parts/nodes).

When positioning components that can have different sizes in one or two dimensions, we make a further distinction between axial control

A 2.3.8 Types of reference
a Axial controlling lines
Here, the component is positioned in such a way that its centre axis coincides with the coordinating lines in at least one dimension, i.e., its position is defined.
b Face controlling lines
Here, the component is positioned between two parallel coordinating lines in such a way that it is aligned with these in at least one dimension, i.e., its size, position and often also its form are defined.
c Combination
When axial and face controlling lines are combined, the component position is defined in one dimension, and its size is defined in the second dimension.
A 2.3.9 Geometrical definitions

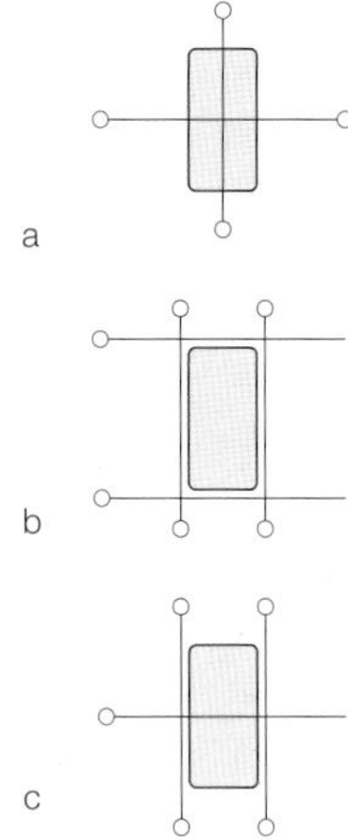

A 2.3.8

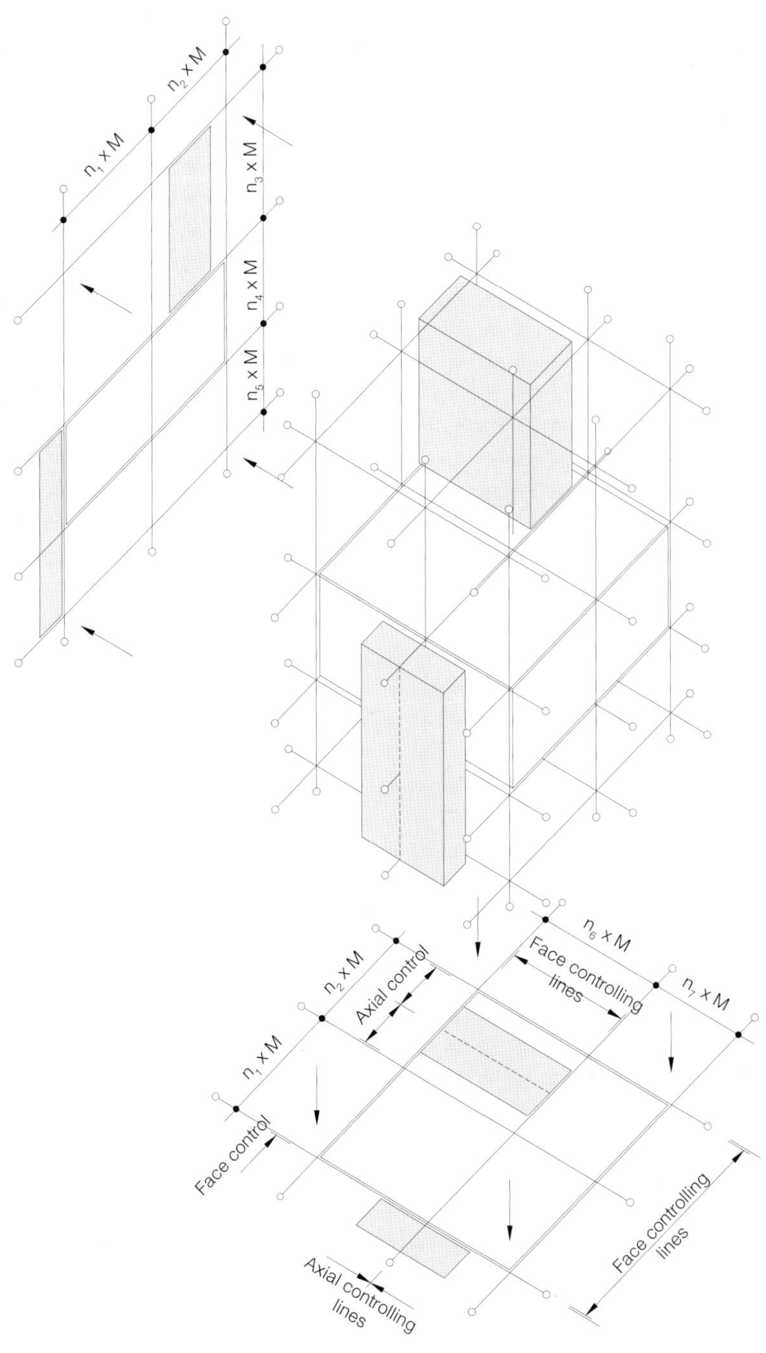

A 2.3.9

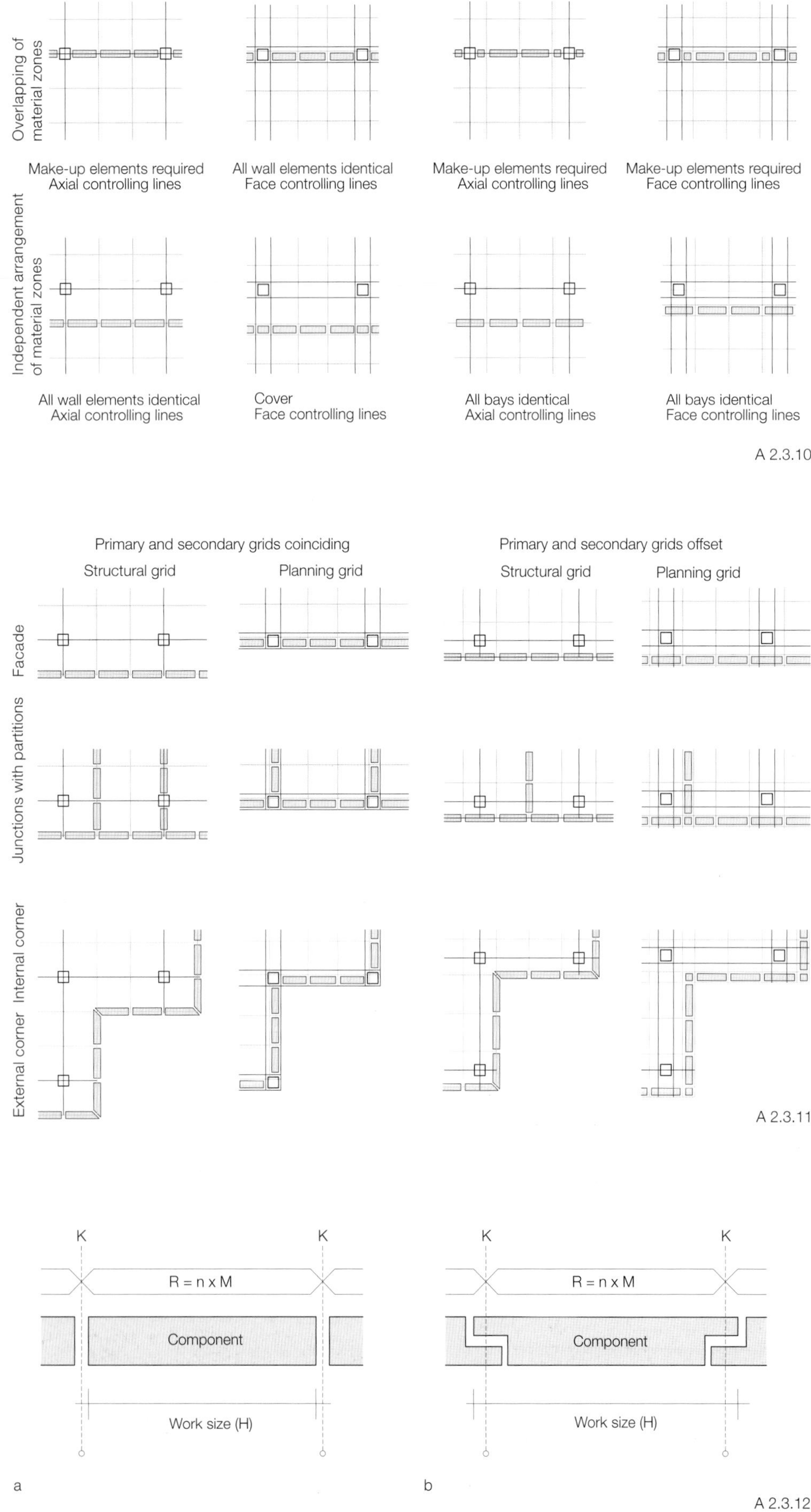

A 2.3.10

A 2.3.11

A 2.3.12

and facial control. Axial control means the component is arranged in such a way that its central axis coincides with the central axis of the modular zone, while with facial control the primary reference face of the component (in terms of dimensions) is aligned with one of the coordination lines. Components with different dimensions can therefore have the same reference plane. As a rule, axial control and facial control are applied in conjunction with axial and face controlling lines. The deviation of the component from the normal position means deviating dimensions for adjoining components, which requires them to have special formats (figure A 2.3.9) [11].

Geometric definitions
Modular systems are created when the spacing of the parallel coordination lines alternates with one or more modules. Modular grids may be based on one or various modules in any one of the three dimensions in space.

Primary and secondary grids
The coordination of individual components calls for the superimposition of reference planes and hence a weighting, i.e., the definition of primary and secondary grids. Normally, the structural grid (for loadbearing elements) is taken as the primary grid and the planning grid (for non-loadbearing elements) as the secondary grid. The most common geometrical relationships between the facade and structural grids are the offset and coincident arrangements. When material zones overlap, as with axial controlling lines, the deviating dimensions of the adjoining fields require special (shorter) formats for the elements. But the separation of material zones enables the loadbearing structure and facade to be arranged independently of each other and allows the design of identical elements (fig. A 2.3.10).

Junctions and corners
The overlapping or independent arrangement of modular zones (material zones for loadbearing structure and envelope/fitting-out) combined with the coincident or offset arrangement of the reference systems results in numerous different construction-related boundary conditions for the dimensions of the components and the junctions between loadbearing structure and envelope. This is particularly true at internal and external corners (figure A 2.3.11).

Sizes of components
As the dimensions established in the modular coordination are only general, the production of specific components calls for coordinating sizes. The coordinating size (R) is the spacing of the reference planes defining the position and size of a component, and is usually a modular dimension (R = n x M). The work size (H) can be derived from the coordinating size by taking into account the joints, the mating faces of a component and the dimensional tolerances: H < R. Depending on the joint design, the work size can extend beyond the modular dimension:

H > R. In this case a joint size has to be taken into account in order to control the dimensions between components (figure A 2.3.12) [12].

Geometrical position in relation to loadbearing structure

Apart from leading to different connection conditions, the position of the facade in relation to the loadbearing structure has consequences for the performance and the appearance of the facade. In principle, we can distinguish between the following positions (considered from outside to inside) in the case of non-loadbearing facades (figure A 2.3.13) [13]:

Position of plane of facade
- in front of the columns (1)
- on the front face of the columns (2)
- between the columns (3)
- on the rear face of the columns (4)
- behind the columns (5).

These geometrical positional relationships determine the role of the loadbearing structure as an architectural element, whether the divisions in the facade are influenced by the loadbearing structure, the detailing of junctions with partitions, the extent to which the facade penetrates column and floor planes, etc.
The incorporation of the horizontal loadbearing elements (floor slabs) into the vertical ones (columns) is another distinguishing criterion. In the case of non-loadbearing facades we can basically distinguish between:

- integrated into rear face of column (A)
- projecting beyond column (B)
- flush with front face of column (C).

The position and allocation of the loadbearing members in relation to the facade is characterised by the emphasising of vertical and/or horizontal elements, i.e., piers, columns or projecting floors, or a grid effect.
In terms of construction, the position and orientation of the columns is important for forming and fixing the facade, i.e., the connections between columns and beams and their three-dimensional form, junctions with partitions, the routing of services – even fire protection. In terms of performance, the position of the columns in relation to the facade results in requirements concerning:

- deformations (changes of length due to temperature fluctuations)
- thermal bridges (heat conduction through adjoining components)
- acoustic bridges (sound transmission between inside and outside)
- weather protection (e.g., protecting steel stanchions against corrosion).

Likewise, the position and orientation of the columns influences the subdivision of the facade. For example, closely spaced columns enable the respective bays to have an identical form, while widely spaced external columns may call for special make-up elements owing to the different dimensions, depending on position and arrangement.

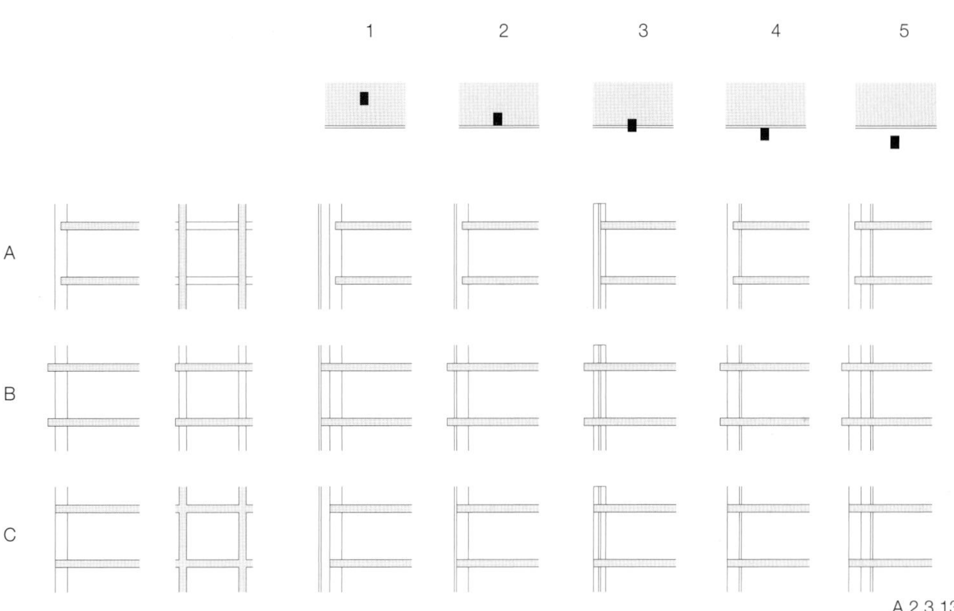

A 2.3.10 Primary and secondary grids (selection)
A 2.3.11 Elements and corners
A 2.3.12 Coordinating and work sizes
Owing to the types of joint possible, components can extend beyond the modular sizes.
A 2.3.13 Geometrical positions of facade in relation to loadbearing structure

Tolerances

"Tolerances are intended to limit the deviations from the nominal sizes, forms and positions of components and structures" [14].

We distinguish between:

- production tolerances
- erection tolerances
- tolerances due to deformation of components.

Joints are spaces between two modular components. One reason for these is dimensional inaccuracies during production and erection. As play in the joints is necessary when erecting adjoining components, there are permissible deviations for determining the (permissible) minimum and maximum sizes. Production tolerances are those permissible dimensional deviations that occur during the production of components and parts of the building. These are the result of the difference between minimum and maximum sizes. Erection tolerances designate the ranges of permissible positional deviations of components during erection and assembly. These may occur as linear, planar or three-dimensional deviations. During design – particularly when establishing method of construction and details – it is vital to ensure that the corresponding tolerances are allowed for depending on each individual case.
Different types of tolerances are often added together at the junctions between neighbouring components. Dimensional deviations must be accommodated, relative movements and seals must be guaranteed permanently, and thermal bridges avoided [1].

Notes

[1] Fundamental and further deliberations in: Herzog, Thomas: Zur Kunst des Fügens oder: Nachdenken über das Standbein; in: Der Architekt, 2/1987, pp. 86–89.
[2] Naredi-Rainer, Paul von: Architektur und Harmonie; Cologne, 2/1984, p. 17.
[3] ibid., p. 130.
[4] Nitschke, Günter: Architektur und Ästhetik eines Inselvolkes; in: Schittich, Christian (ed.): Japan, Munich/Basel, 2002, p. 24ff.
[5] Nerdinger, Winfried: "Das Hellenische mit dem Neuen verknüpft" – Der Architekt Leo von Klenze als neuer Palladio; in: Nerdinger, Winfried (ed.): Leo von Klenze. Architekt zwischen Kunst und Hof 1784–1864, Munich/London/New York, 2000, p. 11.
[6] Wachsmann, Konrad: The Turning Point in Building, Dresden, 1989, p. 54.
[7] ibid. [2], p. 133.
[8] DIN 30798 part 2, 1982.
[9] DIN 18000, 1984.
[10] Bussat, Pierre: Modulordnung im Hochbau, Stuttgart, 1963, pp. 30–33.
[11] ibid. [9].
[12] Project MOSS – OE 06/11 part 1: Grundlagen der Modulordnung, seminar report, Kassel Polytechnic, 1974, p. 26f.
[13] Trbuhovic, L.: Untersuchungen des Strukturschemas und der Fassadenentwicklung beim Stahlbeton-Skelettbau; in: Girsberger, Hans (ed.): ac panel. Asbestzement-Verbundplatten und -Elemente für Außenwände, Zürich, 1967, pp. 46–49.
[14] DIN 18201, 1997

A 3 Planning advice for the performance of the facade

A 3.1 Swiss Re headquarters, London (GB), 2003, Foster and Partners

The conception, design and construction of the facade are crucial. Not only for the external appearance of the building, but also for the serviceability, durability, costs and energy consumption of the entire building, the protection of people and property, and comfortable interior conditions.
The requirements placed on the facade differ according to location and utilisation. Other influencing factors are the shape and height of the building plus the arrangement of areas, rooms and functions, which create terms of reference for the horizontal and vertical subdivision of the facade and the interior layout. Furthermore, there is legislation covering sound insulation, fire protection, smoke control and interior daylighting levels, depending on the use of the building (e.g., offices with computer workstations, atria, foyers, staircases, escape routes). The degree of freedom afforded to the designer also depends on whether the project involves a new building, a conversion, or a refurbishment.
In terms of the type of construction of the facade, it is vital to know whether the building is a "heavyweight" design with loadbearing external walls, or a concrete, steel or timber frame. In addition, the interior temperature and humidity depend on the building services (e.g., with or without air conditioning); that also influences the requirements placed on the facade.
During the design process we have to consider these framework conditions in order to decide which type of facade, and which type of facade structure, is to be chosen for the various areas of the facade:

- loadbearing or non-loadbearing
- one or more leaves
- one or more layers
- post-and-rail or prefabricated.

All the requirements placed on the properties of the facade must be satisfied by choosing suitable materials and components, and by ensuring that they fit – and are fitted – together properly to guarantee long-term durability.
All inhomogeneities and "leakage" points in the facade embody particular risks for the performance and an increased risk of damage. Those weak spots are all joints between facade components and all penetrations, particularly in the form of and around fixings and cables (e.g., for sunblinds, photovoltaic systems). However, these are not the only possible weak spots; junctions between different parts of the building also represent critical performance interfaces between different trades. This latter aspect also applies to junctions between the interior fitting-out (partitions in particular) and the facade. In addition, the ability to create new room layouts to accommodate changing framework conditions, possibly with different requirements, plays a decisive role here.
Special areas of the facade, such as terminations at top and bottom, and vertical and horizontal internal and external corners (especially with offset insulating and sealing layers/leaves), demand special attention in terms of performance issues.
The aspects of airtightness, waterproofing, thermal and acoustic performance, moisture and smoke control, protection against solar radiation, glare and fire, plus the use of solar energy and daylight can generally only be treated as a whole unit and then optimised by taking into account the respective framework conditions. This is because the corresponding measures frequently influence each other.
The various potential solutions often result in, from the functional viewpoint, different merits and demerits in each situation. And from the performance point of view they lead to typical weak spots in the details. Many of the problems identified in practical building situations can be minimised (and the engineering and architectural design freedom retained) when the planning and building work:

- is based on task-specific standards ("system technology") whenever possible,
- based on project-specific standards ("platform strategy") only when unavoidable, and
- is carried out without working to standards only rarely, if at all.

Type of facade

From a construction point of view, we can distinguish between two fundamental types of facade:

- loadbearing external wall
- non-loadbearing facing leaf.

In the first case, windows are formed or incorporated in a loadbearing external wall (fig. A 3.2). These can be in the form of individual openings or can be combined to form continuous bands of windows horizontally (also storey-high) or vertically (also over several storeys). The junctions with the building envelope around the window frames in particular call for careful detailing to ensure acceptable thermal and acoustic performance, and moisture control in the given environment. The areas of the facade between the windows can be clad externally with sheet metal or opaque glass if necessary. Their external appearance then resembles that of a non-loadbearing facade, although they have a totally different construction (fig. A 3.3). Such facades are positioned totally in front of the structure and form an enclosing, additional weatherproof envelope into which glazing, windows (singly or in bands) can be integrated as distinct elements. Experience has shown that the weak spots in terms of performance for this type of construction are to be found at the junctions of floors and walls.

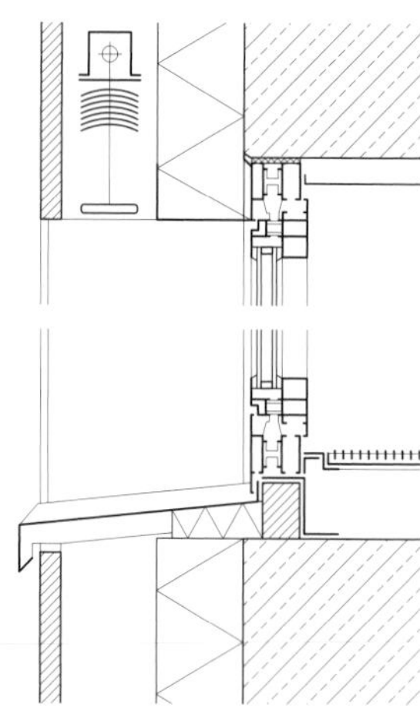

A 3.2

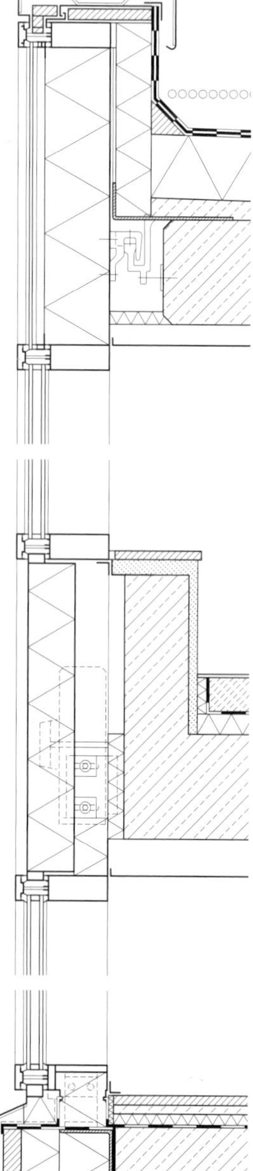

A 3.3

A 3.2 Vertical section through a fenestrated loadbearing external wall

A 3.3 Vertical section through a non-loadbearing, post-and-rail facing leaf (top: parapet; centre: junction with floor; bottom: base)

It is at these points in particular that we face practical problems of sound insulation, fire protection and smoke control between neighbouring rooms when joints are not properly designed or constructed in terms of their insulating and sealing functions. This is especially the case when the following aspects are not given adequate attention and not allowed for in the design and construction:

- deformation of the building fabric, e.g., due to dead and imposed loads
- production-related tolerances
- dynamic, horizontal floor displacements caused by wind pressure/suction or seismic actions
- differences in changes of length due to differing materials and temperatures.

Facade structure

The structural and performance-related properties of single-leaf (monolithic) external walls are determined purely by material and thickness. The material of the wall must therefore satisfy a multitude of functions in this approach. In contrast, in multi-layer or multi-leaf facades the materials and thicknesses of the individual layers and leaves can be optimised to suit their respective functions.
For example, in a multi-leaf facade we can include an air cavity between several leaves which is either fully enclosed or can be connected to the inside and/or the outside. The associated weatherproof layer can either be transparent, translucent or opaque, depending on which functional or architectural characteristics are desired.
The airtightness of the thermal insulation and moisture control layers/leaves must not be interrupted; a suitable sealing system must be used at joints in particular. If this function is on the inside, the material must be more vapour-tight than the outer, weatherproof layer/leaf. In practice it has been shown to be advantageous when the weatherproof layer/leaf is provided with vapour pressure compensation openings at least; moisture within the construction can then escape outwards unimpeded (fig. A 3.6). However, as driving rain, under unfavourable conditions, can enter these openings and penetrate to the air cavity, this water must be able to drain away to the outside directly via suitable openings. So the degree of waterproofing is guaranteed by two coordinated sealing layers/leaves.
If such facades are designed and built properly, they achieve better protection not only against rain, but also against wind, sound and moisture in general. For this reason, multi-layer and multi-leaf facades are used on buildings with high noise or wind loads where a high standard of comfort is also required.

Method of construction

Distinguishing facades in terms of their method of construction concentrates on the issue of whether individual components (e.g., posts, rails) or fully functioning modules (called elements) have to be delivered to and erected on the building site.

Facing leaves using a post-and-rail construction are very popular solutions (fig. A 3.4). This form of construction uses sliding longitudinal and transverse connections between the posts and rails. The infill elements take the form of windows, glazing or panels, to a certain extent "floating" in a rebate, the depth of which has to take account of the tolerances, movements and deformations to be expected. On the building site, the erection work requires scaffolding, is time-consuming, and is at the mercy of the weather.

By contrast, facades made from prefabricated elements mean that the mechanical fabrication and assembly of fully functioning facade elements – including glass, panels, sheet metal and thermal insulation, even stone and sunblinds plus sensors and drives in extreme cases – can take place in a factory (fig. A 3.5). One essential advantage of this is that, contrasting with the situation on the building site, this work takes place under controlled, industrial conditions with a high degree of automation and high level of accuracy. This leads to reliable quality assurance measures and hence a consistently high standard of quality. Completely prefabricated modules are delivered to the site and hung on (adjustable) brackets previously fixed to the structure. This type of construction also includes those prefabricated facades in which the facade sections are assembled to form frames by using T- and/or L-connectors. The edge sections of adjoining facade elements, provided with rubber seals, are interlocked, labyrinth-style, during the erection procedure on site. This means the system can accommodate not only tolerances, movements and deformations, but also guarantee the necessary thermal and acoustic performance plus airtightness and waterproofing requirements in the joints between the elements. Incorrectly designed intersections between facade elements are the typical weak spots of this system.

Prefabricated facades consume more materials and more plant time, and require experienced designers and engineers. Mistakes at the design stage cannot easily be corrected by additional manual measures. These facades require more intensive planning and therefore need appropriate (planning) lead times, something that must be taken into account when awarding contracts. However, they are equally suitable for high-rise buildings and single-storey sheds, and are preferred for those with a regular structural system.

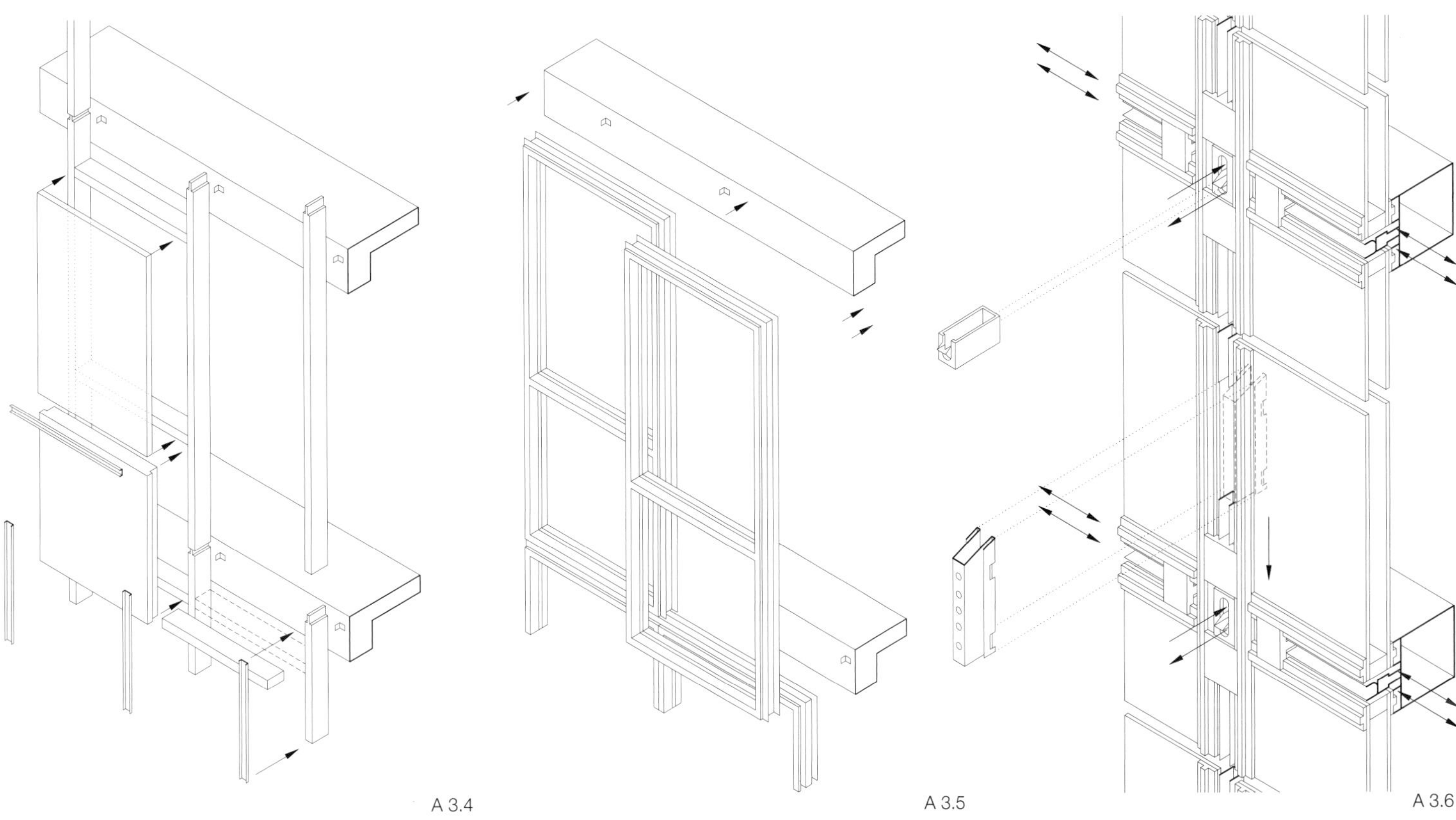

A 3.4 Post-and-rail facade
A 3.5 Prefabricated facade
A 3.6 Vapour pressure compensation in post-and-rail facade

Thermal performance

Good thermal insulation increases the surface temperatures on the inside of the facade, which improves the level of comfort near the facade, reduces the maximum heating requirement and hence lowers the capital outlay. Furthermore, it shortens the operating time of the heating system, which cuts the consumption of heating energy and operating costs. The optimisation of the thermal performance of the facade involves an overall optimisation of frames, glazing and opaque areas using means to reduce heat conduction, convection and the exchange of long-wave radiation. Useful here are framed constructions with varying amounts of thermal insulation, opaque/translucent insulating materials or transparent/translucent insulating glazing units with a thermally insulating gas filling and/or surface coating.

Typical weak spots in terms of thermal performance are the joints, along the hermetic edge seals of glazing and panels, and around fixings, caused by linear or discrete thermal bridges and/or leakage points. Horizontal and vertical internal and external corners, parapets and bases plus offsets in the insulation or sealing layers/leaves, particularly at transitions between different facade types and assemblies, have proved to be especially critical.

Moisture control

Thermal bridges generally represent weak spots in terms of moisture control as well because these are the places where there is an increased risk of condensation on inner surfaces and, possibly, within the facade itself. The same is true for facade details where the developed inner surface area is smaller than the outer area, for instance, at "slender" external corners, or where sections positioned externally act as cooling fins.

The risk of interstitial condensation in a facade is governed by the vapour permeability of the individual components and, in particular, by the actual detailing of sealing measures around joints and fixings.

Effective protection against condensation is a fundamental requirement for the durability of the facade and for a healthy interior climate. According to the latest findings, mould can start to grow even before condensation is visible; DIN 4108 has therefore redefined the critical surface temperatures.

In Central Europe the principle that applies to design and construction is: "inside more vapour-tight than outside". In hot, humid climates this principle must be reversed: "outside more vapour-tight than inside".

Condensation can form in a multi-leaf glass facade when damp interior air in the cavity comes into contact with cold surfaces. Improving the quality of the thermal insulation in the outer layer/leaf and ventilating the cavity reduces this risk. The moisture control requirements to be met by the facade also depend heavily on the use of the building and the services installed. For example, high interior humidities generally prevail in swimming pools (in air-conditioned buildings only in winter), which increases the risk of condensation.

One phenomenon frequently neglected during the design phase is the formation of condensation or hoar-frost on the outer surface of the facade. This risk increases with the quality of the thermal insulation in the facade, especially at panels with a high level of thermal insulation and with triple glazing, where, owing to the low heat transmission, the outermost surface experiences hardly any temperature rise. The result is that the fogged-up glass never completely dries. More attention must be devoted to this aspect in the future.

Acoustic performance

The requirements to be met by the facade with respect to sound insulation to protect against

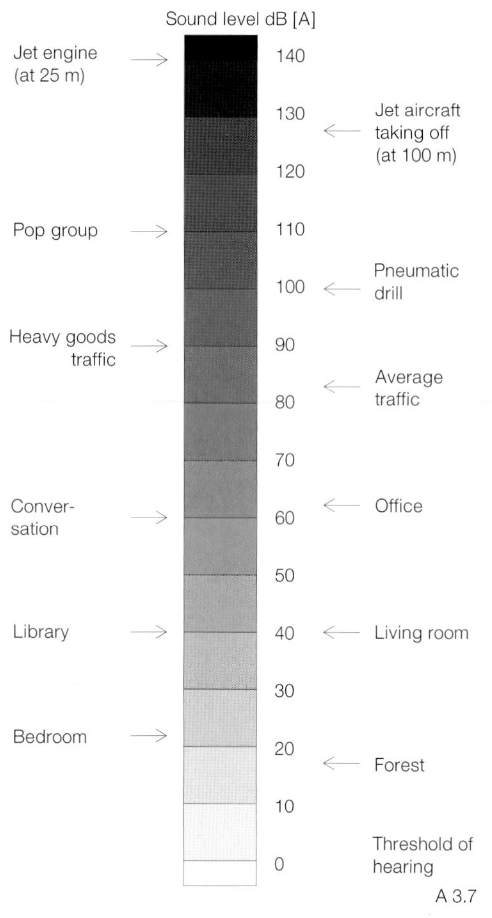

A 3.7

Noise level range	Relevant external noise level dB [A]	reqd. $R'_{W,res}$ of ext. component dB [A] bed-rooms	living quarters	offices[1]
I	≤ 55	35	30	–
II	56–60	35	30	30
III	61–65	40	35	30
IV	66–70	45	40	35
V	71–75	50	45	40
VI	76–80	[2]	50	45
VII	> 80	[2]	[2]	50

[1] No requirements are specified for the external components of rooms in which, owing to the activities performed in those rooms, the external noise level does not contribute significantly to the background noise level internally.
[2] These requirements are to be specified based on the local circumstances.

A 3.8

A 3.7 Sound levels of various sources
A 3.8 Noise level ranges and sound reduction indexes (R') to be maintained
A 3.9 Fire resistance classes to DIN 4102 part 2
A 3.10 Examples of building materials and their combustibility, plus the German building materials and European classes

external noise depend on the prevailing external noise level and the permissible and actual noise levels within the building (fig. A 3.7). DIN 4109 lays down the main requirements for the acoustic performance of the facade. If, compared to the partitions and the junctions between parts of the building and between partitions, there is a disproportionate amount of sound insulation in the facade to protect against external noise (or the background noise level in the building is lower than expected), the subjective disturbing effect of internal noises – high frequencies in particular – from neighbouring rooms can prove problematic. The sound insulation between adjacent rooms is not only the result of the sound insulation in the intervening walls and floors, but also of their junctions with the facade. Flanking transmissions via the facade itself are also present. This effect is much more noticeable with post-and-rail facades than with prefabricated facades when the joints between the elements are positioned adjacent to the floor and partition junctions. VDI Guideline 2719 classifies facades according to acoustic performance classes 1 to 6 depending on their airborne sound insulation index to DIN 52210. The design and construction must ensure that the necessary acoustic properties of the facade are guaranteed in the long term (fig. A 3.8). The sound insulating effect of facades, also junctions with floors and partitions, can be improved, mainly by employing the following constructional measures

- Increasing the weight of the components, also filling with sand or a heavy gas, or adding lead planking
- Increasing the number of successive, separate leaves, e.g., double leaves, preferably with different material thicknesses
- Increasing the elasticity of the components, e.g., the "amination" of several thin sheets of metal or panes of glass, and by isolating them (in acoustic terms) at joints and fixings by means of soft gaskets, etc.
- Increasing the asymmetry of the assembly in terms of the weight of successive layers
- Increasing the distance between surfaces forming the boundary to an air layer
- Increasing the degree of absorption of surfaces forming the boundary to an air layer, e.g., by using porous materials or by creating a labyrinth.

If a facade is to achieve the requirements of acoustic performance classes 4 to 6 of VDI Guideline 2719, insulating glass systems with very thick panes (especially the outer ones) and wide cavities plus a heavy gas filling are one way of achieving this. A reduction in the total glass thickness – resulting in a less expensive facade construction – can be achieved by replacing individual panes with laminated glass (with casting resin or PVB interlayer); one laminated pane achieves class 4, two panes classes 5 and 6. Double-leaf facades, when designed and constructed properly, achieve a noise level reduction of 4–8 dB compared to a single-leaf facade equivalent to the inner leaf; however, this does depend on the size of the ventilation openings in the outer glazing, plus the sound absorption in the ventilation openings and in the cavity between the leaves.

Fire protection and smoke control

The main issues for facades are measures to prevent fires, to prevent/delay the outbreak/spread of fire, and to allow heat and smoke to escape. The fire protection and smoke control properties of a facade are critical for preventive fire protection measures and hence for the protection of life and property.

There are many rules to be observed, and in Germany these can even vary from state to state. The federal state building codes, the regulations of the factory inspectorate, the building authorities, the Technical Monitoring Service (TÜV) and the general DIN and VDE standards are all involved in regulating fire protection. Furthermore, the guidelines of the local fire brigades, the Institute for Building Technology (IfBt) and the Association of Property Insurers (VdS) all have to be taken into account. The fundamental prerequisites of preventive fire protection are fire detection/alarm options and access to the building for the fire brigade. The basic requirements are given in numerous statutory instruments, which essentially regulate the following measures/precautions:

- fire prevention
- preventing/delaying the outbreak of fire
- preventing/delaying the spread of fire
- provisions for fire detection/alarms
- allowing heat and smoke to escape
- fire-fighting
- rescue/safety of occupants and fire-fighters.

The requirements concerning preventive fire protection specified in DIN 4102, the general building code and the federal state building codes are compulsory. Irrespective of those requirements, building authority guidelines regulate the requirements for smoke vents.

Classification, loading classes

Fire-resistant glazing is a light-permeable component comprising a frame, one or more light-permeable elements, glazing beads, gaskets and fixing materials. Such a component can withstand a fire for 30, 60, 90 or even 120 minutes, depending on its classification.

DIN 4102 part 13 subdivides such components into F and G glazing units (fig. A 3.9). Both types of fire-resistant glazing are light-permeable components for use in vertical, sloping or horizontal arrangements. They prevent the spread of fire and smoke in accordance with the duration of their fire resistance rating.

Fire resistance class	Building materials class to DIN 4102 part 1 for materials used in components tested		Designation[1]	Building authority designation[1]
	main parts[1]	other parts not included in col. 2		
F 30	B	B	F 30-B	fh = fire-retardant
	A	B	F 30-AB	fh, and the main parts made from incombustible materials
	A	A	F 30-A	fh, and made from incombustible materials
F 60	B	B	F 60-B	–
F 90	B	B	F 90-B	–
	A	B	F 90-AB	fb = fire-resistant
	A	A	F 90-A	fb, and made from incombustible materials

[1] for explanations see DIN 4102 part 2

A 3.9

Building material	Building materials class to DIN 4102 part 1	European class
Incombustible material (e.g., steel lattice girder)	A 1	A1
Incombustible material with combustible components (e.g. plasterboard as interior finish to timber construction)	A 2	A2
Not readily flammable material (e.g., oak parquet flooring on screed)	B 1	B
Low contribution to fire		C
Flammable material (e.g., glued laminated timber joist)	B 2	D
Acceptable behaviour in fire		E
Highly flammable material (e.g., untreated coconut fibre mat)	B 3[1]	F

[1] not permitted in buildings

A 3.10

Unlike G glazing, F glazing also prevents the transmission of high-temperature heat radiation. F glazing becomes opaque when exposed to fire and forms a heat shield; in terms of fire protection, F glazing behaves like a wall. Consequently, F glazing is ideal for use as walls (or parts thereof) enclosing interior spaces, with no restrictions apart from the dimensions specified in the building authority approval documents. Fire-resistant glazing of fire resistance class G (G glazing), on the other hand, remains transparent during a fire. It reduces the temperature of the escaping heat radiation and is a special component in terms of fire protection. G glazing may be incorporated only at those points where it would not impair fire protection measures, e.g., in windows in the walls to corridors used as escape routes, where the underside of the glazing must be at least 1.80 metres above floor level so that persons using the corridor are still protected in the event of a fire.
The local building authority regulates other uses of G glazing depending on each individual case, e.g., taking into account the heat radiation and the risk of ignition on the side not exposed to the fire when combustible materials are stored or used within the range of the radiation. G glazing must remain effective as an enclosing wall; no flames are allowed to penetrate to the side not exposed to the fire.

When which class of fire resistance is relevant to the facade is usually decided – for each individual case – by the building authority responsible. The authority will take into account the type of building, the storey height, the nature and magnitude of the fire loads, and the other measures of the fire protection concept specific to the building (fig. A 3.10). The general building code calls for measures to be taken to prevent fire spreading from one storey to the storey above in high-rise buildings (FFL of top storey > 22 m). This is achieved by providing incombustible materials of class F 90 (or W 90) in the walls and floors which must extend either 1.0 m in the vertical direction or 1.5 m in the horizontal direction (e.g., by means of a fire-resistant overhang). Sheet metal spandrel panels require additional mechanical fixings in this case. A number of facades of this type – also without a backing of masonry or concrete – have been approved and built in recent years. The same applies to the internal corners of multi-storey office blocks. Providing such details with fire-resistant glazing serves to extend the fire wall and hence helps to protect against fire spreading horizontally to the facade on a part of the building separate in terms of the compartmentation concept. If a low-rise extension is added to a multi-storey building, the separating wall between the two parts of the building must be built as a fire wall right up to the roof of the taller building.

Likewise, staircases used as escape and rescue routes in the event of a fire represent opportunities to achieve the necessary fire protection with a glass facade. When neither spandrel panels and lintels nor cantilever arrangements can fulfil the spread-of-fire requirements, the authority responsible for the fire protection must clarify whether the requirements can be fulfilled with the help of a sprinkler system.
Furthermore, the needs of fire protection require the designer to check that the connections between facade and structure can reliably prevent the passage of smoke. If joints and junctions are not smoke-tight, the smoke and toxic gases associated with fires can quickly spread upwards, and thus put occupants at risk who would otherwise possibly not be directly affected by the fire.

Constructional measures

Smoke vents are either opened automatically in the event of a fire, or opened manually by fire-fighters. Besides the typical smoke and heat vents, the size of which is defined in DIN 18230 depending on the risk group, the cross-sectional area required in each case can also be achieved with openings in the facade (side- or bottom-hung opening lights), which must vent directly to the open air. But this must be discussed with a fire protection specialist.
The effectiveness of the smoke venting essentially depends on correctly sized vents and an adequate supply of fresh air. In specifying the cross-sectional area of the smoke vents, the authorities will distinguish between aerodynamically effective smoke venting and a geometrically calculated opening. The method of opening is important here (e.g., lights in a vertical facade opening outwards at the top approx. 60°). At the same time, a corresponding cross-sectional area for a fresh air supply must be made available (factor of 1.5 x vent opening; with simultaneous opening – e.g., automatic – factor of 1.0). Door openings may also be taken into account. Smoke venting via a vertical facade is currently not covered by the standards and codes; approval case by case must be obtained.

Fire protection and smoke control for weak spots in the facade

Besides the typical thermal bridges within the facade (e.g., gaps between frame and sash, or between frame and facade, or edge fixings of infill panels and the sealing thereof), all the inhomogeneities within the facade represent special risks in terms of fire protection. In facing leaves, slender, continuous posts and rails adjacent to partitions/floors and their junctions with the building/partitions have proved to be additional weak spots in terms of spread of fire. Movements and deformations of the facade, which can increase considerably during a fire due to the high temperatures, must be compensated for in the detailing at the

connections and joints between facade and building/partitions.
Specific measures to improve fire protection characteristics include:

- materials that foam up when exposed to heat and thus form a seal which improves the fire resistance or the mechanical integrity
- materials that vaporise when exposed to heat and thus compensate for the effect of the rise in temperature.

Facades connected with particular risks
Double-leaf facades to multi-storey buildings employ fire-resistant glazing primarily to prevent fire spreading from one storey to the storey above. Vertical routes for the spread of fire must be fitted with class F 30 glazing. The fire resistance class W 90 required for the spandrel panels of high-rise buildings can be integrated into the inner leaf of a double-leaf facade. Fire protection concepts in which the venting of the cavity between the leaves is achieved by means of voids extending like chimneys over several storeys require special testing. It cannot be ruled out that the pressure relationships prevailing during a fire could cause smoke to spread to adjacent floors when the windows are open.

The use of daylight

The daylight availability can be specifically exploited by means of intelligent daylighting concepts. Besides the targeted distribution of the solar radiation entering the interior by means of suitable shading systems, there is a second approach which is based on the fact that only the visible part of the solar spectrum can be utilised for illuminating the interior. As the infrared component in particular increases the thermal load in the interior, systems with specially coated glasses are desirable. These coatings are selective, that is, they are designed to admit the wavelenghts of visible light within the spectrum.

One special form of glass for improved daylight usage is insulating glazing with daylight-redirecting components in the cavity: grids of mirrors in two and three dimensions and aluminium honeycombs, made of specially shaped metal or plastic structures, sometimes with a highly reflective coating. These represent, as it were, a miniaturisation of rigid sunshading systems. Prismatic systems to redirect the incoming light can be employed to improve the illumination of the interior. These redirect primarily near-zenith luminance into the interior. However, prismatic systems interrupt the occupants' view of their surroundings and therefore should be restricted to those parts of openings above eye level.

Variable daylighting systems
Variable systems are much simpler and obviously more widespread form of active daylight control. The advantage of these compared to rigid measures is that their position and degree of opening can be varied. The incidence of light and the view out are therefore not impaired when the sky is completely overcast.

The desire for visual contact with the outside world, even when the sunshading is in operation, plus the demand for maximum transparency in the facade, have led to the development of perforated louvre blinds. The surroundings can still be perceived through these blinds. The proportion of perforations of the products generally available is about 9%. The size of each individual perforation depends on the thickness of the sheet metal and hence on the dimensions of the louvre. Blinds with perforation diameters of 0.6 and 1.1 mm are in use.

The degree of radiation transmission is 8% for a single louvre with the light striking the louvre at 90°. As the perforations mean that the louvre is not opaque, there is also always direct transmission in addition to the transmission of reflected radiation passing between the louvres. On average, considering ambient reflections of 20%, the perforations increase the radiation transmission from 4% to just over 6%. What this means is that the use of a perforated louvre system compared to one which, in terms of construction and surface characteristics, is essentially closed increases the radiation transmission by a factor of 1.6, and this could mean the cooling load has to be increased by the same amount. Louvre blinds in which the louvres can be set at different angles over the height of the blind have been available for some years. The angle of the upper louvres is shallower than that of the lower louvres, which enables shading and light-redirecting characteristics to be achieved simultaneously. The reflectance of the top sides and undersides of the louvres can be optimised to suit different requirements. Bright surfaces improve the light redirection, while dark colours will reduce glare phenomena in the interior. Louvres with different colours or reflectance values on top side and underside are now available.

Large louvres
Variable large louvres are much more robust than foils, textiles and louvre blinds, and are therefore generally wind-resistant. These louvres can be made from opaque materials (e.g., extruded aluminium sections) or partially transparent materials (mirrored or printed glass, perforated sheet metal) and can be installed either horizontally or vertically in sliding or pivoting arrangements. They are fixed to the outside of the building parallel with the facade or cantilevering from it, and hence affect the appearance quite dramatically.

Since the early 1990s fully automatic, sensor-regulated and microprocessor-controlled systems have ensured that the louvres are always positioned ideally for the direction and angle of the sun. During periods of lower lighting levels, e.g., a completely overcast sky, the louvres are moved so that their outer edges point upwards. In this position they act as light-redirecting elements, forcing more daylight into the interior and providing better, more even illumination.

The majority of glass louvres and active solar-control glasses cannot fulfil stricter antiglare requirements. The transmission character of these glasses, virtually unaffected by the angle of incidence, generally reduces the luminance of direct solar radiation only insufficiently. Furthermore, the variability of the degree of transmission of active solar-control glasses is still inadequate for antiglare and daylight usage purposes. The comments regarding perforated louvres apply here as well.

Sunshading and antiglare provisions

The effective intensity of solar radiation at openings in the building exhibits a more or less non-steady-state character. This is due to the variations in the available solar radiation in the open air and the geometrical influencing variables around the openings. The geometry of the building, with projections and returns, plus size, distribution, orientation and inclination of transparent facade components are the main elements relevant in this respect. The interior illumination by means of daylight, the heat load due to solar gains and the visual contact with the outside world are also affected by the arrangement and the properties (radiation transmission and light transmission) of the glazing. The same applies to additional components for shading, controlling glare and redirecting daylight (fig. A 3.11).

Sunshading
The function of rigid components, e.g. overhanging components or fixed louvres, is based on the solar altitude angle as it varies in a defined way over the course of the day and the year. If it were possible to design a system to screen direct solar radiation (also not redirecting it into the interior after being reflected on a surface) and to admit all the diffuse light from the sky (and not partly absorbing or reflecting it to the outside), then such a system would exhibit a reduction factor of 21%. However, this target cannot be fully met by rigid systems because these temporarily allow either the passage of some direct solar radiation or screen some light from the sky, which impairs the interior illumination.

Variable systems do approach this target. These can take into account the effects of the weather, and in some cases redirect the incident daylight on to the ceiling of the room, hence contributing to an even level of illumination in the interior. The shading and light-redirecting effects of variable louvre systems can be optimised by employing:

- high-level and eye-level louvres that can be set to different angles
- different reflectance values for top side and underside of louvres
- louvre surfaces with geometrical textures.

Standard perforated louvre systems can result in a 50% increase in radiation transmission and a corresponding increase in the cooling load compared to a non-perforated system with a similar construction and surface finish. In addition, it must be remembered that every system that does not screen direct solar radiation completely will lead to problems of glare in the interior. The deciding factor for the sunshading effect of the facade is not only the type of shading itself, but also its arrangement: the further away from the building, the better!

Preventing glare

Visual performance and visual comfort must not be impaired by disturbing factors. The recognition/detection of objects and the occurrence of glare are dependent on the absolute level of luminance and its distribution within the field of vision, and the ensuing contrast. We distinguish between physiological glare, which leads directly to a reduction in the visual performance, and psychological glare, which results in premature fatigue and a decrease in performance, activity and comfort. Direct glare is caused directly by the light source, while reflected glare is due to reflections of light-coloured areas on shiny surfaces.

The factors relevant to direct glare are the viewing angle of the observer and the perceivable luminance in the respective viewing direction. The brighter the surroundings, the lower the risk of glare.

Owing to the low luminance of monitor screens (10-100 cd/m^2), rooms with computer workstations are subject to strict antiglare requirements. This is another reason for providing the windows with means to screen direct solar radiation and the associated heat gains; there should be no remaining bright slits which could cause glare problems. In addition, suitable measures should be implemented to prevent glare due to surfaces reflecting direct sunlight. As these requirements are also necessary during high winds, antiglare measures must be protected from the wind, fitted internally or in a facade cavity.

Concluding remarks

Building owners and users will be satisfied with a building envelope in the long term only when the requirements and framework conditions specific to the building are clarified and the relevant technical options, including their specific risks, are evaluated in terms of their practical applicability. At the same time, the targets thereby derived must be rigorously implemented by the designers and builders.
In doing so, it is important to remember that, on the one hand, all the interfaces between different trades and all the inhomogeneities and leakage points within the facade re-present potential weak spots. On the other hand, the different constructional and performance aspects can generally only be treated as a whole because the corresponding measures frequently influence each other.

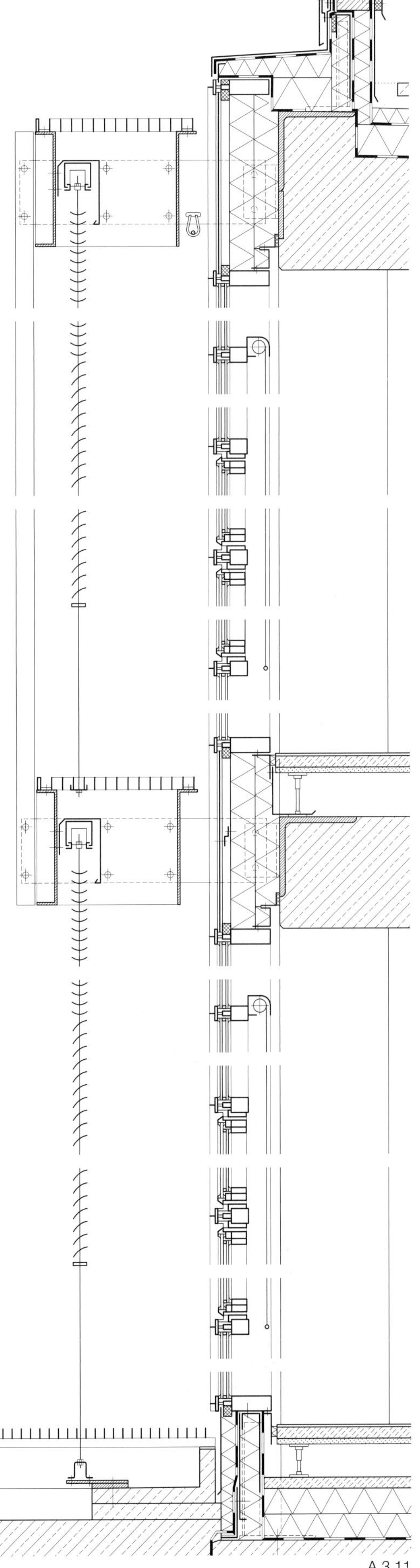

A 3.11 Facade with roller blind inside and louvre system outside, Munich (D), Peter C. von Seidlein

Part B Case studies in detail

1 Facade materials

- 1.1 Stone
- 1.2 Clay
- 1.3 Concrete
- 1.4 Timber
- 1.5 Metal
- 1.6 Glass
- 1.7 Plastics

2 Special topics

- 2.1 The glass double facade
- 2.2 Manipulators
- 2.3 Solar energy

Wrapped Reichstag, Berlin (D), 1995, Christo & Jeanne-Claude

B 1.1 Natural stone

If the "Stone Age" is taken to be the first relevant cultural epoch, then it is because people used the natural materials at hand to produce diverse utensils. Throughout history, the use of natural stone ranges from tools and weapons to tombs and walls and to precision artefacts, such as jewellery.
The stone referred to in this chapter is that obtained directly from the Earth's crust – "natural stone". This natural product can be divided into three main classes, according to its origin:

- igneous rock
- sedimentary rock
- metamorphic rock

These three rock classes are further subdivided into about thirty types, among them granite, sandstone and marble, to name just three. All types of rock occurring naturally on the Earth (about 4500–5000 types) can be classified in one of these groups. There are numerous opportunities for using stone externally (fig. B 1.1.10). Granite, for example, is suitable for applications ranging from solid, heavyweight construction to the cladding of facades.

Ashlar stonework
In order to be able to use natural stone in building projects, it must first be worked and given a particular form by means of, for example, cleaving, sawing or milling. These are the dimension stones used for ashlar stonework.

Depending on its compressive strength, a type of stone is classed as hard or soft (hard rocks: include granite, diorite; soft rocks: include limestone, tuff). Natural stone intended for use in stone masonry must satisfy certain physical requirements, such as minimum compressive and bending strengths and frost resistance [1].

Fig. B 1.1.11 (page 65) shows the most important characteristic values of natural stones: density, thermal conductivity, compressive strength, bending strength. Man-made "stone" is known as cast or reconstructed stone (e.g., clay bricks, concrete). The nature of the production methods leads to prefabricated, modular elements.

The use of natural stone in the facade

In historical terms, the development of stone facades is closely linked with that of masonry. Stone is among the oldest of building materials. Early civilisations such as those of Mesopotamia or Egypt made use of stone to construct loadbearing walls. Today, the applications also include ventilated, non-loadbearing facing leaves. The first stone buildings in the history of the human race were the result of local circumstances. Initially, these were simple additions to natural structures such as caves, made by

B 1.1.2

B 1.1.3

B 1.1.4

B 1.1.5

B 1.1.2 Graveyards, Petra (JOR), 4th century BC
B 1.1.3 Unit comprising stairs, wall, architecture and sculpture, Temple of Athena Nike, Athens (GR), 5th century BC
B 1.1.4 Mountain village in Ticino (CH)
B 1.1.5 Panel below shop window, with petrified ammonites as decoration

B 1.1.1 German Pavilion, Barcelona (E), 1929/1986, Ludwig Mies van der Rohe

B 1.1.6

- Natural stone
  - Igneous rocks
    - granite
    - syenite
    - dionite
    - gabbro
    - rhyolite
    - trachyte
    - basalt
    - diabase
    - lava stone
    - volcanic tuff
  - Metamorphic rocks
    - conglomerate
    - breccia
    - sandstone
    - graywacke
    - clayey shale
    - limestone
    - shelly limestone
    - travertine
    - tufaceous limest.
    - Solnhofener plates
    - dolomite
    - onyx
  - Sedimentary rocks
    - paragneiss
    - quartzite
    - mica slate
    - chlorite schist
    - serpentine
    - marble
    - migmatite
    - phyllite
    - granulite

B 1.1.7

B 1.1.6 "Palazzo dei Diamanti", Ferrara (I), begin: 1493, Biagio Rossetti
B 1.1.7 Rock classes and rock types
B 1.1.8 Cathedral of "S. Maria del Fiore", Florence (I), 1296 (–1887), Arnolfo di Cambio, Filippo Brunelleschi et al.
B 1.1.9 German Pavilion, Barcelona (E) 1929/1986, Ludwig Mies van der Rohe
B 1.1.10 Use of various natural stone types externally [3]

piling up layers of stones. Although these archetypal forms of stone external walls offered mainly permanence and safety, later in history we see certain cultural groups constructing stone facades made from stone cut with maximum precision and satisfying aesthetic demands.

The quarrying of stone for building purposes began around 5000 B.C. However, the accurate working of stone to produce dimension stones only became possible with the appearance of bronze (c. 2500 B.C.) and the availability of correspondingly hard tools.

The techniques for dressing stone and for cutting into this hard material, which had been practised with great precision by the Egyptians – e.g., to produce hieroglyphics and sunken reliefs – were further refined during the height of Greek architecture. The study of entasis and curvature of the dado also bear witness to the desire for visual modulation in the facade with the highest possible accuracy.
The Romans developed the techniques for cutting stone even further and this practical experience was committed to paper for the first time by Vitruvius in his 10-volume treatise on architecture *De Architectura*. So the first technical rules generally applicable on the European continent within the boundaries of the Roman Empire were laid down some 2000 years ago.

The systematic separation of loadbearing elements and cladding resulted in clear principles for the conception of a construction and for the organisation of a building site.

Modular prefabrication, which had been practised for thousands of years in the form of clay bricks, did not become established for ashlar stonework until the early Middle Ages. Brought about by increasingly specific requirements for the construction of large cathedrals, the techniques for the construction of stone facades continued to develop and made possible the prefabrication of large numbers of dimension stones. Construction times were also reduced by the invention of frame and layered forms of construction with continuous bed joints. These construction methods, which evolved during the Romanesque period, continued to be improved, and reached their zenith in the Gothic facades of the 13th century and later [2].

As the Renaissance started to spread, the desire to express global power by way of architecture started to grow. The outward appearance of major secular buildings such as palaces therefore became more and more important – illustrated excellently by the Palazzo dei Diamanti in Ferrara by Biagio Rossetti (fig. B 1.1.6).

In many cases the facade became completely detached from the body of the building itself, for the first time becoming an independent architectural element within the overall structure. Facades produced at great expense, in Italy more than anywhere else, were separated quite distinctly from the loadbearing wall, not only in terms of style, but also in terms of material (fig. B 1.1.8).

In one particular technical variation, the outer layer of thin, cut and worked stone cladding panels are bedded in mortar on the loadbearing external masonry walls – an "incrustation". Such incrustation facades, produced by outstanding artistic craftsmanship using cladding panels of different types of stone, are particularly widespread in Tuscany and Umbria.

Until the arrival of windows with transparent glass panes, stone cladding panels ground thin served as light-permeable protection against wind and weather. A modern example exploiting the translucent properties of natural stone is St Pius Church in Meggen by Franz Füeg (1966) (pp. 72–73).

B 1.1.8

B 1.1.9

	Solid construction	Floor coverings	Steps	Cladding	Sculptures
Basalt	○	○	○		-
Granite	•	•	•	•	•
Marble	-	○	○	○	○
Slate				○	
Sandstone	○			-	
Limestone	•	-	-	○	-

• very suitable
○ sometimes suitable
- low suitability

B 1.1.10

	Density [kg/m²]	Thermal conductivity [W/mk]	Compressive strength [N/mm²]	Tensile bending strength [N/mm²]
Basalt	2700–3000	1.2–3.0	250–400	15–25
Granite	2500–2700	1.6–3.4	130–270	5–18
Marble	2600–2900	2.0–2.6	80–240	3–19
Slate	200–2600	1.2–2.1		50–80
Sandstone	2000–2700	1.2–3.4	30–200	3–20
Limestone	2600–2900	2.0–3.4	75–240	3–19

B 1.1.11

B 1.1.11 Characteristic values of natural stone types [4]

B 1.1.12 Thermal baths, Vals (CH), 1995, Peter Zumthor
B 1.1.13 Kaufmann House ("Falling Water"), Mill Run (USA), 1937, Frank Lloyd Wright

Individual architects have developed new and unusual applications for stone to suit particular projects. At the vine-growing estate in Yountville, California, Herzog & de Meuron used stone baskets made from woven fabric, normally used in landscaping works, as a facade material. This illustrates wonderfully just how the penetrating light can create exciting effects in the interior. The great mass of stone in the facade acts as a temperature regulator but owing to its coarse structure it has a high "permeability" (a refuge for reptiles!), which can be compensated for by additional constructional measures if necessary (see also the example of Mortensrud Church by Jensen & Skodvin, p. 75).

Quarrying stone

Various methods are used to obtain the rough stone blocks from the quarries (fig. B 1.1.14), depending on type, stratification and the abundance or scarcity of the material (figs. B 1.1.15 and B 1.1.16). What all these methods have in common is the aim of obtaining the largest possible flawless blocks, without wastage. The dimension stones required for building are cut to size and shape from the rough blocks by sawing or gang-sawing. Modern computer-controlled cutting techniques mean that it is possible to produce virtually any shape, even rounded forms.

Construction assembly

They are great differences in the various construction options for creating stone curtain walls and their individual appearance. Even at the start of the twentieth century, we saw the predecessors of stone facing leaves, as on the Post Office Savings Bank by Otto Wagner in Vienna. In the second half of the century, this type of construction was one of the most common and most economical forms of stone facade.

The Concert and Congress Hall "Finlandia" (1975) by Alvar Aalto in Helsinki is an example of the aesthetic potential of this technical solution [5].

The construction principle of the facing leaf, known for many centuries, is today becoming popular again with architects. Compared to the "thin" stone curtain wall, it demonstrates distinct advantages in terms of mechanical resistance to horizontal forces. To create the illusion of distinct horizontal layering in a stone facade, using a facing leaf is the simplest solution.

One outstanding example of using a facing leaf of natural stone is Kaufmann House ("Falling Water") by Frank Lloyd Wright. The coarse, layered structure of the external wall seems to reflect the layered structure of the riverbed over which the house is built.

More than six decades later, Peter Zumthor chose the same method of construction (facing masonry leaf) – but with regular stones – to finish off the facade of the thermal baths in Vals.

In the twentieth century, the Modern Movement took up the idea of a distinct outer layer again, but in the form of a ventilated curtain wall, normally secured with metal supports and retaining fixings to withstand the vertical and horizontal forces. The technical approach of treating the leaves of a masonry construction separately according to their function has once again become a feature of modern facades. In these, the natural stonework is detached from the loadbearing wall and serves purely as a cladding material.

The advantages of such constructions in terms of economy and performance have led to this approach being used almost universally for the stone facades of recent years (see p. 33).

B 1.1.12

B 1.1.13

B 1.1.14

B 1.1.15

B 1.1.16

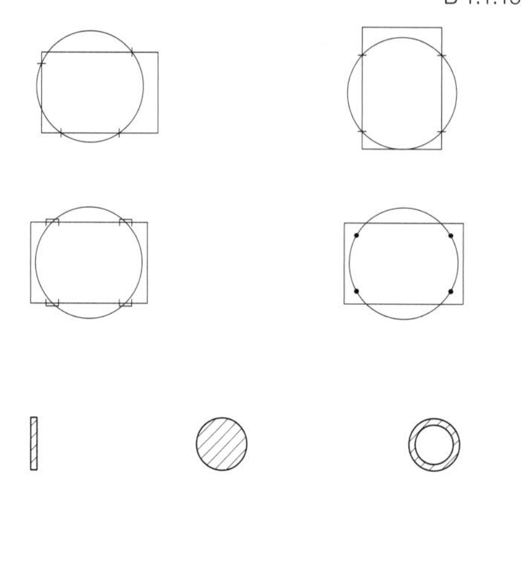

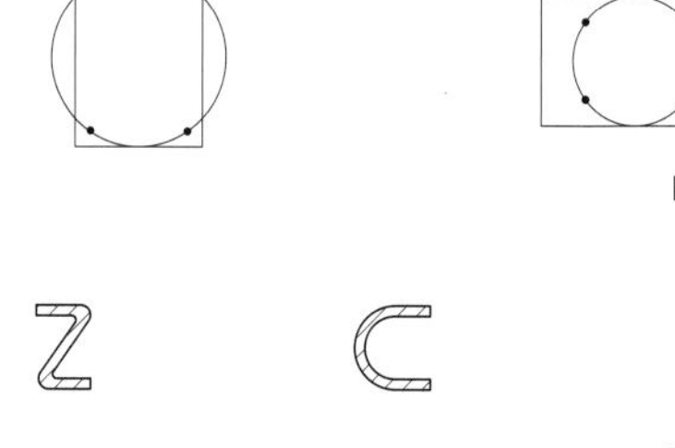

B 1.1.17

B 1.1.18

DIN 18516 part 3 describes external wall claddings of natural stone as follows:

- natural stone cladding panels
- ventilation zone
- thermal insulation layer (when the external wall itself does not provide the necessary thermal performance)
- fixing and anchoring of cladding panels to various external wall constructions

Sizes of natural stone cladding panels

Structural analyses are required to assess the bending strength and the pull-out load for dowels. DIN 18516 part 3 specifies the following minimum thicknesses for natural stone cladding panels:

- angle of inclination with respect to the horizontal > 60°: 30 mm
- angle of inclination with respect to the horizontal 60°: 40 mm

The minimum thicknesses of cladding panels with a higher tensile bending strength are normally also those given in the DIN standard. To take into account permanent actions, vibrations, shocks and dynamic loads, the self-weight of stones with an angle of inclination of 0–15 degrees is increased by a factor of 2.5 owing to the reduction in the bending strength and the pull-out load for dowels.

B 1.1.19

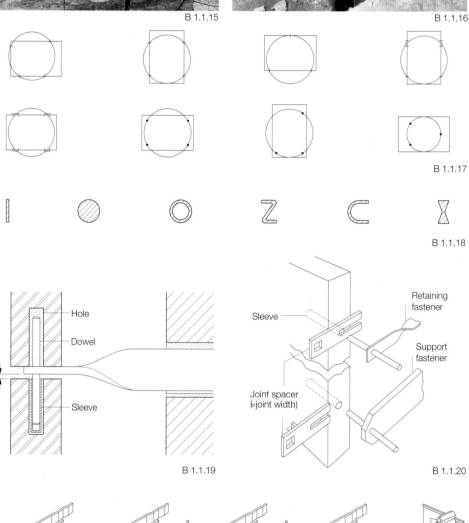

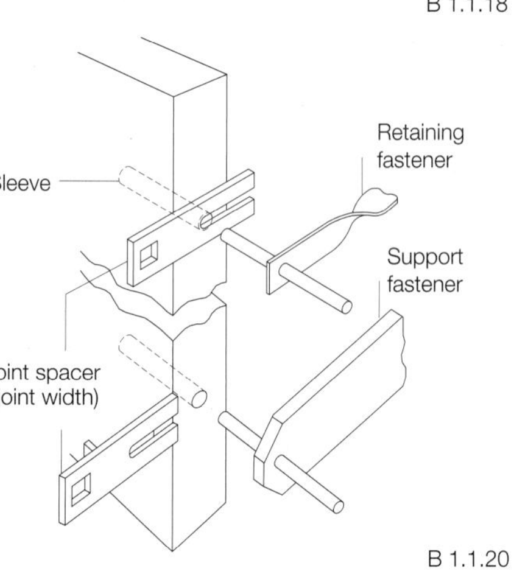

B 1.1.20

Fixing

Stone cladding panels transfer loads to the supporting construction or the substrate individually, i.e., each cladding panel. In facing leaf arrangements without adequate structural strength, the supporting construction (e.g., rail system) must be able to transfer the forces due to dead loads and wind loads to the load-bearing components. Normally, every cladding panel is secured by three or four fixings, whose geometrical arrangement guarantees that the cladding panel is held in position but not subjected to any restraint stresses (fig. B 1.1.17).

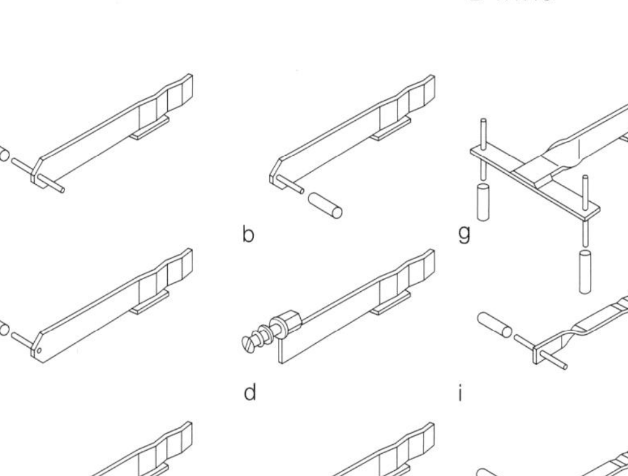

B 1.1.21

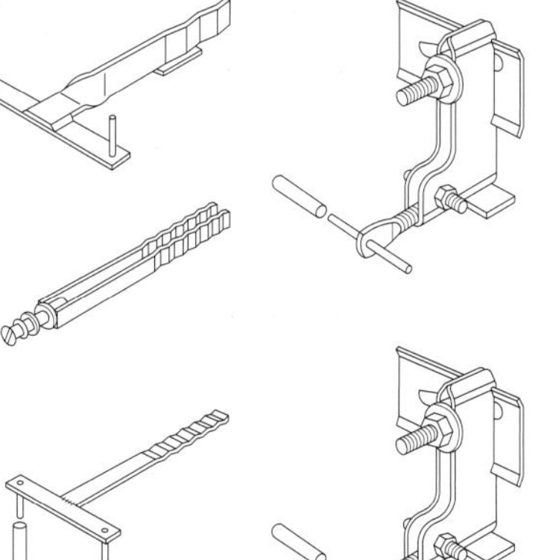

B 1.1.22

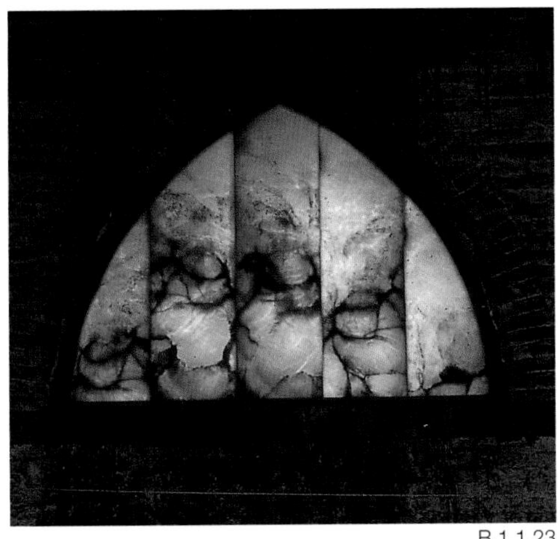

B 1.1.23

mm	2–3	2–5	3–7	5–10	8–15	12–30	Candle in darkened room
	quartzite without mica	limestone	coarse-crystalline marble	fine-crystalline marble	onyx without pigment	alabaster without bitumen	

B 1.1.24

Large cladding panels that require more than four fixings for structural reasons must include appropriate constructional measures to ensure that there are no restraint stresses. The types of fixings can be divided into four main groups:

- dowels
- threaded dowels
- profiled cramps and corbels
- other (e.g., adhesives).

Joints

Joints are required to accommodate movement caused by temperature differentials or static and dynamic actions. A stone cladding to a facade normally requires open (drained) joints 8-10 mm wide. If sealed (closed) joints are required, the permanently elastic sealing material must be able to accommodate the maximum movement calculated. In most cases cladding panels are fixed along their joints. For this reason it is important to ensure that the fixings coincide with the joints in the load-bearing structure, and that in each case allowance for movement of the adjoining cladding panel(s) is possible only on one side of the fixing.

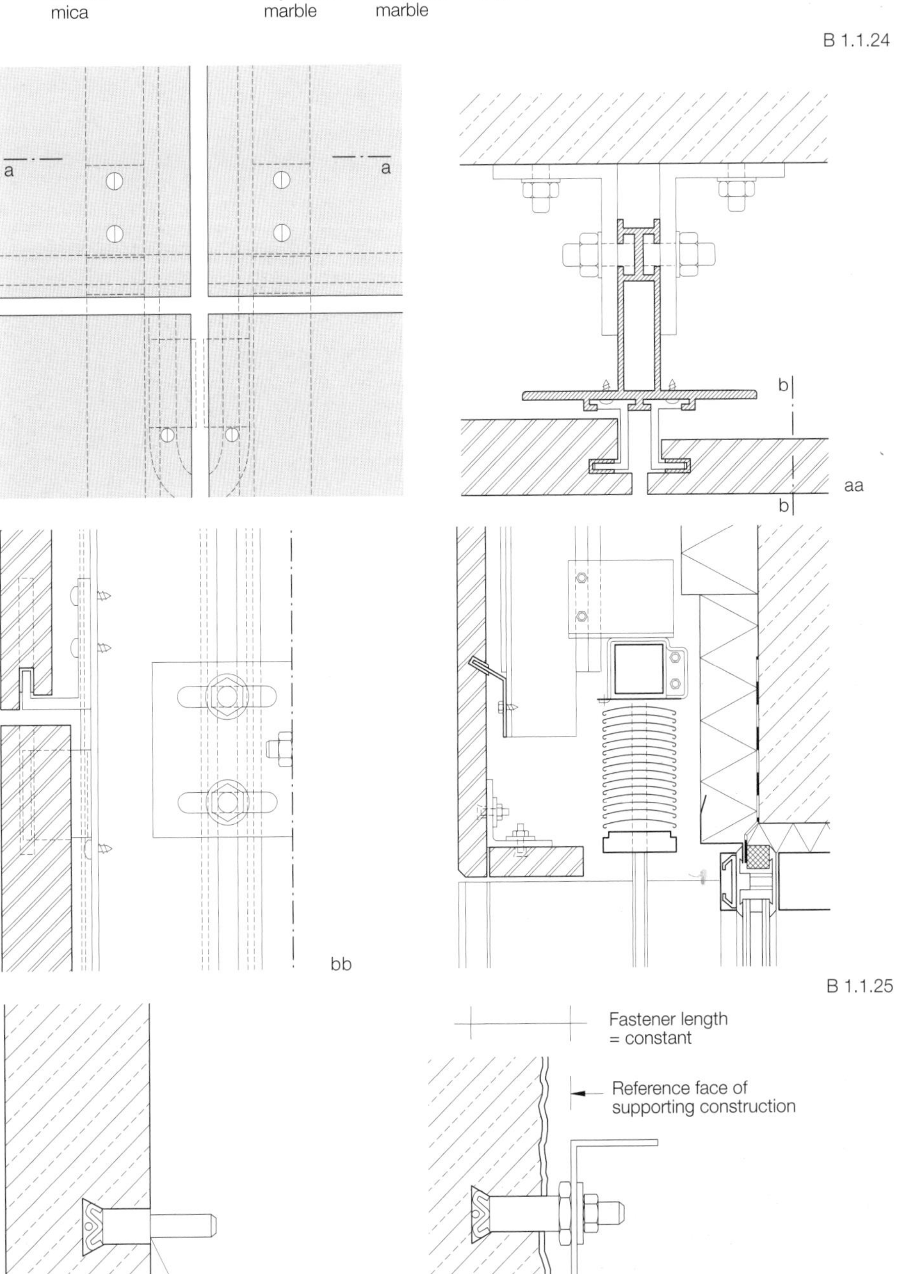

B 1.1.25

B 1.1.26

B 1.1.14 Stone quarry (Fark), 1952
B 1.1.15 Prising apart a block of stone with a long iron bar
B 1.1.16 Wedges can be driven into a plane of weakness to split a block of stone
B 1.1.17 Geometrical conditions for positioning fixings
B 1.1.18 Cross-sectional forms for cramps and corbels
B 1.1.19 Grouted dowel in sleeve to allow movement, horizontal section
B 1.1.20 Axonometric view of support fastener and retaining fastener
B 1.1.21 Support fasteners (a-h) and retaining fasteners (i–l)
B 1.1.22 Dowel with fine adjustment
B 1.1.23 Marble window in the Arsenal at Venice
B 1.1.24 Translucency of light-coloured rocks (light permeability in equivalent material thickness) [6]
B 1.1.25 The use of slots and grooves for fitting support and retaining fixings
B 1.1.26 Undercut fastener for mounting flush or with clearance

B 1.1.27 Hotel, Berlin (D), 1996, Josef Paul Kleihues

The facade of the "Four Seasons" Hotel consists of prefabricated storey-height panels suspended from the intermediate floors. Each panel comprises ground Roman travertine, 30 mm thick, in an overlapping "weatherboarding" arrangement fixed with stainless steel pins. The aluminium frame also carries the window surrounds (separated by a thermal break) in addition to the natural stone cladding panels and thermal insulation (with ventilated air cavity).

B 1.1.28 Office building, Berlin (D), 1996, Jürgen Sawade

This shiny facade consists of highly polished, black granite from Africa. The window elements are fitted flush with the front face of the stone. The basic grid measures 1.2 x 1.2 m and the stone panels are 30 mm thick. The use of a temporary facade hoist allowed the facade to be erected without the need for scaffolding; that shortened the erection time considerably.

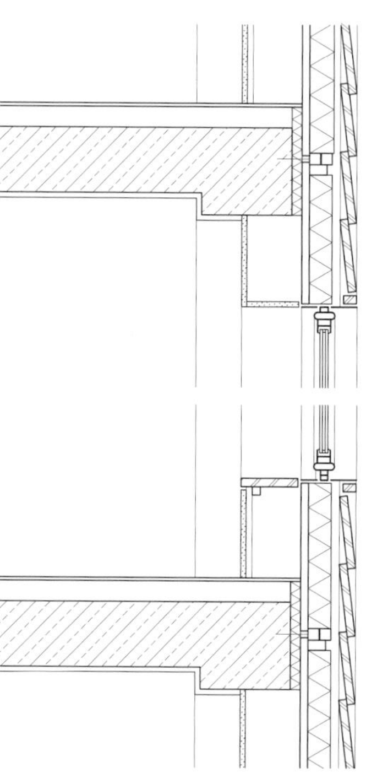

B 1.1.27

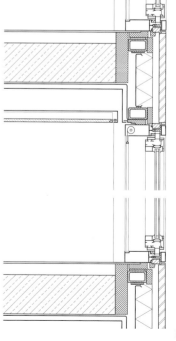

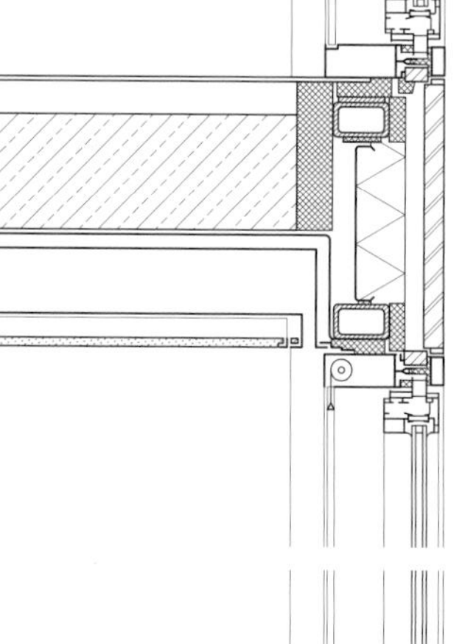

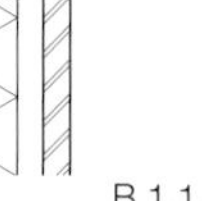

B 1.1.28

B 1.1.29 Office building, Berlin (D), 1997, Klaus Theo Brenner

This strictly regimented stone facade consists of green dolomite with conspicuous stainless steel fixings, which keep the stone panels from falling away from the facade. The shadows of the stainless steel fixings change during the day and over the seasons to add to the buildings its individual character.

B 1.1.30 Mixed commercial and residential development, Berlin (D), 1996, Josef Paul Kleihues

A traditional fenestrated facade. The aluminium windows positioned in the middle of the thickness of the facade construction and the projecting stone window surrounds reinforce the effect of the openings. The frames are made from ground green serpentine, the wall and spandrel elements from ground open-pore yellow travertine.

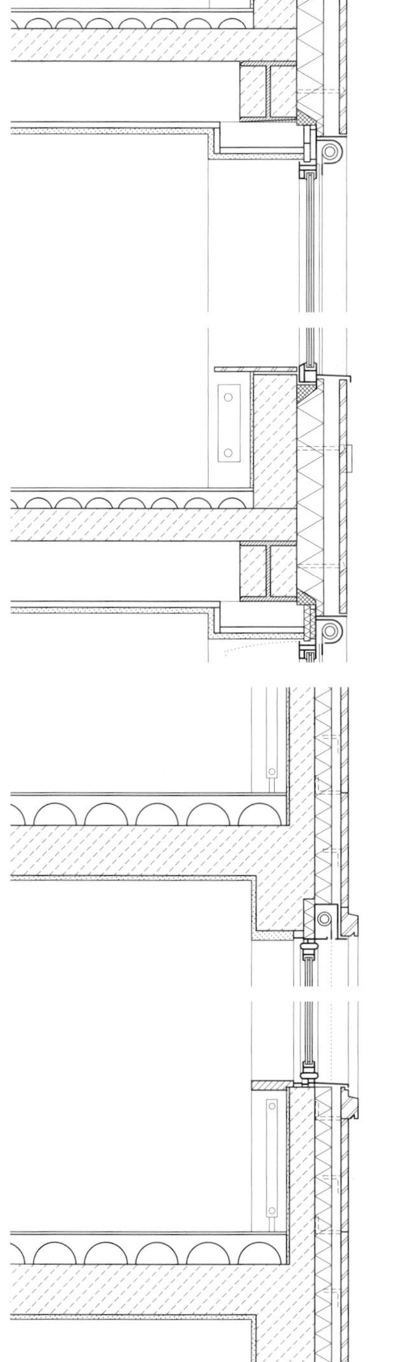

B 1.1.29

B 1.1.30

B 1.1.31

B 1.1.32

B 1.1.33

B 1.1.34

B 1.1.35

B 1.1.36

B 1.1.37

B 1.1.38

Colour and surface texture

The colour and texture of a natural stone material depends on the mixture of the minerals and pigments it contains. In addition, in limestone rocks there is often the visual effect of fossil inclusions as well. Due to physical, chemical or biological contamination, rocks can lose their natural colouring. However, soft and porous types of rock tend to fade even without any polluting effects, especially when used externally. In contrast to this, water on the surface of natural stone often reinforces the intensity of the colouring.

Depending on its hardness and individual properties, the surface of natural stone can be worked either by machine or manually, using mason's tools.

Examples of natural stone types from German quarries:
B 1.1.31 Fürstenstein diorite (igneous)
B 1.1.32 Greifensteiner basalt (igneous)
B 1.1.33 Dorfprozelten sandstone (sedimentary)
B 1.1.34 Mosel slate (sedimentary)
B 1.1.35 Aachener blue stone (sedimentary)
B 1.1.36 Odenwald quartz (metamorphic)
B 1.1.37 Zöblitz garnetiferous serpentinite (metamorphic)
B 1.1.38 Jura marble (metamorphic)
B 1.1.39 The colours of natural stone types [7]
B 1.1.40 Methods of working stone with machinery [8]
B 1.1.41 Methods of working stone with manual tools [8]

Surface finishes of natural stone:
B 1.1.42 Coarse-pointed
The surface is broken away with a pyramid-shaped pointed chisel. The entire surface must be worked. The type of blow determines the difference between a coarse-pointed and a fine-pointed surface finish.
B 1.1.43 Tooth-chiselled
The use of a toothed chisel, working in different directions (straight, curved, or criss-cross) allows a wide variation of surface finishes.
B 1.1.44 Tomb-chiselled in herringbone pattern
Various surface finishes can be achieved by using different tool widths (approx. 80–150 mm) and by varying the striking action.
B 1.1.45 Pointed, bush-hammered, axed and rubbed
These four different types of working result in different surface finishes.
B 1.1.46 Bush-hammered
A method of working that uses a bush hammer with a fine or coarse head. For a fine texture, the head has 7 x 7 pyramid-shaped teeth, for a coarse texture, 4 x 4 teeth.
B 1.1.47 Bush-hammered, brushed and waxed
The wax treatment protects the surface and intensifies the colouring.
B 1.1.48 Polished
Polishing creates a smooth surface with an intensive shine. In order to achieve the best possible effect, any holes should be filled beforehand.
B 1.1.49 Flamed
This exploits the different thermal expansion properties of the particles present in the natural stone. The brief flame treatment to the surface removes flakes of material and leaves a cleaved-type surface. This reduction in the amount of material must be taken into account when calculating the thickness of the stone.

Notes

[1] DIN 18516 parts 1 and 3.
[2] Pfeifer, Günter et al.: *Masonry Construction Manual*, Basel/Munich, 2001, pp. 17–18.
[3] Müller, Friedrich: *Gesteinskunde*, Ulm, 1994, pp. 196–97.
[4] Hugues,Theodor et al.: *Naturwerkstein*, Munich, 2002, p. 72.
[5] *architecture and urbanism* 05/1983: Alvar Aalto, pp. 160–67.
[6] ibid., p. 171.
[7] ibid. [3], p. 169.
[8] ibid. [4], p. 74.

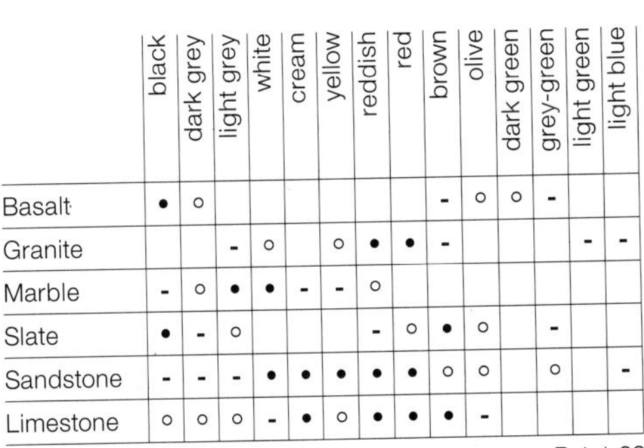

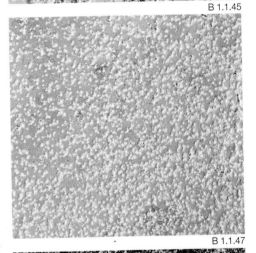

	black	dark grey	light grey	white	cream	yellow	reddish	red	brown	olive	dark green	grey-green	light green	light blue
Basalt	•	○							-	○	○	-		
Granite			-	○		○	•	•	-				-	-
Marble	-	○	•	•	-	-	○							
Slate	•	-	○				-	○	•	○		-		
Sandstone	-	-	-	•	•	•	•	•	○	○		○		-
Limestone	○	○	○	-	•	○	•	•	•	-				

- individual forms
○ some forms
• many forms

B 1.1.39

	sawing	sanding	milling	grinding	honing	rubbing	flaming	polishing
Basalt	•	•	•	•	•			•
Granite	•	•	•	•			•	•
Marble	•		•	•	•	•		
Slate	•		•	•				
Sandstone	•	•		•	•	•		
Limestone	•		•	•	•	•		•

B 1.1.40

	cleaving	split-facing	pointing	bush-hammer.	axing	scabbling	batting	tooth-chiselling	comb-chiselling	sandblasting	rubbing
Basalt	•	•	•	•							
Granite	•	•	•	•							
Marble	•	•	•	•	•	•		•	•	•	•
Slate	•										
Sandstone	•	•	•		•	•	•	•	•	•	•
Limestone	•	•	•	•	•	•		•	•	•	•

B 1.1.41

B 1.1.42

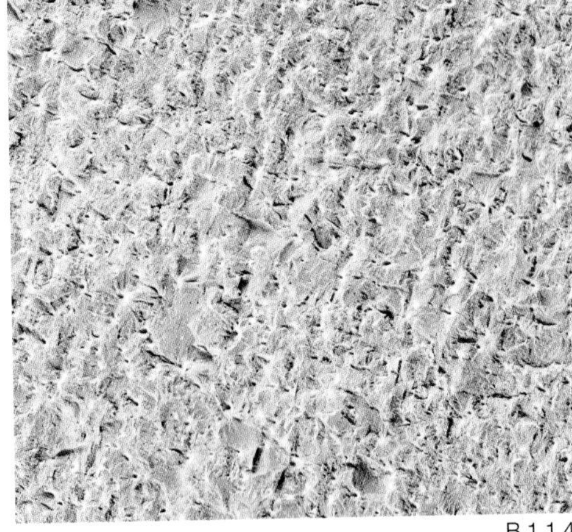

B 1.1.43

B 1.1.44

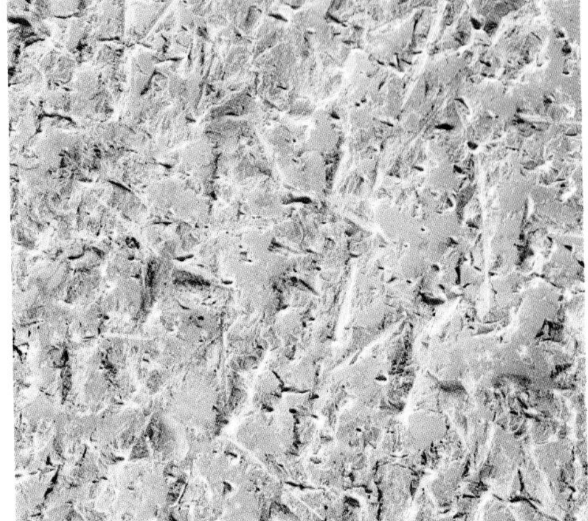

B 1.1.45

B 1.1.46

B 1.1.47

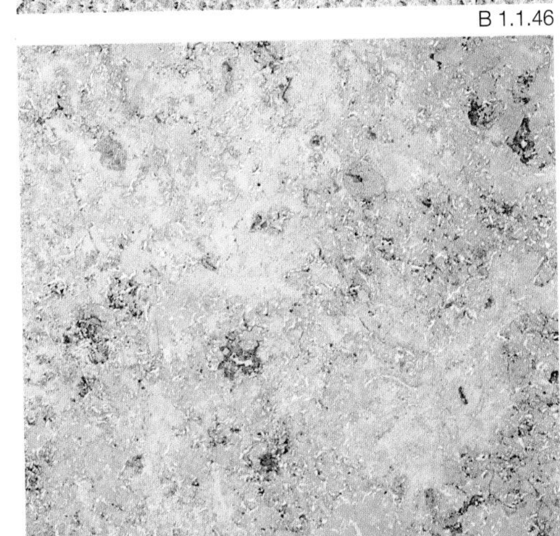

B 1.1.48

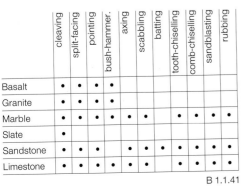

B 1.1.49

St Pius Church

Meggen, CH, 1966

Architect:
Franz Füeg, Solothurn
with Peter Rudolph and Gerard Staub

A+U 11/2003
Bauen + Wohnen 5/1966 and 12/1966
Casabella 677, 2000
Detail 03/1967
Stock, Wolfgang Jean (ed.):
Europäischer Kirchenbau 1950–2000.
Munich 2002

- Steel frame with a basic grid of 1.68 m
- Roof structure of steel circular hollow sections, 63.5 mm dia., span 25.5 m
- Translucent facade of marble panels (1020 mm high x 1500 mm wide)
- Unusually festive interior atmosphere

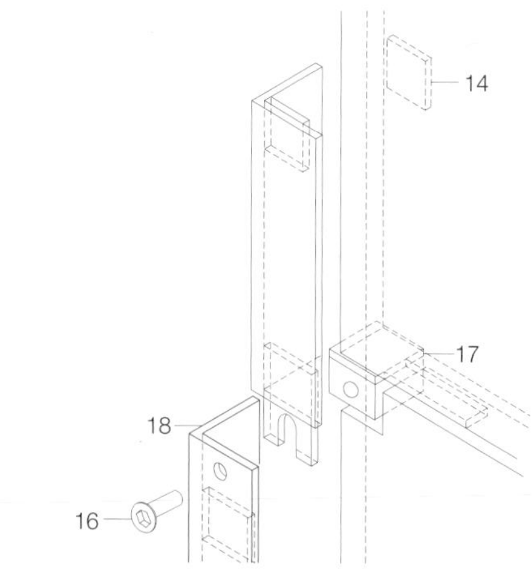

Isometric view, not to scale
Plan • Section, scale 1:750
Vertical section • Horizontal section scale 1:20
Details of panels, vertical and horizontal, scale 1:5

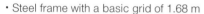

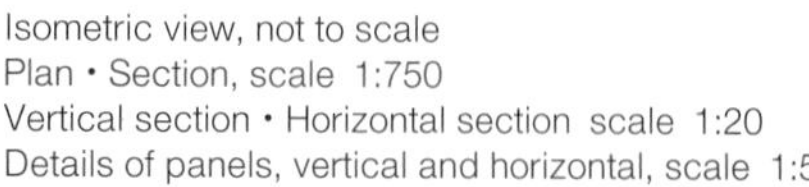

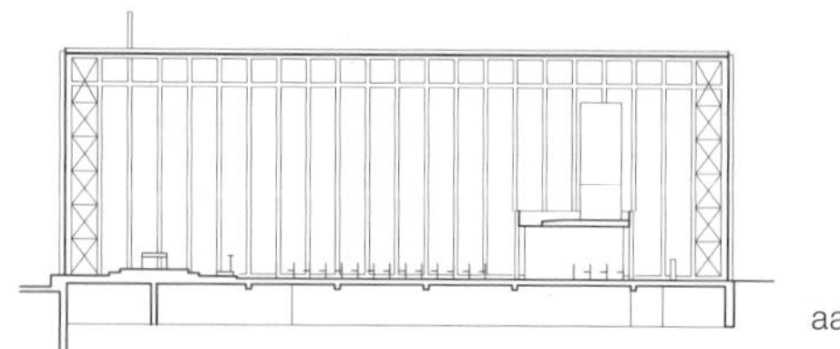

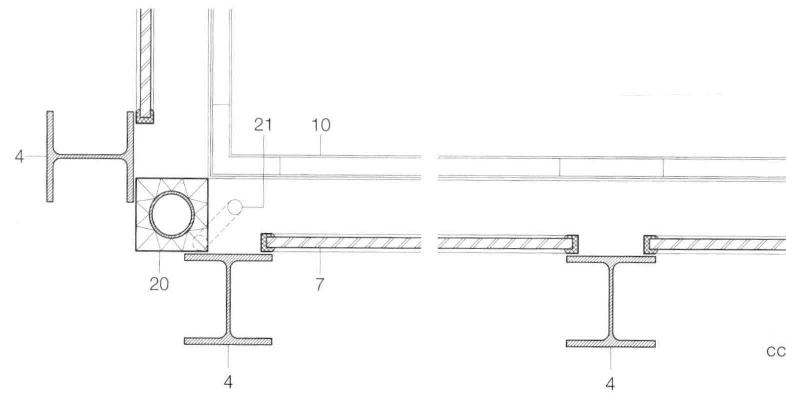

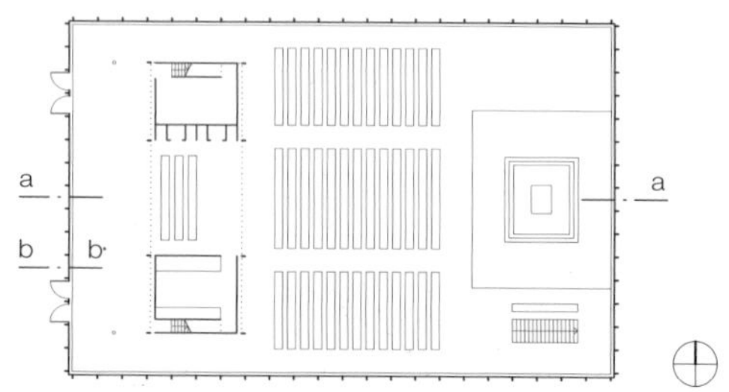

1 Peripheral timber batten
2 Steel flat, 550 x 10 mm
3 Marble panel, 150 x 102 x 21 mm, outside face ground smooth
4 Facade column I IPB 240
5 Lattice beam, steel circular hollow sections, 63.5 mm dia.
6 Steel flat, 260 x 10 mm
7 Marble panel, 150 x 102 x 28 mm, outside face ground smooth
8 Steel flat, 240 x 10 mm
9 Bent sheet steel as channel for condensation water
10 Fresh air inlet
11 Fresh air duct
12 Angle, 35 x 35 x 4 mm
13 Angle, 40 x 25 x 4 mm
14 Spacer, 25 x 25 x 4 mm
15 Spacer, 30 x 30 x 3 mm, with seal
16 M8 hexagon socket screw
17 Panel support 20 x 20 x 15 mm steel flat, with rigid foam bearing pad
18 Angle, 40 x 40 x 4 mm
19 Rigid foam strip to prevent contact between marble and steel
20 Sheet steel box, insulated, for 125 mm dia. rainwater downpipe
21 Drain from condensation water channel

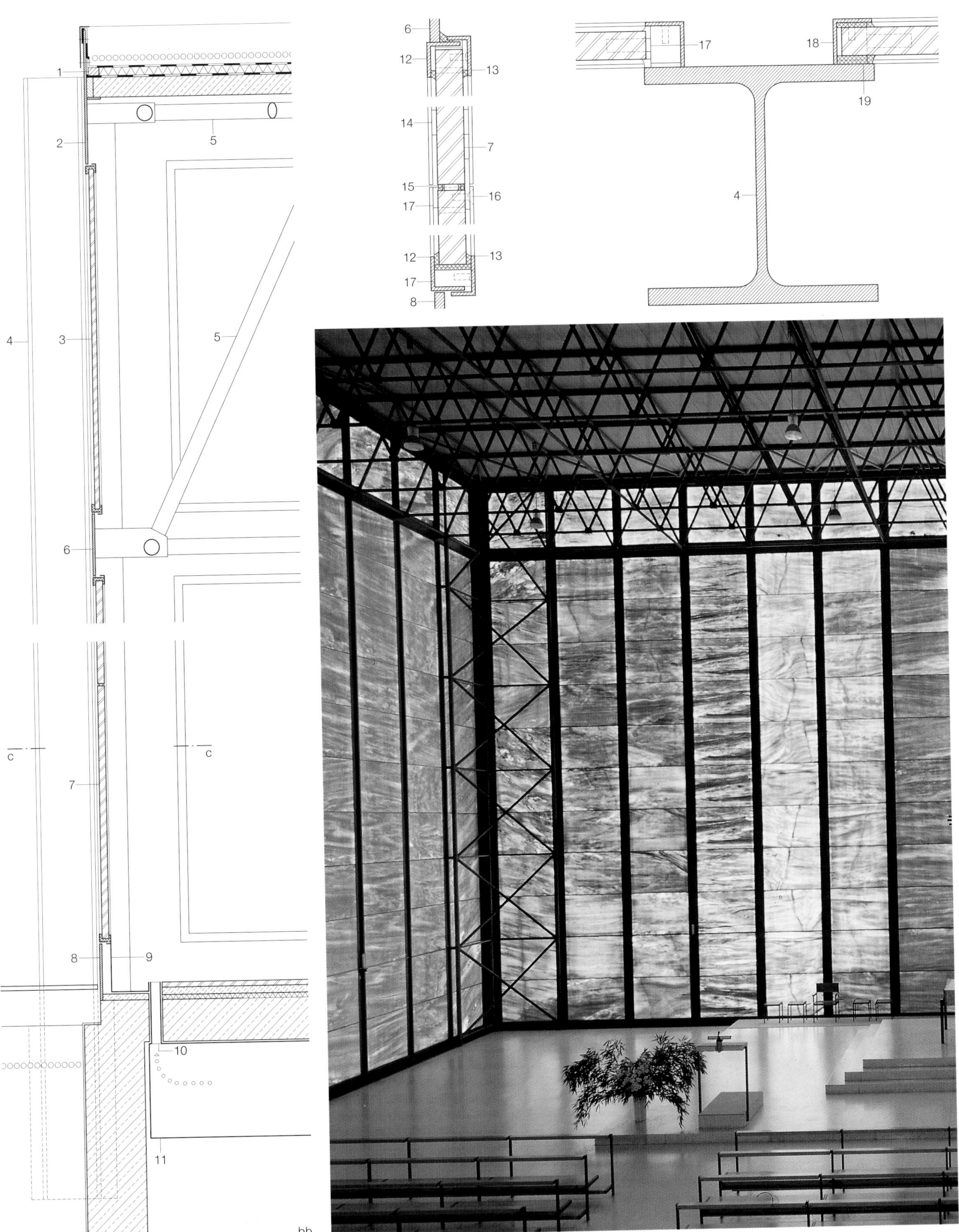

1
2
3
4
5
6
7
8
9
10
11
c
c
bb
6
12
13
14
7
15
16
17
12
13
17
8
17
18
19
4

Private house

Sarzeau, F, 1999

Architect:
Eric Gouesnard, Nantes

l'architecture d'aujourd'hui 320, 1999
A+U 06/1999
LOTUS 105, 2000. special issue:
Aperto over all

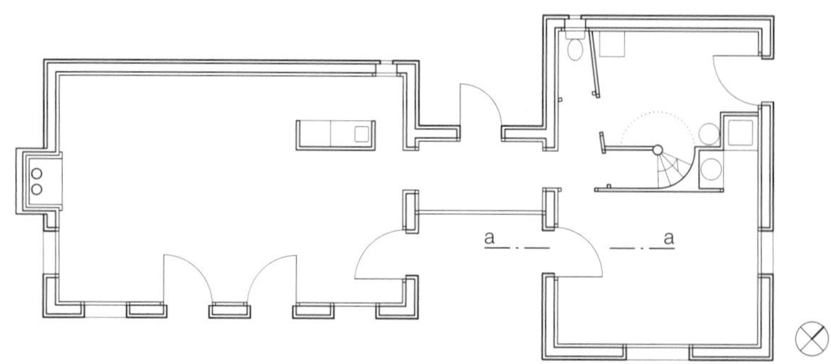

- "Monolithic" effect of building because facade and roof clad with same material
- Dark grey slate panels, 500 x 500 mm
- Concealed rainwater gutters

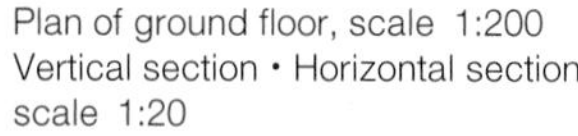
Plan of ground floor, scale 1:200
Vertical section • Horizontal section
scale 1:20

1 Facade construction:
20 mm slate panel
supporting framework of steel Z-sections
20 mm cement rendering
200 mm masonry
closed-pore thermal insulation
vapour check
100 mm composite panel made from plasterboard
2 Sheet aluminium rainwater gutter, concealed
3 Rainwater downpipe

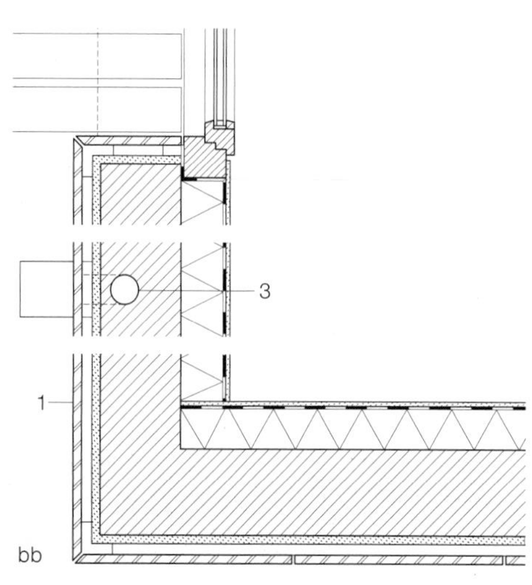

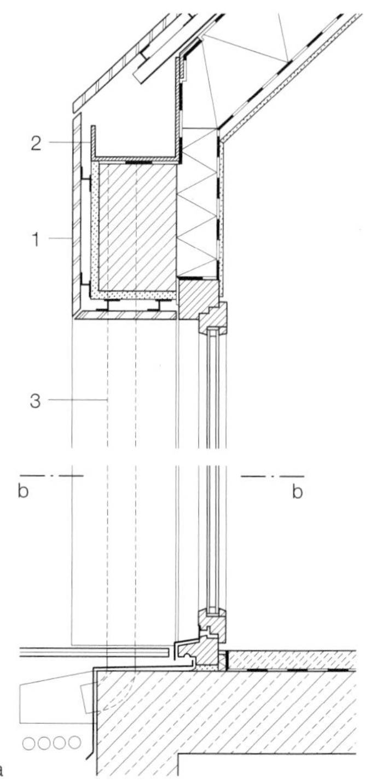

Mortensrud Church

Oslo, N, 2002

Architects:
Jensen & Skodvin, Oslo

Architectural Review 12/2002
Architektur Aktuell 01–02/2003
A+U 08/2002
Byggekunst 04/2002
Detail 11/2003
Living Architecture 19, 2004

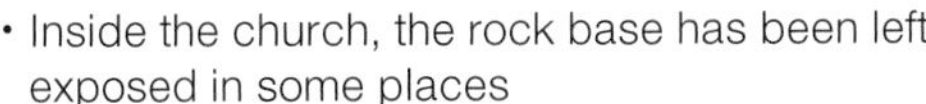

- Inside the church, the rock base has been left exposed in some places
- External glass facade and internal frame of oiled steel
- Dry walling construction – slates laid without mortar
- Rubble filling stabilised by large steel plates between columns at 1 m spacing
- Price per square metre corresponds to that of publicly assisted housing in Oslo

Section • Plan, scale 1:1000
Vertical section through west facade
scale 1:20

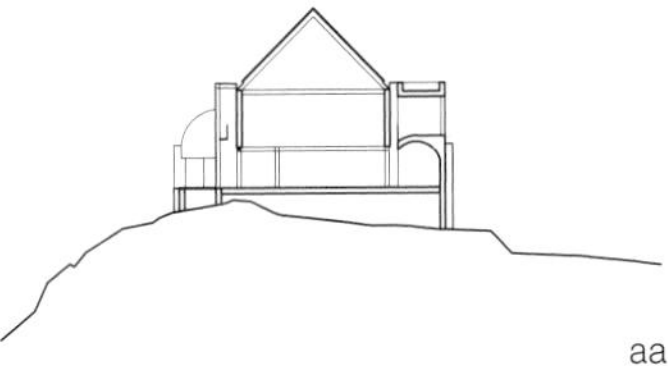

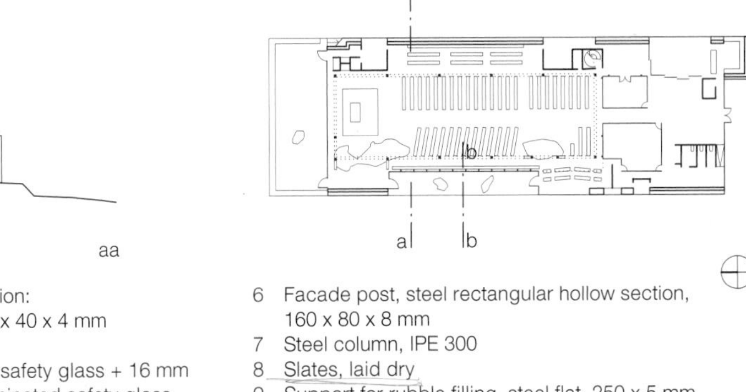

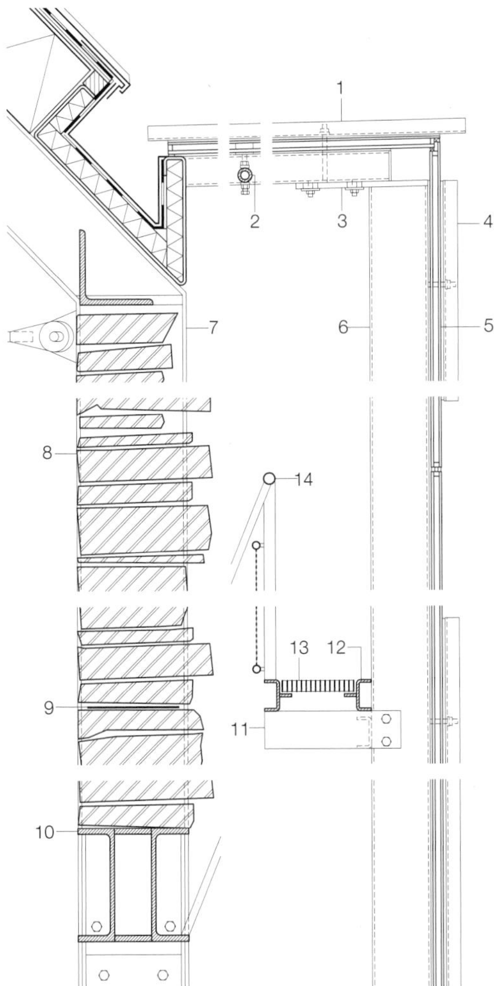

1 Facade construction:
steel channel, 80 x 40 x 4 mm
double glazing:
6 mm toughened safety glass + 16 mm cavity + 8 mm laminated safety glass
steel square hollow section, 80 x 40 x 4 mm
2 Steel circular hollow section, 38 x 5 mm, as central support for glass pane
3 Steel plate, 360 x 80 x 15 mm
4 Steel channel, 80 x 40 x 4 mm
5 Double glazing:
6 mm toughened safety glass + 15 mm cavity + 7 mm laminated safety glass
6 Facade post, steel rectangular hollow section, 160 x 80 x 8 mm
7 Steel column, IPE 300
8 Slates, laid dry
9 Support for rubble filling, steel flat, 250 x 5 mm
10 Steel compound lintel:
2 No. 300 x 100 mm channels +
2 No. 100 x 10 mm flats
11 Steel flats, 2 No. 100 x 10 mm
12 Steel channel, 80 x 40 x 5 mm
13 Steel grating, 30 mm
14 Steel handrail, 30 mm dia.

Museum of prehistory and ancient history

Frankfurt am Main, D, 1989

Architect:
Josef Paul Kleihues, Berlin/Dülmen
with Mirko Baum (project manager)

Arkitektur 08/1989
Baumeister 06/1989
Casabella 481, 1982
Feldmeyer, Gerhard:
The New German Architecture.
New York 1993

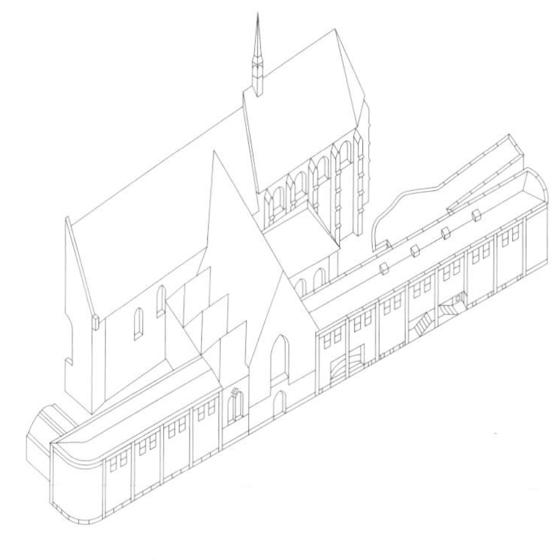

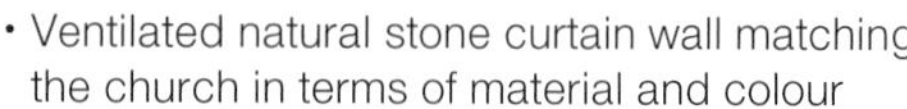

- Ventilated natural stone curtain wall matching the church in terms of material and colour
- Exposed fixings appear as ornamentation for technical reasons

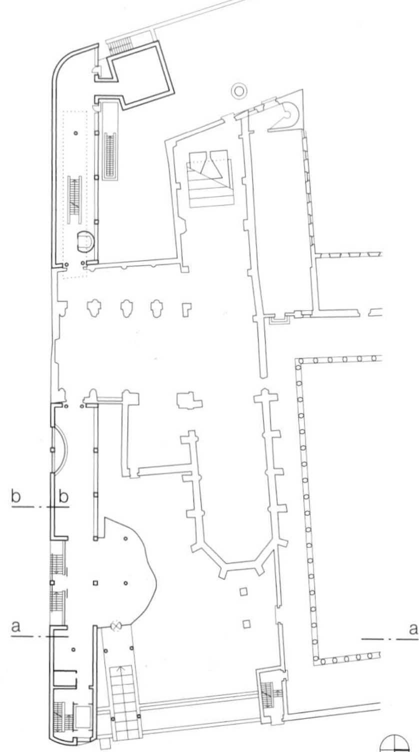

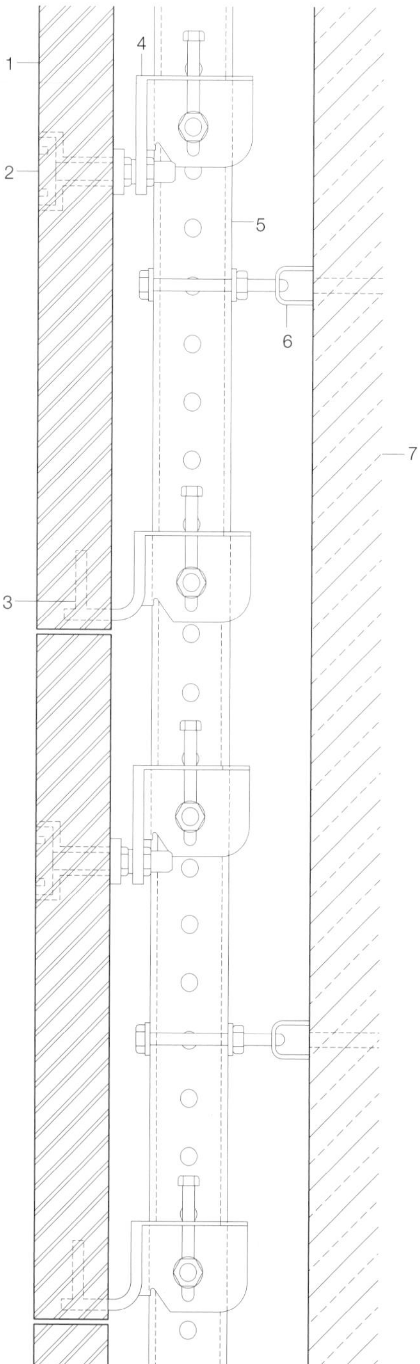

Plan • Section
scale 1:1000
Vertical section scale 1:5

1. Red sandstone without veining, and yellow-green sandstone from the Würzburg region (southern Germany)
2. Spacer with special screw, left exposed externallyr
3. Supporting/retaining bracket, not visible externally
4. Bracket for exposed screw retaining fixing
5. Mounting rail with standard drilling
6. Wall anchor
7. Reinforced concrete

Office of the Federal President

Berlin, D, 1998

Architects:
Gruber + Kleine-Kraneburg, Frankfurt am Main

Detail 06/1999
Burg, Annegret; Redecke, Sebastian: Kanzleramt und Bundespräsidialamt der Bundesrepublik. Boston/Berlin/Basel 1995

- Dark, polished natural stone (Nero Impala)
- Shape of building emphasised by stonework (elliptical cut)
- Windows fitted flush with the outer face of the stone facade

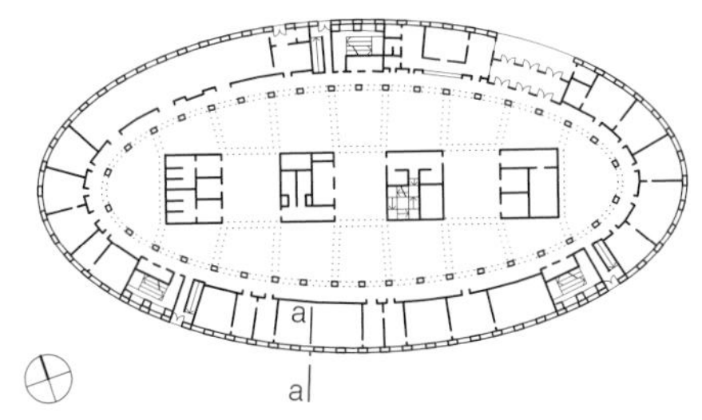

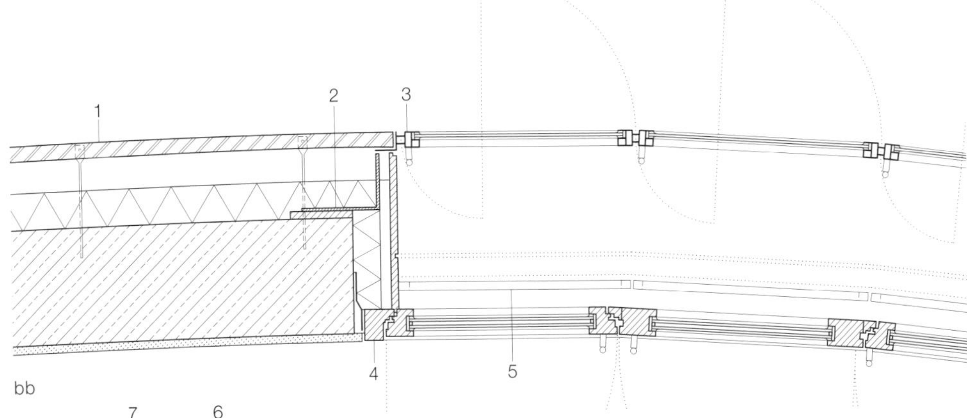

Plan, scale 1:3000
Vertical section • Horizontal section
scale 1:20

1 Facade construction:
40 mm natural stone
85 mm air cavity
100 mm thermal insulation
300 mm reinforced concrete
25 mm gypsum plaster
2 Spandrel panel construction:
aluminium angle on three sides
with plastic wedge below
as thermal break
3 Aluminium window, anthracite stove-enamelling glazing:
ground floor: 16 mm laminated safety glass (comprising 2 panes of toughened safety glass)
1st–3rd floors: 10 mm toughened safety glass
4 Wooden window, oak with dark stain finish
double glazing: 6 mm laminated safety glass + 14 mm cavity + 4 mm toughened safety glass
5 Safety barrier, 20 x 20 mm aluminium sections
6 Parapet construction:
3 mm aluminium capping plate
aluminium fixing, ribbed profile with integral rubber seal, both sides of butt joint
supporting construction: 50 x 3 mm aluminium channel screwed to 40 x 3 mm aluminium channel, and screwed to timber plank
7 Aluminium angle, 50 x 50 x 2 mm
8 Retaining fixing
9 Support fixing
10 Ventilation grille
11 Sunshade, can be lowered down to max. 100 mm above window board (air circulation)

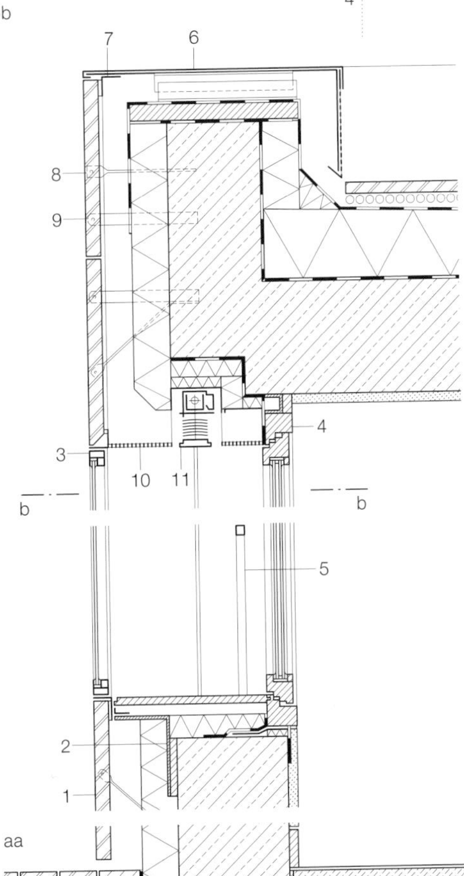

Arts centre

Würzburg, D, 2002

Architects:
Brückner & Brückner, Tirschenreuth
Assistant:
Norbert Ritzer

AV Monografías/Monographs 98, 2002
Bauwelt 14/2002
Detail 10/2002

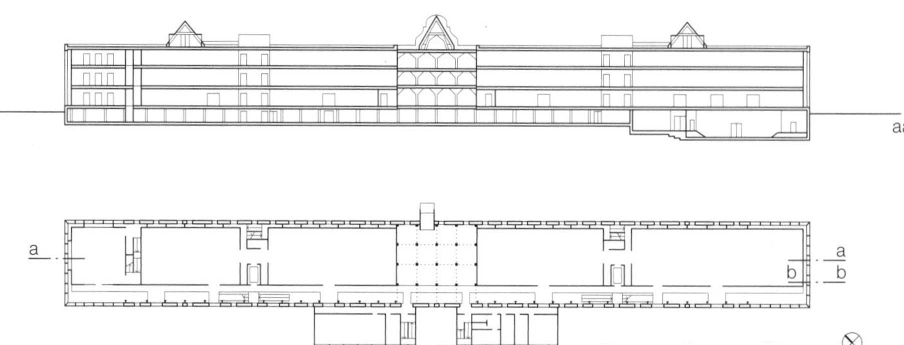

- "Burenbruch" shelly limestone to ground floor and plinth
- Udelfanger sandstone
- Successful "dialogue" between old and new
- Converted building integrated into new function

Section • Plan of upper floor, scale 1:1500
Vertical section • Horizontal section, scale 1:20

1 Facade construction:
100 x 225 mm Udelfanger sandstone louvres
air cavity
insulating render
40 mm thermal insulation
waterproofing
250 mm reinforced concrete parapet
2 Steel column, HEB 300
3 Double glazing: 8 mm toughened safety glass + 16 mm cavity + 10 mm float glass
4 Aluminium square hollow section, 50 x 50 mm
5 Copper heating pipes, Ø 24 mm
6 Ground floor and plinth: 100 x 225 mm "Burenbruch" shelly limestone louvres
7 Steel flat with lugs, 250 mm
8 Steel flat, 500 x 10 mm, welded to 250 x 10 mm steel flat
9 External wall (existing): clay bricks with whitewash finish internally, untreated natural stone externally

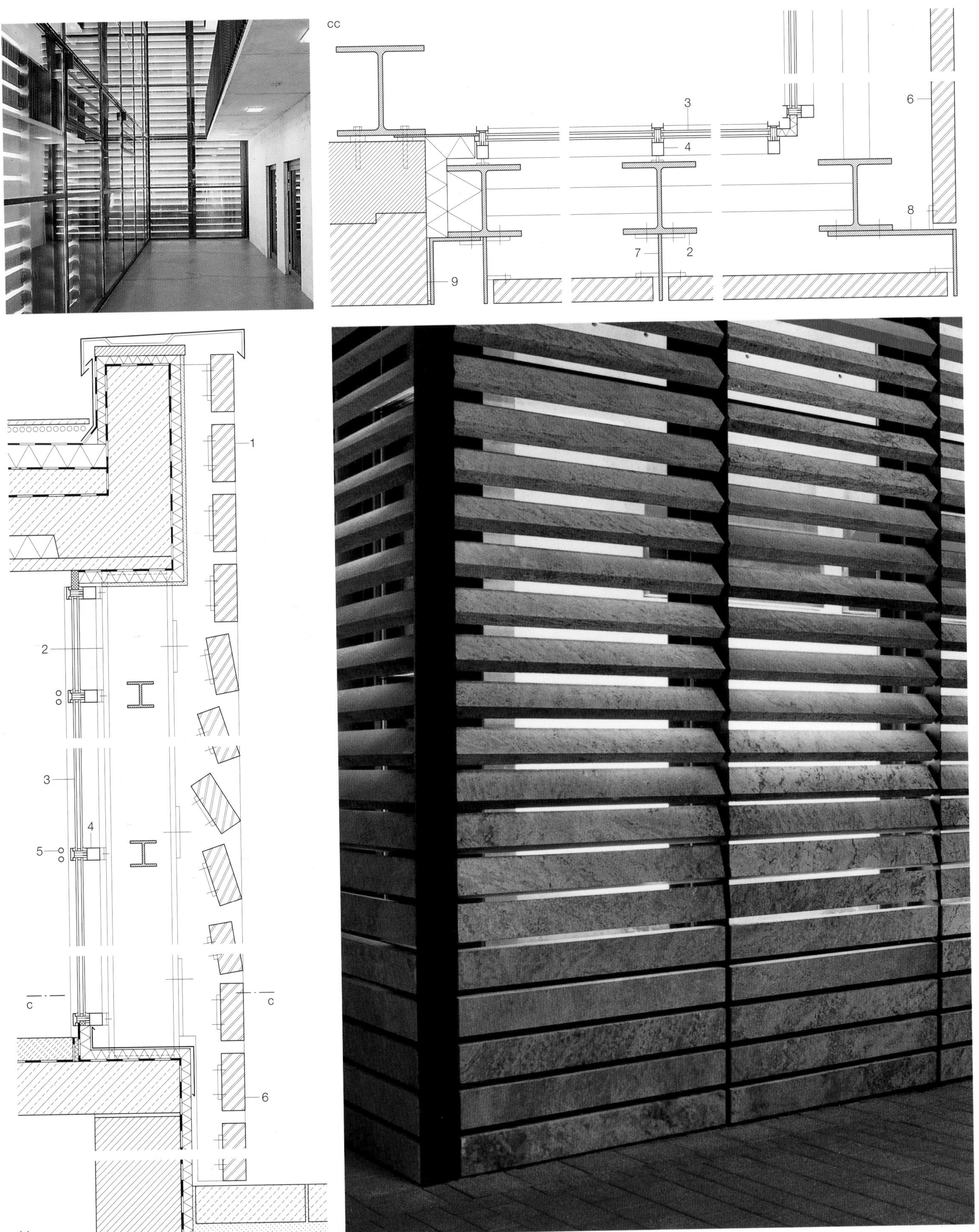
cc
3
4
6
8
7
2
9
1
2
3
4
5
c
c
6
bb

Museum of modern art

Vienna, A, 2001

Architects:
Ortner & Ortner Baukunst, Vienna
with Christian Lichtenwagner
Structural engineers:
Fritsch Chiari & Partner, Vienna

A+U 01/2002
Materia 39, 2002
Dernie, David:
Neue Steinarchitektur.
Stuttgart 2003

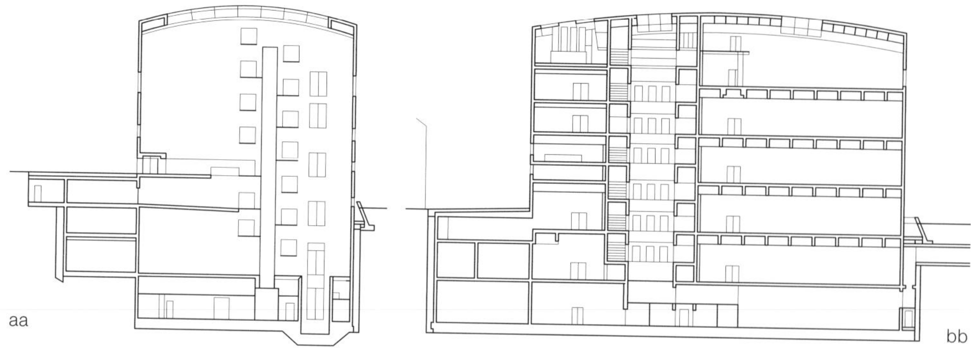

- Ventilated stone curtain wall of basalt lava
- Panel formats increase in size with the height of the building
- Curved roof covered with basalt panels
- Diamond-sawn stone with porous, but smooth surface

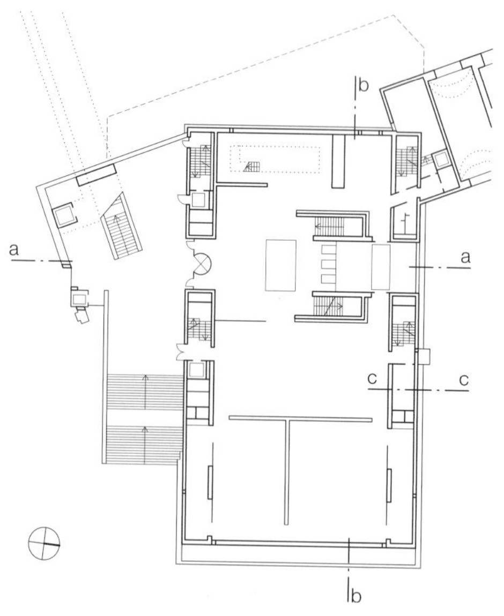

1 Heated rainwater gutter
2 Overflow gutter
3 Retaining fixing
4 Supporting fixing
5 100 mm Mendiger basalt lava stone hung on grouted dowels bed joints filled with permanently elastic material
50 mm ventilation cavity
80 mm mineral wool
300 mm reinforced concrete
50 mm timber battens
25 mm 3-ply core plywood
2 No. 12.5 mm plasterboard
6 Insect screen
7 Limestone, 250 mm
8 Steel angle, 100 x 100 x 10 mm, connected to wall via thermal break
9 Door frame, 100 x 100 x 6 mm steel square hollow sections
10 Frame, 60 x 60 x 4 mm steel square hollow sections, with steel lugs for fixing stonework
11 Door leaf: 40 mm Mendiger basalt lava stone, fastened with dovetail slot fixing
60 mm mineral wool
20 mm rigid polystyrene foam
2 mm sheet aluminium
12 Glazing to double window:
inside: laminated safety glass (comprising 2 panes of toughened safety glass) + cavity + toughened safety glass
outside: toughened safety glass + cavity + toughened safety glass
13 Cover plate, 2 mm stainless steel

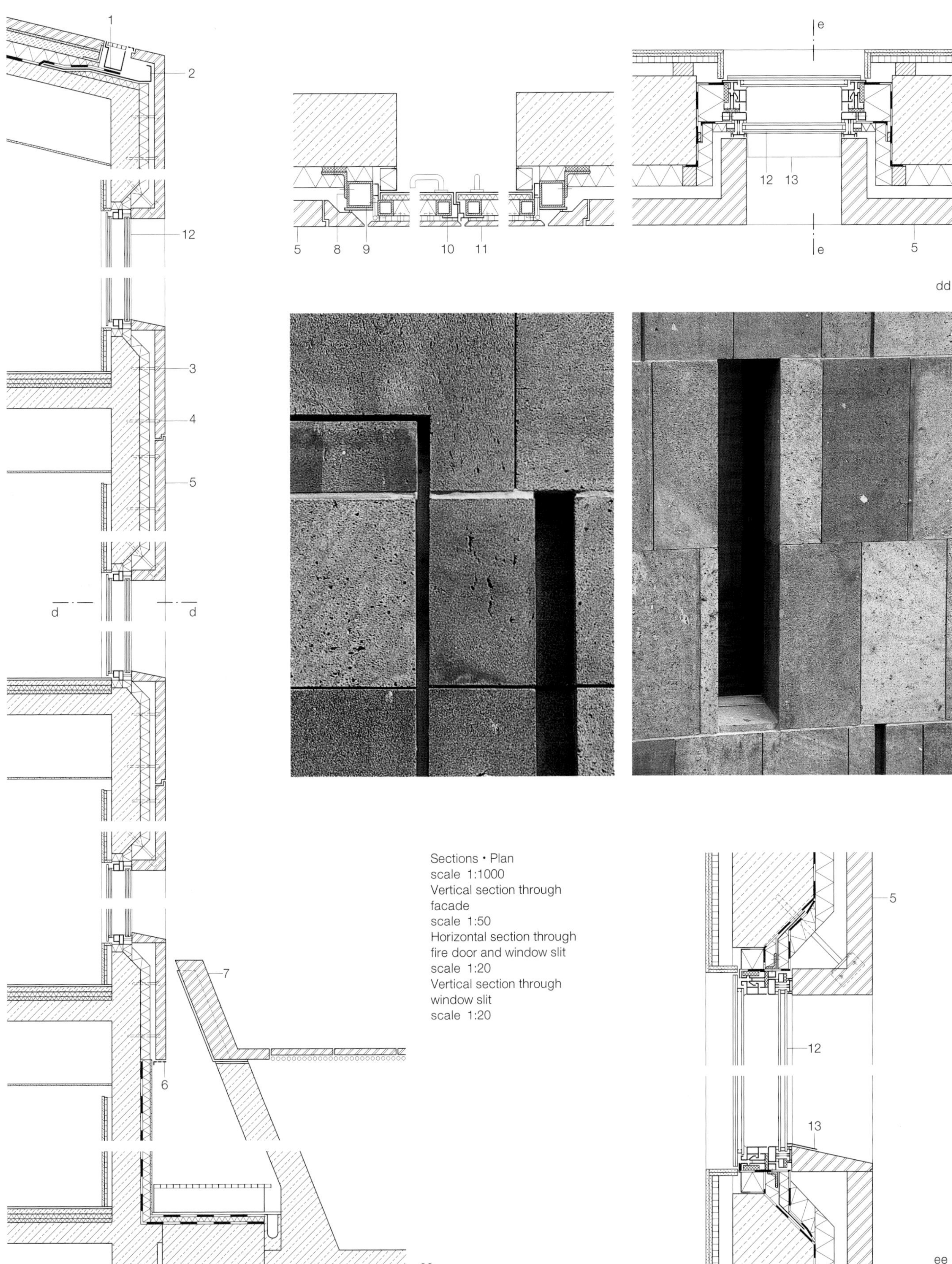

Sections • Plan
scale 1:1000
Vertical section through facade
scale 1:50
Horizontal section through fire door and window slit
scale 1:20
Vertical section through window slit
scale 1:20

B 1.2 Clay

Fired materials made from clay, the main component of all ceramic building materials, have been used in building for more than 7000 years. Although the basic principles of their production have hardly changed in that time, ceramic materials tend to be classed among the "modern" building materials [1].

Manufactured masonry units

The number of manufactured units has multiplied in recent decades; clay bricks are among these. One important reason can be found in the development of the various additional substances, which can have a major impact on the properties (thermal conductance, compressive strength, colour, etc.) of a manufactured masonry unit. Despite the great variety of products available, they can be classified in three groups according to their method of production:

- dried (oldest form of manufactured unit)
- hardened
- fired

The dried types include, primarily, loam building materials, the further development of which has been given a major boost recently owing to their ecological relevance. Calcium silicate, concrete, and lightweight concrete types, among others, form the group of masonry units hardened by means of steam and pressure. Clay bricks, which are available in many formats, hardness grades and colours, fall into the category of fired types. Fig. B 1.2.4 summarises the material properties of a number of masonry unit types.

Clay bricks in the facade

In the valley of the Nile we find traces of buildings made from hand-moulded loam bricks with indications that they originated around 14 000 B.C. Loam buildings are at risk if exposed to the weather without being protected by additional constructional measures. The properties specific to loam (a mixture of clay and quartz sands) make it vulnerable to moisture. Furthermore, loam does not set upon drying out, but rather simply hardens. What this means is that when the material gets wet again (e.g., due to rain, moisture in the ground), it softens and loses its strength. Similar constructional solutions to protect loam buildings against erosion (e.g., by building beneath overhanging rocks, by using natural stone plinths, by cladding with fired bricks, natural stone) are employed everywhere. To create long-lasting masonry with loam bricks, early builders began to fire them starting about 5000 B.C. If firing can be done at a temperature of 1000°C, the material is sintered – and that creates a building material offering good protection against the weather. At that time it was already possible to glaze surfaces or produce bricks with coloured

B 1.2.1 Apartment block, rue des Meaux, Paris (F), 1991, Renzo Piano Building Workshop

B 1.2.2

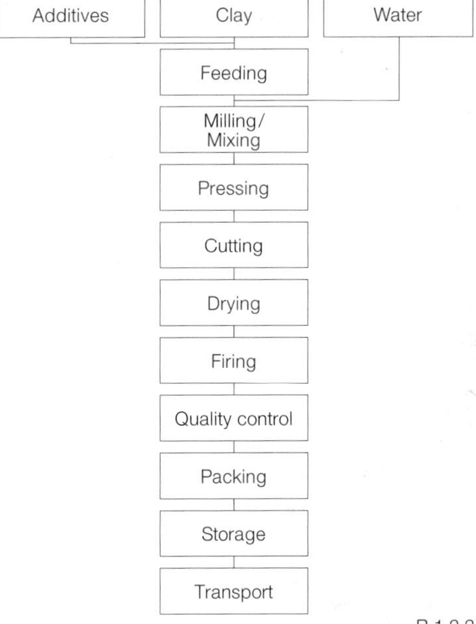

B 1.2.3

	Density [kg/m²]	Thermal conductivity [W/mk]	Compressive strength [N/mm²]	Tensile bending strength [N/mm²]
Loam materials	1800–2000	0.64–0.93	2.40	0.52
Calcium silicate units	600–2200	0.23–0.98	4–6	**
Aerated concrete units	350–1000	0.07–0.21*	2–8	**
Concrete units	500–2400	0.24–0.83	2–48	**
Granulated slag aggregate units	1000–2000	**	6–28	**
Clay bricks	1000–2000	0.18–0.56*	4–60	**
Ceramic materials	1600–2000	**	36–66	7–20

* oven-dry values, 50% fractile
** no figures available

B 1.2.4

B 1.2.2 Traditional loam buildings, Yemen
B 1.2.3 Flowchart for the production of clay bricks [1]
B 1.2.4 The material properties of various manufactured masonry units [1, 3, 5]

B 1.2.5

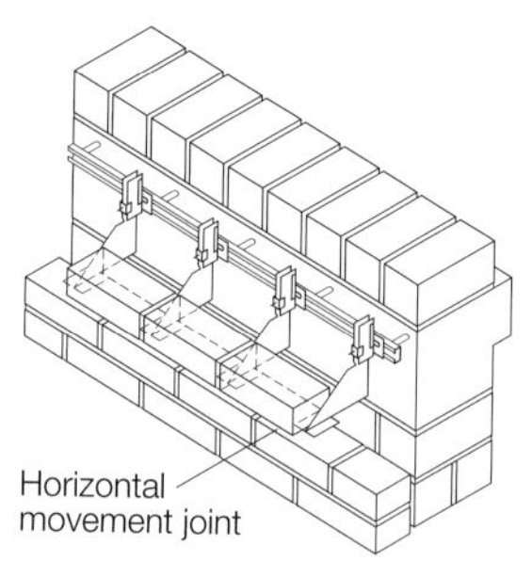

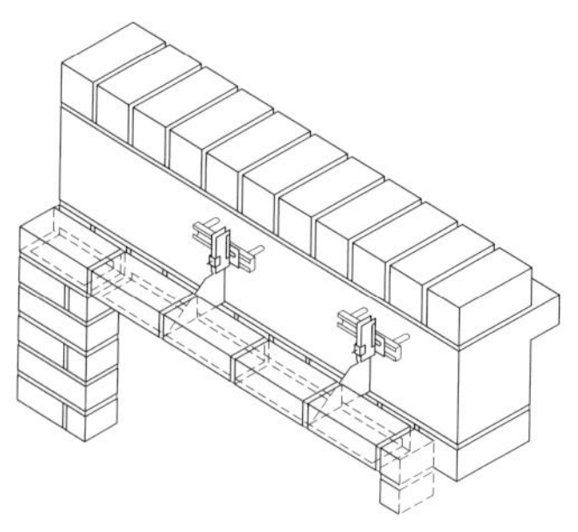

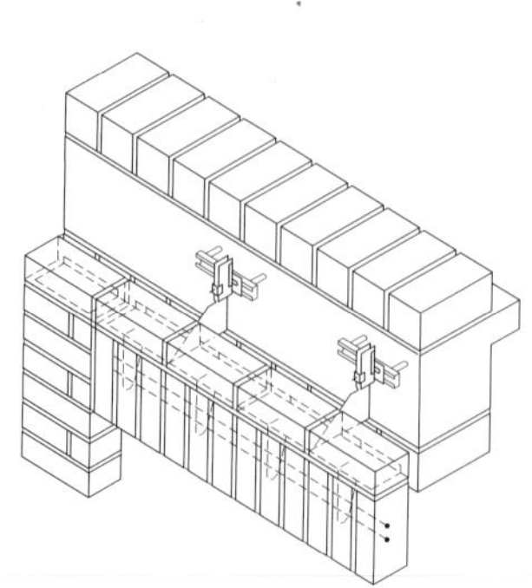

B 1.2.9

B 1.2.6

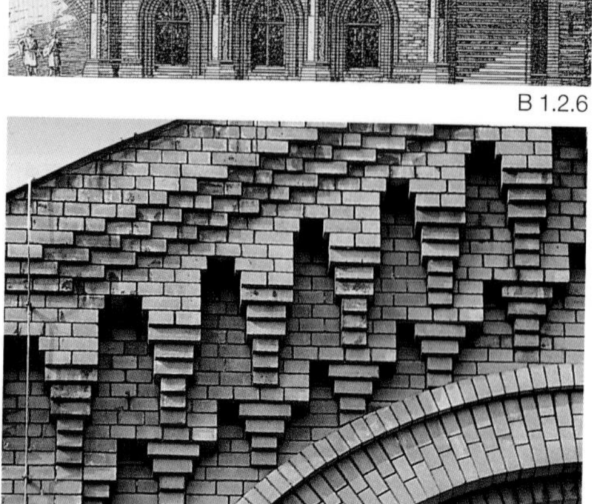

B 1.2.7

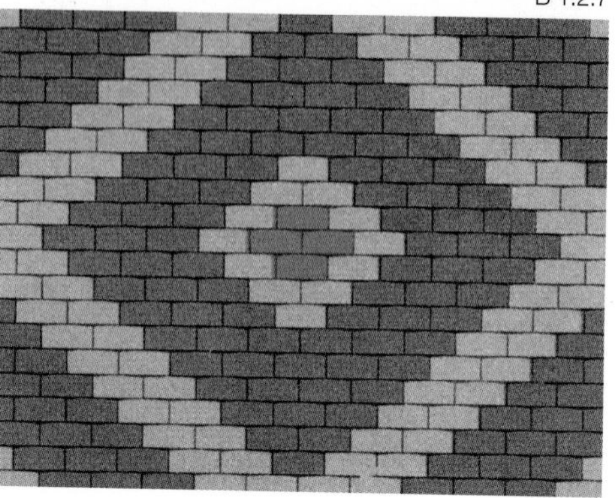

B 1.2.8

pigments (fig. B 1.2.2). Manufactured masonry units have therefore been an everyday building material for many thousands of years. Since that time, these materials have been used for a hugely diverse range of structures, depending on local, climatic and geological circumstances, as well as aesthetic requirements and social contexts. Decisive progress towards mass production of fired bricks was first made by the ancient Romans. Throughout the Roman Empire we find brickworks supplying building materials for all kinds of building projects [1]. In England and in Germany fired clay material took on great significance during the Middle Ages, as shown by the buildings in the impressive and very simplified style of Gothic, called the "Backsteingotik" (fig. B 1.2.6).
The invention of the extruder, the circular kiln and, shortly afterwards, the tunnel kiln in the eighteenth century made possible the mass production of clay bricks. Thanks to the firing process, the original readily water-soluble clay aquires a high physical and chemical stability. Its high resistance to soiling, flue gases, mould growth and frost make this material ideal for external use [1]. During the founding years of the modern German state, towards the end of the nineteenth century, hard-burned facing bricks gradually became the standard weatherproof material for the outer face of masonry facades in many places; almost without exception – at least on the side facing the street – also with diverse historicising decoration and ornamentation, which in those days could be ordered straight out of a catalogue. "Stony Berlin" with its great blocks of rented apartments was built of clay bricks. And the use of clay bricks was a matter of course for architects of the Modern Movement such as Alvar Aalto, Ludwig Mies van der Rohe and others. From the middle of the twentieth century onwards, others such as Eladio Dieste, continued the Iberian tradition of magnificent architectural creations in which – like the church in Atlantida, Uruguay – fired clay is an important element in the loadbearing structure. At the same time, the material conveys the sensation of a lightweight, undulating envelope (fig. B 1.2.14). Today, ceramic cladding with a thickness of just a few centimetres is possible, and owing to its weather resistance, is particularly suitable for protecting thermal insulation materials.

Ceramic facades

When masonry is used for the external walls of buildings, the loadbearing walls simultaneously act as the building envelope. Both these aspects are reflected in a wide spectrum of alternatives and methods of construction developed in different cultural spheres over the centuries. Innumerable publications deal in detail with the corresponding design approaches for walls and openings [1, 2].

The examples later in this chapter illustrate essentially non-loadbearing external wall constructions which mainly serve as the outermost protective mantel for the building within. Selected examples also show how to use "clay brick elements" to create light- and air-permeable wall surfaces that act as privacy screens and sunshades.

Design of facades with facing brickwork
Owing to their similar appearance, at least at first glance, there is sometimes a danger of confusing faced walls and facing masonry. This can lead to misunderstandings, in terms of the construction, when planning a faced wall, which is these days normally a non-loadbearing ventilated facade cladding. This outer leaf must therefore be permanently connected to the loadbearing structure. In contrast to other types of facade cladding, in a masonry leaf the individual components (facing clay bricks) are joined together with mortar to form a complete system within a short time. This leaf must satisfy various requirements, which depend on the orientation, height and colour of the facade. Besides load-carrying characteristics, the accommodation of movement due to moisture-related and thermal influences is critical.

Fixings
As with every type of facade cladding, the main loads to be carried are those due to the self-weight of the components plus positive and

B 1.2.10

negative wind pressure. Owing to the relative heaviness of faced walls, the weight issue is of primary importance in their design. Parts of the building with structural functions (e.g., columns, floors and loadbearing walls) are suitable for carrying the loads. In practice, the loads due to self-weight are mainly transferred to the floors, storey by storey. At openings in the facade, structurally effective fixings direct the dead loads of the respective facade segments via a lintel into the loadbearing components. These days, many different types of prefabricated lintels are available off the shelf. Fixings that extend into the backing masonry provide the necessary stability to cope with wind loads. But these fixings must still be sufficiently flexible to accommodate the different movements of the inner and outer leaves. The number of fixings required per metre varies depending on the position in the facade; five are required in the middle and nine at corners and openings [3].

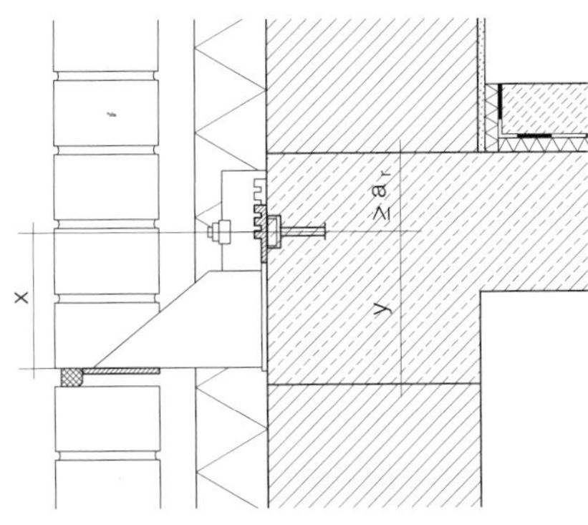

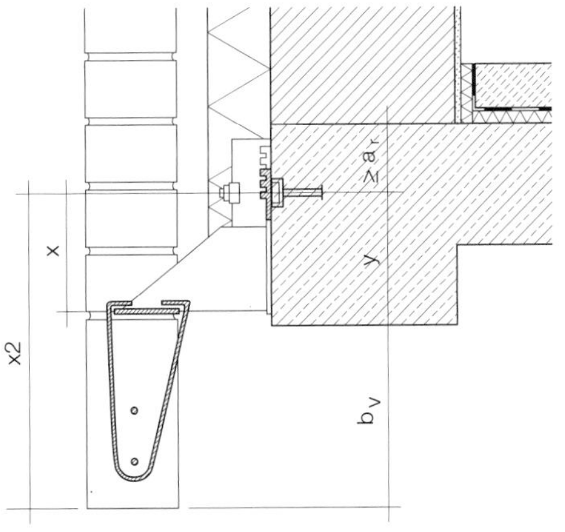

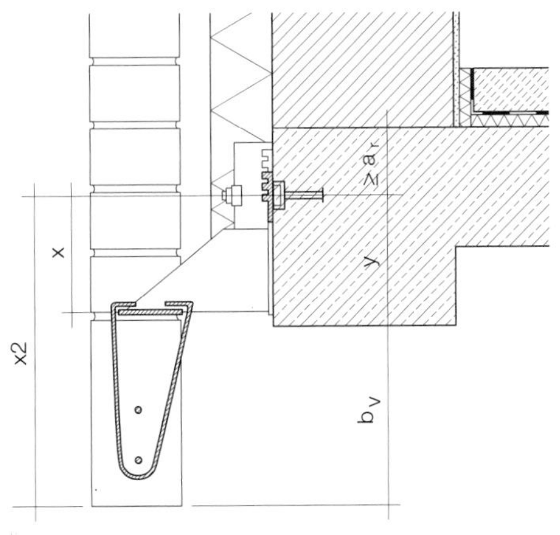

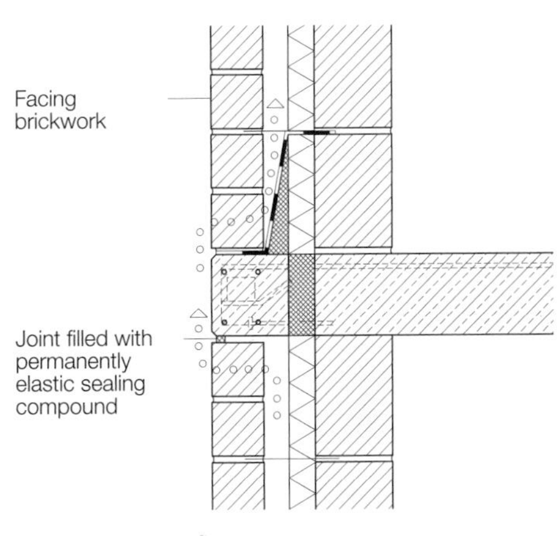

B 1.2.11

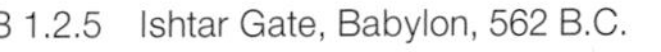

B 1.2.5 Ishtar Gate, Babylon, 562 B.C.
B 1.2.6 Town Hall, Tangermünde (D), 1430
B 1.2.7 Ornamentation, Berlin (D), 1891, Franz Schwechten
B 1.2.8 Industrially produced, coloured clay bricks, c. 1880
B 1.2.9 Brackets for supporting facing brickwork
B 1.2.10 Clay brick production, Pakistan, 1999
B 1.2.11 Brackets for supporting facing brickwork in a plain area of the facade, at an opening, and with the floor slab projecting but with a thermal break
B 1.2.12 Chile House, Hamburg (D), 1924, Fritz Höger
B 1.2.13 School, Hamburg (D), 1927, Fritz Schumacher
B 1.2.14 Church, Atlantida (ROU), 1959, Eladio Dieste

B 1.2.12

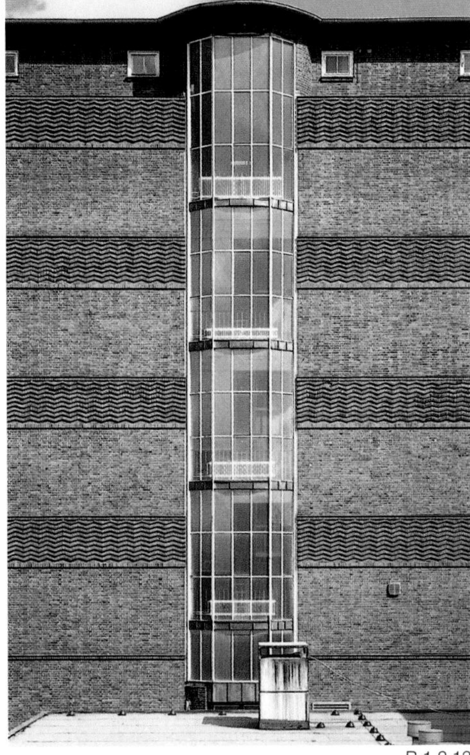

B 1.2.13

B 1.2.14

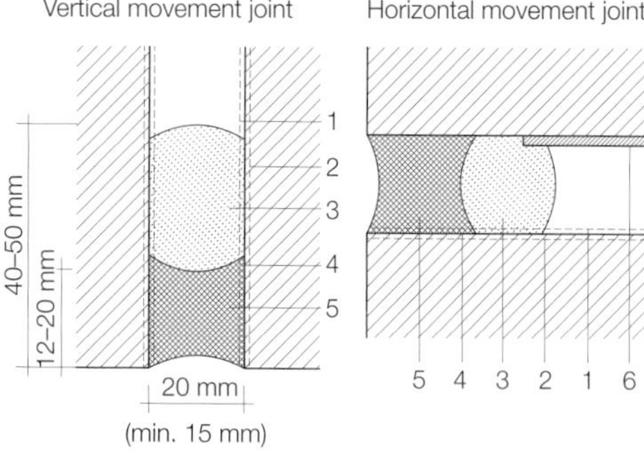

1 Joint compressed
2 Joint expanded
3 Closed-cell foam backing strip
4 Bonding coat
5 Elasto-plastic joint sealing compound
6 Support bracket

B 1.2.15

Direction of movement 1/2 $L_R \leq 4.00$
VF
VF
Stationary point
Direction of movement 1/2 $L_R \leq 4.00$
VF
VF

B 1.2.16

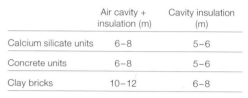

	Air cavity + insulation (m)	Cavity insulation (m)
Calcium silicate units	6–8	5–6
Concrete units	6–8	5–6
Clay bricks	10–12	6–8

B 1.2.17

B 1.2.18

B 1.2.19

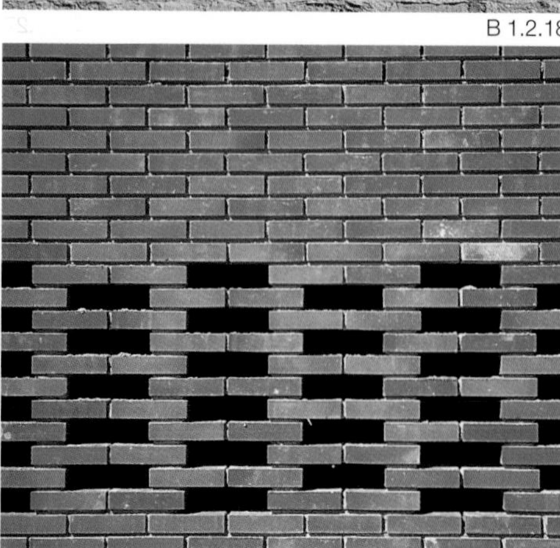

B 1.2.20

B 1.2.21

B 1.2.22

B 1.2.23

Joints

We basically distinguish between horizontal and vertical movement joints. These are 10 to 20 mm wide and are normally sealed with a permanently elastic material. The maximum spacing for vertical joints is 15 m for a continental climate, 25 m for a marine climate [3]. According to Eurocode 6 the distance between movement joints should not exceed 12 m, although the colour and orientation of the facade play a crucial role here.
Horizontal joints also have to take into account the height of the building. Up to a height of 12 m it may not be necessary to include any movement joints at all. However, on taller buildings a horizontal movement joint at least every 9 m is a mandatory requirement. In practice the normal solution is to provide a movement joint at every storey or every two storeys, directly below the course where the structural fixings are positioned. Spandrel panels below windows, corners, changes in the facade cladding or places where expansion within the overall building system is expected represent special cases which require additional movement joints. An air cavity within the facade can be ventilated by means of open vertical joints between individual elements.

Visual effect

Many factors contribute to the aesthetics of a clay brick facade. One of the most important of these is the bond, which in turn is very much dependent on the basic module of the masonry units themselves. Furthermore, the material (raw materials, firing, pigments, glaze) and the pattern (combination of different units, layout) have a major influence on the appearance of the facade.
Joints are an engineering necessity, but also have a major effect on the appearance of a building. The colour, width and depth of joints are fundamental factors determining the visual impression of a facade – to the same extent as the formats and colours of the masonry materials (figs B 1.2.18 to 1.2.23).
Artistic differentiation through the use of relief is rarely exploited, although the small sizes of clay bricks would seem to present a good opportunity for varying the elements. To animate an

B 1.2.15 Joint design, proposal by the German Masonry Association
B 1.2.16 Positioning of vertical movement joints at a corner
B 1.2.17 Recommended spacings for movement joints
B 1.2.18–23 Decorative masonry bonds for rectangular formats
B 1.2.24 Facade construction, axonometric view Ceramic panels on aluminium supporting construction with clip fixings only
B 1.2.25 Dimensionally coordinated heights for rectangular panels
B 1.2.26 Rectangular panels fixed vertically
B 1.2.27 Vertical section through standard panels
B 1.2.28 Rectangular panels fixed horizontally
B 1.2.29 Large-format system
B 1.2.30 Drying chamber
B 1.2.31 Printing works, Munich (D), 1993, Walter Kluska

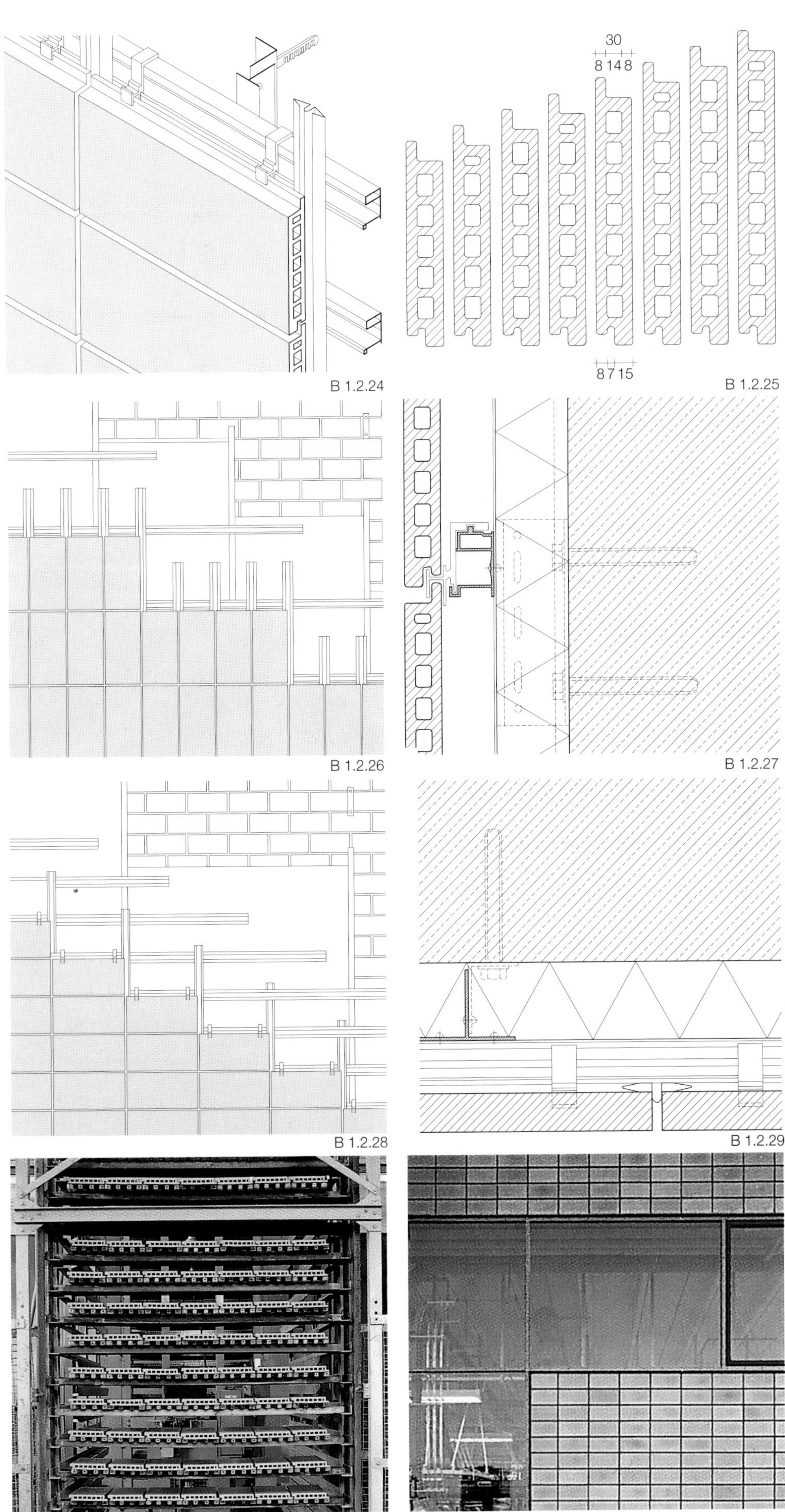

B 1.2.24 B 1.2.25 B 1.2.26 B 1.2.27 B 1.2.28 B 1.2.29 B 1.2.30 B 1.2.31

otherwise monotonous facade surface it is often sufficient to allow individual bricks to protrude minimally from the plane of the facade.

Ceramic panel facades

Newer systems using ceramic panels are available only in the form of ventilated curtain walls, which exhibit distinctive advantages in terms of performance. We distinguish between small-, medium- and large-format systems, with the small-format systems offering the big advantage that they can be more readily adapted to match the geometry and nature of the building. According to DIN 18516 standard panels may not exceed 0.4 m^2 in size and 5 kg in weight if a separate structural analysis is not carried out.

Supporting construction
Ceramic panel facades require a supporting construction that can transfer the loads due to self-weight, wind pressure/suction and thermal movement to the loadbearing structure without setting up restraint forces. As these constructions are normally made from stainless steel or aluminium [4], the connection to the loadbearing structure also represents a thermal bridge. Timber supporting constructions are also permitted provided they are suitably treated, but only up to a certain height.

Facade panels
Several methods are suitable for production of facade panels. Cyclic pressing in negative moulds requires tapering sides. This method does not permit any undercuts. In the extrusion process the shape of the die determines the cross-sectional form (figs B 1.2.25 and 1.2.30). Such panels are attached individually, which permits a limited amount of movement. This means that only a few additional joints, matching the requirements of the loadbearing structure, are necessary. The facade run-off water can be handled in various ways:

- in the horizontal joints by overlapping (like shingles) or by forming a rebated joint,
- in the vertical direction by designing the joints to drain the water.

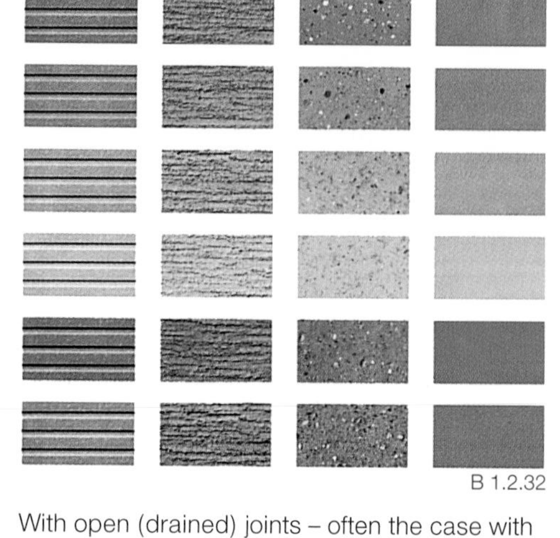
B 1.2.32

B 1.2.33

B 1.2.34

B 1.2.35

B 1.2.36

B 1.2.37

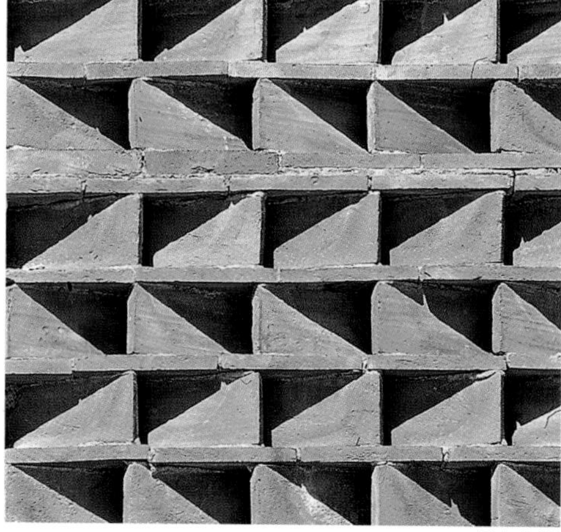
B 1.2.38

With open (drained) joints – often the case with fine ceramic stoneware – it is important to provide the right size of air cavity, and the area of the ventilation inlets and outlets must comply with the stipulations of DIN 18516, part 1. Another important aspect to be considered when designing ceramic panel facades is the ease of replacing damaged panels; the form of the panels and the supporting construction must allow for this (fig. B 1.2.46).

Colour and surface finish
Ceramic panels are mainly available in their own "natural" colours. Generally, the firing temperature, the oxygen content of the kiln atmosphere, the form and level of the iron content plus the raw materials and any additives all influence the final colour of ceramic building materials. In the customary method of producing ceramic panels the only possibility for adjusting the surface finish lies in the firing process. However, in the extrusion process the shape of the die also has an effect. Owing to the extra work and costs involved, the introduction of other colours (glazes) is on the decline.

Small-scale openings in ceramic facades admit air and light, protect against glare and sunlight, and also help to maintain a visual link between inside and outside. This effect has been used in the past and is a major factor affecting the appearance.

A more recent example of a non-loadbearing ceramic facade of clay bricks can be seen on the multi-storey car park in Genoa (Renzo Piano Building Workshop, 1992). Here, each facing brickwork panel with its peripheral steel frame is fixed to two round steel bars with metal discs acting as spacers (figs B 1.2.40 and 1.2.41). Another innovative example is the BP Studio showroom in Florence by Claudio Nardi (2001); in this case the long, extruded cross-sections are slid onto metal rails (figs B 1.2.44 to 1.2.46).

Notes

[1] Pfeifer, Günter et al.: *Masonry Construction Manual*, Munich/Basel, 2001, pp. 8–51.
[2] Acocella, Alfonso: *L'architettura del mattone faccia a vista*, Rome, 1990.
[3] DIN 1053.
[4] Please refer to the restrictions on materials given in DIN 18516, part 6.2.2.
[5] Rauch, Martin: *Tamped Clay-Construction*. in: Detail 06/2003.

B 1.2.32 Colour scale (extract)
B 1.2.33–38 Decorative openings in masonry
B 1.2.39 Large-format system
B 1.2.40–41 Multi-storey car park, Genoa (I), 1992, Renzo Piano Building Workshop
B 1.2.42 Fine ceramic facade panel with bonded, concealed fixings
B 1.2.43 Fine ceramic facade panel with mechanical, exposed fixings
B 1.2.44–46 Extruded, linear components for semi-permeable facade designs
B 1.2.44 Detail of facade, showroom, Florence (I), 2001, Claudio Nardi
B 1.2.45 Details
B 1.2.46 Showroom during construction, Florence (I), 2001, Claudio Nardi

B 1.2.40

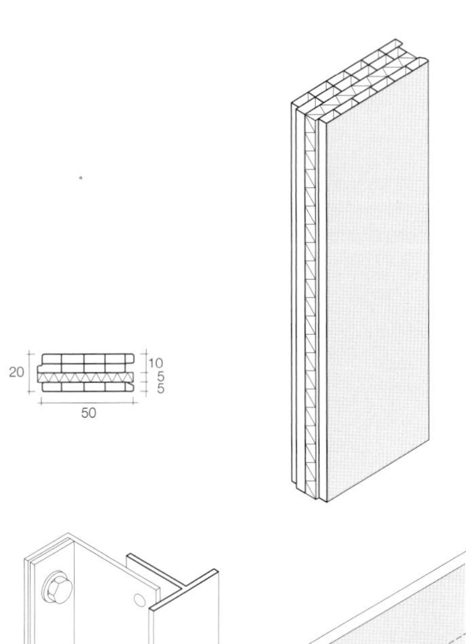

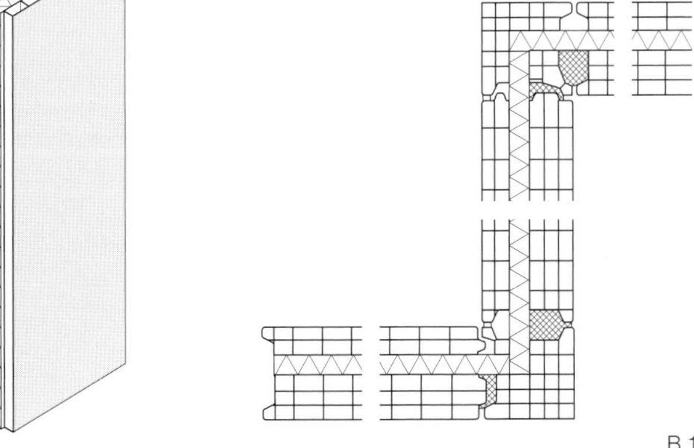

B 1.2.39

B 1.2.41

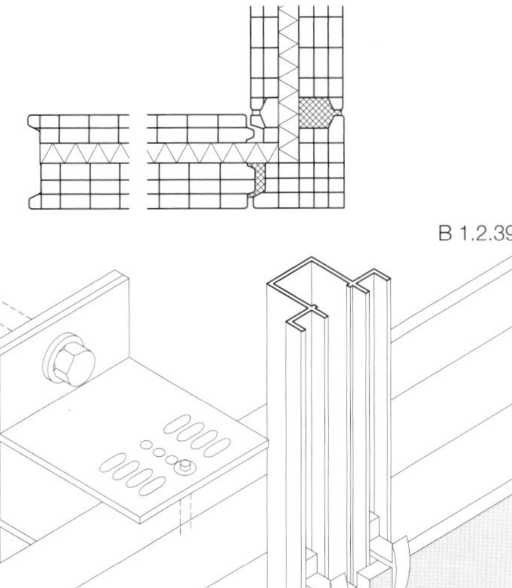

B 1.2.42

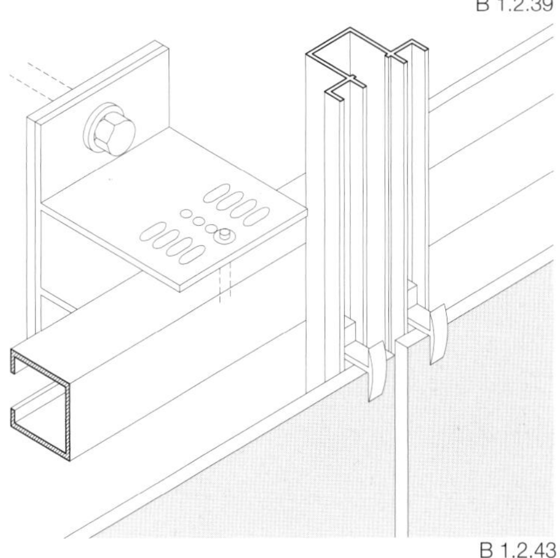

B 1.2.43

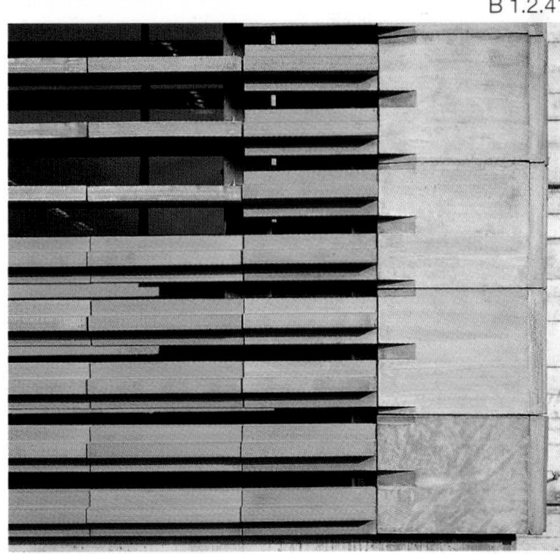

B 1.2.44

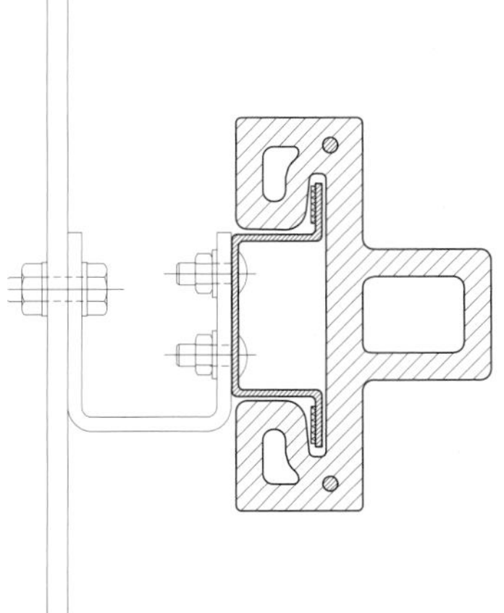

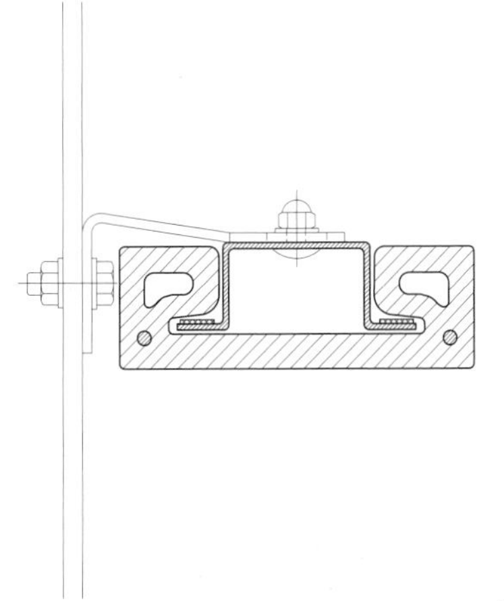

B 1.2.45

B 1.2.46

Chapel of rest

Batschuns, A, 2001

Architects:
Marte.Marte, Weiler
Structural engineers:
M+G, Feldkirch

l'architecture d'aujourd'hui 346, 2003
Detail 06/2003
Waechter-Böhm, Liesbeth (ed.):
Austria West Tirol Vorarlberg.
Neue Architektur.
Basel/Berlin/Boston 2003

- Tamped loam without chemical additions
- Constructed between formwork in approx. 120 mm-high courses without joints
- Compacted with hand-held plant
- Minimal erosion of surfaces during rain is easily accommodated by minimal oversizing of the loam components

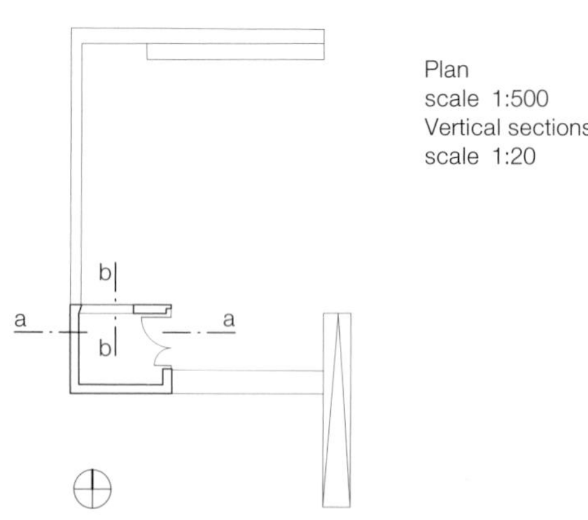

Plan
scale 1:500
Vertical sections
scale 1:20

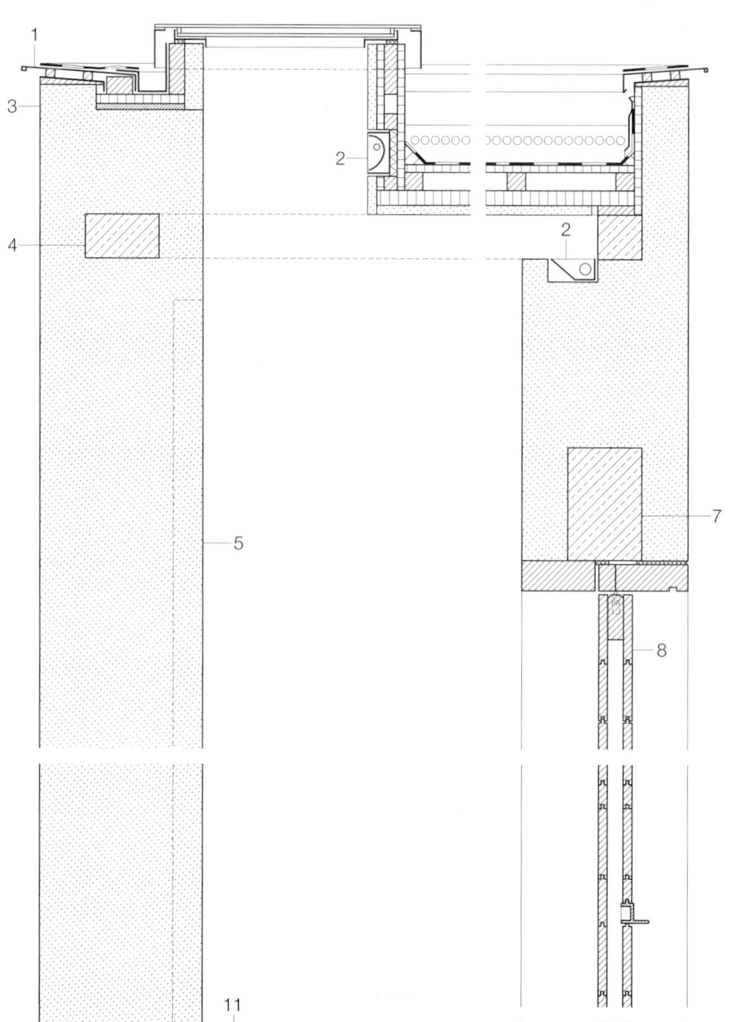

1 Sheet steel, 3 mm
2 Lamp
3 Tamped loam wall, 450 mm
4 Reinforced concrete rail, 205 x 120 mm
5 Squared timber section, oak, 80 x 80 mm, symbolising a cross with the horizontal lines of the layers of loam
6 Tamped concrete, containing pigment to match colour of loam
7 Reinforced concrete beam, 300 x 200 mm
8 Oak door leaf, 2 No. 24 mm thick
9 Oak door threshold, on steel rectangular hollow section, 200 x 100 x 7 mm
10 Sheet stainless steel, 240 x 10 mm
11 Steel beam made from steel flats, 380 x 15 mm + 2 No. 180 x 20 mm, welded
12 Float glass, 8 mm, bonded to sheet steel frame
13 Steel angle, 220 x 150 x 10 mm, and waterproofing
14 Anti-capillary hard-core to prevent rising damp

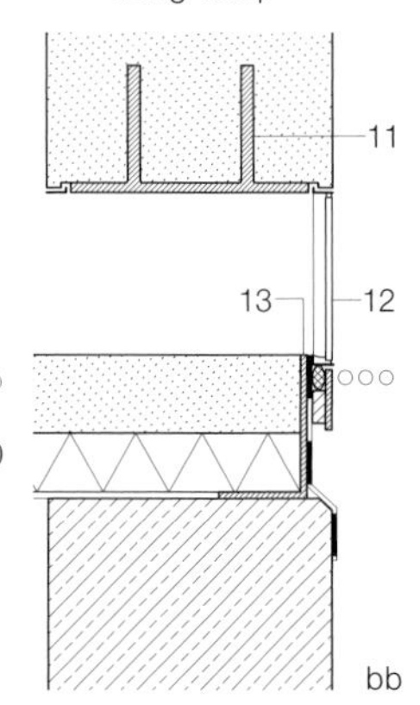

Private house

Ealing/London, GB, 2001

Architects:
Burd Haward Marston, London

Architectural Design 01/2002

- Small-format clay brick slip facade on timber supporting construction
- Pattern of joints adheres very closely to the form of the building
- All corners treated with the same material

Isometric (not to scale)
Plan, scale 1:200
Vertical section • Horizontal section
scale 1:20

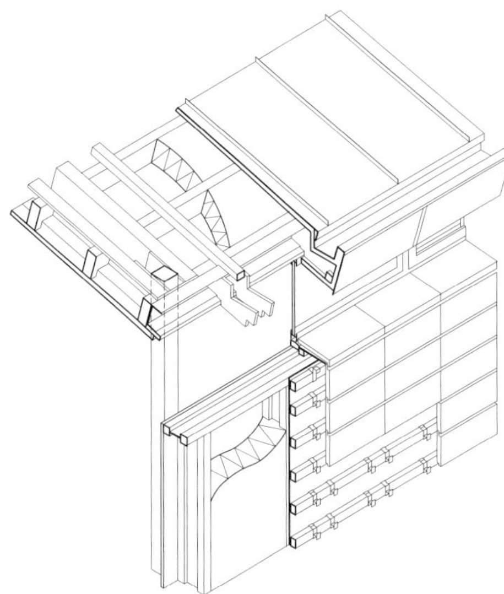

1 Aluminium gutter concealed in eaves detail
2 Continuous ventilation opening and insect screen
3 Sliding window, double glazing in aluminium frame
4 Clay tile window sill on sheet lead as water proofing, fixed with clips screwed through the lead and sealed with silicone sealant
5 Facade construction:
30 mm clay brick slips
40 mm supporting construction, horizontal support battens with slip fixings
9 mm wood fibreboard, vapour-permeable and moisture-resistant
170 mm thermal insulation between timber frame members
12 mm plasterboard, moisture-resistant both sides
6 Steel column with matt paint finish
7 Steel column with intumescent paint finish (for fire protection)
8 Clay brick slip make-up piece under lintel
9 Aluminium window with double glazing
10 Timber I-beam (incorporated horizontally)
11 Reinforced concrete pad foundation
12 Clay brick slip make-up piece at corner

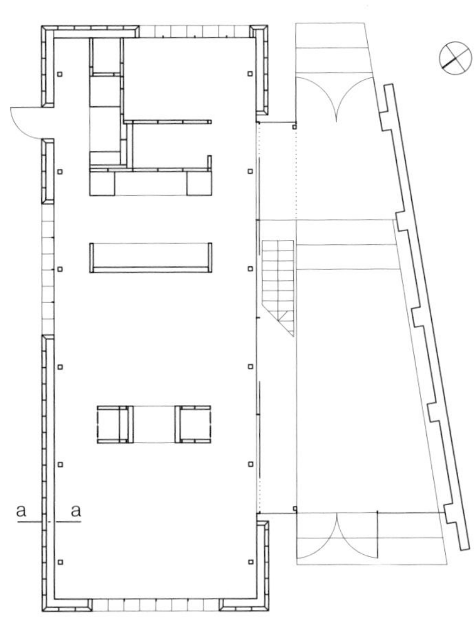

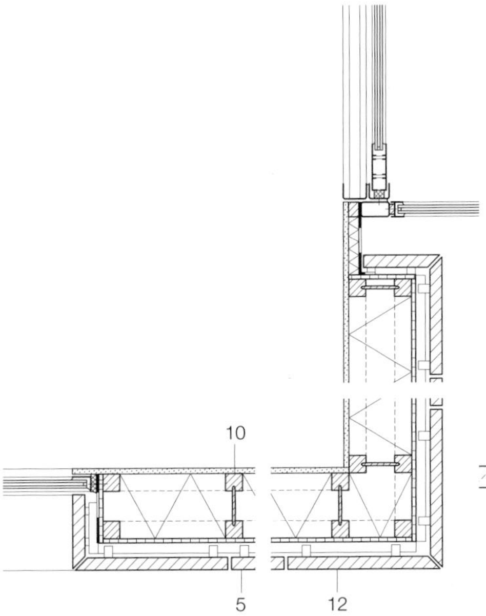

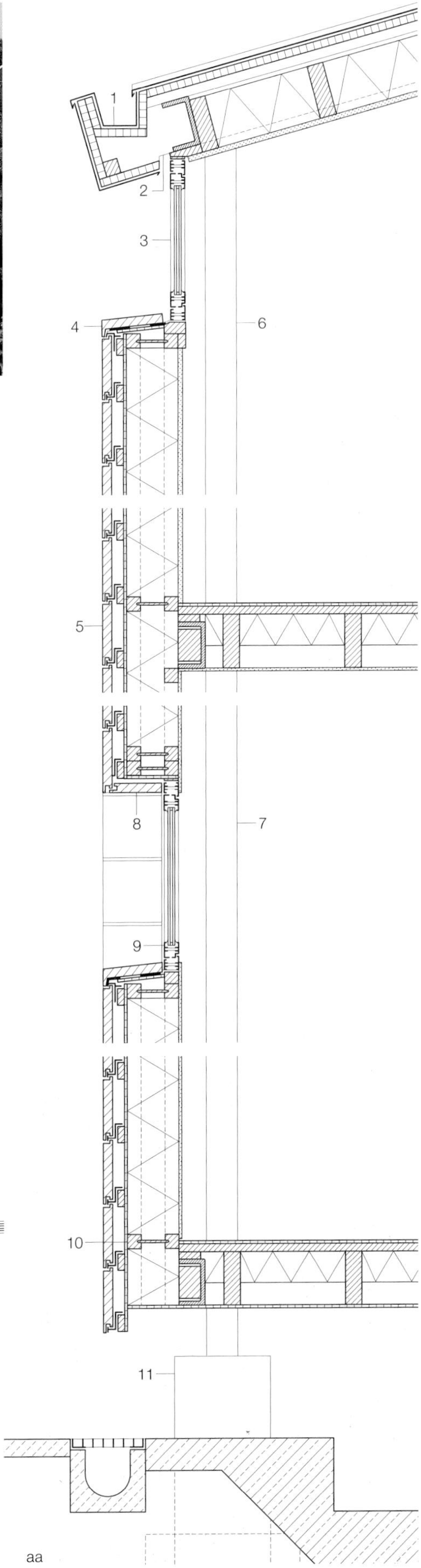

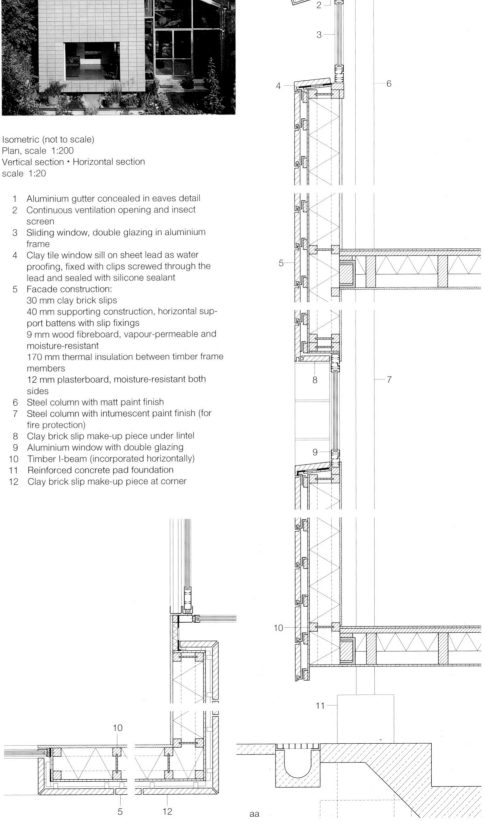

Trade fair grounds administration tower

Hannover, D, 1999

Architects:
Herzog + Partner, Munich

Architectural Review 01/2001
modulo 10/2002
Gissen, David: Big & Green.
Washington DC 2003
Herzog, Thomas (ed.):
Nachhaltige Höhe – Sustainable Height.
Munich/London/New York, 2000

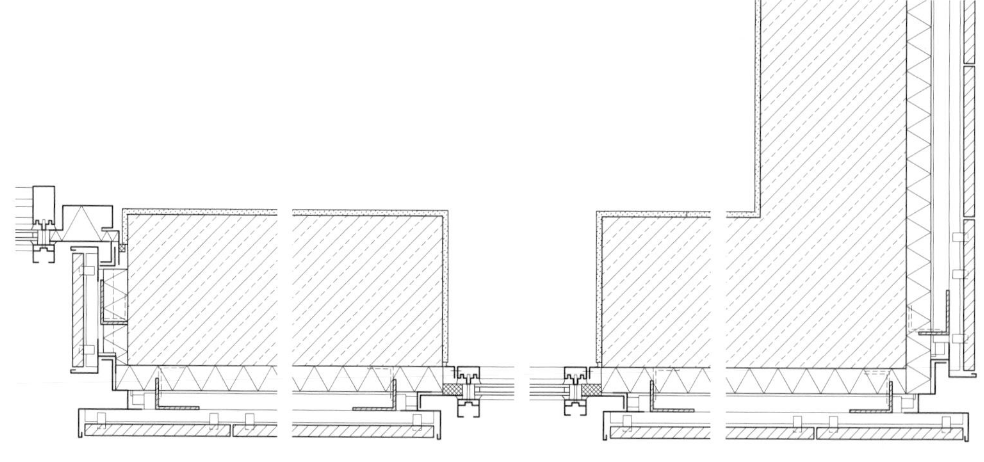

- Ventilated clay brick curtain wall system on aluminium supporting construction
- Light pearl-grey is the natural colour of this ceramic material (no superficial colouring)
- Facade panels with horizontal grooves: for checking facade run-off water driven upwards during rainfall, and for reducing the peak stresses during the manufacturing process

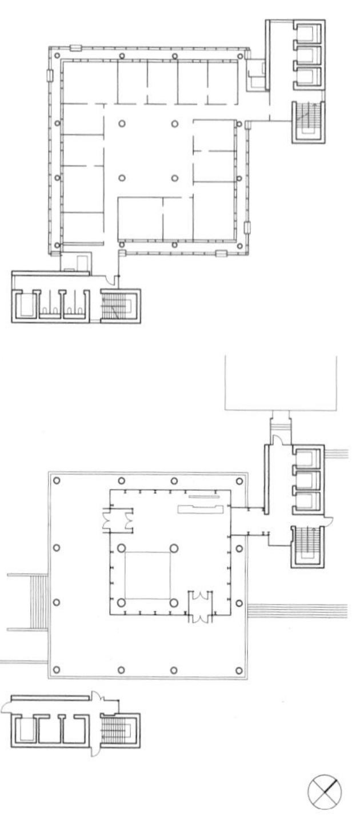

Horizontal section, scale 1:20
Plans
Ground floor • Standard upper floor
scale 1:1000
Vertical section • Horizontal section
scale 1:5

1 Aluminium capping plate, 3 mm, with anti-drumming (anti-noise) coating
2 Extruded aluminium section
3 Clay brick panels, with grooves, 200 x 400 mm
4 Aluminium trim
5 Thermal insulation, 60 mm
6 Reinforced concrete, 300 or 400 mm
7 Sheet stainless steel, positioned to match glass-and-steel facade

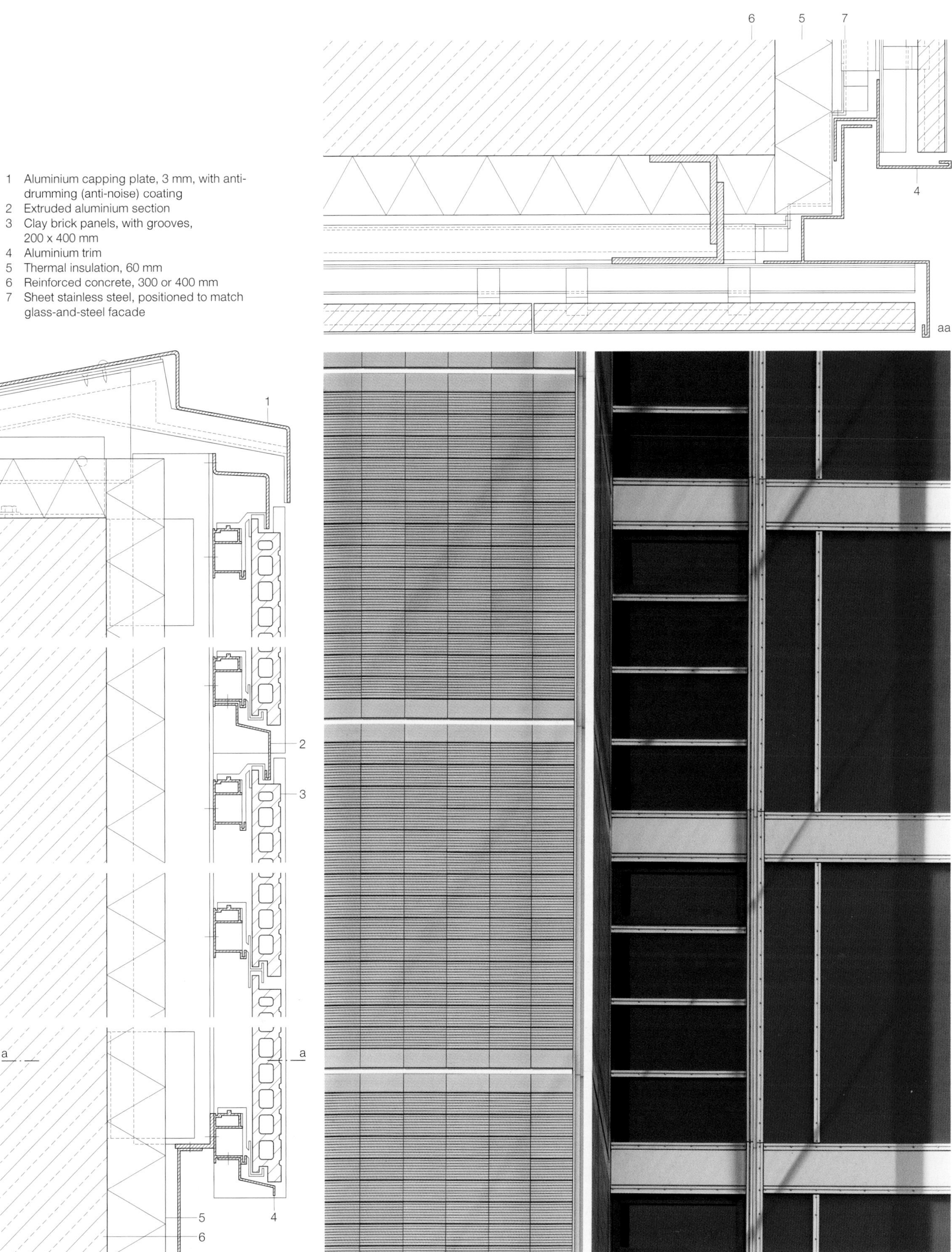

Daimler Benz / Debis C1 building

Potsdamer Platz, Berlin, D, 1998

Architects:
Renzo Piano Building Workshop, Paris
with Christoph Kohlbecker
Facade design:
Emmer Pfenninger Partner, Münchenstein

📖 Architectural Record 10/1998
Architectural Review 01/1999
Bauwelt 43–44/1996
Fassade/Façade 04/1997

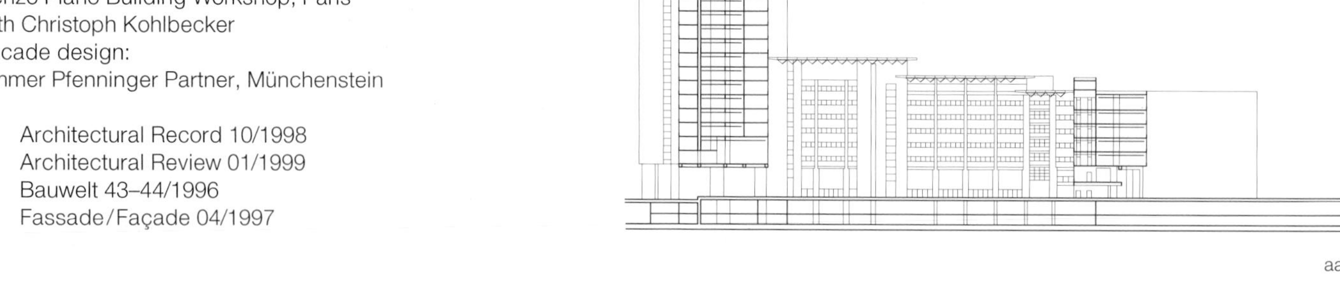

aa

- Facade developed as part of an EU research project
- First type of facade in the form of a rainproof leaf made from clay panels in front of a highly insulated external wall with standard windows
- Second type of facade is a similar construction but with an additional outer leaf of glass louvres to protect against wind, rain and traffic noise
- Very long, extruded ceramic components
- Vertical lesenes with horizontal square tubes ("baguettes") spanning between them, with an internal steel tube as a safety measure in case of breakage

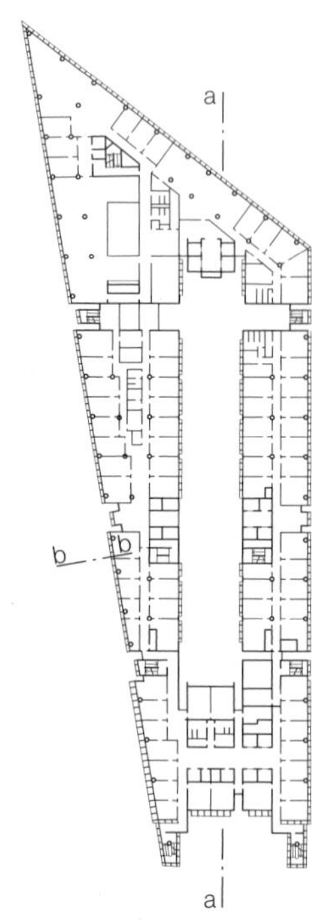
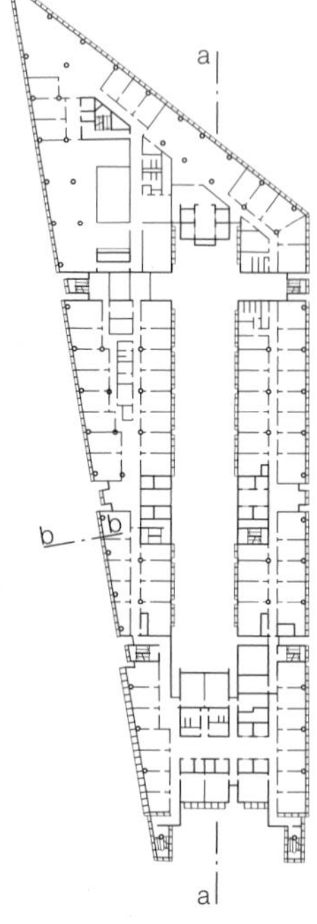

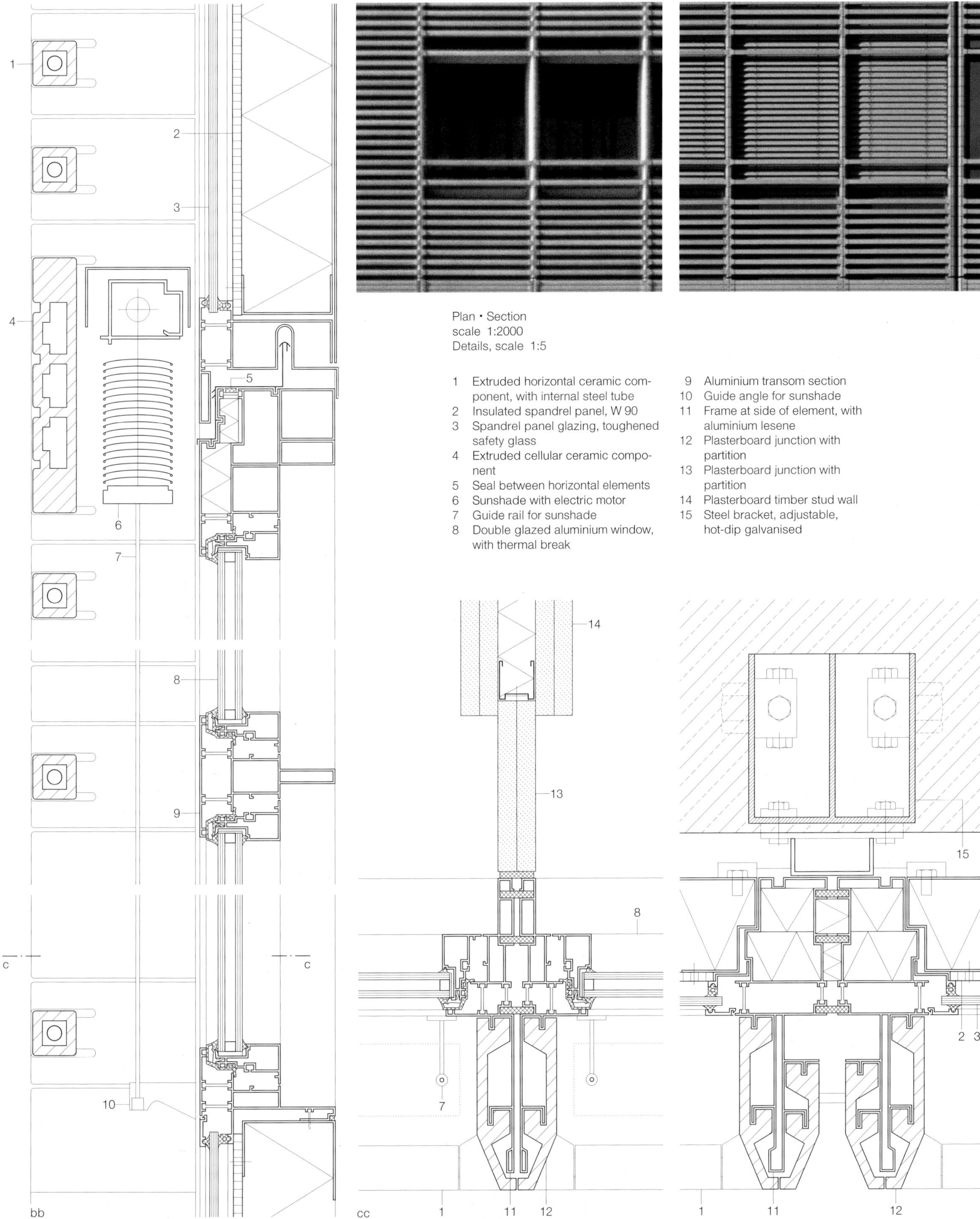

Plan • Section
scale 1:2000
Details, scale 1:5

1 Extruded horizontal ceramic component, with internal steel tube
2 Insulated spandrel panel, W 90
3 Spandrel panel glazing, toughened safety glass
4 Extruded cellular ceramic component
5 Seal between horizontal elements
6 Sunshade with electric motor
7 Guide rail for sunshade
8 Double glazed aluminium window, with thermal break
9 Aluminium transom section
10 Guide angle for sunshade
11 Frame at side of element, with aluminium lesene
12 Plasterboard junction with partition
13 Plasterboard junction with partition
14 Plasterboard timber stud wall
15 Steel bracket, adjustable, hot-dip galvanised

Private house

Brühl, D, 1997

Architect:
Heinz Bienefeld, Swisttal-Ollheim

A+U 10/2001
Baumeister 11/1997
Pfeifer, Günter et al.:
Masonry Construction Manual,
Munich/Basel, 2001

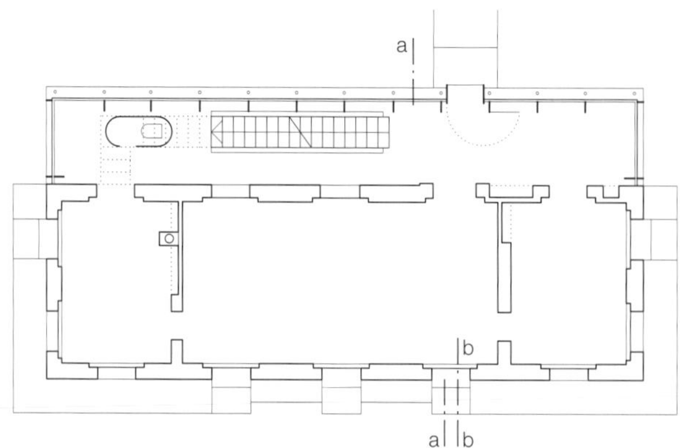

- Massiveness of cubic form emphasised by choice of material and roof lifted clear of the walls
- Faced single-leaf brickwork
- "Rustic" bond
- Wall almost 500 mm thick, with multi-level lintel feature

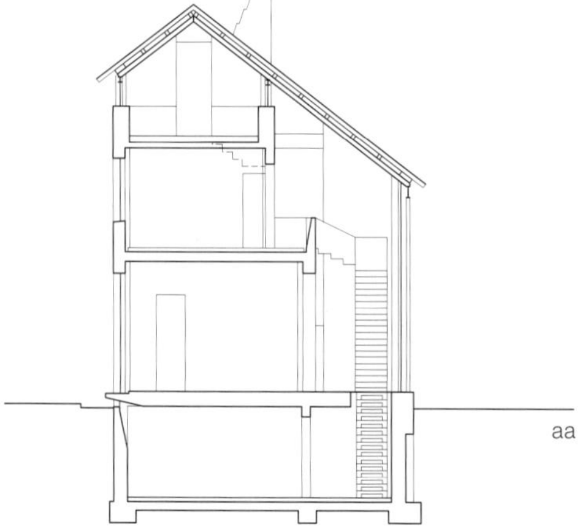

Section •
Plan of ground floor
scale1:250
Horizontal section •
Elevation on glazed door
scale 1:20

Vertical section through facade
scale 1:20

1 Gutter, 140 mm channel section
2 Steel purlin, 2 No. 80 x 80 x 10 mm angles
3 Galvanised steel window, micaceous iron oxide finish, double glazing
4 Steel channel, 40 x 35 mm, on steel rectangular hollow section, 50 x 25 mm
5 Sheet metal window sill, bent up and tucked into masonry bed joint at ends
6 Wall construction:
facing bricks of Taunus stone, NF 115 mm, "rustic" bond, 20 mm bed joints
"Poroton" lightweight clay bricks
25 mm lime plaster
neat lime finish with marble dust
7 Cambered arch lintel, 15 mm rise
8 Galvanised steel glazed door, micaceous iron oxide finish, double glazing
9 Galvanised steel channel, 120 x 40 x 8 mm, micaceous iron oxide finish
10 Precast concrete step

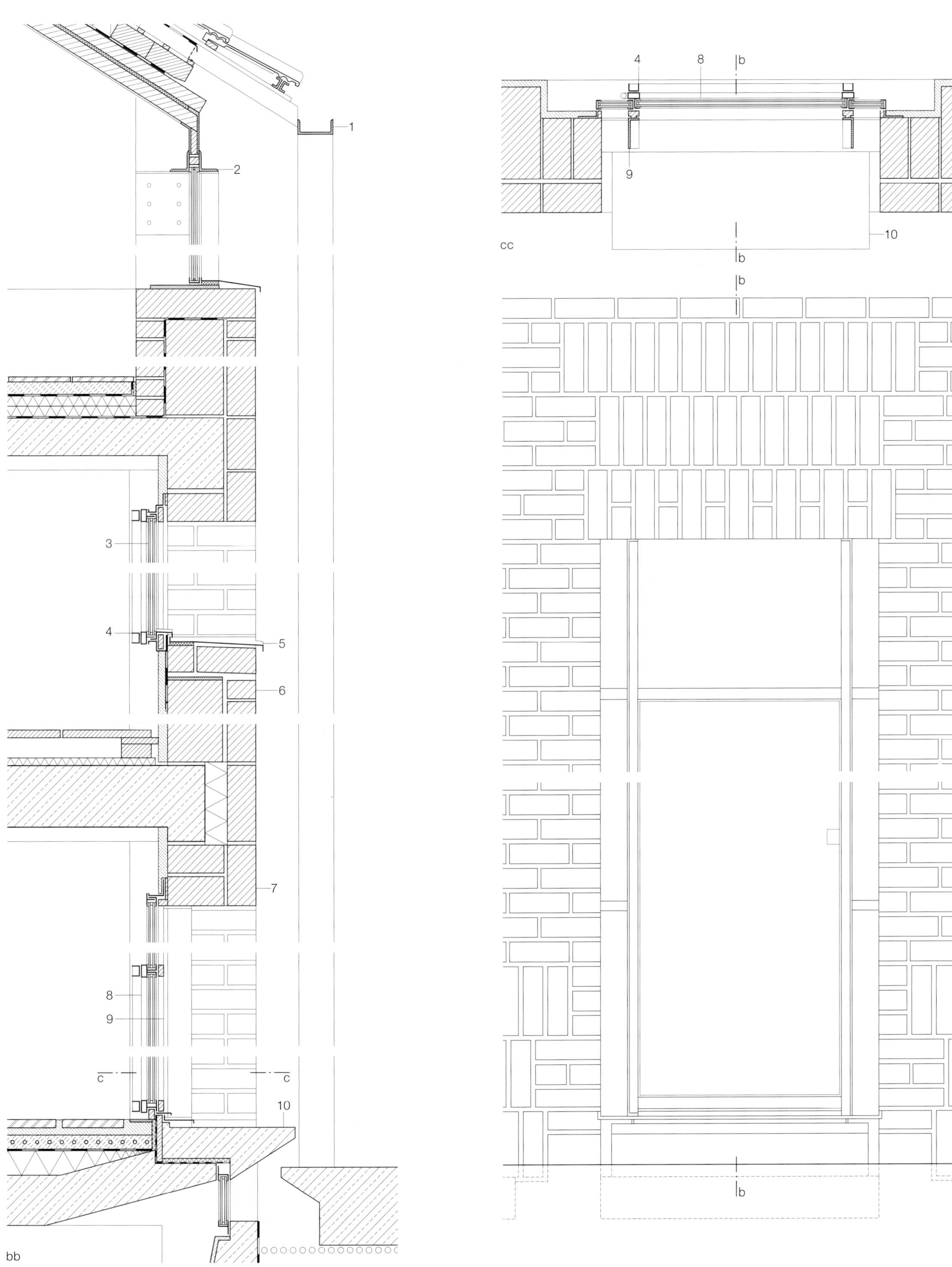
1
2
3
4
5
6
7
8
9
c
c
10
bb
4
8
b
9
10
6
cc
b
b
b

Museum of porcelain

Herend, H, 1999

Architects:
Turányi + Simon, Budapest
Gábor Turányi
Facade consultant:
Gábor Becker, Budapest

Construire in Laterizio 03–04/2003
Gall, Anthony; Kerényi, József:
Porcelánium. A mü, a mester és a mester-jelöltek.
In: Új magyar építömüvészet.
Budapest 2002
Slapeta, Vladimir (ed.):
Baustelle: Ungarn (Neuere ungarische Architektur). Berlin 1999

- Combination of materials such as clay brick-work, timber and a stone curtain wall
- Shape of building resembles an old kiln
- Glass blocks incorporated between clay bricks
- Mortar coloured to match bricks

Section • Plan, scale 1:800
Horizontal section • Vertical sections
scale 1:20

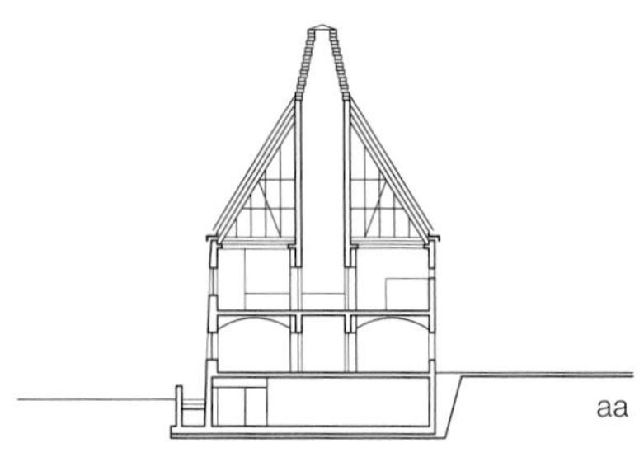

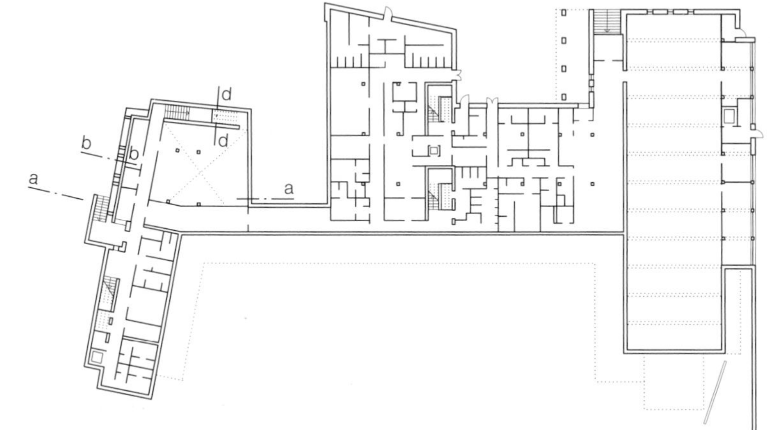

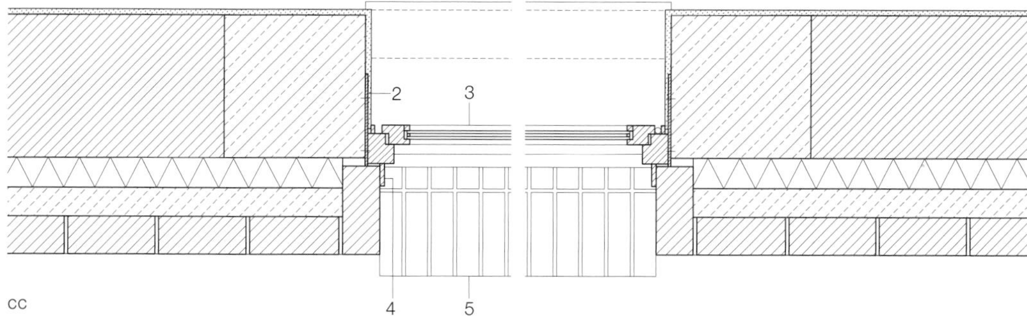

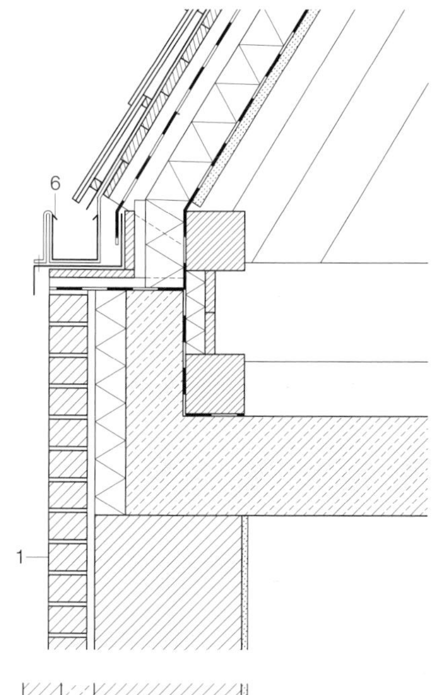

11
10
1
12
13
dd

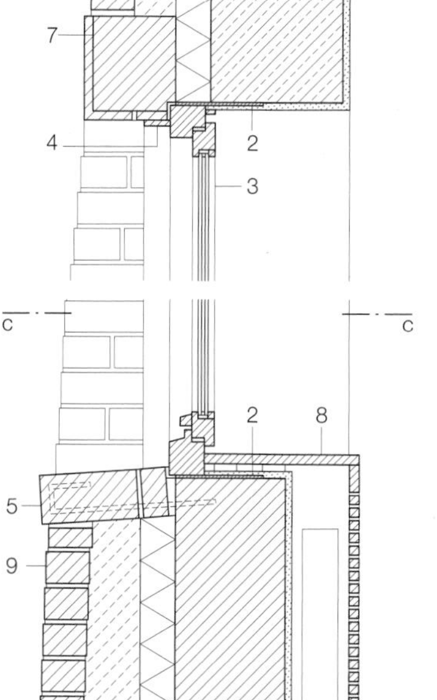

1 Wall construction
120 mm facing bricks
20 mm air cavity
380 mm hollow clay block masonry between reinforced concrete frame
15 mm plaster
2 Steel flat, 200 x 100 x 6 mm
3 Wooden window with double glazing
4 Pine cover strip, 60 x 10 mm
5 Brick-on-edge course (facing bricks) as window sill
6 Sheet zinc gutter
7 Insulated stone lintel, 200 x 200 mm, with facing brickwork on outside
8 Beech window board
9 Spandrel panel
120 mm facing bricks
100–200 mm lightweight concrete
80 mm thermal insulation
250 mm hollow clay block masonry
15 mm plaster
130 mm radiator behind
24 mm perforated pine panelling
10 Glass block, 195 x 195 mm
11 Steel flat, 400 x 400 x 20 mm
12 Slate, 10 mm
13 Insulated shallow brick lintel

B 1.3 Concrete

Concrete, the very first artificial and heterogeneous building material, marked an important stage of development in the history of building. Concrete is extremely durable, easy to work with, easy to connect and, in conjunction with steel, has a high loadbearing capacity. Reinforced concrete was therefore widely adopted for structural purposes and, thanks to its mouldability, opened up new avenues to (completely) different forms of construction.

There are also diverse opportunities for using concrete in the facade which, however, are distinctly less widely used, mostly because of a simplifying pragmatism. As a "monolithic" building material which can be worked "as if from one mould", seamless transitions between elements are easily produced. Besides *in situ* fair-face concrete facades, there is also a huge structural and architectural repertoire available, ranging from large panels down to small masonry units. In connection with the subject of "concrete facades" we shall generally restrict our survey to applications of cement-based and cement-bonded building materials. In doing so, we shall concentrate mainly on the successful aesthetic effects which can be obtained; these can be divided into five categories:

- fair-face concrete
- precast concrete
- reconstructed stone panels
- facing concrete masonry units
- cement-bonded sheets

These various options presuppose very different production techniques and statutory requirements. There is also wide scope for material-related adjustments by means of almost limitless colour and surface texture options, such as:

- lightweight/heavyweight
- insulation/thermal mass
- dense microstructure/open pores

From *opus caementitium* to (reinforced) concrete

Concrete has exerted a permanent influence on the evolution of modern architectural forms [1]. In terms of its constituents, concrete is a very old material. As long ago as 12 000 B.C. lime mortar was being used as a building material, and experience with this led to the invention of *opus caementitium* in the second century B.C. This was the material that helped the Romans achieve masterly feats of architecture and engineering, such as the Pantheon in Rome (A.D. 118). But with the downfall of the Western Roman Empire, *opus caementitium* lost its significance as a building material for nearly 1500 years.

It was the invention of Portland cement in 1824 that marked the beginning of the modern development of concrete.

Trials with the reinforcement of concrete were carried out in the mid-nineteenth century in France and England. The initial aim of these experiments was to replace timber and natural stone because the new material promised better protection against moisture penetration. A patent was granted in England in 1854 for a composite floor slab reinforced with iron. At about the same time, François Coignet developed the tamped concrete method, based on the idea of tamped loam, the "béton aggloméré"; he built a three-storey house using this technique.

These pioneering construction projects were accompanied around 1900 by numerous experimental investigations into the behaviour of the material and the further development of methods of analysis towards a general theory of reinforced concrete construction. All this gradually opened up new applications – especially for long-span loadbearing structures.

Concrete in the facade

The new material became established around 1900, primarily for industrial and commercial buildings such as factories and wholesale markets. However, it is the linear framework of columns and beams that really distinguishes these structures. One pioneer was Auguste Perret, whose apartment block in the Rue Franklin in Paris (1903) demonstrated the use of this material in the facade of a residential building for the first time, at least in principle. From about 1910 onwards, more structured approaches started to influence reinforced concrete construction. Concepts from Tony Garnier (plans for the ideal city, the "Cité Industrielle", 1901–17), Le Corbusier's "Dom-Ino System" (1914), and the reinforced concrete office building (1922) and country house (1923) designs of Ludwig Mies van der Rohe illustrate the use of plates, slabs and continuous spandrel panels.

In situ concrete

Concrete is regarded as a modern building material, and around 1900 both architects and contractors were hoping that *in situ* concrete would bring advantages. However, the degree of mechanisation of the concreting operations and the formwork heavily influenced the economic success.
The external walls of this time were mainly constructed as conventional fenestrate facades and their surfaces finished with render just like clay masonry. Three religious buildings and one "amateur project" symbolise those early years in terms of the impression made by this material. Frank Lloyd Wright used numerous plastic forms in his Unity Church at Oak Park, Illinois (1904–06). Even at this early stage, he was already incorporating special aggregates to exploit the opportunities for adding colour to fair-face concrete surfaces.

B 1.3.1 Art and Architecture Building, Yale University, New Haven (USA), 1964, Paul Rudolph

B 1.3.2

B 1.3.3

B 1.3.4

B 1.3.5

In 1922 Auguste Perret left the surfaces of the frame exposed on the church in Rainy near Paris and transformed the enclosing walls, essentially resolved into individual members, into a light-permeable, tracery-like concrete grid. Karl Moser chose a strict cubic architectural language with an as-struck finish for the concrete surfaces of the St Antonius Church in Basel (1922–27), thereby expressing this material to the full both on the facade and in the interior.
The Goetheanum in Dornach (1928) by Rudolf Steiner is a structure in which concrete is utilised in a virtuoso fashion in the forming of the facades. However, the realisation of such plastic-organic forms calls for a high labour input and great manual skills in constructing the formwork.

During the 1950s concrete evolved into an all-purpose building material, being used for every conceivable building task. Le Corbusier provided much impetus here, striving to show concrete in its immediate, "raw" quality – the "béton brut". He used this skilfully as an artistic medium in the formation of the facade surfaces, making use of reliefs and/or plastic forms, for example at the Sainte-Marie-de-la-Tourette monastery (1957–60) in Eveux near Lyon (fig. B 1.3.2). While the Swiss practice Atelier 5 was also using rough exposed concrete forms in (small) domestic properties at the Halen Estate near Bern (1955–61), Louis Kahn preferred surfaces finished as smooth as possible for the Jonas Salk Institute at La Jolla (1959–65). And it was also Kahn who used recessed joints and carefully positioned formwork ties to add an orthogonal pattern of lines to the concrete facade and at the same time render visible the production process.

In the 1960s and 1970s many architects made increasing use of the options surrounding the three-dimensional mouldability of the external wall and building envelope as well as the opportunities for adding artistic expression to the surface. Unique structures from this period are the pilgrimage church in Neviges (1963–68) and Bensberg town hall (1963–69), both by Gottfried Böhm. He sculpts – especially in his church structure – a plastic, jagged volume, where the fine textures provided by the formwork result in powerful, opaque surfaces without any effect of monotony (fig. B 1.3.3).

While Carlo Scarpa was sounding out the mouldability of concrete in an almost (artistic) craftsman-like fashion – particularly at the Brion family cemetery in San Vito d'Altivole near Asolo (1969–75) – Paul Rudolph was using industrial, structured formwork on the Art and Architecture Building at Yale University in New Haven (1958–64) (fig. B 1.3.1). The fluting-like profiling on the coloured surfaces yields a differentiated play of light and shadow across the alternating smooth grooves and rough, broken ribs. The addition of materials found locally and/or the structuring of the green surface opens up further artistic options, even new socio-cultural relationships with the surroundings, as can be seen on the ESO Hotel at Cerro Paranal by Auer + Weber (2001) (see example on p. 123) or at the "Schaulager" art storage facility in Basel by Herzog & de Meuron (fig. B 1.3.8).

More recently, architects have been attempting to express this monolithic method of construction more comprehensively, right down to the tiniest detail. The avoidance of all forms of construction joints, concealing the marks left by formwork ties and extremely minimal component cross-sections coupled with the use of new visual effects has created enormous building technology challenges for this extremely versatile material.

Prefabrication

As the casting of concrete elements on the building site proved to have disadvantages in terms of construction and production, attempts were made to divide the construction into identical, transportable elements which could then be produced in series in precasting plants. Once freed from the constraints of the weather, better quality, better accuracy and a better standard of surface finish could be achieved.

The first instance of a temporary precasting yard for concrete elements was in France in the early 1890s. And in 1896 the French stonemason François Hennebique achieved the first industrially produced building, comprising transportable room modules of 50 mm thick, reinforced concrete panels.
From 1920 onwards the industrialisation of reinforced concrete construction started to gain significance. Architects like Ernst May, who used his own system with different sizes of wall blocks in a number of housing estates around Frankfurt am Main (including Praunheim, 1927), or Walter Gropius, who made use of a small-format system with hollow clinker blocks in Dessau-Törten (1926–27), worked on concepts involving considerable prefabrication. Even though such systematic approaches were not able to establish themselves, neither in terms of technology nor economics, these experiments represent an important (preliminary) stage on the way to the industrialisation of building [2]. During the 1950s and 1960s, construction with large-format, loadbearing wall panels became very widespread. While these building systems led to the realisation of masses of very "mechanical" facades, in the course of the Postmodern era these approaches resulted in almost the opposite situation, in that the prefabrication and plastic mouldability of concrete elements was used for an arbitrary play of shapes and colours.

Architects like Angelo Mangiarotti (see example on p. 114), Bernhard Hermkes (Faculty of Archi-

B 1.3.2 Monastery of Sainte-Marie-de-la-Tourette, Eveux (F), 1960, Le Corbusie
B 1.3.3 Pilgrimage church, Neviges (D), 1968, Gottfried Böhm
B 1.3.4 Faculty of Architecture, Berlin Technical University (D), 1967, Bernhard Hermkes
B 1.3.5 Headquarters for a welfare organisation, Berlin (D), 2003, Léon Wohlhage Wernik
B 1.3.6 John Storer House, Hollywood (USA), 1924, Frank Lloyd Wright
B 1.3.7 Office building, Centraal Beheer, Apeldoorn (NL), 1972, Herman Hertzberger
B 1.3.8 "Schaulager" art storage facility, Basel (CH), 2003, Herzog & de Meuron

B 1.3.6

B 1.3.7

tecture, Berlin technical University, 1965–67, fig. B 1.3.4), Gottfried Böhm or Eckhard Gerber formulated architectural answers. Böhm's office building for Züblin AG in Stuttgart (1982–84) is a formal yet differentiated use of precast elements. Gerber, on the other hand, attached orthogonal, planar reinforced concrete facade elements to an office building in Dortmund (1994) in a structured, clear fashion, as cladding to the columns and spandrel panels. Even "heavyweight prefabrication" is an option again today, both from an engineering and an architectural viewpoint. Architects like Thomas von Ballmoos and Bruno Krucker (Stöckenacker Estate, Zürich, 2002), or Léon Wohlhage Wernik (headquarters for a welfare organisation, Berlin, 2003, fig. B 1.3.5) use storey-height, multi-leaf precast concrete elements in such a way that moderate variations to the dimensions achieve harmonious results.

One form of plain (i.e., unreinforced) facade cladding is the small-format reconstructed stone panel. Panels attached with mortar have been used in building for the past 100 years or so. Particularly at the base of the wall , they represent a robust, easily handled building material. One of the earliest examples in Germany is the town hall in Trossingen (1904), where reconstructed stone panels were used to clad the plinth and splayed door jambs. In particular, the numerous surface finish options, the mouldability and the combination with different aggregates leads to the creation of ornamental elements such as (demi-)columns, balusters, gables, and rosettes.
The use of reconstructed stone panels in the form of a ventilated curtain wall of small-format cladding units is widespread, as in the red-pigmented facade of the German Embassy School in Beijing (2001) by Gerkan Marg + Partner.

Concrete masonry units
Concrete masonry units offer the advantages of small formats and lightweight construction, together with a wide range of colour and surface finish options.
Frank Lloyd Wright worked with various systems after 1914. His "textile block" system was an attempt to find an alternative to large-format panel systems. Based on an initial square module, he worked with a number of shaped blocks. Buildings like John Storer House in Hollywood (1923–24) are examples of richly decorated facade surfaces with alternating patterns of smooth and textured blocks (fig. B 1.3.6) [3].

Egon Eiermann themacised the motif of a light-permeable wall by using concrete grid blocks with (coloured) glass infills both on the Matthäus Church in Pforzheim (1952–56) and on the Kaiser Wilhelm Memorial Church in Berlin (1957–63). Another use of facing masonry blocks is as opaque masonry infill panels on a reinforced concrete structure, a particular feature of the work of Herman Hertzberger. In buildings like the Centraal Beheer offices in Apeldoorn (1968–72, fig. B 1.3.7), the Vredenburg Music Centre in Utrecht (1973–78) or the Apollo Schools in Amsterdam (1980–83), the exposed, untreated facing masonry forms an effective contrast – both inside and outside – to the smooth fair-face concrete and glass (block) surfaces thanks to the slightly porous surfaces, and the different coloured textures [4].

The Ticino-based architect Mariano Botta has also used concrete masonry units in a number of private houses. Their small formats and colouring are a deliberate allusion to the granite rubble stone walls of the regional building tradition.

Cement-bonded sheet materials
Fibre-cement sheets represent a totally different way of using mineral-bonded building materials [5]. Around 1900 asbestos-cement – a composite material made from asbestos fibres and cement – was the subject of a patent application in Austria – and the Eternit company has been marketing its sheets of the same name since 1903.

By the mid-1970s it had become clear that asbestos fibres are carcinogenic. Sprayed asbestos was banned in 1979, and in 1990 also the use of asbestos-cement sheets (fibre content approx. 10%). Once asbestos had to be replaced, cement-bonded sheets with safe substitutes for the asbestos fibres, such as wood chips, started to appear. This material exhibits high mechanical strength and good fire resistance despite its minimal thickness,

B 1.3.8

and can be produced in various sizes and formats.
Originally developed as a lightweight roofing material, small-format slates and large-format sheets quickly became popular as facade cladding. Small-corrugation sheets were available starting in 1912 and large-corrugation versions after 1923. Besides the positive material properties and ease of use, this composite material was suitable for large-scale industrial production right from the outset, which turned it into an inexpensive building material.

One pioneer of the intentional architectural use of this material for facades was Marcel Breuer. For instance, in the early 1930s he employed fibre-cement corrugated sheets for a shop front in Basel.
In Germany millions of asbestos cement sheets were used, also in facades, particularly during the 1950s and 1960s. Renowned architects such as Ernst Neufert, who published a Well-Eternit manual in 1955, and Egon Eiermann, employed fibre-cement sheets for industrial structures but also residential and office buildings; likewise Rolf Gutbrod, on an office and commercial development in Stuttgart (1949–52). Newer projects demonstrate that fibre-cement sheets represent a lightweight, robust facade cladding material still favoured for many applications. For example, on the Ricola warehouse in Laufen (1987) by Herzog & de Meuron, with its three-dimensional, graded, band-like arrangement (fig. B 1.3.9), or the technology centre in Zürich (1989–92) by Itten and Brechbühl, with its two-dimensional design and exposed fixings [6].

Concrete technology

Concrete is an artificial stone produced by hardening of cement-water mixture (cement paste) to form hydrated cement. The aggregate is bonded into this to form a solid matrix.
EN 206 part 1 is the most important standard for the design and construction of concrete elements. The individual constituents of concrete are:

- binders
- aggregates
- additives
- admixtures

Cement is used as the binder; this is obtained by firing lime and clay or marl and subsequently grinding the fired material. The most important type is Portland cement, with a gypsum or anhydrite content of 3–5%. Cement sets after being mixed with water. The resulting hydrated cement is water-resistant and exhibits a high strength.
EN 197 part 1 divides standard cements into five main classes (CEM I to V), with the twenty-seven different products distinguished in each case by means of their main constituents. The most common type of cement currently in use is CEM II, a Portland composite cement containing at least 65% Portland cement clinker by weight and one other main constituent in each case.

Concrete is made up of about 70% aggregate by volume. Limestone, quartz, granite or porphyry can be obtained in rounded or angular form as sand or gravel from rivers or gravel pits. Crushed rock fine aggregate, chippings or fine chippings in the form of crushed, broken materials are obtained from stone quarries. Additives like plasticisers, superplasticisers, air entrainers or stabilisers affect the material properties by way of chemical or physical actions. Additives like pigments – rare stone dusts – are used to add virtually any colour to the concrete.

Concrete has high compressive strength and good durability even at an early stage of hardening. On the other hand, its tensile strength is always fairly low. We compensate for this by incorporating reinforcement – usually steel bars. This turns concrete into an excellent composite material whose material properties can be determined with great accuracy. These properties govern the use and function of the concrete and hence its potential applications. Generally, requirements regarding strength, corrosion protection, frost resistance, etc. are specified, with exposure classes distinguishing between the effects on the concrete and on the reinforcement. External components made from concrete that, on the one hand, are subjected to frost with moderate saturation, but whose reinforcement, on the other hand, has to be protected against carbonation with changing saturation, are classified as exposure classes XC4 and XF1. The concrete used must comply with the requirements of grade C25/30 at least, and must have a water/cement (w/c) ratio of 0.60 and a cement content of 280 kg/m³.
With regard to the properties of wet concrete, fair-face concrete has to satisfy requirements regarding ease of workability, i.e., it should be stable, non-bleeding and non-segregating, which is specified in consistency class F3. In order to guarantee a constant content and a consistent granulometric make-up, i.e., grain sizes and shapes, sufficient numbers of the finest particles of cement and aggregate represent important parameters for the workability.

Types of concrete

The two critical properties of hardened concrete are density and compressive strength. Depending on the production method and the aggregates, concrete can be given very specific properties. For example, good load-carrying capacity and good sound insulation require a dense concrete, while porous aggregates improve the thermal insulation behaviour. We classify concrete according to its oven-dry density as follows:

B 1.3.9

B 1.3.9 High-level racking warehouse, Laufen (CH), 1987, Herzog & de Meuron
B 1.3.10 Classification of "concrete in the facade"

- Heavy concrete: > 2600 kg/m³
  Aggregates: e.g., iron ore, iron granulate, baryte
  Applications: radiation protection etc.
- Normal-weight concrete: > 2000–2600 kg/m³
  Sand, gravel, chippings, blast furnace slag; this is the type of concrete used for the majority of construction tasks; providing no confusion can arise, normal-weight concrete is usually simply called concrete.
- Lightweight concrete: 800–2000 kg/m³
  The features of this type of concrete are essentially determined by:
  – the properties of the lightweight aggregates, e.g., expanded shale/clay
  – the nature of the concrete microstructure – no-fines porosity or dense
  – the proportion of pores – aerated or foamed concrete
  No-fines porous lightweight concrete is primarily used for thermal insulation tasks and, compared to normal-weight concrete, has a lower load-carrying capacity, which is, however, still adequate for many structural purposes.

Furthermore, concrete is divided into compressive strength classes designated with the letter "C" and two figures. These are a result of the standardisation in DIN EN 206 part 1 and designate the compressive strengths in N/mm² according to the cylinder and cube tests:
- normal-strength concretes (C8/10 to C50/60)
- high-strength concretes (C55/67 to C100/115)
- lightweight concretes (LC8/9 to LC50/55)

Lightweight concretes are further subdivided into six density classes from D1.0 to D2.0, which must be specified by the designer, depending on the application.

High-performance and textile-reinforced concretes

In the realm of concrete production, numerous research activities are currently being undertaken to improve the performance of concrete.

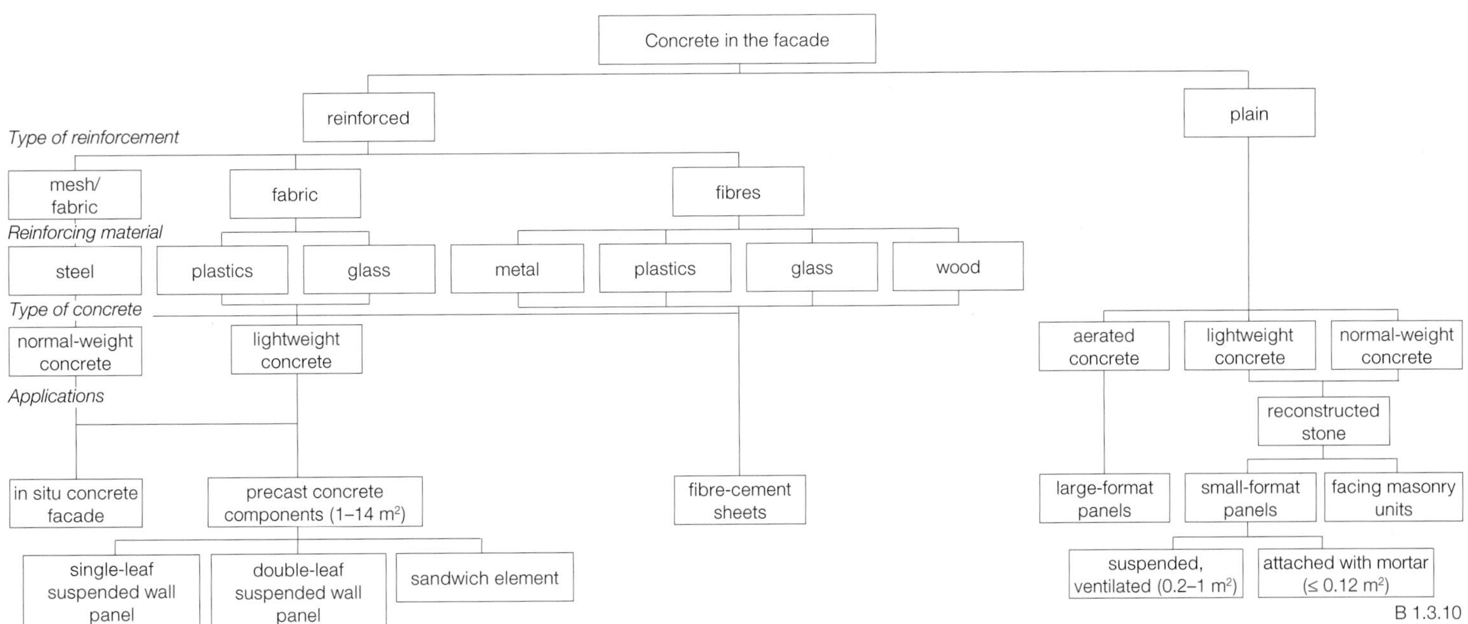

B 1.3.10

Self-compacting and high-strength concretes are just two of the focal points of these research activities. The objective is to add building chemicals to produce concretes with high flowability and exactly the toughness required to compact and bleed, themselves, without the need for any mechanical plant. This would improve the production of particularly slender components with closely spaced reinforcement that would demand geometrical configurations, but would at the same time result in high-quality, dense, fair-face concrete surfaces. Further developments include improvements to strength and protection against the ingress of moisture.
In the meantime, high-strength concretes with compressive strengths > 125 N/mm^2 (cylinder test) have been used for structural purposes. These concretes have a low w/c ratio, and a very dense microstructure, due to the use of ultra-fine fillers such as microsilica or ultra-fine cement, and surfaces with an extremely low porosity.

Besides these construction "superlatives", the combination of concrete and corrosion-resistant textile fibres as the reinforcement is growing in significance. The new composite material "textile-reinforced concrete" brought about through the use of aramid, glass or carbon fibre fabrics enables the production of relatively thin-walled concrete components because the concrete cover on the reinforcement is very much reduced. The results to date confirm that this new material opens up new applications for concrete and lightweight construction which go beyond the simple substitution of conventional composite materials and existing constructions.
In the meantime, it has been shown that cement-bonded materials can achieve extremely high strengths and it is possible to produce extremely dense fair-face concrete surfaces. Both these new developments have given new impetus to concrete construction, particularly in precast work [7].

Generally, however, the use of self-compacting concrete is not yet regulated by standards or directives. And the use of fabrics or fibres as the reinforcement also generally requires the approval of the authorities in one way or another.

Construction aspects

Despite the diverse types of concrete available, the majority of mixes are based on normal-weight concrete. Back in the 1970s the various types of lightweight concrete promised to deliver an enormous boost to the use of concrete in terms of quantity and quality. But the expectations were not fulfilled, and so the use of (structural) lightweight concrete for external walls was restricted almost exclusively to detached and semi-detached houses or commercial properties, and apart from a few exceptions was rendered over, and joined with mortar or adhesive in the form of masonry units or precast components.

The diverse processing options result, to some extent, in very different boundary conditions for the construction of facades. On the one hand, there are pure material-specific requirements, but, on the other, there are forms of construction based on standards and directives that also apply to other building materials (fig. B 1.3.10).

Fair-face concrete

When we speak of concrete facades, then we are usually referring to *in situ* concrete – and we usually mean fair-face concrete. Such concrete surfaces must comply with certain requirements governing their production and their appearance. Experience shows that the specification of architectural features calls for specialist knowledge about these particular requirements – also on the part of the architect.

The final appearance of the exposed face of the concrete wall can be accomplished in many ways by:

- adjusting the formwork
- adjusting the concrete mix

The concrete surface

One critical boundary condition is the formwork. The surface of the concrete consists of a layer of mortar made up of cement paste and the finest grains of aggregate, and therefore produces a copy of the face of the formwork employed. The absorption properties of the formwork can affect the finish of the exposed face:

- absorbent (e.g., rough-sawn, unplaned boards; unfaced chipboard)
- low absorbency (e.g., multi-ply boards with improved surface finish)
- non-absorbent or very low absorbency (e.g., sheet steel, plastic lining, chipboard)

The frequency of use and cleanliness of the formwork affect the occurrence of pores, mottled effects, blushing and discolouration. Other essential aspects affecting the appearance are the position and layout of:

- joints
- formwork joints
- formwork ties

The use of trapezoidal or triangular fillets (e.g., 7, 10 mm) can emphasise construction or dummy joints especially well, or disguise them by relegating them to the shadows. However, adequate concrete cover on reinforcement must be ensured at these positions as well.

Joint spacing L [m]	Recommended joint width b 1) based on +10 °C [mm]	Required minimum joint width min b [mm]	Thickness of joint sealing compound t_F 2) [mm]	perm. dev. [mm]
up to 2	15	10	8	±2
> 2 to 3.5	20	15	10	±2
> 3.5 to 5	25	20	12	±2
> 5 to 6.5	30	25	15	±3
> 6.5 to 8	35	30	15	±3

1) Permissible deviation: 5
2) The figures given here apply to the final condition; the volumetric shrinkage of the joint sealing compound must also be taken into account.

B 1.3.11

B 1.3.11 Recommended values for joint design and permissible minimum joint width to DIN 18540 parts 1-3
B 1.3.12 Formwork tie holes:
a with fibre-cement plug
b plugged with filling compound and finished flush
c recessed plug of filling compound
d recessed plug of filling compound, chisel-finish concrete
B 1.3.13 Minimum thicknesses and lengths of precast concrete elements
B 1.3.14 Fixing of large-format precast concrete elements:
a using cast-in components
b using anchors
c using cast-in rails

As it is impossible to produce completely watertight formwork joints and that means the w/c ratio may alter around the joints (which can lead to discolouration), special attention should be paid to such details. Besides the regular positioning of formwork ties, the form of the cones at the surface has an effect on the overall look of the surface of the facade. Experience shows that filling the depressions flush with the finished surface frequently leads to unsatisfactory results (fig. B 1.3.12).

Sharp edges must be carefully protected against damage – a fact that must be taken into account at the design phase. Spalled concrete must be patched and that usually results in colour differences.

Another important parameter is the thickness of the external wall, which depends on the arrangement of the reinforcement but also the correct placing and compacting of the concrete. Taking into account the diameter of a poker vibrator (approx. 40 mm) and the necessary minimum spacing between reinforcing bars, this gives us a tried-and-tested thickness of 160 mm, or better still 240 mm, for facing leaves.

a

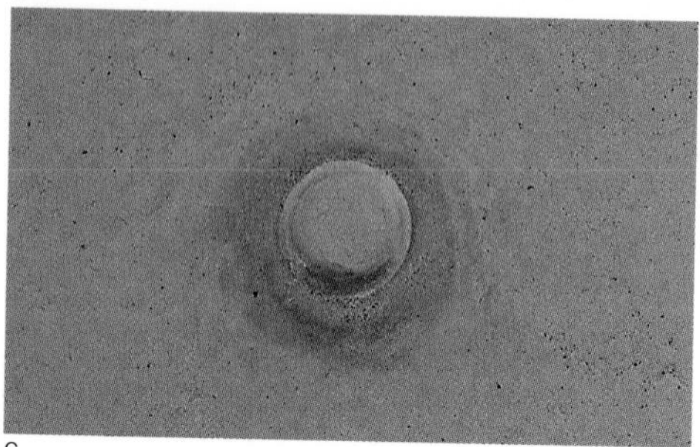

c

On the whole, fair-face concrete facades require detailed preliminary planning with respect to quality of workmanship, evaluating and guaranteeing the quality of the surface, and costs. It is therefore advisable to draw up a formwork specimen schedule showing the particular architectural features (e.g., partitioning of the surface, texture, details). The production of the specimen surfaces – comparable in terms of scale plus position and production conditions – represents an important control medium for attaining the desired quality.
In terms of their appearance and their production, *in situ* fair-face concrete facades are in fact unique, and can never be reproduced exactly because the result depends on many factors. Further, on-site production limits the number of options for finishing the surface [8].

Precast concrete components
The production of precast concrete components [9] offers a number of advantages over *in situ* concrete because the production and processing of the concrete is carried out by one team. Horizontal casting beds result in very good compaction of the concrete, and that in turn leads to lower porosity at the surface. However,

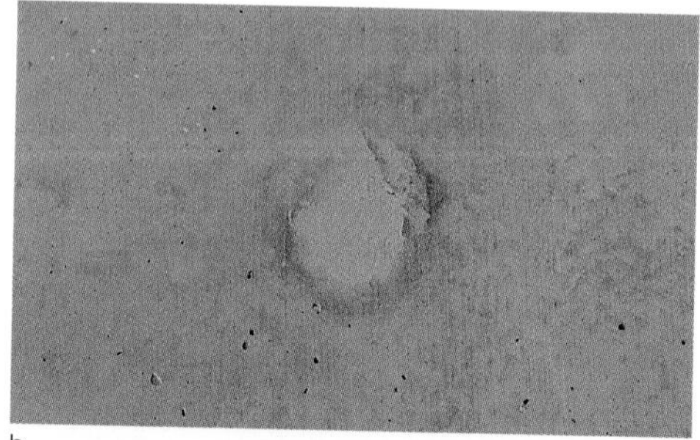

b

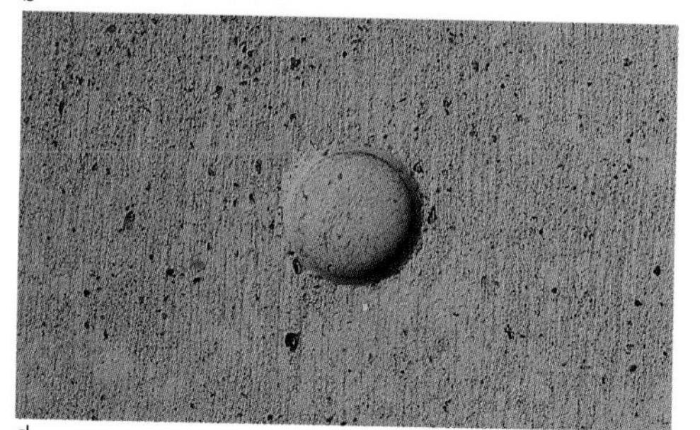

d

B 1.3.12

the transportation and erection options limit the economic use of heavy and large precast components. In terms of the thickness, similar conditions to those for *in situ* concrete apply. However, much slimmer dimensions – from 70 mm to 140 or 160 mm – are feasible. Precast components can have a surface area up to 14 m², although the maximum length of facing leaf components should not exceed 5 m.

At the precasting works a whole range of economical methods for finishing the concrete surface are possible. The surface can given a more three-dimensional form by incorporating projections and depressions, or the surface can be divided up by means of dummy joints.

There is also the possibility of producing facings of (frost-resistant) rubble stone, clay facing bricks, natural stone and ceramic panels. To do this, the materials are positioned with the face to be exposed on the bottom of the formwork and then bonded to the precast components with several layers of concrete. Normal-weight concrete is ordinarily used in precasting plants. However, self-compacting concrete is being used more and more; the very soft consistency – and hence good workability – of this concrete is ideal for fair-face concrete surfaces. The arrangement and form of the joints represents an important detail in working with precast concrete components, where the minimum dimension of 10 mm depends on the respective panel length. Dark surfaces, which are particularly sensitive to temperature changes, will require joints 10–30% wider (fig. B 1.3.13).

Numerous preformed strips, permanently elastic sealing compounds and cast-in plastic sections are available for use in joint details. Vertical joints at corners need to be detailed especially carefully. Generally, it is regarded as more economical to reduce the number of joints by providing larger formats. The use of precast concrete components calls for more work during the preliminary design stage and – despite a number of improved production techniques – generally larger batch sizes as well, if the components are to prove economical.

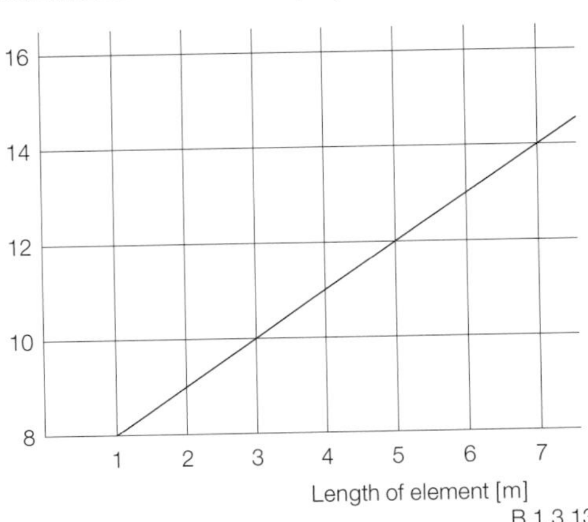

B 1.3.13

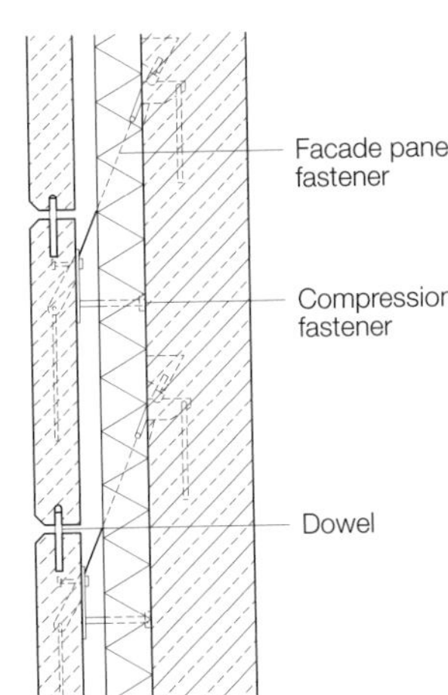

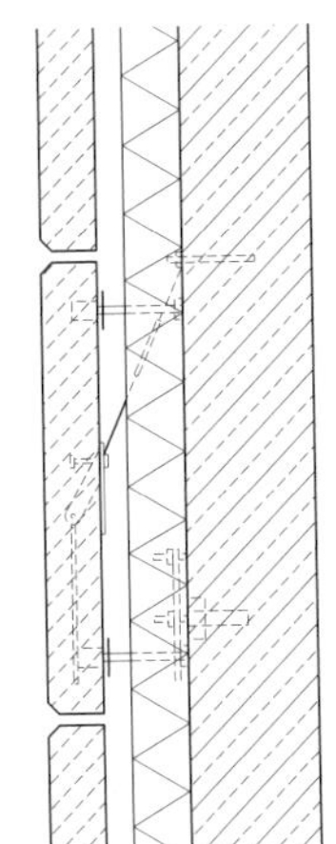

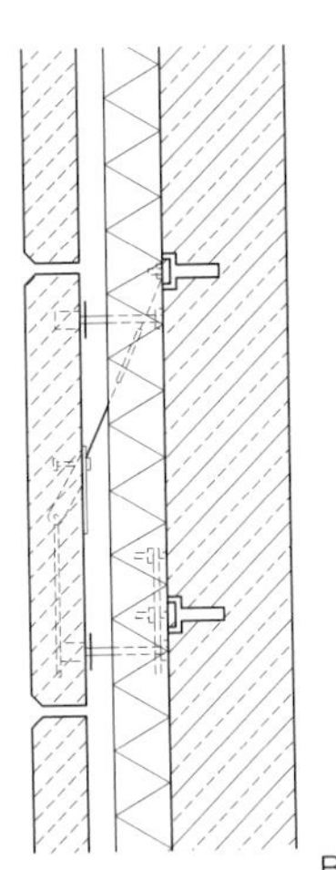

B 1.3.14

Three types of precast concrete elements can be considered for building facades:

- single-leaf curtain walls
- double-leaf curtain walls
- sandwich elements

Large-format, single- and double-leaf precast concrete components are suspended from the loadbearing construction by means of special fixings; restraint forces must be avoided. Depending on the system, the elements are suspended from cast-in loadbearing brackets or screwed/bolted to anchors or cast-in rails provided on site. In terms of ease of erection and pattern of joints, cast-in rails offer greater scope for adjustment. Adjustable horizontal fixings (e.g., compression fasteners, wind load fasteners) resist pressure and suction forces, define the position relative to the loadbearing leaf and, together with dowels, guarantee precise integration into the plane of the facade – all during erection. All fasteners must be made from stainless steel (figs B 1.3.14 to 1.3.16).

Normally, sandwich elements combine loadbearing and insulating layers in one component, together with the exposed concrete face. The facing layer should not be less than 70 mm (to ensure adequate cover on the reinforcement) and not more than 100 mm thick (to avoid larger alternating deformations); the longest element length (as with wall panels) should not exceed 5 m (fig. B 1.3.13).

Connections between individual elements are achieved by way of support and retaining fasteners for carrying the vertical and horizontal loads respectively. Cramps and face or concealed fasteners are used to resist wind loads and temperature fluctuations. The number of thermal bridges increases with the number of fasteners and cramps.
Sandwich elements can be used both as loadbearing and as non-loadbearing components.

Reconstructed stone panels
One form of plain facade cladding is the small-format, suspended reconstructed stone panel [10]. This category includes panel-like building materials whose area does not exceed 0.2-1.0 m². As a rule, this cladding is fitted in front of a ventilated air cavity and connected to a supporting construction. One advantage of the small-format panels is that they can be fixed to masonry as well. The fixing options include (figs B 1.3.5 to 1.3.7):

- individual fasteners fitted into the mortar joints
- individual fasteners fixed with anchors
- a supporting construction of rails.

The thickness of the panel depends on the strength of the concrete and is generally 40 mm, although 20 mm is possible, depending on the dimensions. Size and fixing is governed by the same requirements as for natural stone panels (see DIN 18516).

It is true that with such types of facade the costs rise with the number of metal elements in the supporting construction. They generally offer no economic advantages over natural stone facades.

A great variety of surface finishes are possible with reconstructed stone panels, and a vast range of colours, depending on the aggregate.

Facing masonry units
Facing masonry units [11] are part of the tradition of masonry construction. The combination options, with (no-fines porous) normal-weight concrete, different aggregates (including fine chippings) and pigments, provide chances to optimise the material properties with respect to durability and appearance.

Basically, we distinguish between facing bricks and facing blocks, with the height of a course being the determining factor (125 mm = bricks, 250 mm = blocks). But as this distinction is not consistently adhered to even in DIN 18513, the designation "facing masonry unit" has become established in practice. These units exhibit high dimensional stability and – owing to densities between 1800 and 2200 kg/m³ – also good sound insulation and fire protection properties.

There are a number of coordinated masonry unit formats, which allow a facade to be divided up according to masonry bond, colour and surface finish. External walls are mainly constructed as two leaves (with a non-loadbearing facing leaf) in areas with a central European climate. Depending on the height of the building and the performance requirements, unit thicknesses of 90 or 115 mm are employed.

There are two systems for masonry unit formats:

- modular formats (to DIN 18000)
- "ocametric" formats (to DIN 4172).

The "ocametric" formats are based on 1/8 m (= 125 mm) and essentially reflect the customary masonry unit formats. The modular formats are based on 1/10 m (= 100 mm) and allow greater format variety; wall thicknesses of 90, 115, 140, 190 and 240 mm can be produced using different unit sizes. The two types of formats can be combined.

In terms of surface finishes there are – in addition to different colours – four customary options: smooth, porous, sand-blasted and split. White cement is generally used, which emphasises the colour of the units themselves. In the case of specific project-related solutions, there is also the option of extending or individually adapting the range of colours.

Fibre-cement sheets
When we talk of cement-bonded sheets these days we are mostly referring to the combination of wood fibres (52%) with Portland cement as the binder (38%) plus water (9%) and wood mineralisation substances.

These sheet-type building materials exhibit a number of advantages, which are suitable for ventilated curtain walls: good moisture resis-

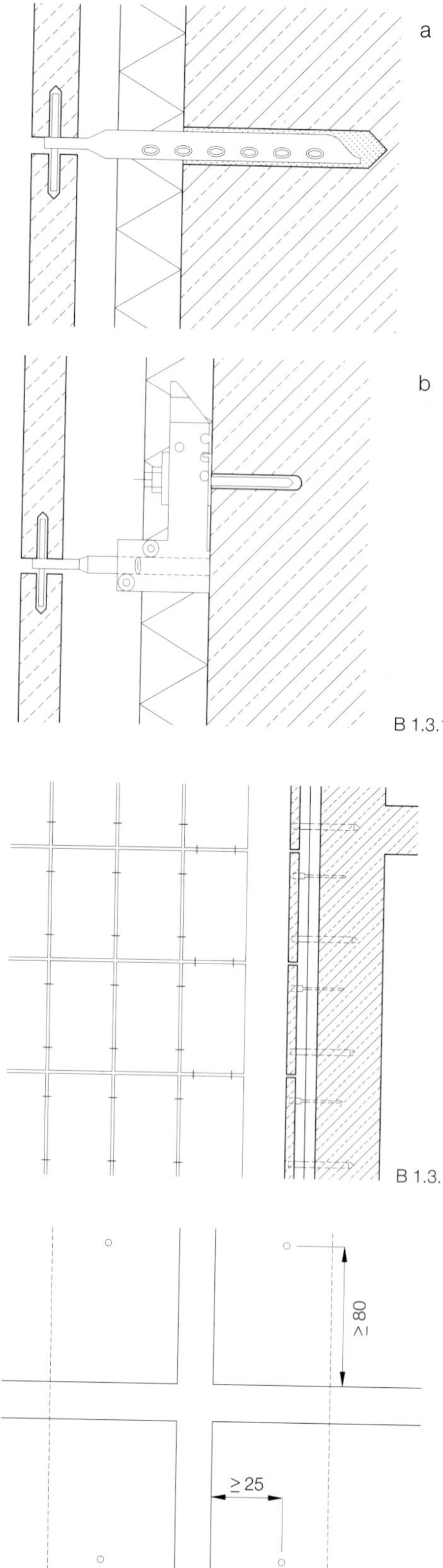

B 1.3.15

B 1.3.16

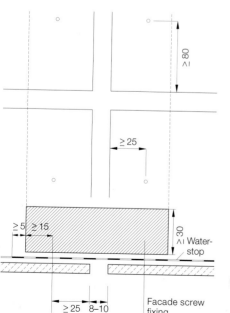

B 1.3.17

tance, frost resistance and little swelling. In addition, fire protection requirements can also be satisfied, depending on material composition.

Cement-bonded sheets [12] are available in the most diverse range of formats. The maximum standard sheet dimensions are 3100 mm long x 1250 mm wide; the thickness varies between 12 and 18 mm normally.

One advantage of these lightweight facade elements is the ease with which they can be cut to suit – even difficult geometrical shapes. The material can be sawn, drilled and milled. However, the – usually – untreated edges must be handled carefully during installation.

The sheets are screwed to a supporting construction of battens and counter battens, or in combination with metal spacers. This type of construction can be used for the tallest of skyscrapers.

Joints can be concealed behind battens and strips, designed as open (drained) joints, or provided with plastic or metal waterstops or water bars. A width of 10 mm has proved to be suitable for the open joints between large-format sheets; spacings ≤ 8 mm are not permitted and those > 12 mm are not advisable (fig. B 1.3.18). Fibre-cement sheets are available with a colour primer and an industrially applied colour coating that does not require any further treatment on site.

Surface finishes

Besides the formwork finish options for concrete surfaces, the exposed faces can be "worked" or "treated" to create other surface finishes. There are fundamental differences between these two approaches. Green or hardened concrete surfaces can be mechanically, thermally and/or chemically worked, or we can apply a waterproofing, coating or sealing treatment to the hardened concrete surface. Furthermore, diverse colour options are also available [10, 13].

Working the surface

We can work the concrete surface to reveal the colour of the aggregate and thus provide the dominant overall colour. DIN 18500 describes the various techniques, which can also be applied in combination.

The most frequent methods are brushing and washing (≥ 2 mm), and light brushing and washing (≥ 2 mm), in which the uppermost layer of fine mortar is removed. This can also be carried out by applying a retarder to the formwork, which can remove the topmost (max.) 1 mm of the surface. The effect of these methods is to allow the aggregate and its natural colour to dominate the surface.

The concrete surface can be roughened by acid-etching, sand-blasting or flame-cleaning. In these methods the hydrated cement and the surfaces of the aggregate are exposed to an equal extent, which leads to a semi-matt surface finish.

The methods based on traditional stonemason techniques (e.g., bush hammering, pointing, chiselling, splitting) produce new surfaces, either manually or by machine. The removal of the uppermost layer exposes parts of the hydrated cement matrix and the aggregate. Special effects can be achieved by using white cement, coloured aggregates or pigments.

There are other mechanical methods for working the surface, primarily used in the production of precast concrete components. Here, we distinguish between production-related textures (grinding plus sawing and splitting of blocks) – without additional measures – and fine working (fine-grinding, polishing), which produces especially smooth or shiny surfaces.

The appearance of the worked concrete surface is governed to a very large extent by the colour of the aggregate(up to 80%). The other factors influencing the final overall shade are the colour of the cement or the finest particles, and any pigments used.

Treating the surface

Silane, siloxane or acrylate coatings are applied for the purpose of:

- waterproofing
- coating
- sealing
- repelling dirt and oil

The "wet effect" can alter the colour of the finished concrete. Products must be "non-yellowing", which makes it essential to carry out preliminary trials on specimen surfaces. Further, surface treatments generally remain effective for only a limited period of time.

Colour

Apart from coloured sealing products and coatings, which can be applied as glazes or opaque finishes, there are also various ways to add coloured accents during the production of concrete, for example by using:

- cements with a particular coloration (white cement or Portland burnt shale cement)
- aggregates with a particular coloration (red granite, Carrara marble, etc.)
- pigments (including iron oxide yellow, chrome oxide green)

Essentially, it is the cement colour that influences the appearance: a relatively high iron content leads to a Portland cement with a darker shade of grey. White cement is obtained by using low-iron raw materials (limestone and kaolin). Portland burnt shale cement contains, besides cement clinker, fired oil shale, which lends the product a reddish colouring. On the whole, the use of grey cement leads to more subdued, darker shades, while white cement makes the colours appear brighter and purer.

The colour of the aggregate becomes effective only after working the surface. The size of the grains leads to different intensities of colour in combination with the method of working the surface. This is why equal amounts of ultra-fine particles and finest sand particles must be used.

Pigments are a very easy way of giving concrete a certain colour. Red, yellow, brown and black shades are primarily achieved by using iron oxide pigments, green shades by adding chromium oxide and chromium oxide hydrate pigments. Pigments based on mixed crystals (e.g., cobalt-aluminium-chromium oxide pigments) create blue shades.

It is usually sufficient to add a small amount (2–3% by wt of cement content) to achieve the desired colour. Minimal profiling of the surface reinforces the effect of the colour. The pigmentation of concrete is both permanent and weatherproof (fig. B 1.3.19). A new form of (coloured) surface finish is photoconcrete. Here, photographic templates are attached to the surface via a screen. The intensity of the effect depends on the different degree of hardening of the concrete (fig. B 1.3.20).

Ageing of facades due to the effects of the weather is a material-related problem and is frequently caused by defects in the detailing. The appearance changes due to the local environmental pollution and the way in which rainwater drains over the facade; wind direction and shelter from the wind determine the quantity of run-off water. The position and orientation of the facade are important factors determining the degree of self-cleaning. In particular, deep surface textures and their direction (horizontal, vertical) together with their cross-sectional geometry (ribs, grooves) have a major effect on the accumulation of dust and the draining of run-off water.

B 1.3.18

B 1.3.19

Notes

[1] Fair-face concrete data sheet; pub. by Deutscher Beton- und Bautechnik-Verein, Bundesverband der Deutschen Zementindustrie, Düsseldorf, 1997.
[2] Junghanns, Kurt: Das Haus für alle Fälle, Berlin, 1994, pp. 113, 116–145.
[3] Ford, Edward R.: Die Pionierzeit des Betonsteins. "Textile-Block"-Häuser von Frank Lloyd Wright; in: Detail 04/2003.
[4] modul, Series of publications on the use of modular concrete masonry units in modern architecture, Rheinau-Freistett, 05/1992.
[5] Eternit Schweiz. *Architektur und Firmenkultur seit 1903*, Zürich, 2003.
[6] Grimm, Friedrich; Richarz, Clemens: Hinterlüftete Fassaden. Konstruktionen vorgehängter hinterlüfteter Fassaden aus Faserzement, Stuttgart/ Zürich, 1994.
[7] Hegger, Josef; Will, Norbert: Bauteile aus textilbewehrtem Beton; in: DBZ 04/2003.
[8] Kling, Bernhard; Peck, Martin: Sichtbeton im Kontext der neuen Betonnormen; in: Beton 04/2003.
[9] Döring, Wolfgang et al.: Fassaden. Architektur und Konstruktion mit Betonfertigteilen, Düsseldorf, 2000.
[10] Fassaden aus Stein; pub. by Dyckerhoff Weiss Marketing und Vertriebs-Gesellschaft, Wiesbaden, 2004.
[11] Technical manual: Kann-Sichtmauersteine; pub. by Kann GmbH Baustoffwerke, Bendorf, 2003.
[12] Product information: Großformatige Fassaden. Fassaden mit Holzzement, Eternit AG, Berlin, 2001.
[13] Kind-Barkauskas, Friedbert et al.: Concrete Construction Manual, Munich/Basel, 2001.

B 1.3.15 Fixing of small-format reconstructed stone panels:
a using individual fasteners cast into mortar
b using anchors
B 1.3.16 Fixings can be positioned in both horizontal and vertical joints
B 1.3.17 Minimum edge distances for fixing fibre-cement sheets to a timber supporting construction
B 1.3.18 Church and parish hall, Cologne (D), 2003, Heinz and Nikolaus Bienefeld
B 1.3.19 Library, Eberswalde (D), 1999, Herzog & de Meuron
B 1.3.20 Different worked surface finishes with the same concrete mix:
top: smooth formwork finish
bottom (left to right): sand-blasted, lightly brushed and washed, acid-etched, finely ground, chiselled, pointed

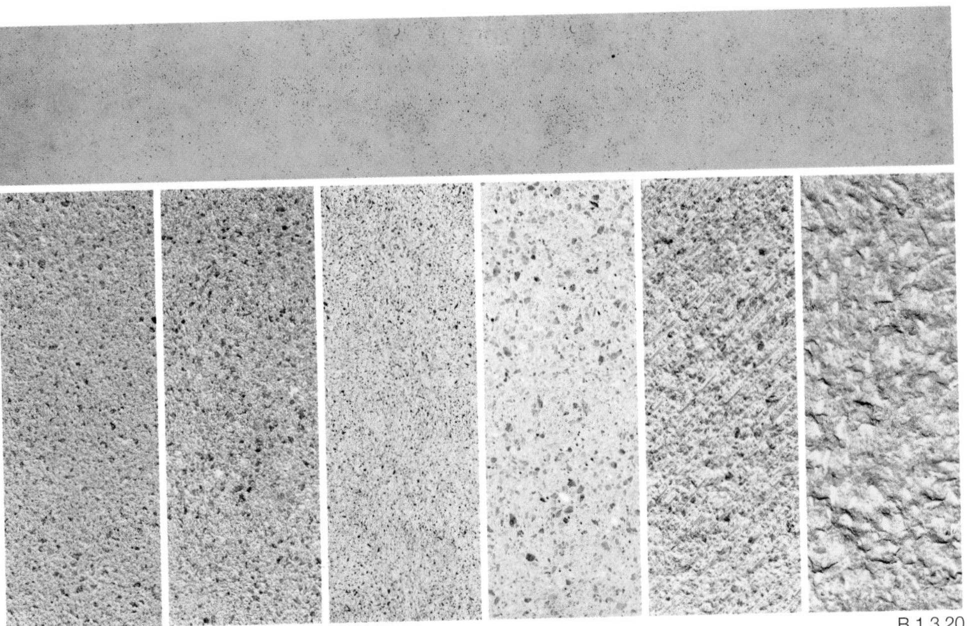
B 1.3.20

Gallery of modern art

Munich, D, 2002

Architect:
Stephan Braunfels, Berlin/Munich
Structural engineers:
Seeberger Friedl + Partner, Munich
Walther Mory Maier, Munichstein, CH
Facade design: R+R Fuchs, Munich

domus 853, 2002
Braunfels, Stephan: Pinakothek der Moderne. Basel 2002
Herwig, Oliver: 6 neue Museen in Bayern. Tübingen/Berlin 2002

- New gallery for four collections (modern art, drawings, architecture, design)
- Seamless, fair-face concrete facades, 16 m high in some places, with cavity insulation, large-format (formwork) grid of 5 m
- Connections between inner and outer leaves can accommodate movement, outer leaf prestressed with horizontal steel strands
- Construction joints positioned just above horizontal triangular fillets – shadow conceals any inaccuracies

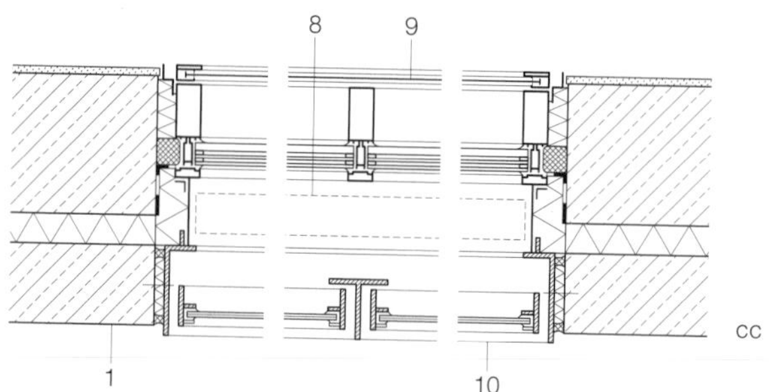

Plan of ground floor • Section
scale 1:2000
Horizontal section • Vertical section south elevation
scale 1:20

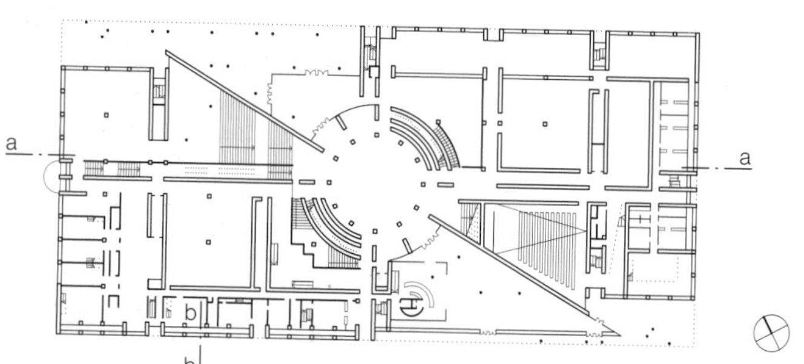

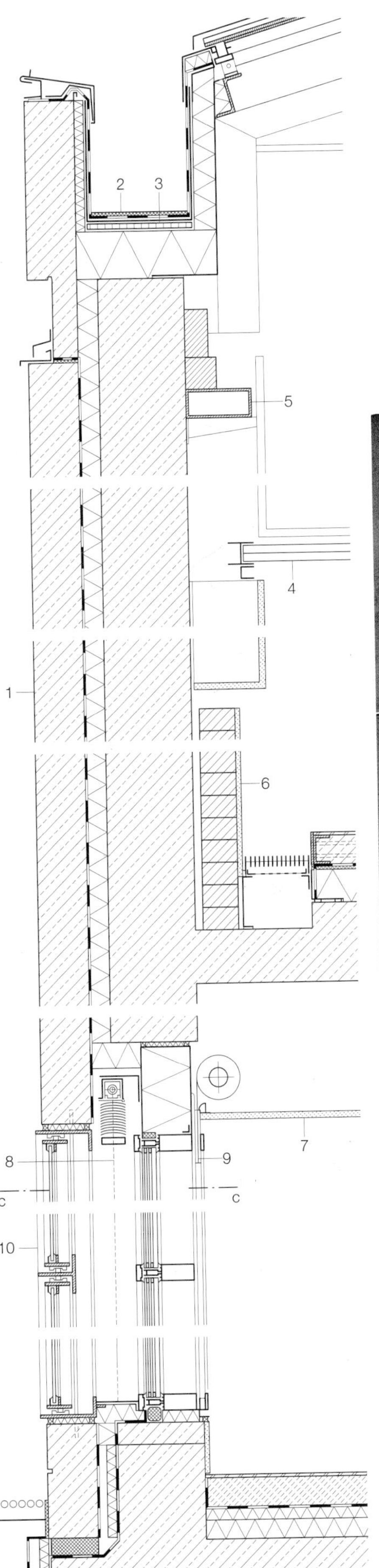

1 Facade construction:
160 mm fair-face concrete, glazed finish foil as slip plane
60 mm Styrodur polystyrene thermal insulation
280 mm reinforced concrete
15 mm plaster
2 Rubber granulate board on separating layer and gutter heating
3 Bonded plywood, waterproof
4 Light-scattering suspended ceiling, laminated safety glass with matt finish
5 Steel rectangular hollow section
6 Facing masonry:
115 mm with integral retaining system
15 mm plaster
7 Plasterboard suspended ceiling
8 Blind with plain-edge louvres
9 Blackout roller blind
10 Double window:
steel frame construction
outside: 12 mm extra-clear toughened safety glass
inside: B1 double glazing, extra-clear laminated safety glass

"House of Tranquillity" –
"monastery for a time"

Meschede, D, 2001

Architect:
Peter Kulka, Cologne/Dresden
with Konstantin Pichler
Structural engineer:
Dieter Glöckner, Düsseldorf
Site manager:
Hans Hennecke, Meschede

Bauwelt 31/2001
domus 849, 2002
Schwarz, Ullrich (ed.): Neue Deutsche Architektur. Ostfildern, 2002

- Visitor accommodation for the Königsmünster Benedictine monastery
- Building split into two units: the narrower one ("accessible wall") acting as an entrance and providing access to the wider one ("His House") with 20 accommodation units plus common rooms, offices, cloister and chapel
- Expression of asceticism by way of reduced language and fair-face concrete surfaces of outstanding quality

Plan of 2nd floor • Section •
scale 1:500
Vertical section • Horizontal section
scale 1:20

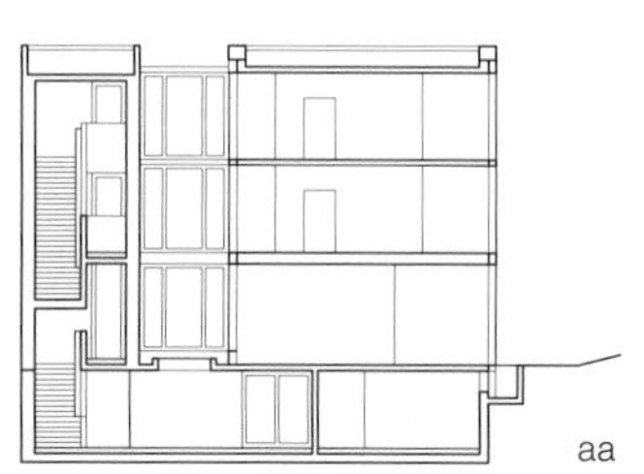

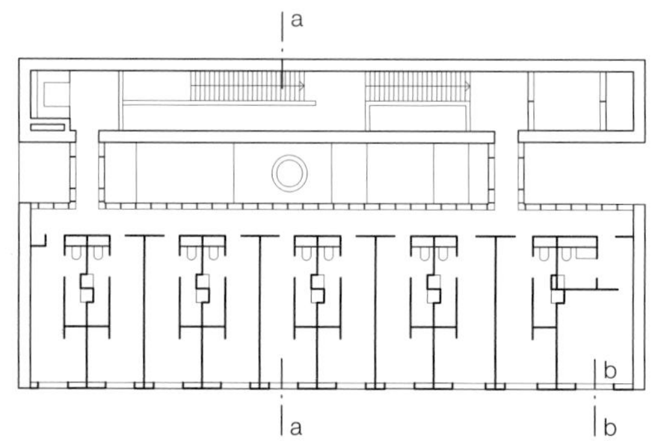

1 Powder-coated sheet aluminium parapet capping plate, bent to suit
2 Compressible preformed sealing strip
3 Waterproofing/vapour check
4 Aluminium angle, anthracite, vapour-tight seal
5 Aluminium angle, anthracite, cork strip in joint
6 Aluminium hollow section post, 50 x 140 mm, anthracite
7 Aluminium hollow section rail, 50 x 140 mm, anthracite
8 Patent glazing cap, 50 mm wide
9 Double glazing, light transmittance 66%, reflection to outside 11%
10 Steel angle, 90 x 90 mm
11 Powder-coated sheet aluminium, 3 mm, bent to suit with anti-drumming coating Styrodur polystyrene board insulation waterproofing/vapour check
12 180 mm external fair-face concrete wall 100 mm extruded polystyrene cavity insulation 200 mm internal fair-face concrete wall
13 100 mm cellular glass insulation embedded in bitumen waterproofing extends 300 mm above ground level
14 100 mm perimeter insulation waterproof concrete
15 Aluminium angle, 30 x 50 x 3 mm
16 Aluminium angle, 40 x 30 x 3 mm
17 2 mm sheet aluminium 15 mm Styrodur polystyrene 2 mm sheet lead

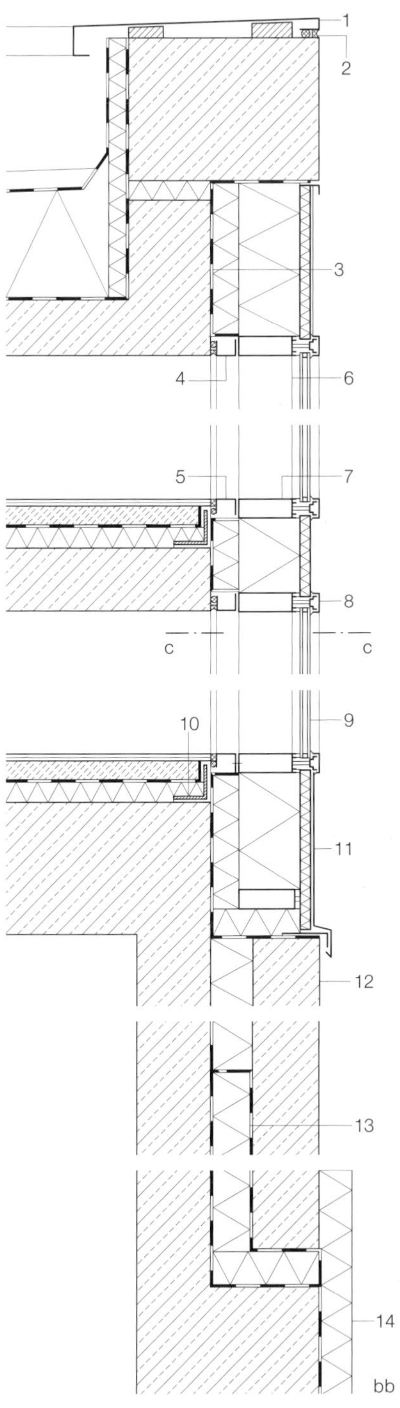

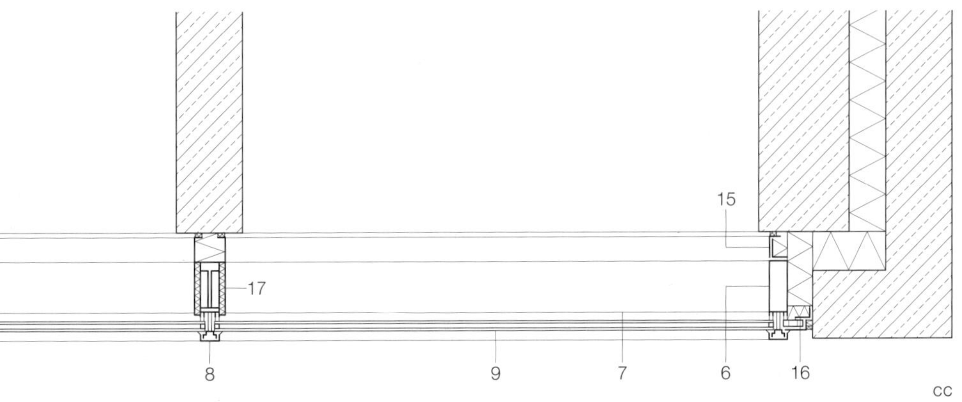

Apartment blocks

Monza, I, 1972

Architect:
Angelo Mangiarotti, Milan

A+U 12/1978
Bona, Enrico D.: Angelo Mangiarotti. Il Processo del Construire. Milan 1980
Finessi, Beppe (ed.): Su Mangiarotti, catalogue of the Milan Triennial for Architecture and Design. Milan 2002
Herzog, Thomas (ed.): Bausysteme von Angelo Mangiarotti. Darmstadt 1998

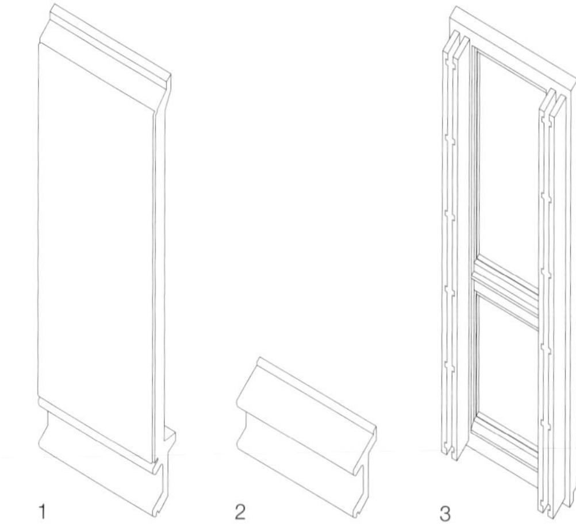

- Storey-height precast concrete sandwich elements
- Used for two different apartment blocks: in Monza and in Arioso, Como, (1977, five-storey building with facade broken up even more than in this building by projections and returns)
- Flexible arrangement leaves open areas for tenants to use as they wish

Isometric view (not to scale)
Plan of 1st, 2nd and 4th floors scale 1:500
Vertical sections, scale 1:20
Detail of roof-facade junction scale 1:5

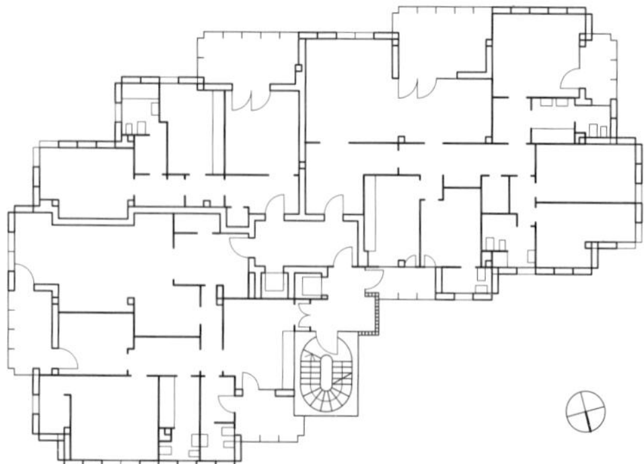

1 Wall panel: storey-height precast concrete, with integral 120 mm rigid polystyrene thermal insulation
2 Precast concrete roof edge element
3 Wooden window element, fir
4 Wooden window, fir, with double glazing: 4 mm toughened safety glass + 9 mm cavity + 4 mm toughened safety glass
5 Folding wooden shutter
6 Sheet copper, bent to suit, 8/10 mm
7 Precast concrete parapet element
8 M12 bolt + nuts
9 Steel angle, 60 x 60 x 8 mm, for fixing roof edge element, welded to steel flat cast flush into reinforced concrete roof slab
10 Steel angle, 60 x 120 x 8 mm, for fixing wall panel, welded to steel flat cast flush into reinforced concrete roof slab

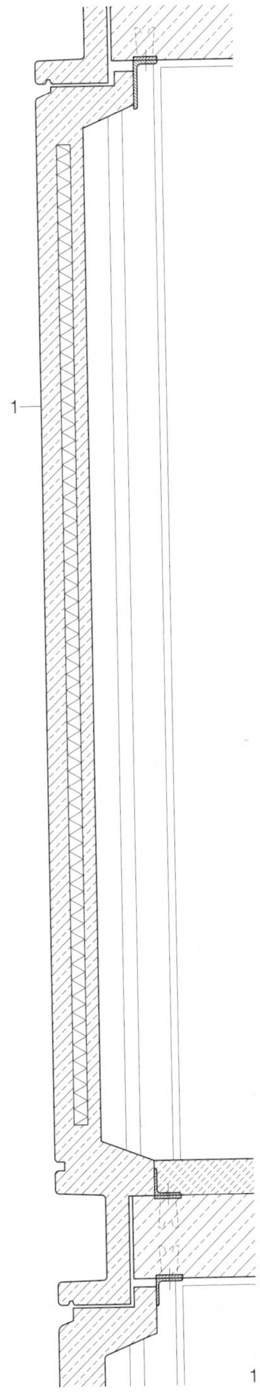

1

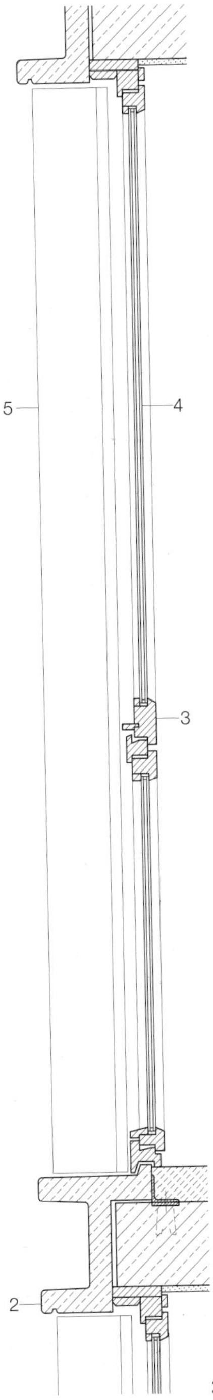

2+3

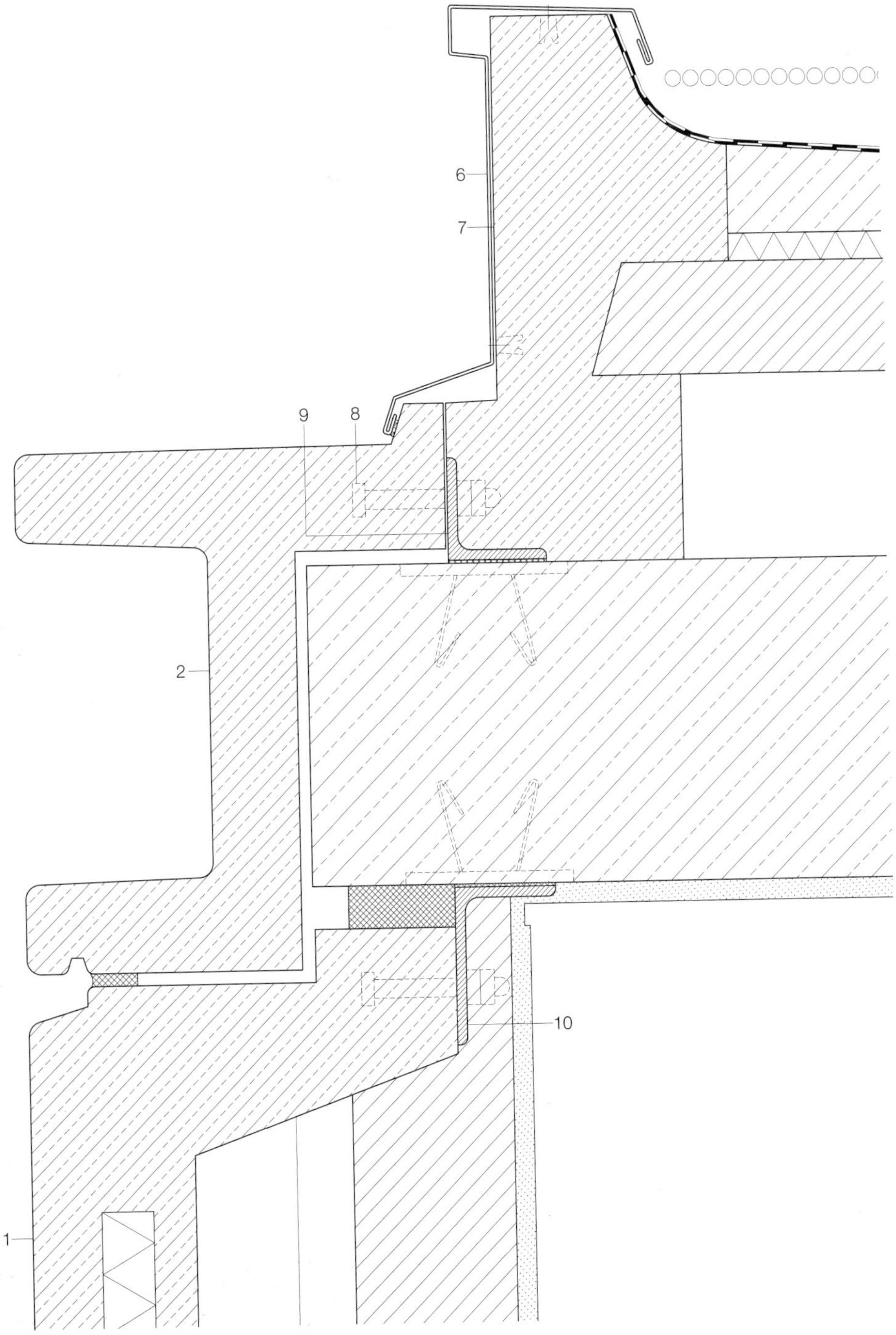

Mixed office and residential block

Kassel, D, 1999

Architect:
Alexander Reichel, Kassel
Structural engineers:
Hochtief, Kassel

Byggekunst 06/2001
Detail 04/2001
Kind-Barkauskas, Friedbert et al.:
Concrete Construction Manual,
Munich/Düsseldorf, 2001

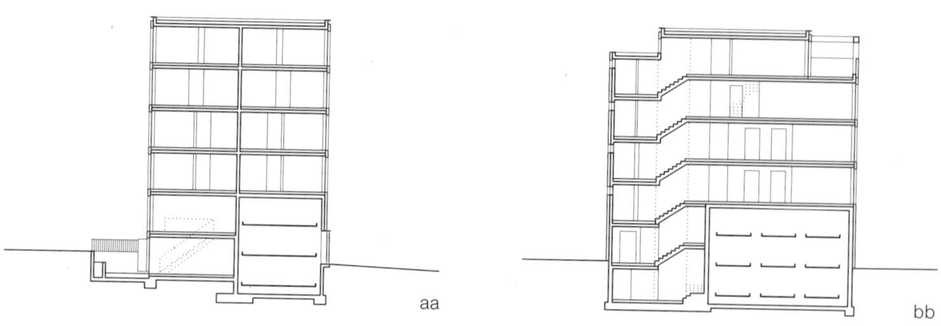

- Town house with columns set out on a 3.00 x 3.50 m grid
- Reinforced concrete frame and large parts of the facade clad with suspended precast concrete components made from glass-fibre reinforced concrete
- Fine-aggregate concrete with aggregates < 4 mm, alkali-resistant glass-fibres as tension and anti-crack reinforcement, approx. 2–4 mm, hydrophobic fluid coating for a water-repellent outer surface
- Storey-height windows and bi-fold shutters made from untreated larch wood provide functional and aesthetic accents

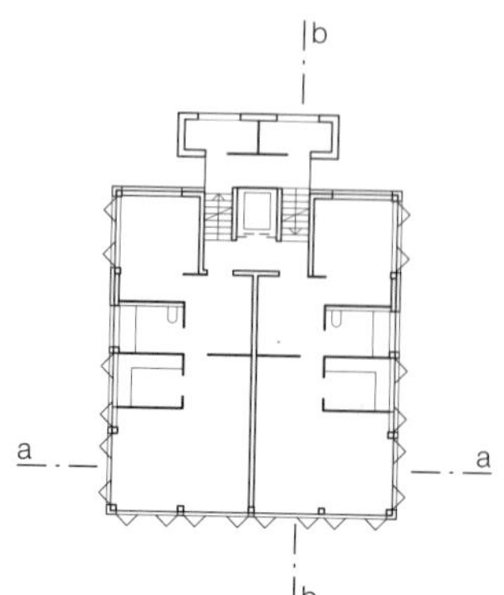

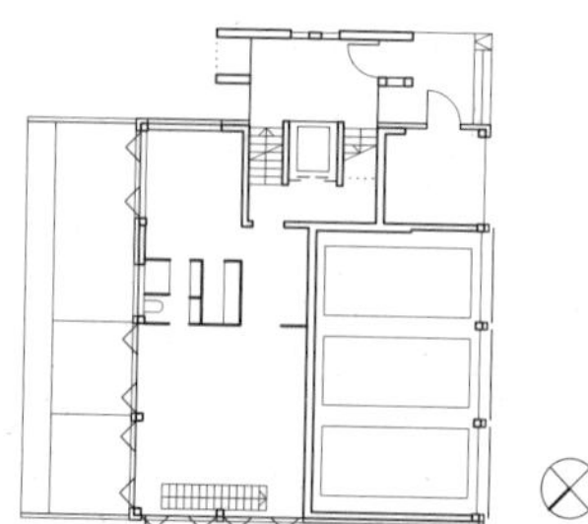

Sections • Plans of basement and 1st/2nd floors
scale 1:500

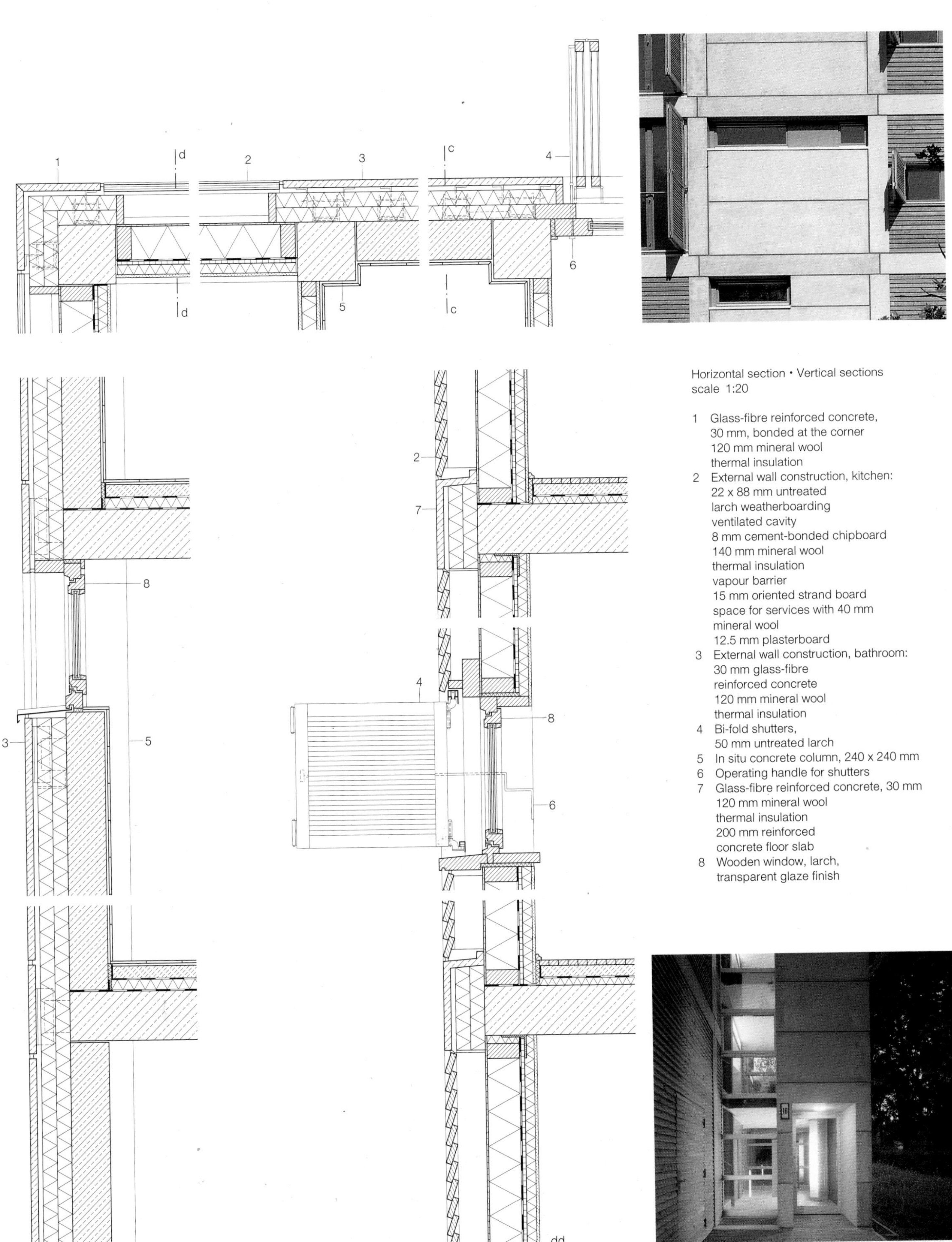

Horizontal section • Vertical sections
scale 1:20

1 Glass-fibre reinforced concrete, 30 mm, bonded at the corner
120 mm mineral wool thermal insulation
2 External wall construction, kitchen:
22 x 88 mm untreated larch weatherboarding
ventilated cavity
8 mm cement-bonded chipboard
140 mm mineral wool thermal insulation
vapour barrier
15 mm oriented strand board
space for services with 40 mm mineral wool
12.5 mm plasterboard
3 External wall construction, bathroom:
30 mm glass-fibre reinforced concrete
120 mm mineral wool thermal insulation
4 Bi-fold shutters, 50 mm untreated larch
5 In situ concrete column, 240 x 240 mm
6 Operating handle for shutters
7 Glass-fibre reinforced concrete, 30 mm
120 mm mineral wool thermal insulation
200 mm reinforced concrete floor slab
8 Wooden window, larch, transparent glaze finish

School of engineering

Ulm, D, 1994

Architects:
Steidle + Partner, Munich
Project architects:
Otto Steidle, Johann Spengler, Siegwart Geiger, Alexander Lux, Peter Schmitz, Thomas Standl

Arkitektur 05/1993
GA document 42, 1995
Feldmeyer, Gerhard G.: The New German Architecture. New York 1993
Sack, Manfred et al.: Steidle + Partner. Universität Ulm West. Fellbach 1996

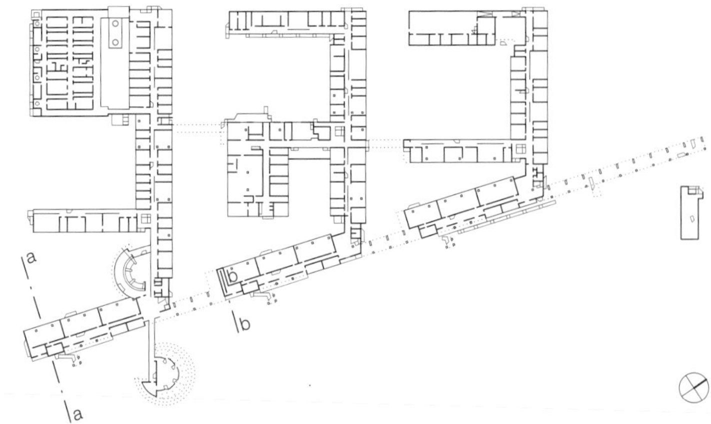

- Designed as a "town" with two- and three-storey structures
- Primary planning parameters: simple design, easy construction, low cost, fast to build
- Two different types of facade: ventilated with up to 7.20 m long panels, non-ventilated spandrel elements with large areas of glazing as prefabricated facade
- Fibre-cement sheets with colour coating prefabricated off site

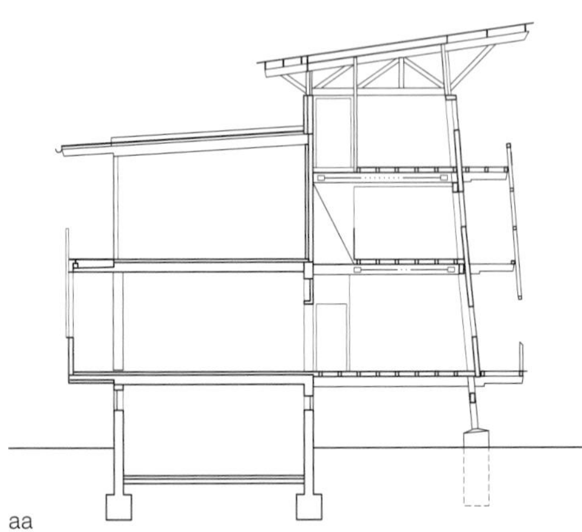

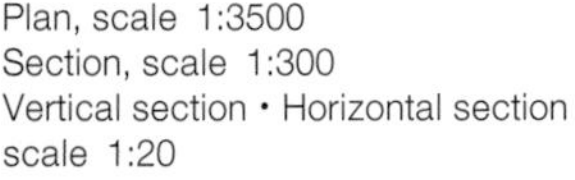
Plan, scale 1:3500
Section, scale 1:300
Vertical section • Horizontal section
scale 1:20

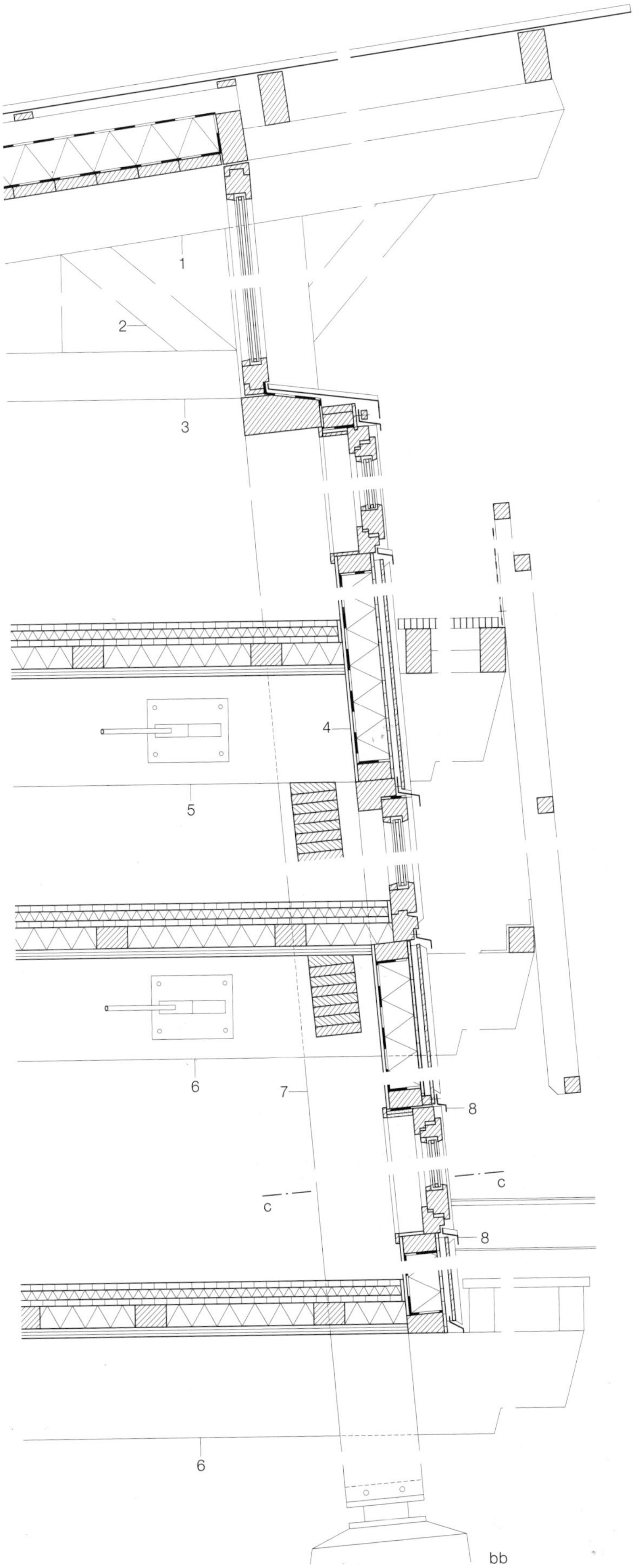

1 Gluelam upper chord, 150 x 200 mm
2 Gluelam diagonal, 150 x 120 mm
3 Gluelam bottom chord, 150 x 150 mm
4 Ventilated facade:
66 x 22 mm timber cover strip
10 mm fibre-cement sheets, painted
20 x 30 mm timber battens
10 mm chipboard
100 mm mineral fibre insulation
vapour barrier
10 mm maritime pine board, oiled
5 Gluelam 2nd floor beam,
150 x 340 mm,
on main grid lines
6 Gluelam ground floor
and 1st floor beams, 150 x 320 mm,
on main grid lines
7 Glulam column, 150 x 250 mm
8 Sheet aluminium bent to suit

bb

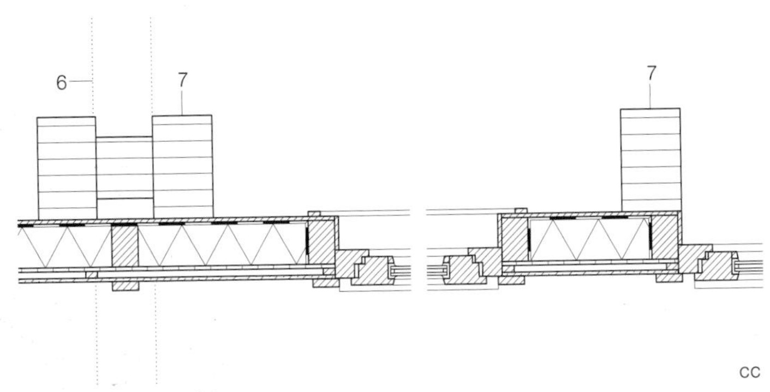

cc

School

Lauterach, A, 2000

Architect:
Elmar Ludescher, Lauterach

A & D 19, 2002
Detail 07/2001

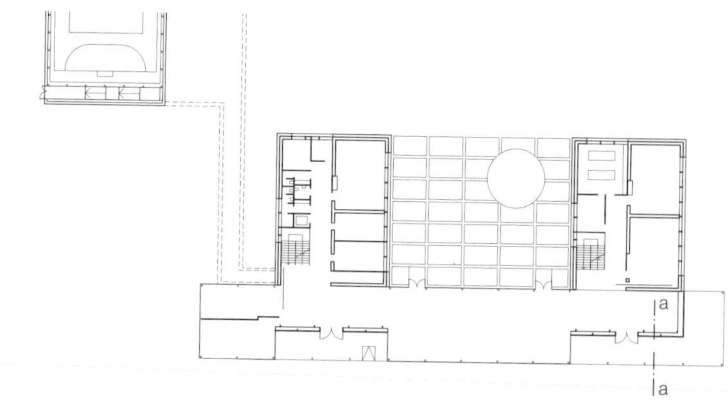

- Extension to and refurbishment of a school building dating from the 1960s
- Facades to the two three-storey structures finished with opaque and semi-transparent fibre-cement sheets (colour: anthracite)
- Perforated facade elements act as climate buffer and light filter

Plan, scale 1:1000
Vertical section • Horizontal section
scale 1:20
Detail of fixing of
fibre-cement sheets
scale 1:5

1 Aluminium window, anodised, with fixed glazing
2 Column, steel rectangular hollow section, 150 x 100 x 8 mm
3 Fibre-cement sheets, 8 mm, perforated
4 Steel square hollow section, 40 x 40 mm
5 Steel rectangular hollow section, 40 x 60 mm

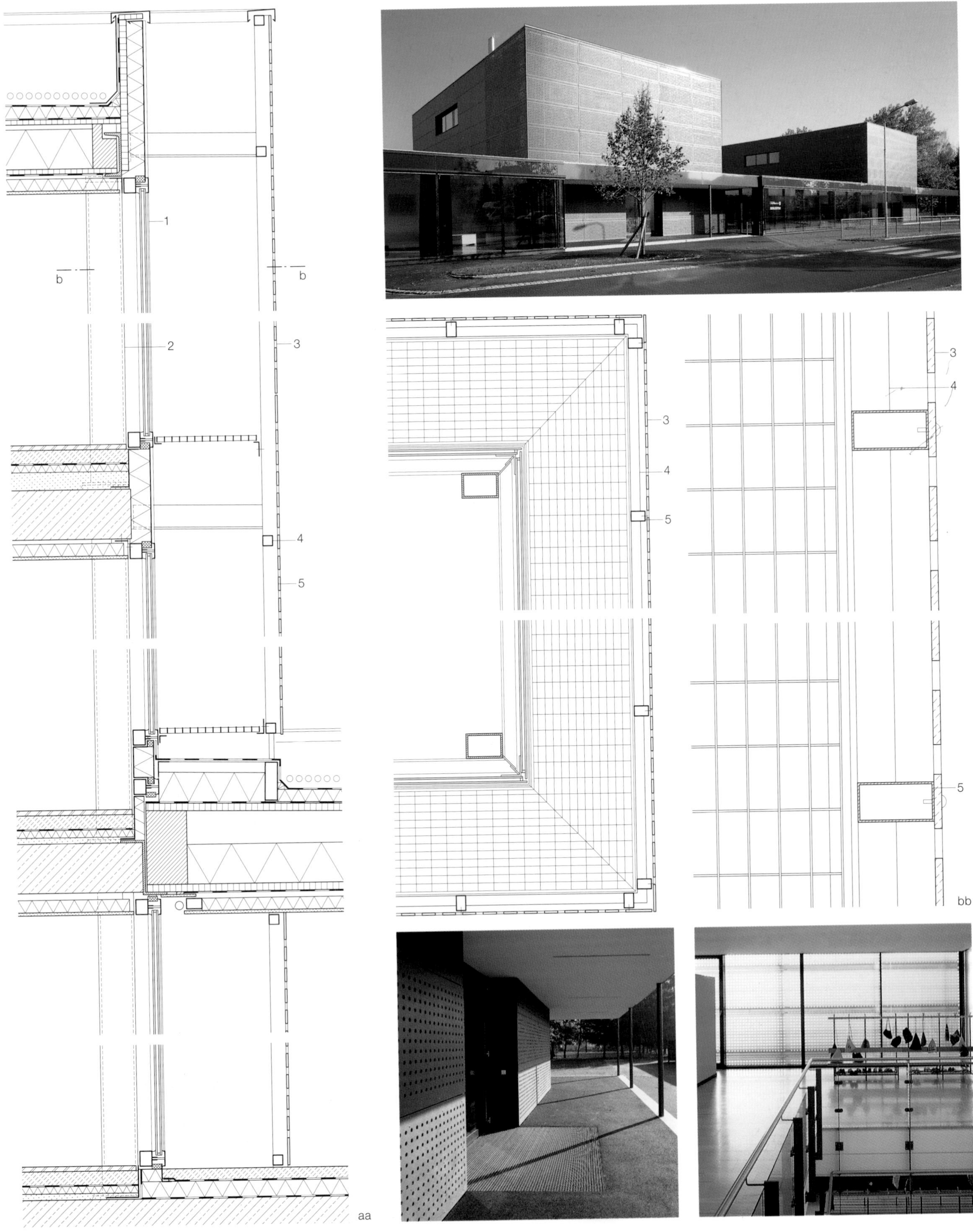
1
2
3
4
5
b
b
3
4
5
3
4
5
bb
aa

Student accommodation

Coimbra, P, 1999

Architects:
Aires Mateus e Associados, Lisbon

Architectural Review 12/2000
Casabella 691, 2001
Detail 07–08/2003

- Plain concrete facade with facing leaf of prefabricated matt white hollow blocks
- Autonomous reference to existing concrete facades on the campus
- Narrow window slits admit light into common rooms
- Small-format blocks, network of joints and surface finish resembling split stones give the surface an animated, structured expression

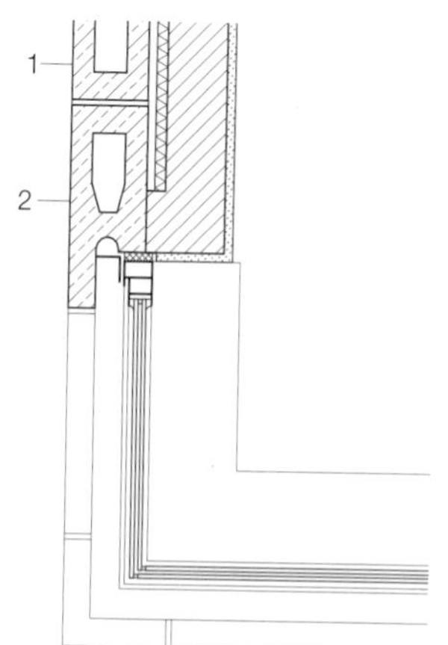

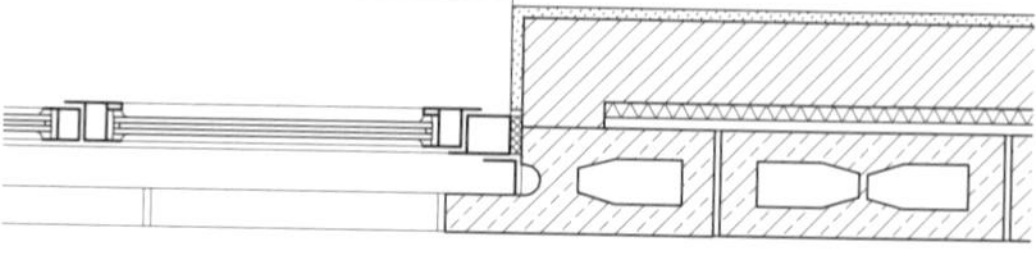

Section • Plan, scale 1:1000
Horizontal section • Vertical section, scale 1:20

1 Wall construction:
hollow concrete blocks,
390 x 140 x 190 mm, white
15 mm ventilated air cavity
20 mm insulation
110 mm masonry
15 mm smooth-finish plaster
2 Special reveal block
3 Special lintel block,
390 x 140 x 190 mm

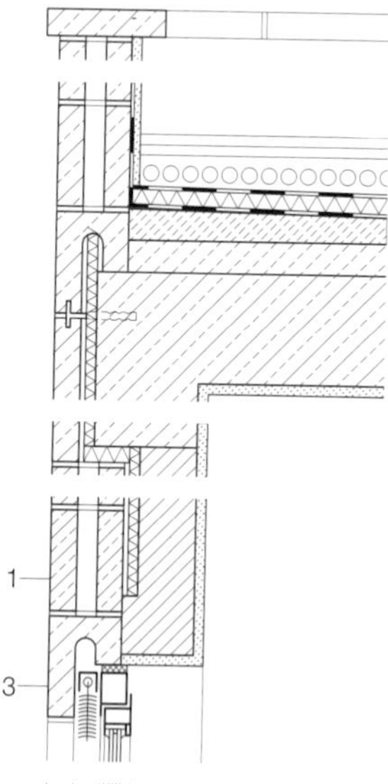
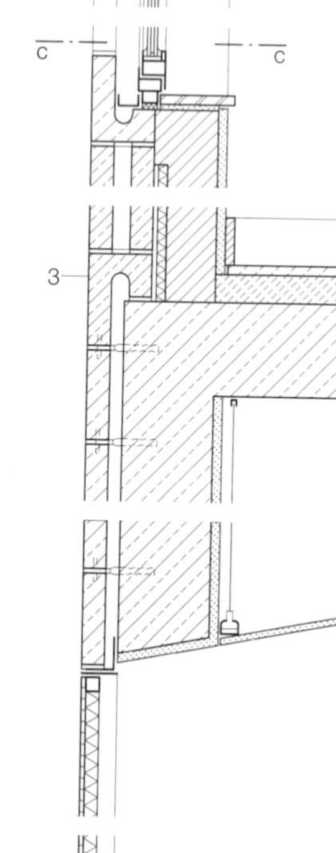

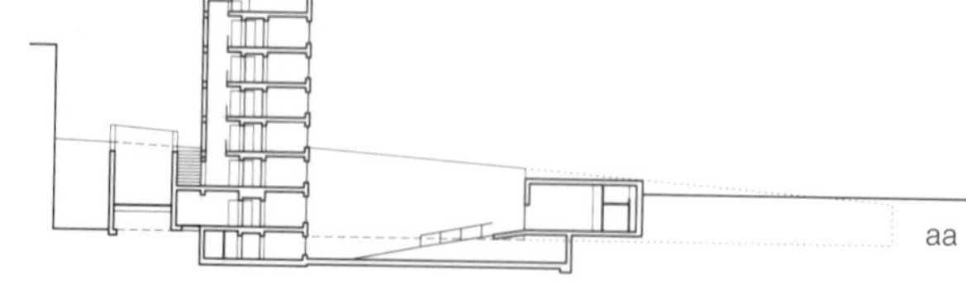

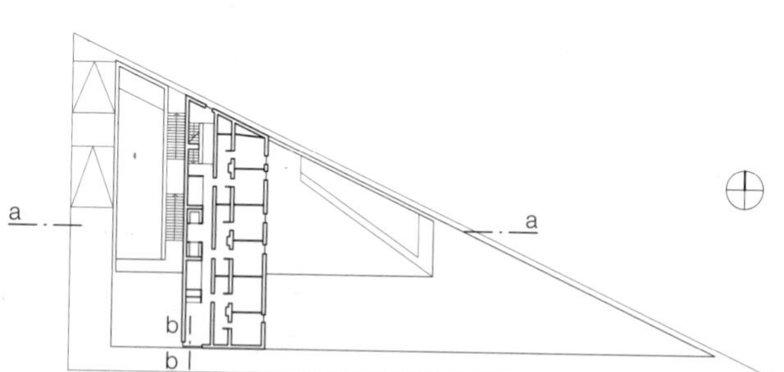

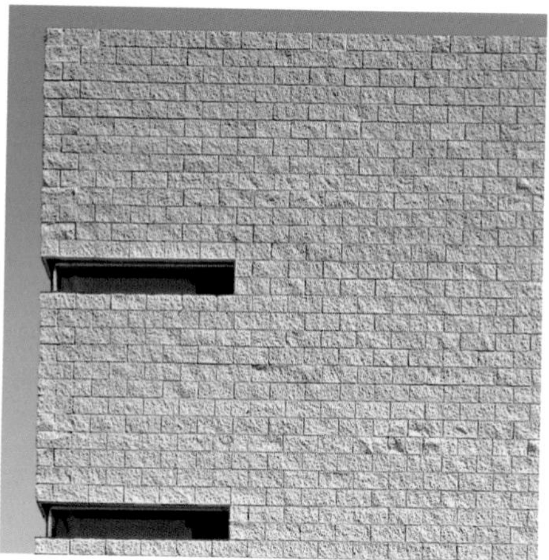

ESO Hotel

Cerro Paranal, RCH, 2001

Architects:
Auer + Weber, Munich

l'architecture d'aujourd'hui 343, 2002
Architectural Review 06/2003
Bauwelt 25/2002
Casabella 704, 2002
Intelligente Architektur 09–10/2003

- Hotel complex for employees of the European Southern Observatory (ESO) located on Cerro Paranal, Chile, at an altitude of 2600 m
- Concrete facade to hotel rooms provides effective protection against sunshine and overheating
- Reinforced concrete as a thermally inert mass for buffering the daily temperature fluctuations (approx. 20 K)
- Ventilation via windows, small additional radiators for extreme, low temperatures
- Fair-face concrete surfaces coloured with iron oxide pigments which create an allusion to the colours of the Atacama Desert

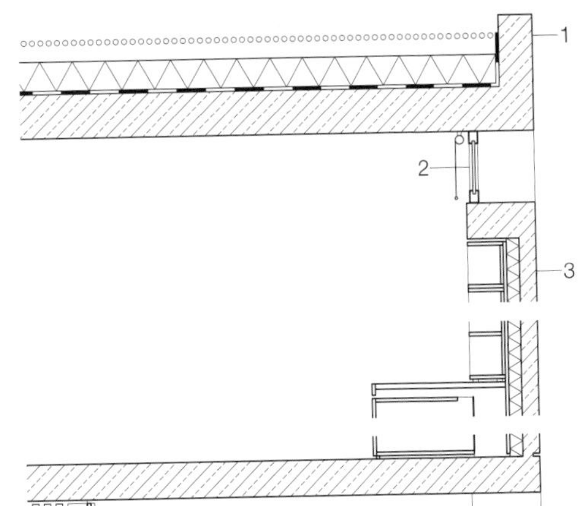

1 Fair-face concrete parapet, 200 mm, rust-red colour
2 Aluminium window with fixed glazing
3 Facade construction:
100 mm fair-face concrete, rust-red colour
75 mm insulation
built-in furniture, veneered chipboard
4 Safety barrier rails, steel hollow sections, 50 x 20 mm, painted, fixed at sides
5 Safety barrier posts, steel hollow sections, 50 x 20 mm, painted, inserted into steel pockets cast into concrete slab
6 Reinforced concrete, sealed
7 Anti-glare blind
8 Aluminium glazed door with double glazing

Section, scale 1:500
Plan of 1st floor,
Vertical section, scale 1:50

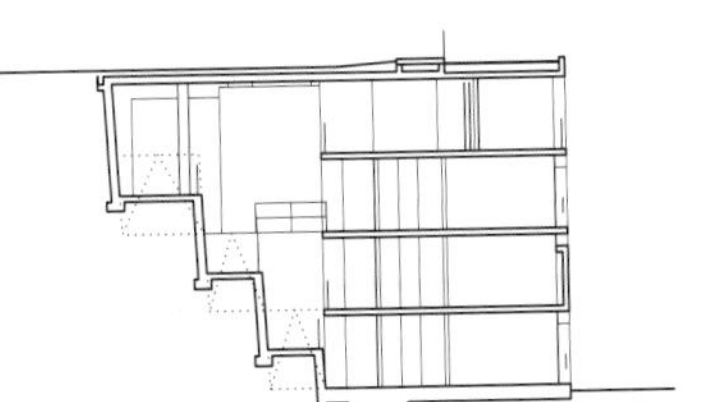

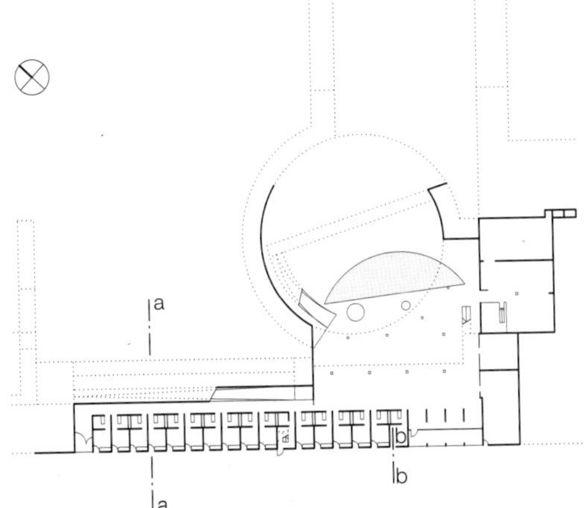

B 1.4 Timber

As a building material, wood can be employed almost universally. The first complex timber buildings date back to the Neolithic period and make use of trunks and branches (the varied natural "timber products") which were joined together to create frameworks and woven to produce walls (fig. B 1.4.3). Much, much later, frameworks made from squared logs appeared, used for buildings, but also for long spans, plus timber constructions made from round logs (e.g., log cabins and stave churches). The main methods of working timber today, e.g. sawing and chipping, are products of the past thousand years and were not used widely before the Industrial Revolution. The forerunners of modern timber engineering can be found essentially in the nineteenth and early twentieth centuries. The precursors of the development of our modern use of timber walls and facades are the glass-houses of England and the timber frame buildings of Central Europe. One outstanding example was the Crystal Palace in London (1851), whose frame construction made use of 17 000 m² of timber in conjunction with cast iron, especially for the roof structure. The current status of timber facade technology stretches from craftsman-like fabrication and erection to – in terms of engineering and performance – very advanced solid timber wall elements manufactured in fully automatic production plants and which, in the form of large panels and room modules, are assembled very rapidly to form complete buildings.

Material properties

Of the many technical properties of wood, the following are relevant to facade construction:

- high strength with low weight
- good working options and advanced techniques
- high thermal resistance
- hygroscopic behaviour promotes moisture equilibrium on the inside of the facade
- species of wood with a high resistance are suitable for use externally without any coatings

Constituents and growth

Wood is a natural product and can be fully recycled. Water containing nutrients from the soil and carbon dioxide react with the help of chlorophyll and sunlight to form starch and in doing so release oxygen. Starch is the basic building block of cellulose, which, at approximately 50%, is the primary constituent of wood. The other constituents are approximately 25% hemicellulose and lignin plus small amounts of pigments, tannins and resins.

Structure of wood

The basic building block of wood is the cell. We distinguish between different types of cells, according to their functions within the living tree, e.g., support, conduction and storage. Most of the cells have an elongated form. They are therefore also known as fibres and lie almost exclusively in the longitudinal direction within

B 1.4.2

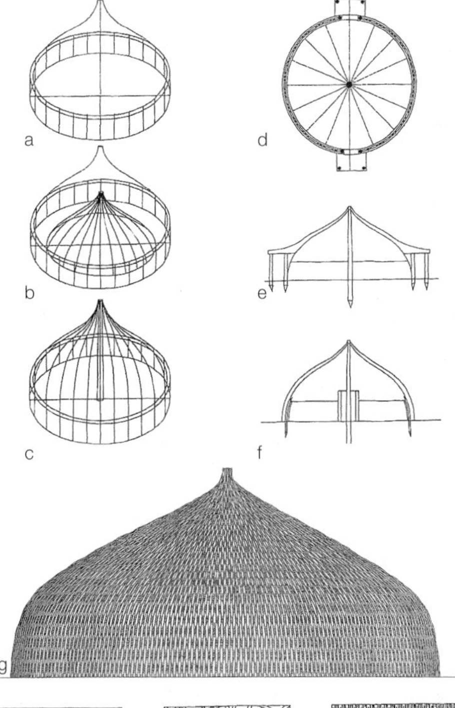

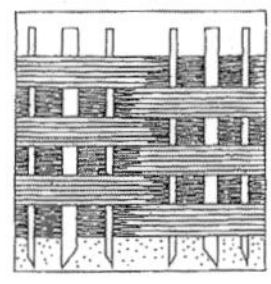

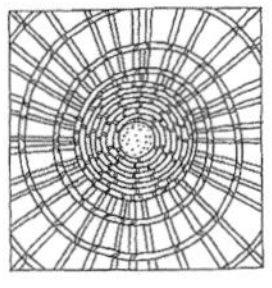

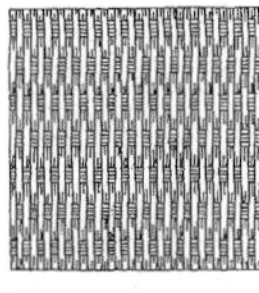

B 1.4.3

B 1.4.1 Komyo-Ji Pure Land Temple, Saijo (J), 2000, Tadao Ando

B 1.4.2 Gokstad ship, Bygdoy (N), c. A.D. 900
B 1.4.3 Sidamo House, Hagara Salam (Ethiopia)
a cylindrical construction
b roof built on the ground
c roof placed on cylinder and waterproofed with leaves from bamboo plants
d plan
e, f sections
g elevation
h detail of cylinder
i roof detail
j detail of covering

B 1.4.4

B 1.4.5

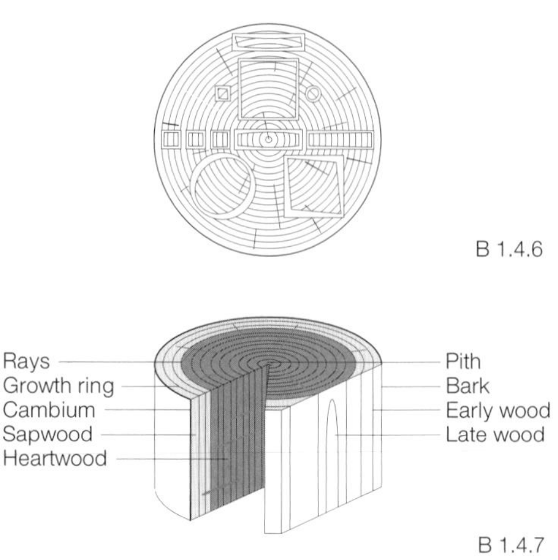

B 1.4.6

B 1.4.7

the trunk cross-section. The exceptions are the rays, whose cells lie in the radial direction. The – in evolutionary terms – older coniferous wood has a simpler structure; it consists mainly of one type of cell, which transports water and nutrients while providing support. In deciduous wood the cells are more specialised, and vessels form. The position and direction of the cells and vessels with respect to each other, in combination with the growth rings, are responsible for giving the wood its grain structure, that important characterising, distinctive feature of each species of wood. The elementary structure of the cell walls is instrumental in determining the strength and elasticity of the wood. The walls have four layers, which essentially consist of lignin for withstanding compressive forces and microfibrils for withstanding tensile forces. Together, the lignin and the microfibrils form an efficient composite structure.

Anisotropy
Wood consists of millions of such cells with their walls and cavities (pores). For simplicity we can consider wood as a bundle of tubes offset from each other in the longitudinal direction. This gives wood its distinctly different properties in different directions (anisotropy), i.e., parallel or perpendicular to the grain. The consequence of anisotropy is the completely different appearance of the various sections (transverse, tangential, radial) and the equally diverse behaviour of the wood parallel or perpendicular to the grain (figs B 1.4.6 and B 1.4.7). This affects, for example, the permissible stresses. The permissible stresses for spruce are:

parallel to the grain
- up to 11 N/mm^2 in compression
- up to 9 N/mm^2 in tension

but perpendicular to the grain only
- up to 2.5 N/mm^2 in compression
- up to 0.05 N/mm^2 in tension

Another consequence of anisotropy is the different swelling and shrinkage in the three planes (parallel to the grain, and perpendicular to the grain in the radial or tangential directions). In spruce the degree of swelling and shrinkage for every 1% change in the moisture content of the wood is:

- longitudinally < 0.01%
- transversely in the radial direction 0.15–0.19%
- transversely in the tangential direction 0.27–0.36%

Density
The density of the pure cell wall substance is about 1.5 g/cm^3 for all species of wood. In contrast, the thickness of the cell wall and the size of the cell cavity vary from species to species, and also within a species.

Furthermore, the cells of the early wood generally have larger cavities than those of the late wood. The ratio of cell wall to cell cavity determines the density and ranges from over 90% cell cavities in balsa wood with a density of 0.1 g/cm^3 to about 10% in lignum vitae with a density exceeding 1.3 g/cm^3. The volume of cell cavities in spruce is 70%, the average density 0.45 g/cm^3; that of oak is less than 60%, its density correspondingly > 0.60 g/cm^3. Density has a considerable influence on the load-carrying capacity of the wood.

Thermal aspects
Owing to its porous structure, Central European building timber with its average density exhibits very good thermal insulation properties. The change in length of the wood under the action of heat is extremely small and in practice is virtually negligible. The coefficients of thermal expansion depend on the species of wood. These are:

- parallel to the grain 2.55 to 5 x 10^{-6} K^{-1}
- in the radial direction 15 to 45 x 10^{-6} K^{-1}
- in the tangential direction 30 to 60 x 10^{-6} K^{-1}

However, an increase in volume does not usually occur because as the temperature rises the wood starts to dry out, causing shrinkage and hence a decrease in volume. The strength of the wood diminishes as the temperature climbs. However, this fact can be ignored in buildings with a normal range of ambient temperatures.

Moisture
The living tree contains water in its cell walls (bound moisture) and cell cavities (free moisture). The moisture content of the wood can amount to around 70% of the mass. At the maximum moisture absorption exclusively in the cell walls we talk of fibre saturation; this lies in the range 22–35%. Regardless of its use, wood remains hygroscopic, i.e., it absorbs water and releases it again, depending on the ambient humidity conditions. The following equilibrium moisture contents tend to become established in timber in use:

- heated structures enclosed on all sides 9±3%
- unheated structures enclosed on all sides 12±3%
- roofed structures open on all sides 15±3%
- constructions exposed to the weather on all sides 18±3%

Inside the building, wood's ability to absorb and release moisture can have a favourable influence on the interior climate. However, during design and construction this moisture absorption property must be carefully considered owing to its possible consequences. The absorption and release of moisture leads to swelling and shrinkage of the wood respectively, i.e., to dimensional changes. The load-carrying capacity of wood decreases as its moisture content increases; the risk of damage by fungi and insects increases too. The disadvantages can be ruled out by in-stalling the timber with a moisture content matching that expected in the long term at a particular location. All timber building components in which an alternating moisture content is to be expected, e.g., components exposed to the weather, must allow for the inevitable associated dimensional changes. This applies, for example, to the timber outer leaves of facades exposed to the changing effects of sunshine and rain. Rapid changes in moisture content bring a great risk of splitting and cracking.

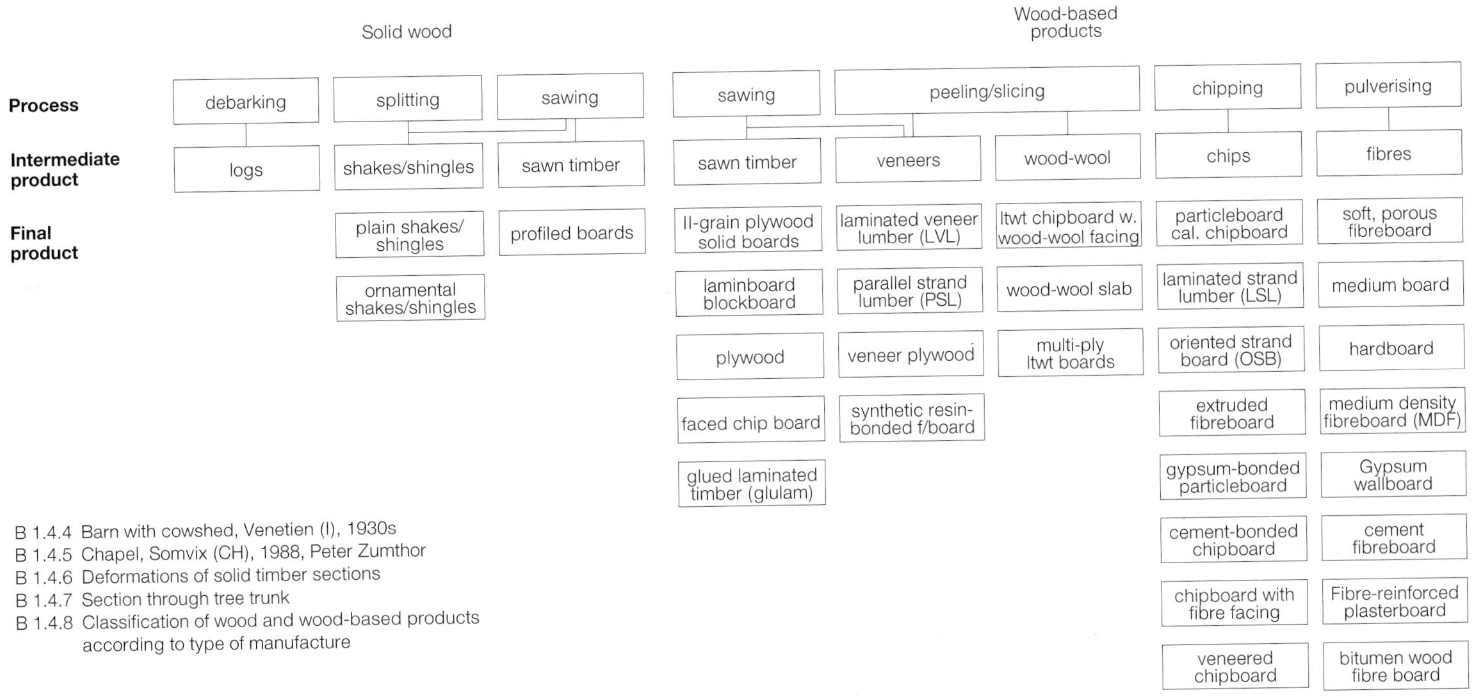

B 1.4.4 Barn with cowshed, Venetien (I), 1930s
B 1.4.5 Chapel, Somvix (CH), 1988, Peter Zumthor
B 1.4.6 Deformations of solid timber sections
B 1.4.7 Section through tree trunk
B 1.4.8 Classification of wood and wood-based products according to type of manufacture

B 1.4.8

Other features
Depending on the species of wood and the location of the individual tree, other features such as knots, slope of grain, pith, width of annual rings, fissures, bark pockets, resin pockets, distortion, discoloration, compression wood and insect damage can affect the wood. These characteristics lead to a wide scatter in the grading of solid timber and play a decisive role in the issue of where and how to use individual pieces of timber in a building.

External wall cladding

Besides protecting against moisture (especially driving rain), temperature effects (summer and winter thermal performance), incident solar radiation and wind, external wall cladding has a very particular influence on the architectural aspects of the building. Wood can be used to produce small-format cladding materials (e.g., shingles, boards) but also medium- to large-format panels. Of course, the choice of external wall cladding affects the outward appearance.

Wood and wood-based products
In recent years we have seen the introduction of numerous new solid wood and wood-based products, which has led to the provision of timber sections with less corruption of the wood but at the same time a more reliable quality (fig. B 1.4.8). The wood-based products have tended to focus on the optimisation of strength and quality of surface finish. Presented below is a selection of solid wood and wood-based products suitable for use in timber facades.

Logs and round sections
These are trunks or trunk segments. The stages of manufacture are as follows:

- Debarking
- If required, machining the cross-section to size over the length of the trunk
- If required, routing of relieving grooves in larger cross-sections (to help prevent splitting and cracking)
- Air drying, followed by kiln drying if necessary
- Visual strength grading

Sawn solid timber made from hardwood (LH) and softwood (NH)
Sawn timber is obtained from round sections by converting or profiling. The manufacturing sequence is as follows:

- Conversion, e.g., using frame saws or bandsaws
- Air and/or kiln drying
- Visual strength grading
- If required, finger jointing and glueing sections together
- If required, planing and chamfering
- If required, further profiling (e.g., rebates, grooves, tongues)

Glued laminated timber (glulam)
Glued laminated timber is an improved form of solid timber in which the growth-related defects in the wood that tend to reduce the strength have been partly eliminated. Glued laminated timber consists of at least three dried softwood boards or laminations glued together with the grain parallel. Besides simple, straight components, forms with a variable cross-section and/or in single or double curvature or twist about the longitudinal axis are also possible. Production sequence:

- Kiln drying of softwood boards to attain a moisture content of about 12%
- Visual or machine strength grading, if necessary with removal of larger defects
- Finger jointing of boards to form laminations
- Planing the laminations and cutting to length
- Application of adhesive to the wide face of the lamination
- Bonding the laminations in a straight or curved press
- Boards of different grades can be arranged within the depth of the cross-section
- Curing under pressure
- Normally, planing, chamfering and cutting to length after curing

Profiled boards
Planed and profiled sections are sawn from round sections, planed and routed. Besides the forms and dimensions given in DIN 4072 and DIN 68126 part 1, numerous modified profiles with different dimensions are possible, depending on the tools available. These profiled boards are manufactured to order in the planing shop and can be purchased from builders' merchants (fig. B 1.4.17).

Shakes and shingles
Shakes are usually split by hand from a bolt (trunk segment) using a mallet and froe. They have a tapering cross-section and the butt (i.e., bottom) end is often given a 45° chamfer. Shingles sawn from the bolt are also suitable as a wall cladding material. However, in sawn shingles all the fibres are cut through and so these shingles are less resistant and are worn by the weather more rapidly.

Soft-woods	Douglas fir DGA	Spruce FI	Scots pine KI	European larch LA	Pine PIP	Fir, silver fir TA	Western Hemlock HEM	Western red Cedar RCW
Colour of wood, sapwood/ heartwood	yellowish white/red-dish brown darkening late wood dark	early wood yellowish white, late wood red-dish yellow sapwood / heartwood indistin-guishable	lt yellowish white/red-dish white, turning to brown, late wood darker	yellowish/ reddish brown, darkening late wood, very dark brown	yellowish/ reddish/ yellow to reddish brown, late wood dark	early wood nearly white, late wood pale reddish sapwood and heart-wood in-distinguish-able	early wood lt brownish grey, late wood darkening, sapwood & heartwood indistin-guishable	white/ red-brown, darkening, late wood darker
Resistance to fungal attack	moderate, sapwood vulnerable to blue stain	low, vulnerable to blue stain	low to mod. sapwood highly vulnerable to blue stain	moderate to low	sapwood low, heartwood moderate	low, vulnerable to blue stain	low to moderate	very high
Resistance to insect attack	moderate	low	low	moderate to high	low to moderate	low	low	high
Growing regions	west coast of N. Am., cultivated in Europe	Europe	Europe, north-west Asia	central Europe	southern/ south-east N. America, Central Am.	central and southern Europe	north-west. North Am., cultivated in Europe	north-west North-America

Hard-woods	Maple AH	Ekki (azobé) AZO	Beech (Europ. b.) BU	Oak EI	Dark-red meranti MER	Merbau MB	Robinia ROB	Teak TEK
Colour of wood, sapwood/ heartwood	yellowish white, satin shine, sapwood & heartwood hardly distinguish-able	light red-brown/ very dark red-brown with pale violet shade	lt yellowish to reddish grey, sapwood & heartwood hardly distinguish-able	grey/grey-yellow, darkening to light to dark brown	yellowish grey to pink-grey/ reddish brown	yellowish white/light brown to reddish brown darkening	light yellow to greenish yellow/ greenish yellow to olive yellow, later shiny gold-brown	grey/gold-yellow, later medium to dark brown often narrow black streaks
Resistance to fungal attack	very low, also with respect to blue stain	high	very low	high	high to moderate	very high	very high	very high
Resistance to insect attack	very low in some instances	very high	low	high	moderate to high	high to very high	high	very high (termite-resistant)
Growing regions	Europe through to Asia minor	west Africa	Europe	Europe	south-east Asia	south-east Asia, Mada-gascar, Papua New Guinea	south-east North Am., cultivated in Europe and elsewhere	S.E. Asia, cultivated in other tropical regions

a

Wood-based product	3-ply and 5-ply core plywood	laminated veneer lumber (LVL)	chipboard, particleboard	facade-quality plywood	wood fibreboard	cement fibreboard
Species of wood/ materials	Softwoods, preferably spruce and Douglas fir, synthetic resins, wood preserva-tives if required	Pine (Kerto product), Douglas fir, southern pine (Microlam product), synthetic resins SVL: Oregon pine, Douglas fir to DIN 68705 pt 3	Adhesive-bonded particle-board wood chips: pine, beech, birch, alder, etc., wood-like fibres, synthetic resins; Cement-bonded particleboard: wood chips: spruce, fir as reinforcement, mineral binder: Portland cement, magnesia cement	Plywood with thin, blemish-free outer plies specially for facades	Primarily of spruce, fir, pine, beech, birch, poplar, eucalyptus; wood-like fibres from annual plants with or without addition of binder; synthetic resins, natural resins, waterproofing agent (waxes/ paraffin) and substances to protect against insect attack and fire	Cellulose-reinforced calcium-silicate boards, made from Portland cement, silicaceous aggregates and cellulose fibres

b

B 1.4.9

Wood-based products

The timber industry supplies a large number of wood-based products, mainly in the form of boards. They are optimised for their particular use in building, exploiting the properties of the wood to best advantage. The main optimisation approaches are:

- Size, in terms of length, width and thickness, for manufacturing larger components and for covering larger areas. Such wood-based products made from boards or veneers usually achieve much higher strengths than a solid piece of timber of the same species.
- Strength, with the aim of achieving greater load-carrying capacity.
- Surface finish, with the aim of achieving maximum compatibility, e.g., in terms of appearance (surfaces of components) or weathering (facades).

Defects (e.g., knots, fissures, twisted growth), which can substantially reduce the strength, are unavoidable in naturally grown wood. However, such defects play at best only a minor role in wood-based products because neighbouring timber components have a neutralising effect. It is for this reason that wood-based products exhibit improved homogeneity, and the swelling and shrinkage is generally much less than in solid timber. Basically, the anisotropy, i.e., the directional behaviour, of wood-based products diminishes with increasing sectioning.

Synthetic resin-bonded wood-based products

Linear or planar products made from crushed wood particles with binding agents (phenolic, resorcinol and other resins).

3- and 5-ply core plywood

These boards consists of a stack of three or five plies glued together, with adjoining plies always at an angle of 90° to each other. The boards of the outer plies are parallel to each other. The strength properties cover a very wide range. They can be controlled through the quality of the wood used and the relationships between the thicknesses of the individual plies.

Laminated veneer lumber (LVL) and structural veneer lumber (SVL)

Laminated veneer lumber (LVL) is produced by bonding together dried softwood veneers about 3 mm thick. We distinguish between the following types:

- Type S: the grain of all plies runs in the same direction, parallel to the direction of production, for primarily linear components and linear stresses.
- Type Q: the grain of most plies runs in the same direction but some run in the trans-verse direction, for planar components and in-plane stresses.
- Type T: is the same as type S in terms of grain direction but is made from lighter

veneers (lower densities) with correspondingly lower load-carrying capacities. The veneers of each ply are generally joined together by a scarf joint or simple overlap.

Structural veneer lumber (SVL) is for essentially linear-type components and consists of the outer plies of LVL laminations glued together. The laminations are made from 2.5 mm thick veneer plies with the direction of grain parallel to the longitudinal direction of the board. Finger joints are employed for the longitudinal joints of the laminations (fig. B 1.4.14).

Particleboards
Particleboards are produced by pressing together small timber particles with adhesives or mineral binders. The particles preferably lie parallel to the surface of the board and are generally arranged in several layers or with a gradual transition within the structure.

Wood fibreboards
Medium density fibreboards (MDF) are pressed with binders in the dry process. Medium boards are pressed without binders in the wet process. Hardboards can be manufactured using either method. The bond is based on the felting (interlocking) of the fibres as well as their own adhesive properties. When used for load-sharing and bracing purposes, hardboards must exhibit a minimum density of 950 kg/m^3, medium boards and medium density fibreboards a minimum density of 650 kg/m^3. Hardboards have virtually identical behaviour in both directions in the plane of the board. The properties can be altered by changing pressure, temperature and binder.

Cement fibreboards
These calcium silicate boards reinforced with cellulose consist of Portland cement, silica aggregates and cellulose fibres (see chapter B 1.3 "Concrete" for applications).

Wood-based materials with new application options

- oriented strand board (OSB) combined with composite materials to form building elements
- wood, hemp, linen and jute fibres, e.g., in the automotive industry for producing interior components for vehicles
- wood/plastic composites (WPC) by means of extrusion and injection methods

Fixings
We distinguish between exposed and concealed fixings. The main requirement is that these should fix the facade components securely and reliably, and – especially in the case of solid timber cross-sections such as boards – prevent them from twisting. However, to prevent cracking and splitting, they must permit the anticipated swelling and shrinkage [1].

Solid wood products	Logs, round sections	Sawn timber (solid hardwood and softwood sections)	Profiled boards	Shakes/shingles
Species of wood	Spruce, fir, pine, larch, Douglas fir, other species to DIN 1052 pt 1/A1, tab. 1	Spruce, fir, pine, larch, Douglas fir, beech (species group A), oak (species group A), ekki (species group C), teak (species group A)	Spruce, fir, pine, larch, Douglas fir	Western red cedar, larch, oak
Surface finish grades	from debarked to smooth	rough-sawn, if required plain and chamfered	scraped or planed	split, rough-sawn

B 1.4.10

B 1.4.9 a Properties of softwoods and hardwoods
b Wood-based products and their constituents
B 1.4.10 Solid wood products and their constituents

Fasteners
Of the various timber connectors and fasteners, it is primarily nails and screws that are used for facades. In the case of nails, adequate penetration depth is important; the recommended value is 35 mm. The head of the nail should neither project above the surface of the timber element nor damage it.
Screws have the advantage that the parts can be detached – useful in the case of refurbishment work. The minimum penetration depth is 25 mm. Again, the head of the screw should neither project above the surface of the timber element nor sit too deep in the wood. Only cross-head or Torx screws with a shank are permitted. Those with a drill tip or low-friction coating reduce the risk of splitting and can therefore be fitted closer to the edge.

Facade elements can also be installed with clips or special hooks. The surface should be coated and smeared with resin in such cases (to increase the pull-out resistance). One hardly avoidable disadvantage of this type of fixing is the squashing of the surface of the wood. Fixing hooks and patent clips can be employed for concealed fixings (fig. B 1.4.36). These elements are nailed or screwed to the supporting construction and engage in tongue and groove profiles. However, the extra work during erection is a disadvantage. Fasteners must be permanently protected against corrosion in order to prevent surface discoloration caused by rusting metal parts or chemical reactions with substances. Only stainless steel fasteners are permitted for heartwood species like oak and larch.

Spacing of fasteners
The number of fixing points within the width of a board depends on its size. One fixing is sufficient up to a width of 120 mm. Boards wider than this require two fixings at the third-points across the width. Along the length of a board, fixings should be spaced no more than 1000 mm apart. The edge distance perpendicular to the grain should be at least 15 mm, parallel with the grain at least 50 mm.

The spacing can be reduced when using self-drilling screws (depending on the species of wood). The denser and hence also the harder the wood, the more likely it will be that a pilot hole will be necessary. The same is true for small edge distances. Alternatively, use self-drilling screws.

Supporting construction
The supporting construction represents the permanent connection to the loadbearing structure. It must also compensate for any unevenness in the wall. Occasionally, the supporting construction also has to carry the thermal insulation. Timber facades with and without ventilated cavities are possible.

In a non-ventilated facade, the rear face of the timber elements should be coated and a rainproof material open to diffusion should be included.

A continuous ventilated cavity (20–40 mm) is advisable owing to the risk of saturation. Air inlets and outlets must be properly protected with screens to prevent insects gaining access and possibly causing damage to the organic material. With open (drained) joints in the facade it is necessary to incorporate rainproofing measures for the underlying construction. A structural analysis will be required for the supporting construction [2].

Surface finish
Untreated timber will gradually turn grey upon exposure to the elements and the effects of ultraviolet light. What happens here is that the lignin in the wood is degraded by photo-oxidation and washed out by the rain. The fibres in the outermost layers become detached and, depending on the species of wood, fungal growth discolours the wood.

Chemical treatment
Chemical wood preservatives can be used as a precaution against fungal growth and insect damage. The preservatives used fall into four categories: water-soluble (primarily inorganic salts), oil-based (e.g., coal tar oil), solvent-based and emulsion concentrates. Chemical wood preservatives usually contain active toxic substances in the form of biocides. The pas-

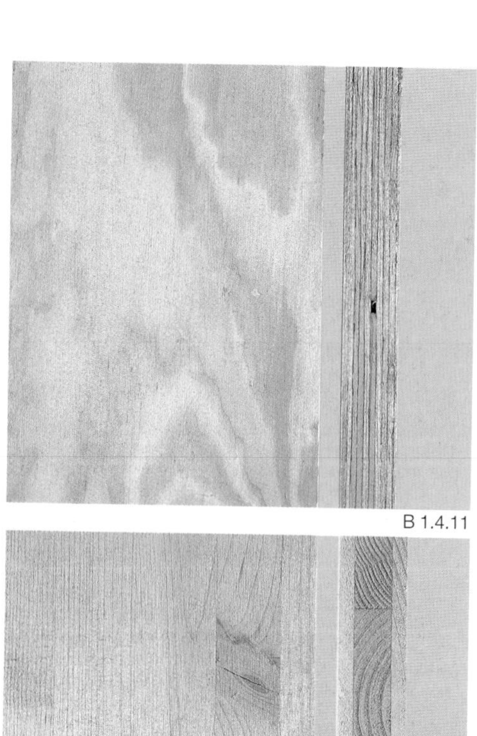
B 1.4.11

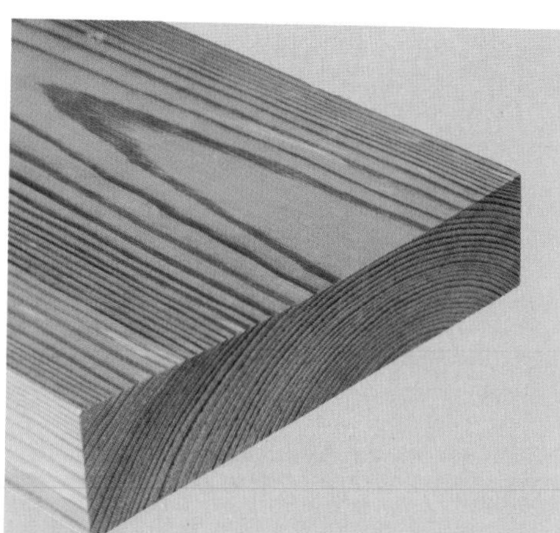
B 1.4.12

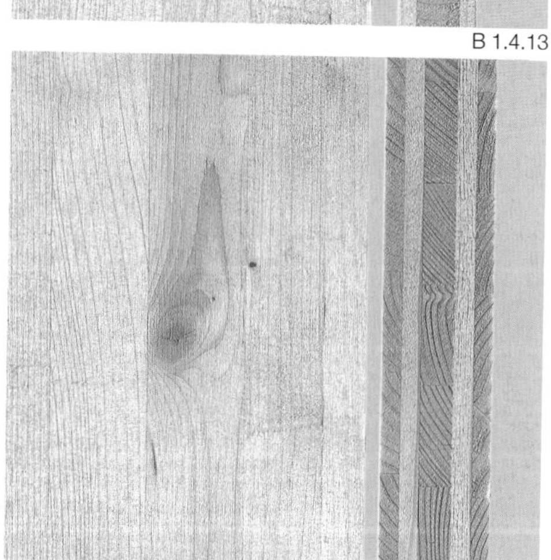
B 1.4.13

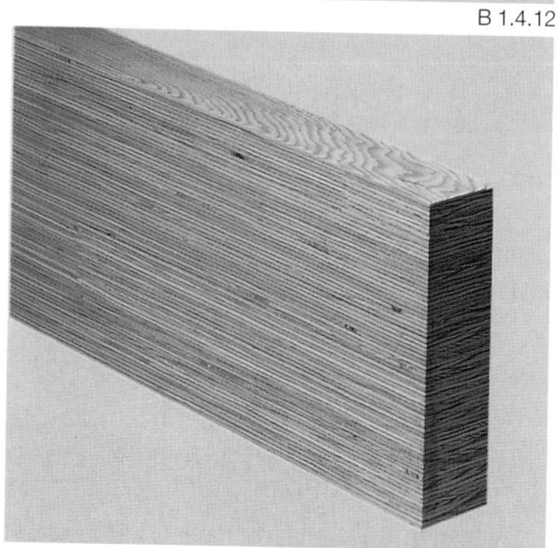
B 1.4.14

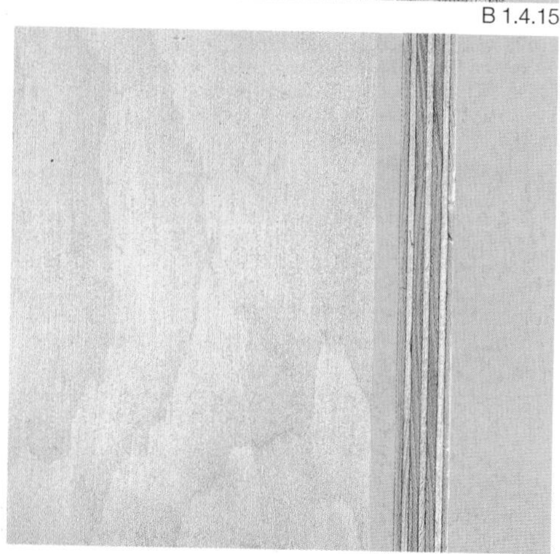
B 1.4.15

B 1.4.16

B 1.4.17

B 1.4.18

B 1.4.11 Laminated veneer lumber (LVL)
B 1.4.12 Board
B 1.4.13 3-ply core plywood
B 1.4.14 Structural veneer lumber (SVL)
B 1.4.15 5-ply core plywood
B 1.4.16 Profiled boards
B 1.4.17 Building-grade plywood
B 1.4.18 Extruded wood fibreboard sections

sive protection measures should be exhausted before the use of chemical preservatives is considered. Basically, such preservatives are only required when there is a risk of an infestation by wood-destroying insects. If it can be guaranteed that the moisture content of the timber will not exceed 20%, there is generally no risk of wood-destroying fungi gaining a hold. And if the moisture content is below 10%, attack by insects is unlikely. Timber constructions that remain exposed are easy to inspect and thus render chemical wood preservatives generally superfluous, apart from loadbearing components.

Biological treatment
Impregnation with water-soluble boron salt (borax mixtures, boric acid), waxes (hard wax, balsam, solutions), natural resin products (lacquers, oils, glazes), oils, wood vinegar, wood tar, pitch, preparations containing citrus oils or extracts from naturally resistant species of wood can also be used to treat timber. However, the problem is that there are currently no building authority approvals for biological wood preservative measures. The definite effectiveness, apart from boron compounds, has not been altogether proved. Longer drying times for such preservatives and re-treatment at intervals are necessary in some cases.

Surface treatments
a) Impregnation
This form of treatment creates a water-repellent surface and protects against insects and microorganisms by incorporating biocides. The substances used are porous, do not form a film and do not penetrate. It is possible to use pigments to indicate the presence of impregnation.
b) Glazes
These represent a compromise between impregnation and lacquer due to the checked penetration capacity and the formation of a relatively thin surface film. Again, the presence of this form of preservative can be indicated by the use of pigments. Protection against UV light can be adjusted with the density of the pigment. Good vapour diffusion capability.

B 1.4.19–22 Examples of timber panels
B 1.4.23–26 Examples of special forms of timber facade

B 1.4.19

B 1.4.20

c) Lacquers
These form a closed surface that repels water and resists abrasion. The vapour permeability can be very severely reduced so that the take-up and release of moisture between the wood and the air is prevented almost completely. We distinguish between colourless, glaze-like lacquers (which form a film, exhibit little penetration, have a gloss to semi-gloss and smooth surface finish; inadequate UV light protection, no fungicidal action) and opaque lacquers (which form a film, exhibit hardly any penetration, have a usually gloss, smooth surface finish; good UV light protection).
d) Dispersion paints
These form an opaque coating with water as the solvent. The pigmentation can be varied from a glaze to an opaque coating; forms a film, does not penetrate, noticeable water-swelling problems, which leads to the vapour diffusion being considerably impaired; matt surface finish, emphasises the texture of the substrate when applied thin; good UV light protection, seldom with fungicide.
e) Stains
These add colour to the wood through the application of pigments (pigment or colour stains) or through chemical processes (chemical stains). The grain of the wood remains visible and, depending on the method of application, can even be intensified. Stains do not provide any protection. Stained surfaces are therefore very sensitive to moisture and in the case of pigment stains also sensitive to light. In contrast to glazes and lacquers, stains can only be removed by planing or sanding off.
f) Waxes
Pores and small fissures are filled; the vapour diffusion capability remains at a high level. In comparison to lacquers and glazes, this finish is less scratch-resistant and less resistant to the effects of heat and water; as a rule, it is advisable to impregnate prior to waxing. Especially suitable for smooth, dry surfaces in areas protected from the weather.
g) Oils
These are the simplest, cheapest and, in ecological terms, the best way of treating the surface. However, their resistance is low (particularly against mechanical damage).

B 1.4.21

B 1.4.22

B 1.4.23

B 1.4.24

B 1.4.25

B 1.4.26

B 1.4.27

B 1.4.28

B 1.4.27–30 Examples of horizontal boarding
B 1.4.31–34 Examples of vertical boarding
B 1.4.35 Horizontal weatherboarding
B 1.4.36 Special retainers
B 1.4.37 Horizontal profiled boards and shakes/shingles
B 1.4.38 Vertical boarding options
B 1.4.39 Vertical profiled board options
B 1.4.40 Detail of joint between vertical boards
B 1.4.41 Detail of joint between large-format timber panels
B 1.4.42–45 Examples of archetypal forms and light/air-permeability
B 1.4.46–49 Examples of shakes and shingles

B 1.4.29

B 1.4.30

B 1.4.31

B 1.4.32

B 1.4.33

B 1.4.34

Boiled linseed oil and herbal varnishes, also wetting oils, are the main contenders. In comparison to wax, they provide better protection against wetting and soiling.

h) Coverings

In addition to being veneered, wood-based products can also be covered with various materials. We distinguish here between decorative and rolled coating materials but also foils and linoleum.

General remarks on surface finishes

External timber components exposed to direct solar radiation tend to require lighter and more intense to opaque pigmented surface finishes (e.g., glazes) in order to minimise the surface stresses due to thermal effects (swelling and shrinkage). The temperature of components painted black can reach about 70°C when exposed to strong sunlight, whereas those painted white reach only about 40°C.

Hardwoods contain less resin and are therefore more suitable for glazes than softwoods. No dark glazes should be applied to species of wood with a high resin content (especially pine and larch) when these are to be used in areas exposed to direct sunlight (bleeding of resin, appearance of blemishes). To improve the durability of surface finishes, all edges should be rounded. Interior surface finishes should be more vapour-tight than the exterior finishes (varnish glaze inside/thin-film glaze outside) in order to prevent the exterior finish flaking off due to water vapour diffusion [3].

Notes

[1] Volz, Michael: The Material; in: *Timber Construction Manual*, Munich/Basel, 2003, pp. 31–46.
[2] Scheibenreiter, Johann: Befestigung; in: *Holzfassaden*, pub. by Holzforschung Austria, Vienna, 2002, pp. 34–39.
[3] Herzog, Thomas; Volz, Michael: Protecting Wood; in: *Timber Construction Manual*, Munich/Basel, 1996, pp. 60–63.

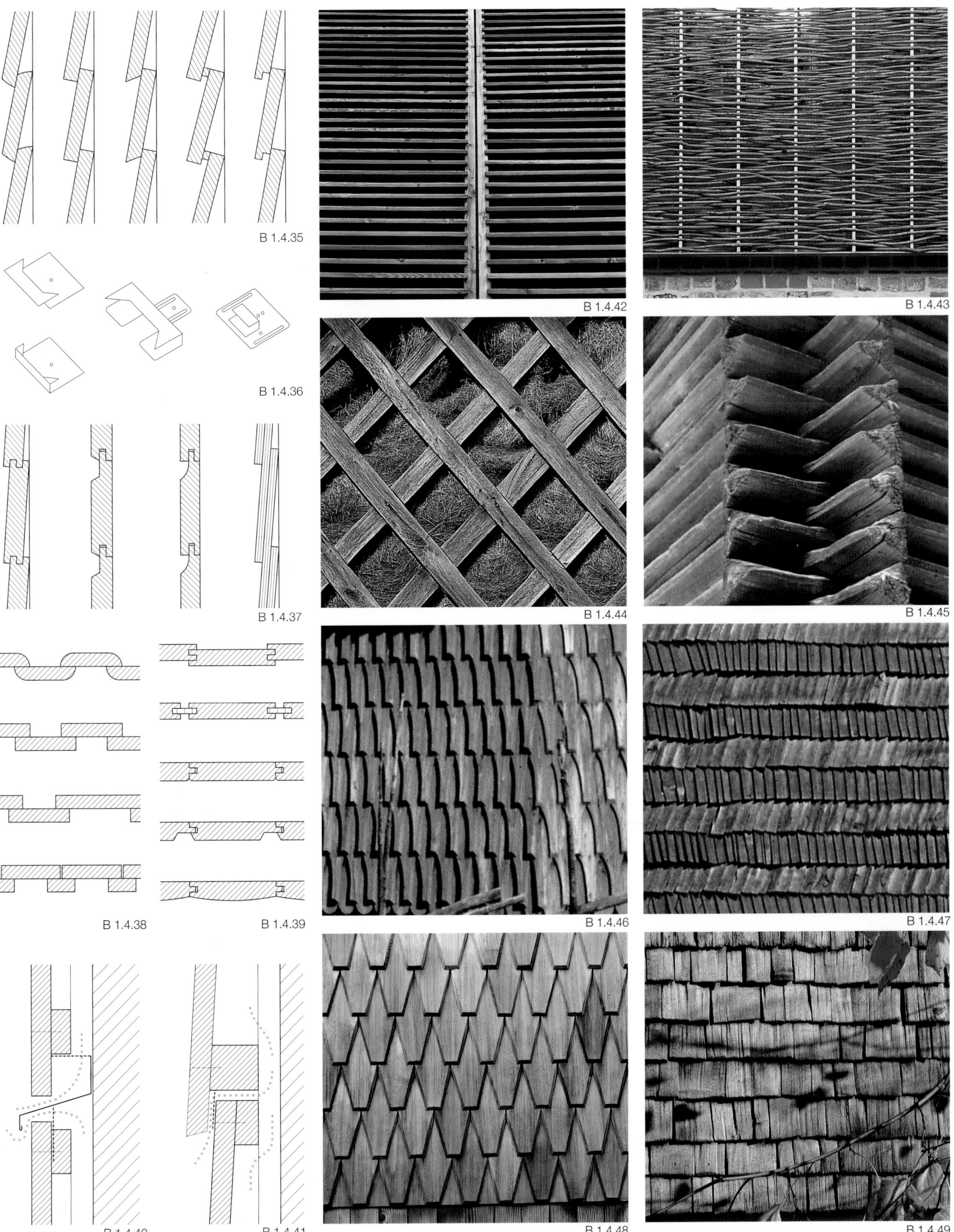

B 1.4.35

B 1.4.36

B 1.4.37

B 1.4.38

B 1.4.39

B 1.4.40

B 1.4.41

B 1.4.42

B 1.4.43

B 1.4.44

B 1.4.45

B 1.4.46

B 1.4.47

B 1.4.48

B 1.4.49

Sea Ranch

California, USA, 1965

Architects:
Moore Lyndon Turnbull Whitaker, Berkeley
Structural engineers:
Davis & Morreau, Albany

A+U 09/1989
DBZ 02/1994
Marrey, Bernard: Des Histoires de Bois. Paris 1994
MLTW/Moore Lyndon Turnbull and Whitaker: Sea Ranch. GA No. 3, Tokyo, 1981

- Simple, robust timber-frame construction with columns of rough-sawn fir
- Profiled redwood boards
- Maintenance-free because impregnation was unnecessary
- Roof overhang inadvisable owing to constant strong winds
- All beams connected through side or placed on top of support
- Bracing diagonals for wind and seismic loads made from 4" x 4" (102 x 102 mm) squared timber sections connected with quarter-circle metal plates left exposed
- Intensive weather-induced colour changes
- No requirements regarding winter thermal performance

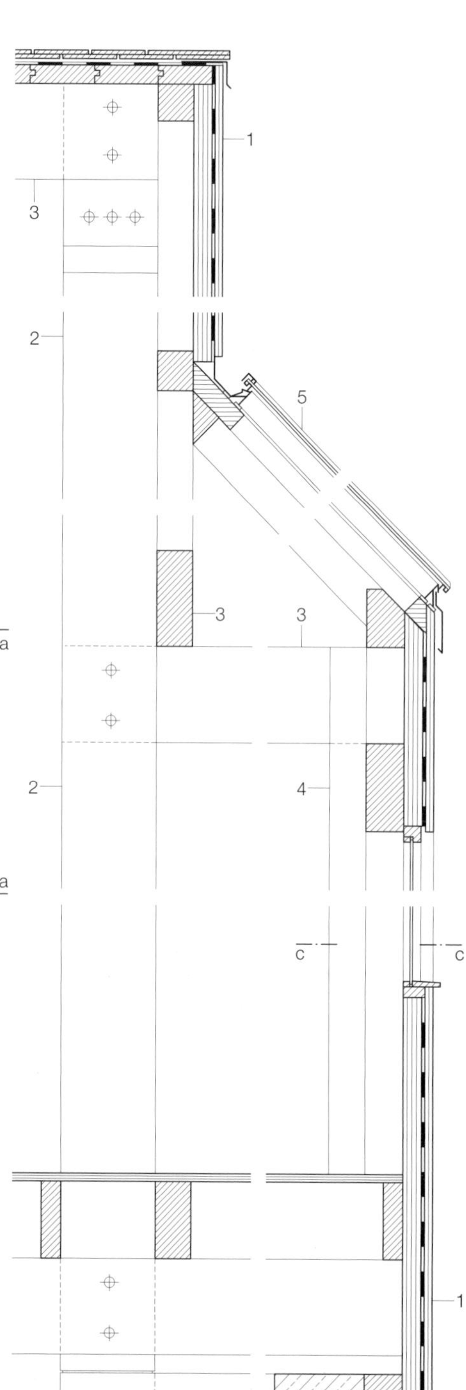

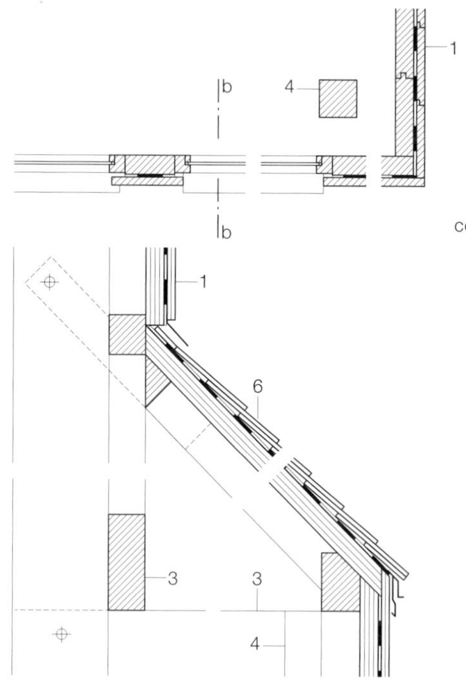

Plan • Section
scale 1:500
Vertical section • Horizontal section
scale 1:20

1. External wall: 1" x 8" (25 x 203 mm) vertical redwood boards, with rebate waterproofing
   vertical rough-sawn fir boards, 2" x 8" (51 x 203 mm), with tongue and groove, opaque paint finish internally in some areas
2. Column, 10" x 10" (254 x 254 mm)
3. Beam, 4" x 10" (102 x 254 mm)
4. Additional posts adjacent to windows, 4" x 4" (102 x 102 mm)
5. Rooflight, aluminium frame
6. Roof covering: 1" x 8" (25 x 203 mm) redwood boards waterproofing, rough-sawn fir boards, 2" x 8" (51 x 203 mm), with tongue and groove

Residential and artist studios complex

Paris, F, 1983

Architect:
Roland Schweitzer, Paris
Assistant:
Alexandre Levandowsky, Paris

AC 110, 1984
Herzog, Thomas et al.:
Timber Construction Manual,
Munich/Basel, 2003

- Vertical boards
- Glaze finish to external boards
- 600 mm grid, prefabricated wall elements
- Very low building costs

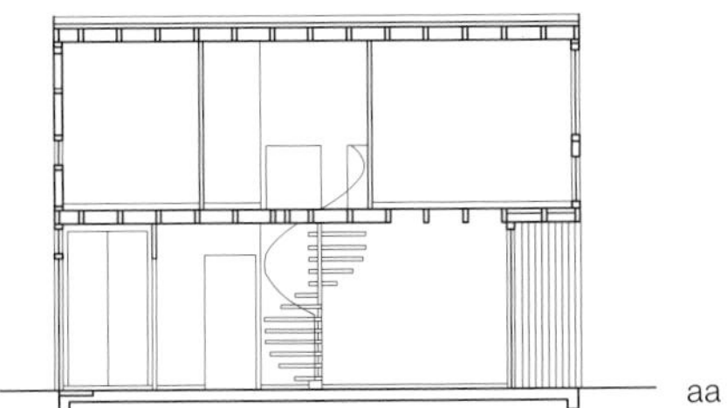

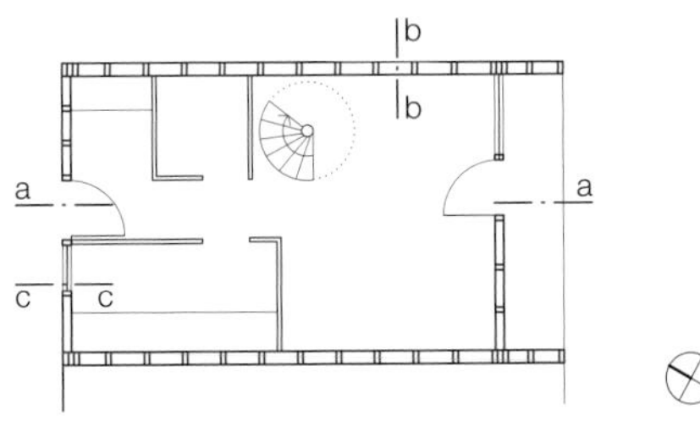

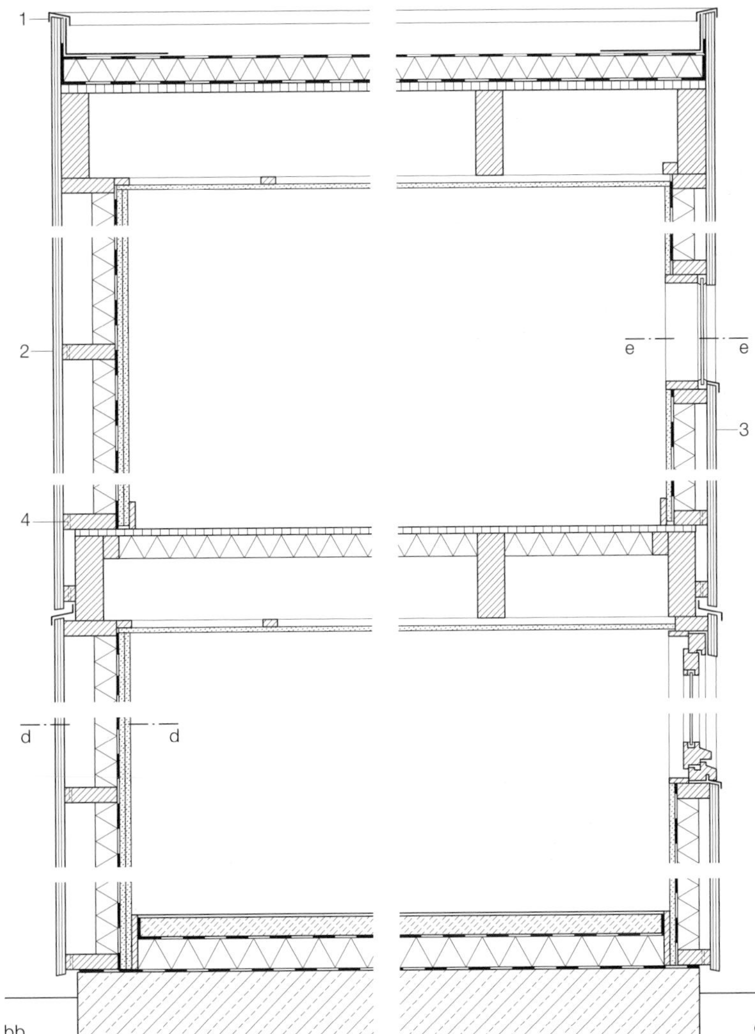

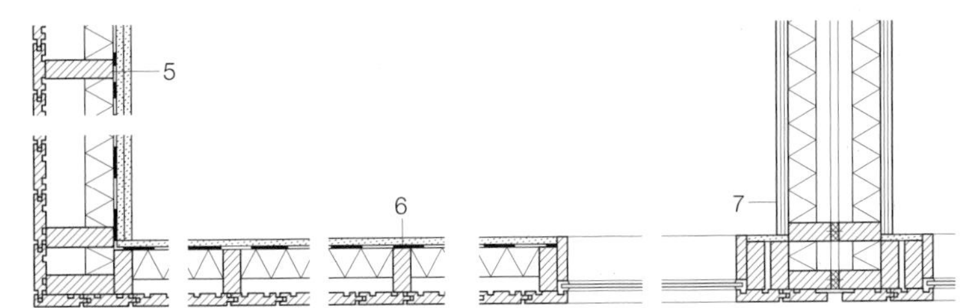

Plan of ground floor • Section
scale 1:200
Horizontal sections • Vertical sections
scale 1:20

1 Sheet metal capping plate, 75 x 100 mm, bent to suit, painted black
2 100 x 25 mm vertical tongue and groove boards
38 x 142 mm rails, with
82 mm air cavity in between
60 mm thermal insulation
vapour barrier
2 No. 15 mm plasterboard
3 100 x 25 mm vertical tongue and groove boards
38 x 90 mm rails, with
30 mm air cavity in between
60 mm thermal insulation
13 mm plasterboard
4 10 mm dia. holes at 150 mm centres for equalising vapour pressure and for drainage
5 Post, 38 x 142 mm
6 Post, 38 x 90 mm
7 Party wall

Criminal courts complex

Bordeaux, F, 1998

Architects:
Richard Rogers Partnership, London
Structural engineers:
OTH Sud-Ouest, Bordeaux

architecture 01/1999
Bauwelt 27/1998
Lemoine, Bertrand: Frankreich 20. Jahrhundert. Basel/Berlin/Boston 2000

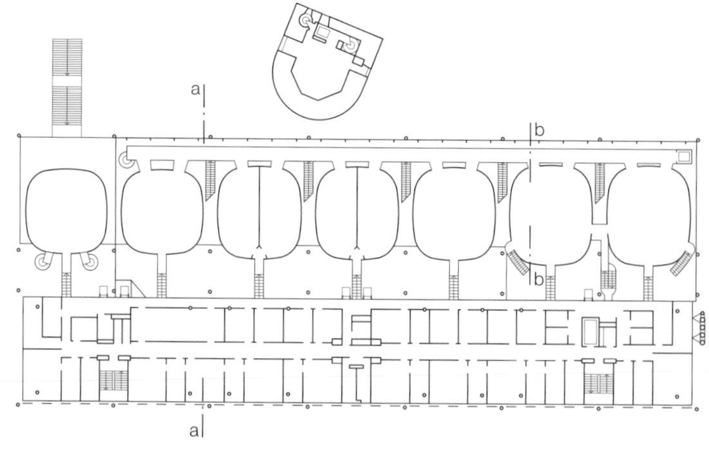

- Seven courtrooms positioned within an open single-storey shed
- Air cooled by means of an external water cascade and water basin; water is pumped into the building and, as it heats up, it flows upwards due to the shape of the building envelope
- "Bottle" shape allows a relatively large amount of daylight to enter through an opening at the top of the building envelope
- Diagonal cedar boarding
- Acoustic panels in the courtrooms cut down the amount of noise entering from outside and limit internal reverberation

Plan • Section scale 1:1000
Vertical section • Horizontal section scale 1:20

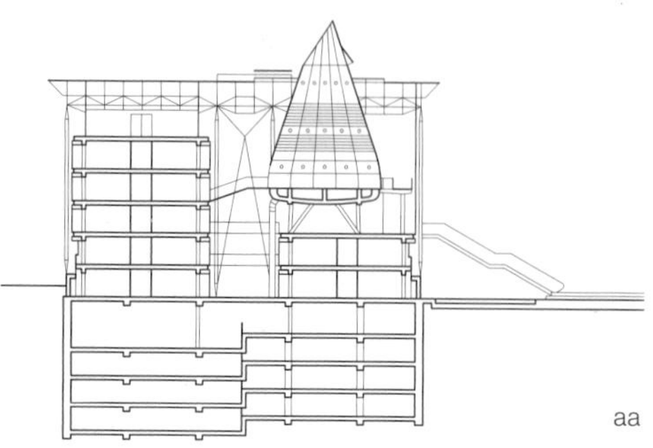

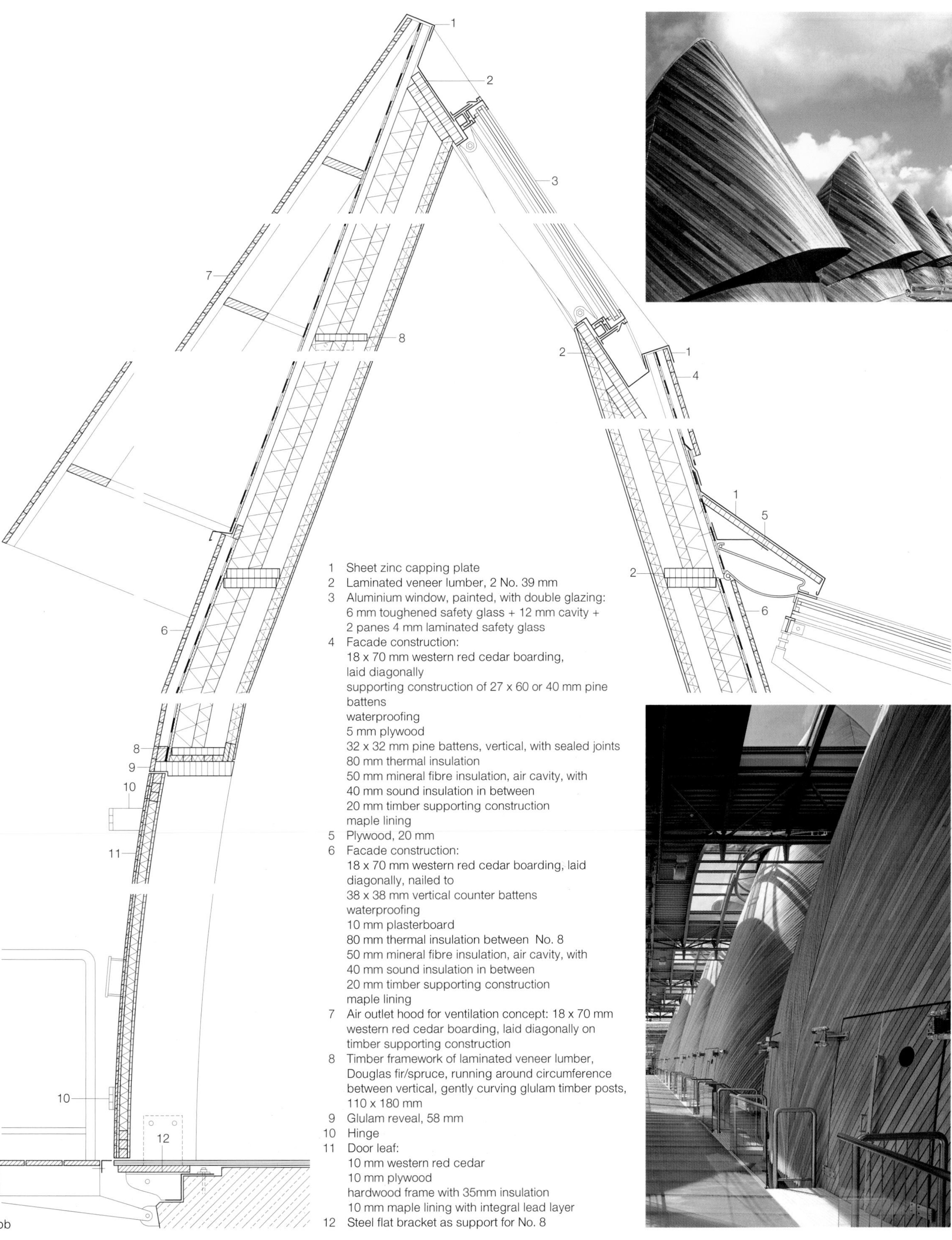

1 Sheet zinc capping plate
2 Laminated veneer lumber, 2 No. 39 mm
3 Aluminium window, painted, with double glazing:
6 mm toughened safety glass + 12 mm cavity +
2 panes 4 mm laminated safety glass
4 Facade construction:
18 x 70 mm western red cedar boarding,
laid diagonally
supporting construction of 27 x 60 or 40 mm pine battens
waterproofing
5 mm plywood
32 x 32 mm pine battens, vertical, with sealed joints
80 mm thermal insulation
50 mm mineral fibre insulation, air cavity, with
40 mm sound insulation in between
20 mm timber supporting construction
maple lining
5 Plywood, 20 mm
6 Facade construction:
18 x 70 mm western red cedar boarding, laid diagonally, nailed to
38 x 38 mm vertical counter battens
waterproofing
10 mm plasterboard
80 mm thermal insulation between No. 8
50 mm mineral fibre insulation, air cavity, with
40 mm sound insulation in between
20 mm timber supporting construction
maple lining
7 Air outlet hood for ventilation concept: 18 x 70 mm western red cedar boarding, laid diagonally on timber supporting construction
8 Timber framework of laminated veneer lumber, Douglas fir/spruce, running around circumference between vertical, gently curving glulam timber posts, 110 x 180 mm
9 Glulam reveal, 58 mm
10 Hinge
11 Door leaf:
10 mm western red cedar
10 mm plywood
hardwood frame with 35mm insulation
10 mm maple lining with integral lead layer
12 Steel flat bracket as support for No. 8

Olympic Art Museum

Lillehammer, N, 1993

Architects:
Snøhetta, Oslo

Architectural Review 04/1993
Byggekunst 04/1993
Techniques + architecture 408, 1993

- Extension to existing art gallery dating from the 1960s
- Design of external cladding reminiscent of boatbuilding

Plan • Section scale 1:1000
Horizontal section • Vertical section scale 1:20

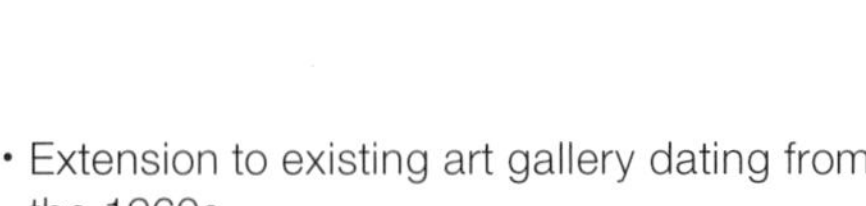

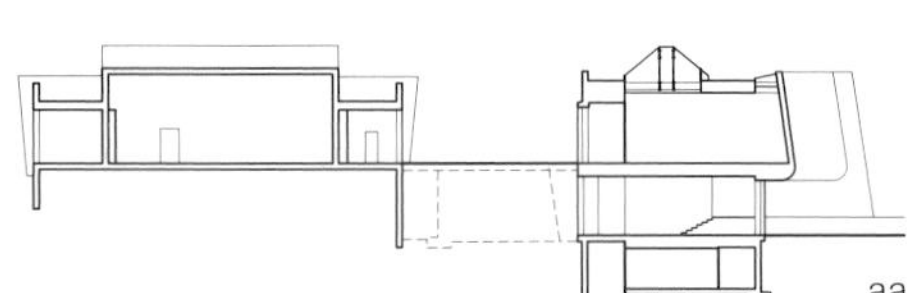

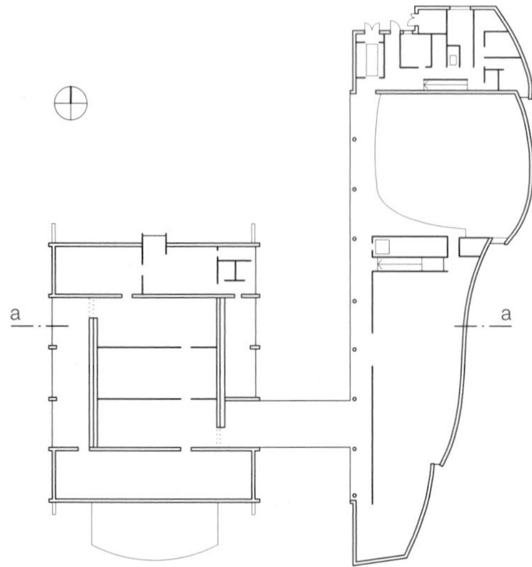

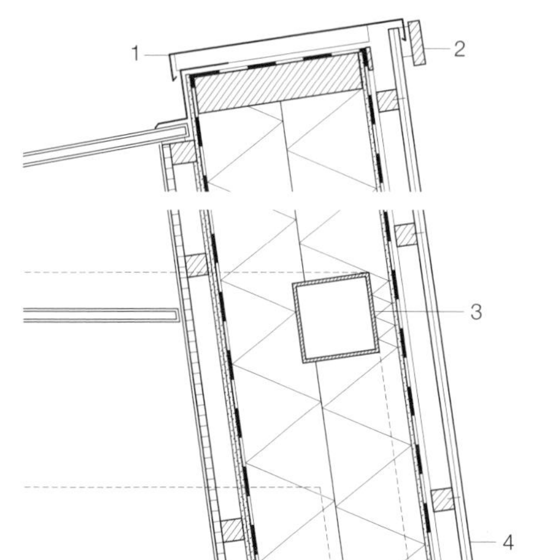

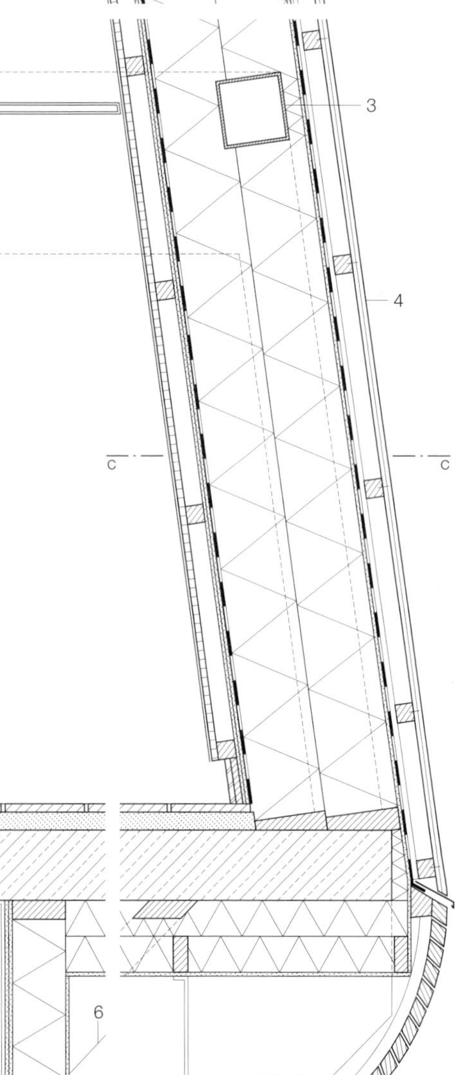

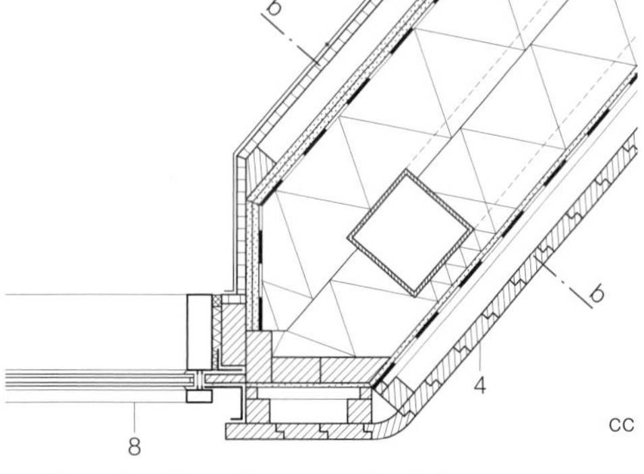

1 Sheet aluminium-zinc capping plate
2 Peripheral larch batten, 23 x 98 mm
3 Steel square hollow section, 180 x 180 mm
4 Facade construction: vertical timber boards with rebate, larch, 28 x 75 mm
48 mm battens
23 mm counter battens
waterproofing
9 mm plasterboard
2 No. 198 mm thermal insulation
vapour barrier
2 No. 12.5 mm plasterboard
48 mm battens
18 mm wood-based board with textile covering, white
5 Horizontal timber boards with rebate, larch, 40 x 40 mm
6 Timber bracing, 98 x 48 mm
7 timber frame with 148 mm thermal insulation 2 No. 12.5 mm plasterboard
8 Aluminium window with double glazing

Café

Helsinki, FIN, 2000

Architect:
Niko Sirola, Woodstudio 2000,
Helsinki University of Technology
Structural engineers:
Nuvo, Espoo

Architectural Review 12/2000
Detail 05/2002
Herzog, Thomas et al.:
Timber Construction Manual,
Munich/Basel, 2003

- Glued laminated timber, spruce
- 620 mm wide panels assembled on site
- Flame-treated surface finish, impregnated with creosote
- Adequate rainproofing thanks to treatment twice a year

Plan • Section scale 1:500
Vertical sections • Horizontal section scale 1:20

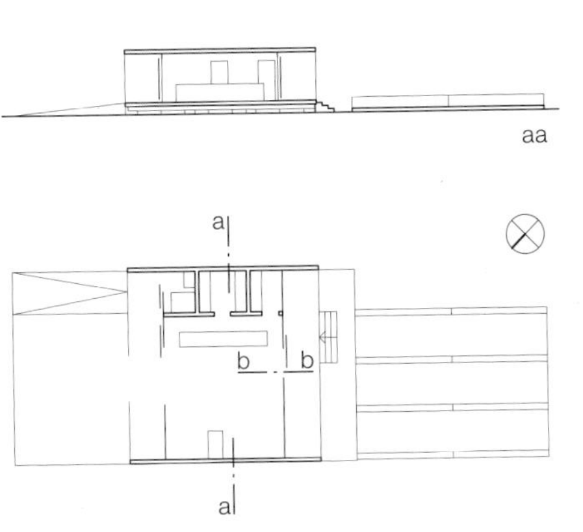

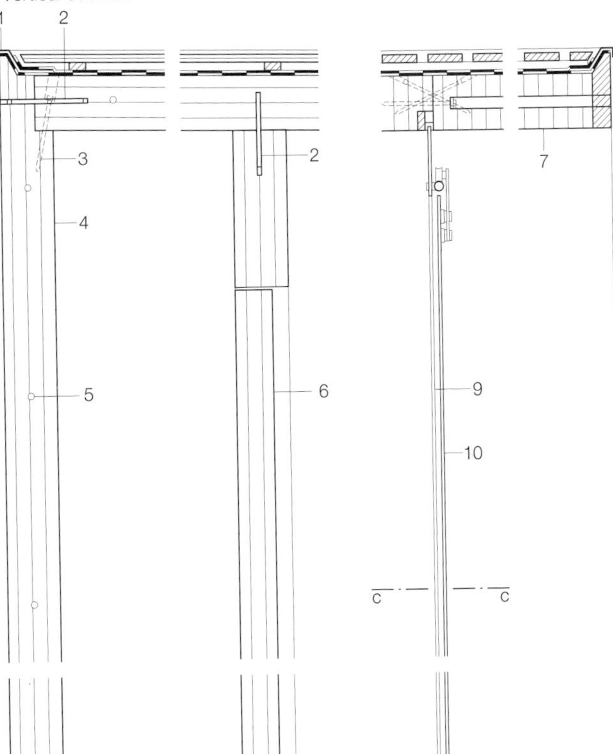

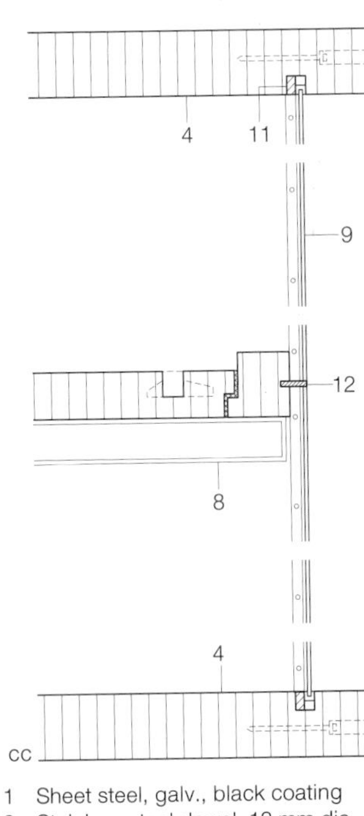

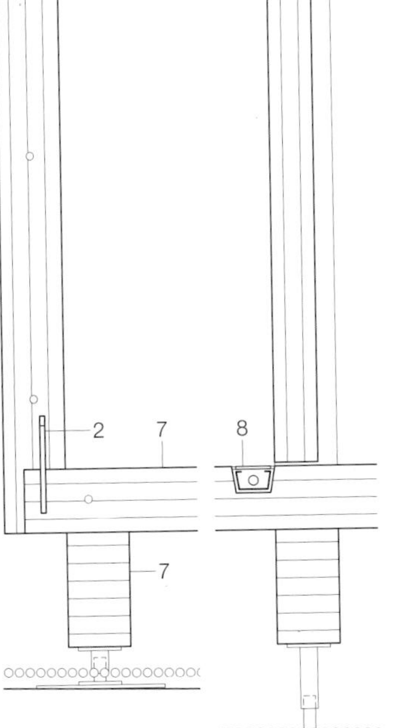

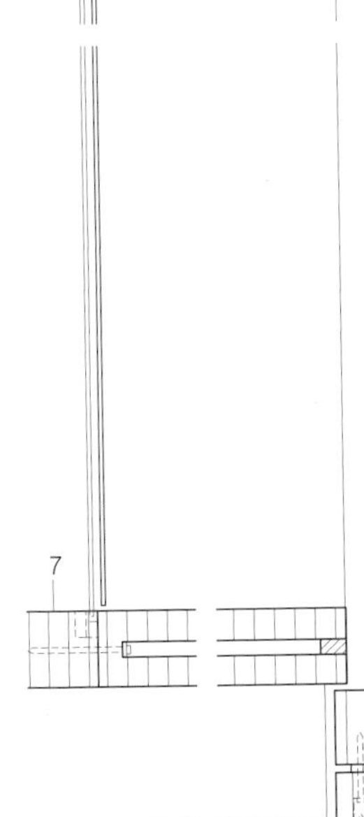

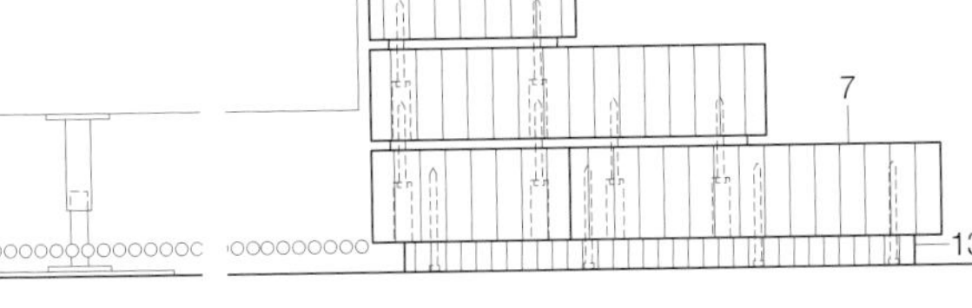

1 Sheet steel, galv., black coating
2 Stainless steel dowel, 12 mm dia.
3 Galvanised steel pin, 10 mm dia.
4 Glulam element, spruce, burned off externally, impregnated with creosote, sanded internally, 145 mm
5 Bolt, 10 mm dia.
6 Door leaf, 100 mm glulam element
7 Glulam element, spruce, glued, 145 mm
8 Lighting unit recessed in floor
9 Fixed light, 10 mm toughened safety glass
10 Sliding door, 10 mm toughened safety glass
11 Timber packing, 25 x 35 mm, planed
12 Steel flat, 10 x 50 mm, painted black
13 Waterproof plywood, 16 mm

Forestry depot

Turbenthal, CH, 1993

Architects:
Burkhalter Sumi, Zürich

DBZ 07/1996
Detail 03/1995
gta (ed.): Marianne Burkhalter, Christian Sumi. Die Holzbauten, Zürich, 1996
Herzog, Thomas et al.: Timber Construction Manual, Munich/Basel, 2003

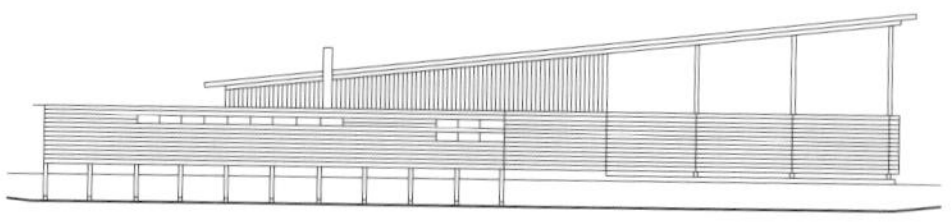

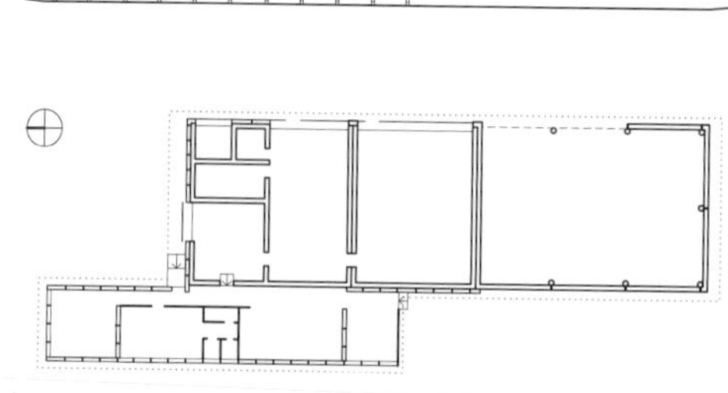

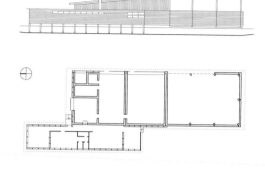

- Prototype for forestry depots, using a modular system consisting of three parts with offices, garage and open shed
- High degree of prefabrication
- Garage in concrete for fire resistance purposes

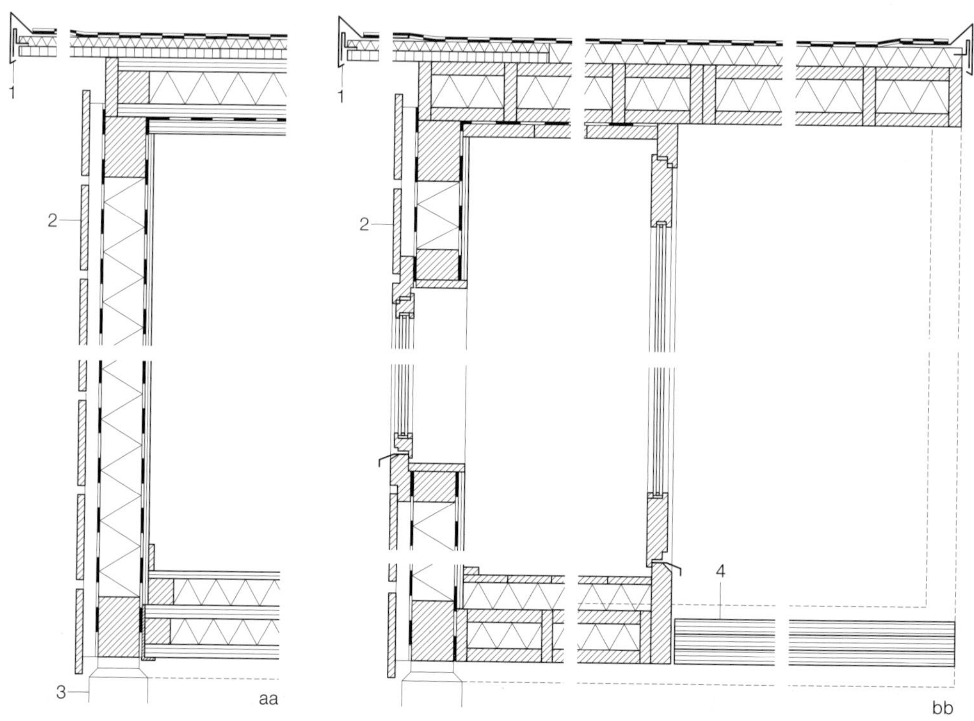

Elevation • Plan scale 1:750
Vertical sections • Horizontal section
scale 1:20

1 Sheet metal capping, bent to suit
2 External wall to offices:
 21 x 230 mm horizontal timber boards
 40 x 80 mm battens
 airtight barrier, vapour-permeable paper
 120 mm thermal insulation between timber studs
 vapour barrier
 19 mm pine board
3 Log, 380–300 mm dia.
4 Solid larch section, 120 mm
5 External wall to garage:
 21 x 230 mm vertical timber boards
 40 x 80 mm battens
 airtight barrier, vapour-permeable paper
 80 mm thermal insulation (where necessary)
 200 mm concrete wall

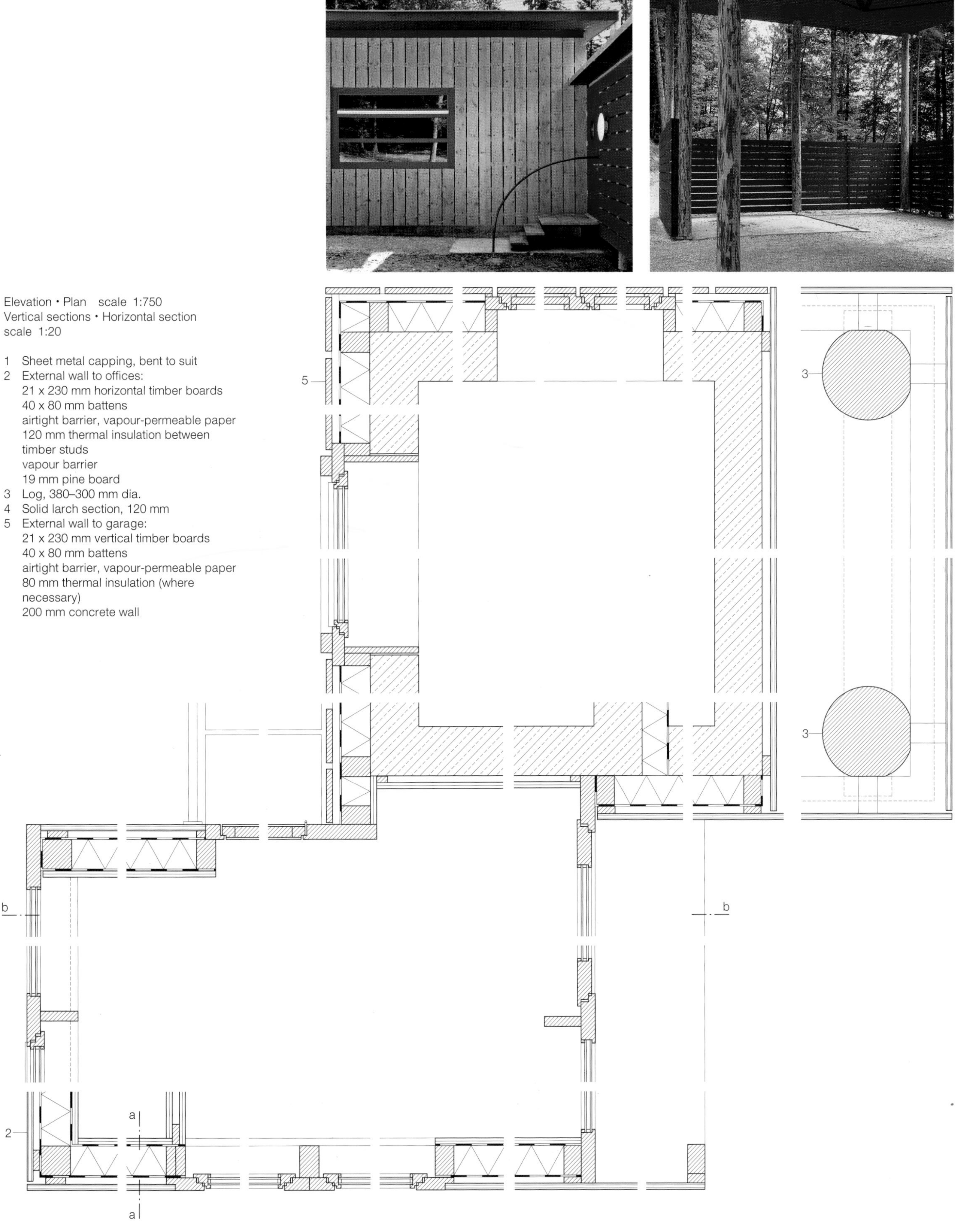

Housing project

Regensburg, D, 1996

Architects:
Fink + Jocher, Munich

A+U 04/1997
Bauwelt 25/1997
DBZ 03/1999
Detail 01/1997
Pfeifer, Günter et al.: Der neue Holzbau. Aktuelle Architektur – Alle Holzbausysteme – Neue Technologien, Munich, 1998

- Horizontal timber battens, larch
- Superstructure completed in four months
- Part of a pilot project of the Bavarian Building Authority for developing low-cost housing which can be built in large numbers with a high degree of prefabrication

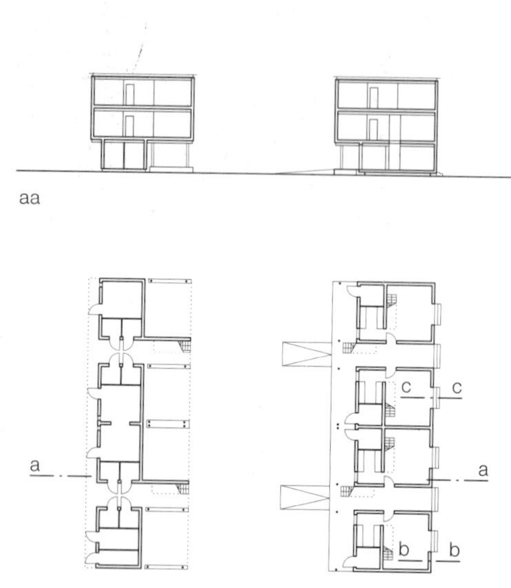

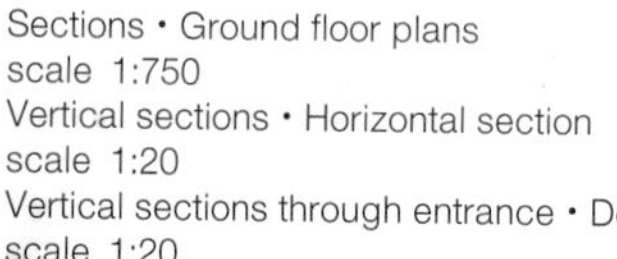
Sections • Ground floor plans
scale 1:750
Vertical sections • Horizontal section
scale 1:20
Vertical sections through entrance • Door
scale 1:20

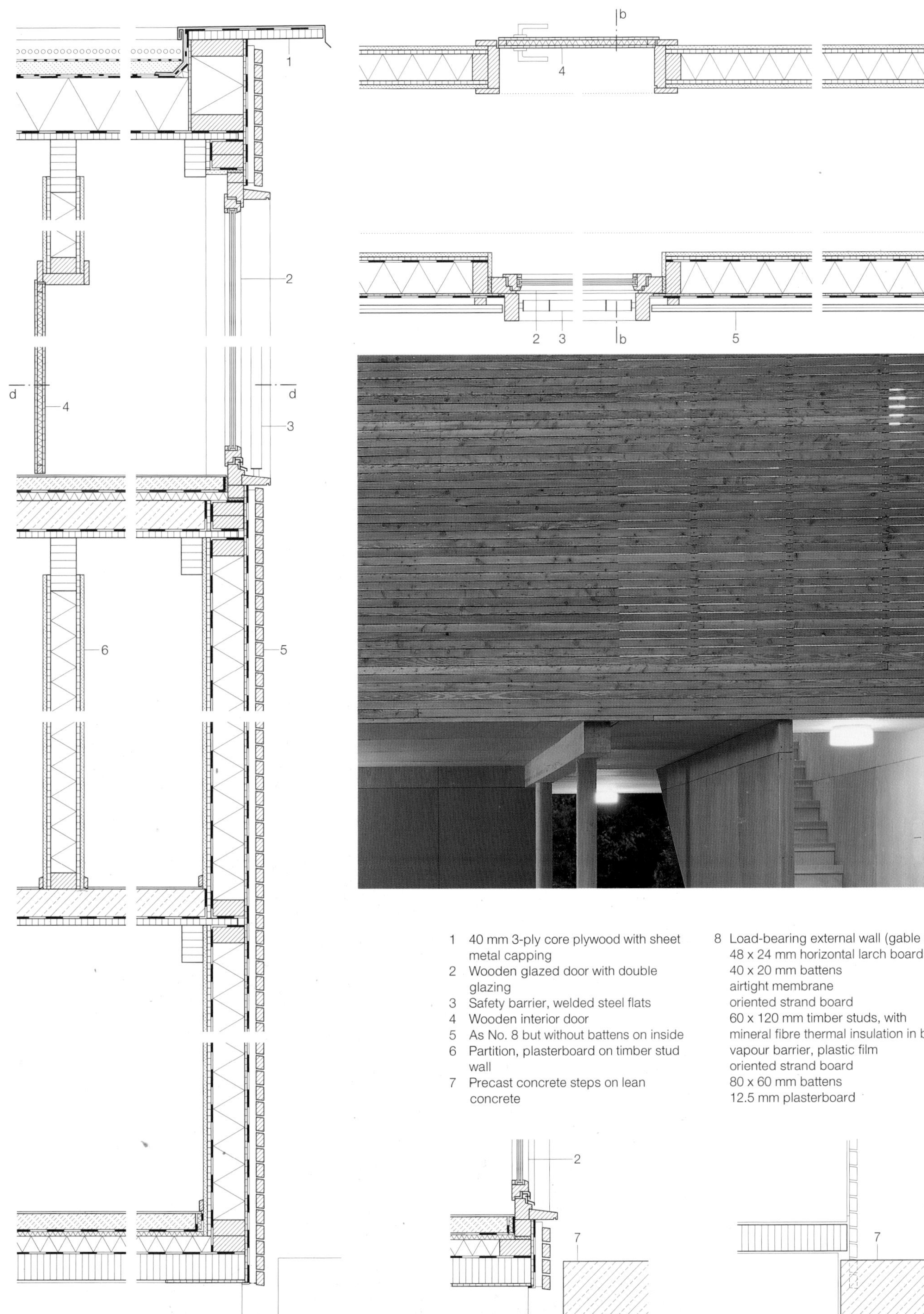

1 40 mm 3-ply core plywood with sheet metal capping
2 Wooden glazed door with double glazing
3 Safety barrier, welded steel flats
4 Wooden interior door
5 As No. 8 but without battens on inside
6 Partition, plasterboard on timber stud wall
7 Precast concrete steps on lean concrete
8 Load-bearing external wall (gable sides):
48 x 24 mm horizontal larch boards
40 x 20 mm battens
airtight membrane
oriented strand board
60 x 120 mm timber studs, with mineral fibre thermal insulation in between
vapour barrier, plastic film
oriented strand board
80 x 60 mm battens
12.5 mm plasterboard

House and studio

Tsukuba, J, 1995

Architects:
Naito Architect & Associates, Tokyo

l'ARCA 12/1995
Bauwelt 38/1997
Detail 04/1996
The Japan Architect 46/2002

- Double-leaf wall construction:
  outer leaf: cedar planks, gaps closed with acrylic glass panels
  inner leaf: cedar planks with cover strips
- Sliding timber elements

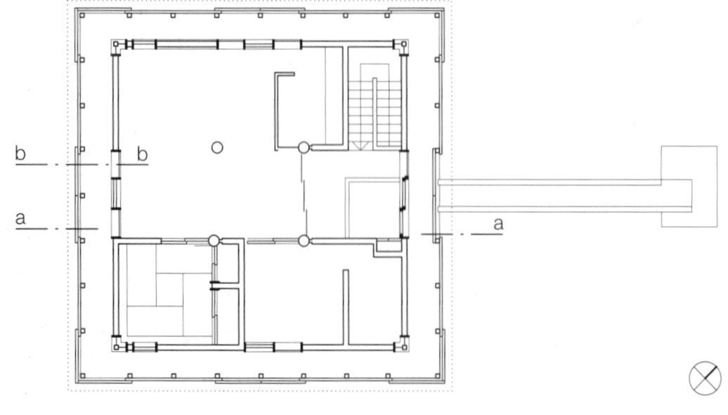

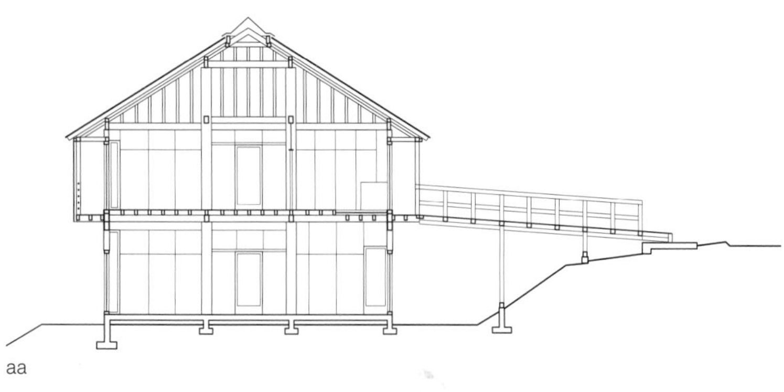

aa

1 Sliding element to close off balcony, Japanese cedar planks, 12 x 150 mm
2 Wooden window with fixed glazing
3 Guide track for sliding door, bent steel flat, 6 mm
4 Facade construction: Japanese cedar planks, 12 x 150 mm, with cover strips, 12 x 10 mm airtight membrane timber wall, with 105 mm thermal insulation in between 6 mm plywood
5 Column, Japanese pine, 105 x 105 mm
6 Stainless steel cable, 3 mm dia.
7 Sliding glazed element to inside of balcony
8 Japanes cedar, 12 mm
9 Gap, 10 mm, closed with acrylic glass panel, 2 mm

Plan of upper floor • Section
scale 1:250
Vertical section • Horizontal section
scale 1:20
Detail of sliding door element
scale 1:5

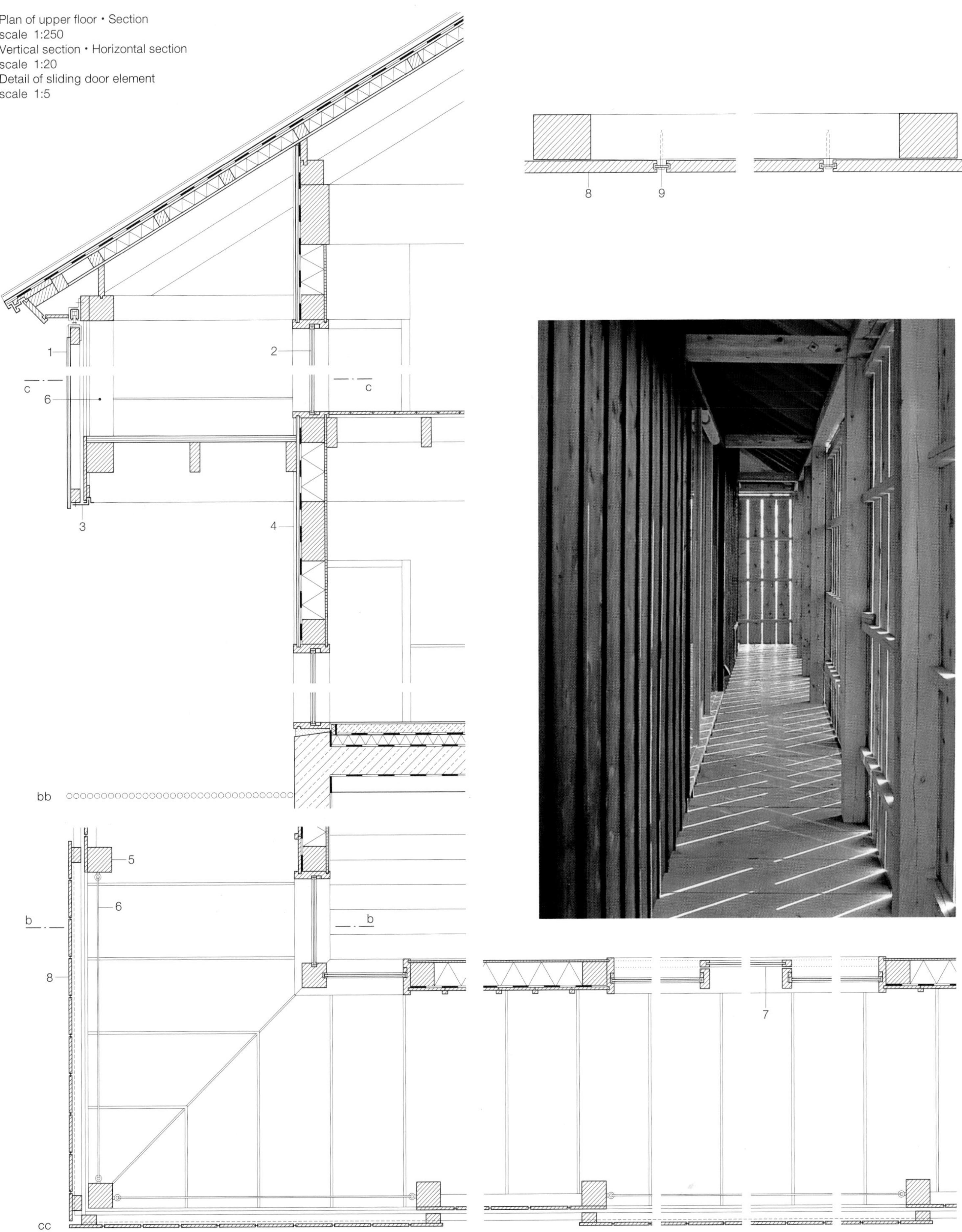

Multi-storey car park

Heilbronn, D, 1999

Architects:
Mahler Günster Fuchs, Stuttgart

A+U 03/2001
Bauwelt 06–07/2000
Casabella 691, 2001

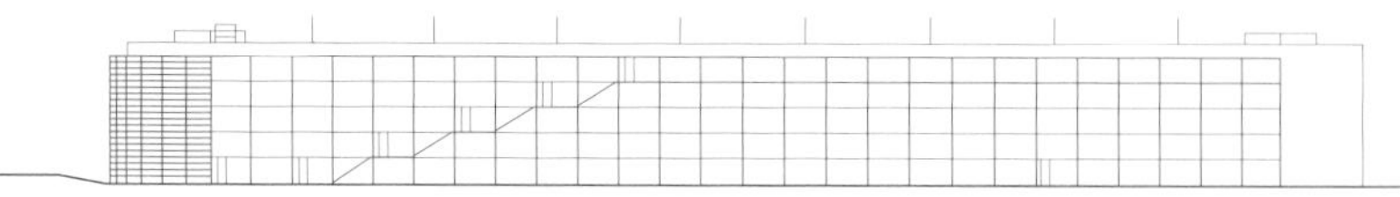

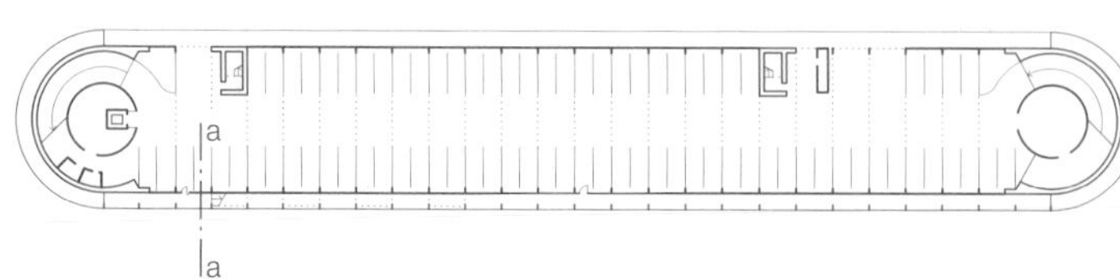

- Squared timber sections, 40 x 60 mm, 15 m long
- Non-insulated building envelope
- Timber facade modules
- Details exposed both internally and externally
- All fixings to squared timber sections are concealed

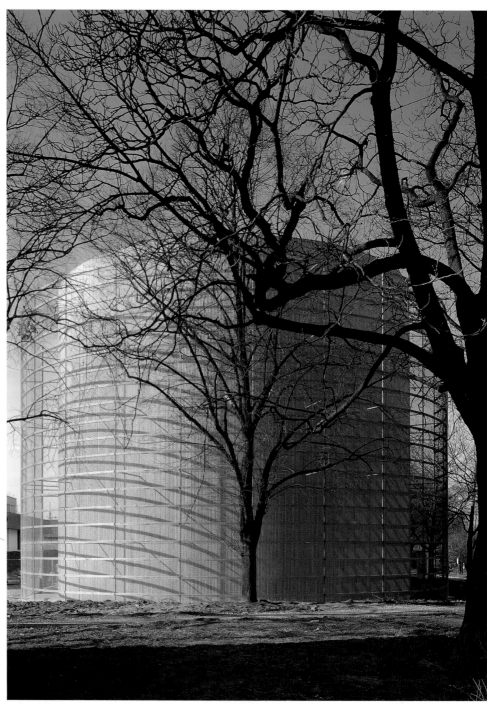

Elevation • Plan of ground floor
scale 1:1500
Horizontal section • Vertical section
scale 1:20

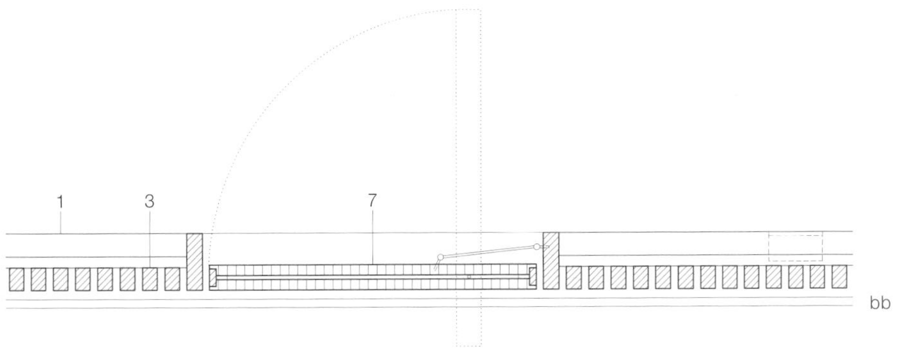

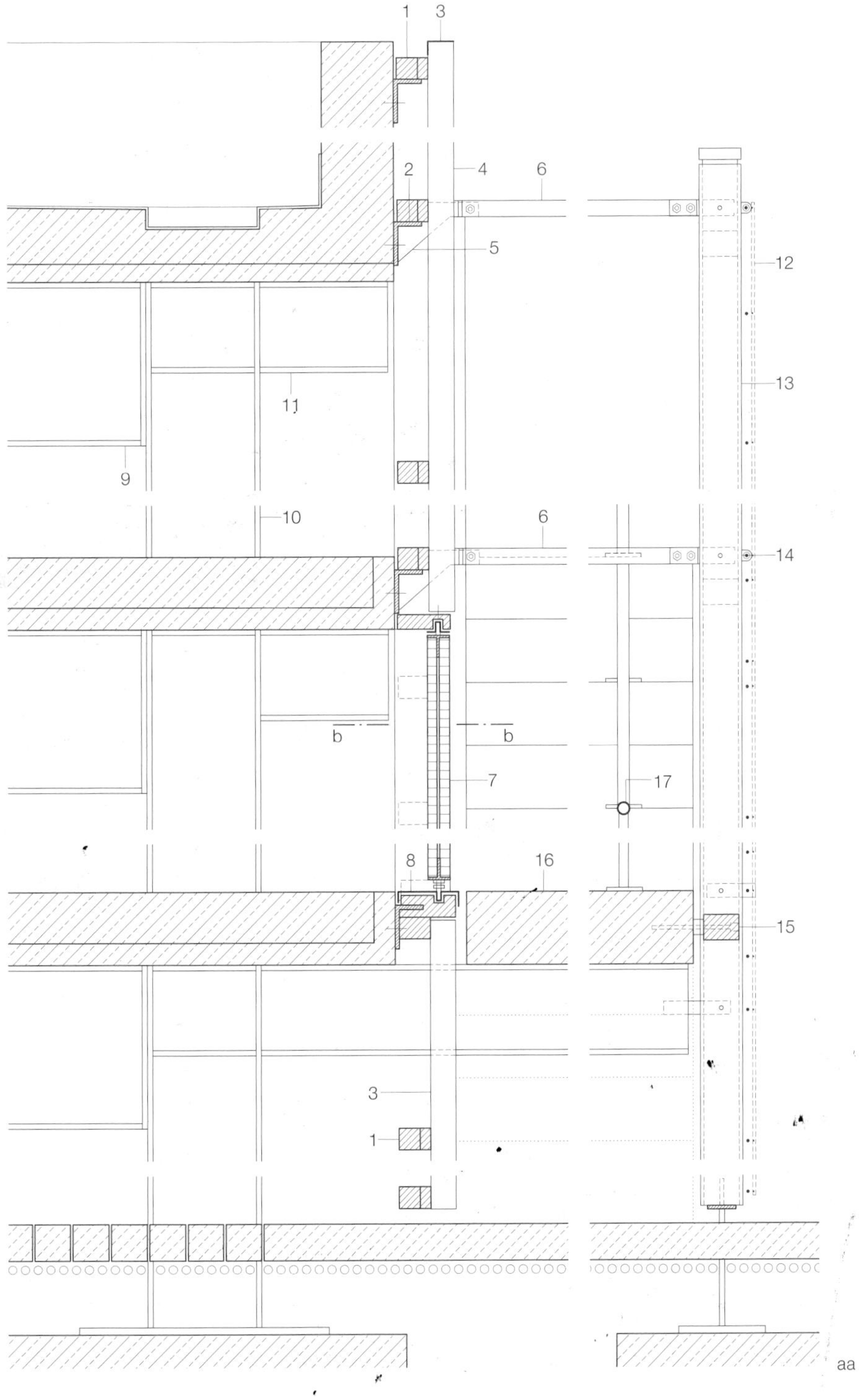

1 Battens, squared timber sections, Douglas fir, 60 x 60 mm and 30 x 60 mm
2 Fixing for timber facade, steel angle, 120 x 80 x 12 mm
3 Galvanised steel angle, 70 x 70 mm
4 Timber facade members, untreated Douglas fir, 40 x 60 mm
5 Steel flat bracket
6 Strut, galv. steel circular hollow section, 44 mm dia.
7 Door fitted on pivot pin, leaf: 2 No. 28 mm 3-ply core plywood, veneered
8 Bent non-slip sheet metal threshold
9 Steel beam, HEB 450
10 Steel col., HEB 320
11 Steel bracket, HEA 260
12 Wire mesh, galv., 40 x 40 x 3.1 mm
13 Round glulam facade post, Douglas fir, 120 mm dia.
14 Stainless steel pre-stressing cable, 10 mm dia.
15 Douglas fir section, 70 x 100 mm, steel anchors concealed behind timber plugs, steel tube spacer, 40 mm dia.
16 Precast concr. stairs
17 Handrail, galvanised steel tube, 22 mm dia.

Multi-storey apartment block

Innsbruck, A, 1996

Architects:
Kathan Schranz Strolz, Innsbruck

Architecture today 05/1998
AV Monografías/Monographs 67, 1997
Bauwelt 15/1997

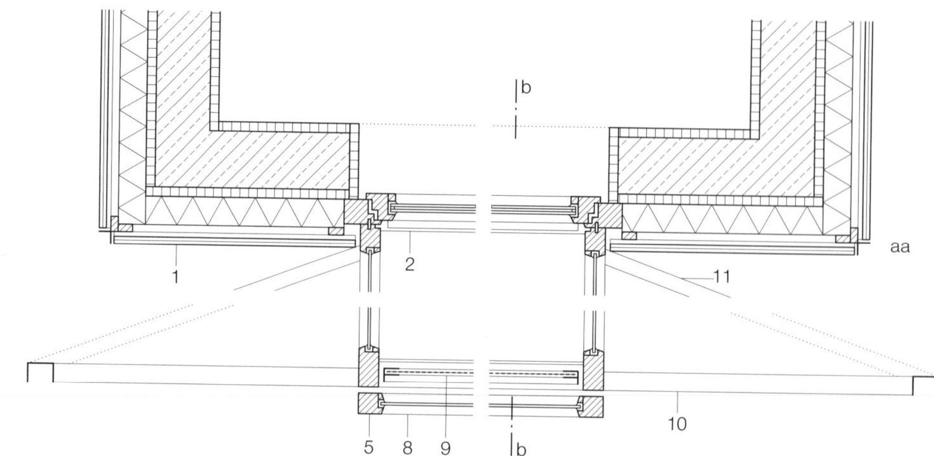

- Oak weatherboarding
- Weatherboarding plus battens, insulation and windows used as permanent formwork for concrete containing recycled aggregate
- High degree of prefabrication

Section scale 1:500
Horizontal section • Vertical section
scale 1:20

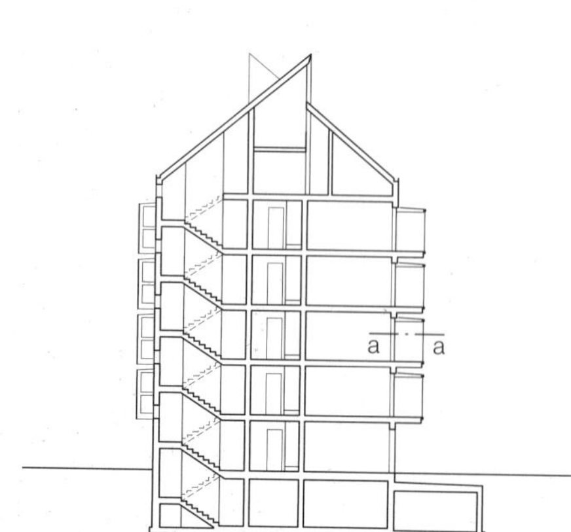

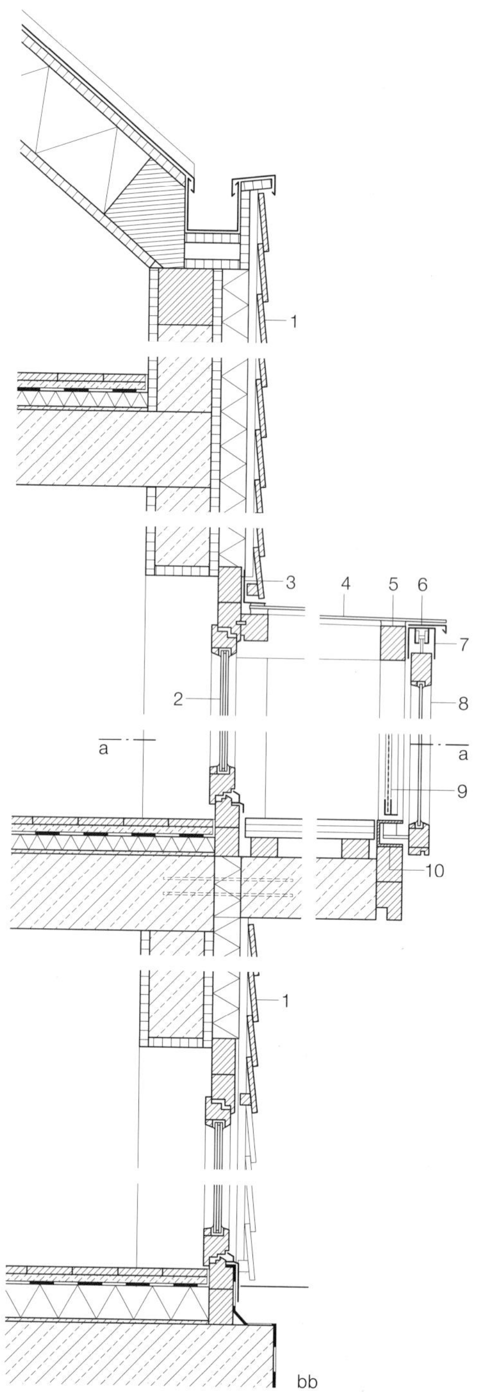

1 Facade construction:
15 x 150 mm oak weatherboarding, 17 mm overlap
20 x 40 mm vertical battens
40 x 80 mm horizontal counter battens, with 80 mm rockwool insulation in between
25 mm chipboard
150 mm reinforced concrete
25 mm chipboard with skim plaster coat ready for painting
2 Wooden window with double glazing
3 Sheet zinc glued in place
4 Single glazing, 2 No. 6 mm laminated safety glass glued to wooden frame, 30 mm fall
5 Timber frame construction, 68 x 90 mm
6 Sheet zinc with drip
7 Galvanised guide track, 80 x 80 x 4 mm channel, screwed to frame, extruded section with EPDM seal
8 Sliding wooden window, 61 x 90 mm, with single glazing, 6 mm toughened safety glass
9 Expanded metal in steel angle frame, 45 x 45 x 5 mm
10 Guide track, 70 x 50 x 4 mm channel
11 Steel circular hollow section, 38 mm dia.

House for school caretaker

Triesenberg, FL, 1994

Architect:
Hubert Ospelt, Vaduz
Assistant:
Marcus Freund

Herzog, Thomas et al.:
Timber Construction Manual,
Munich/Basel, 2003

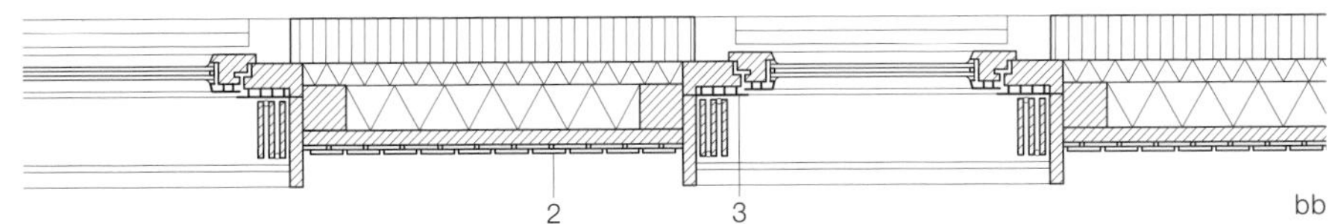

- Four-storey timber construction
- Fine shingle texture on clear stereometric building envelope
- Floors, walls and roof of edge-nailed timber elements
- Glued laminated timber beams (beech) within depth of floor for transferring floor loads to columns without further assistance

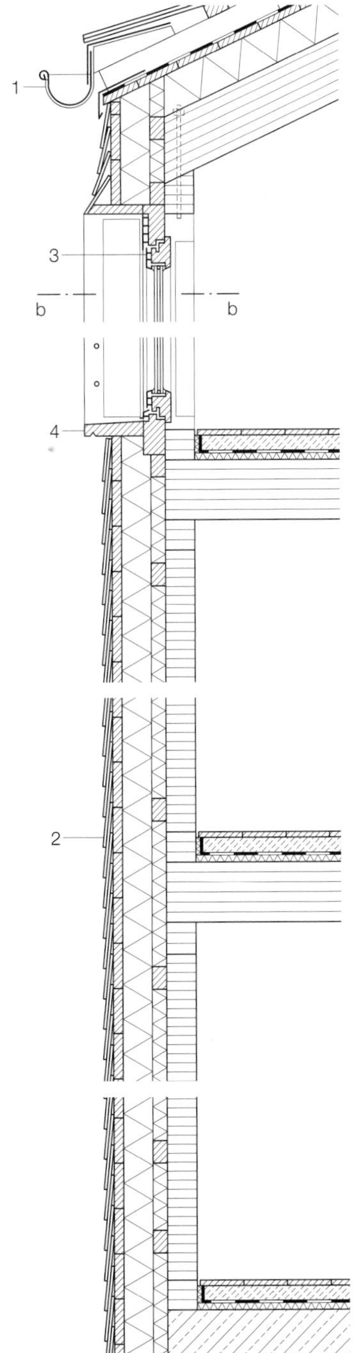

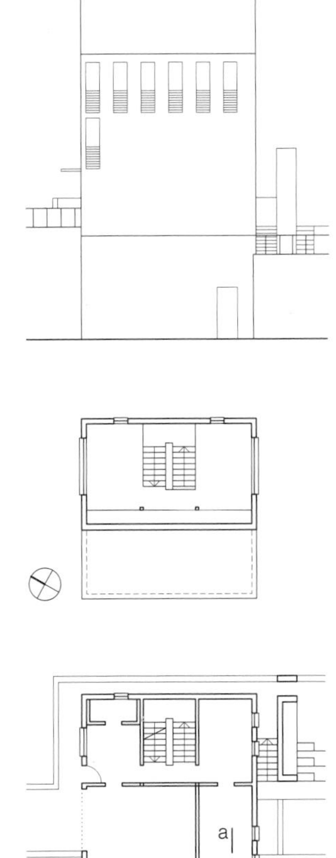

Elevation • Plan of ground floor and gallery
scale 1:400
Horizontal section • Vertical section scale 1:20

1 Sheet copper gutter
2 Facade construction:
larch shingles, double-lap arrangement
80 x 80 mm vertical battens
40 x 60 mm horizontal boards
80 mm edge-nailed timber elements
3 Aluminium/wooden window with double glazing
4 Larch window sill

Private house

Hohen Neuendorf, D, 1997

Architects:
Heinz und Nikolaus Bienefeld, Swisttal-Ollheim
Structural engineer:
Rainer Mertens, Cologne

Architektur Wettbewerbe 09/1998
Baumeister 01/1998
Herzog, Thomas et al.:
Timber Construction Manual,
Munich/Basel, 2003

- Laminated veneer lumber, spruce
- Timber block panel construction, d = 110 mm
- System open to diffusion
- High degree of prefabrication

Section • Plan of ground floor
scale 1:400
Horizontal section • Vertical section
scale 1:20

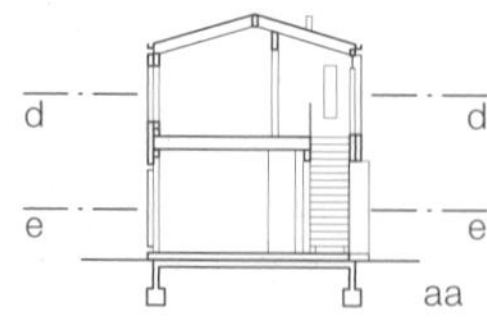

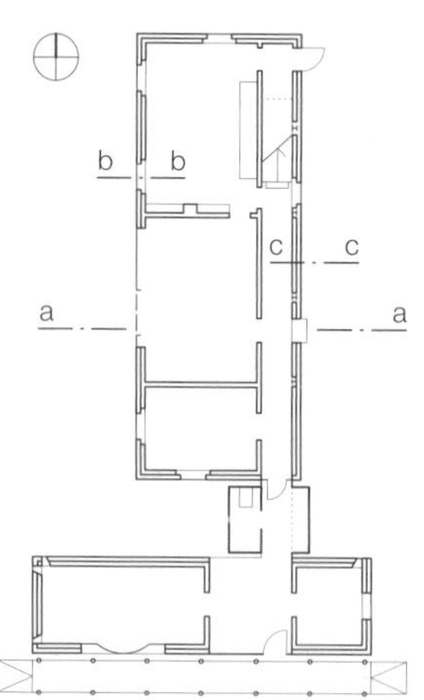

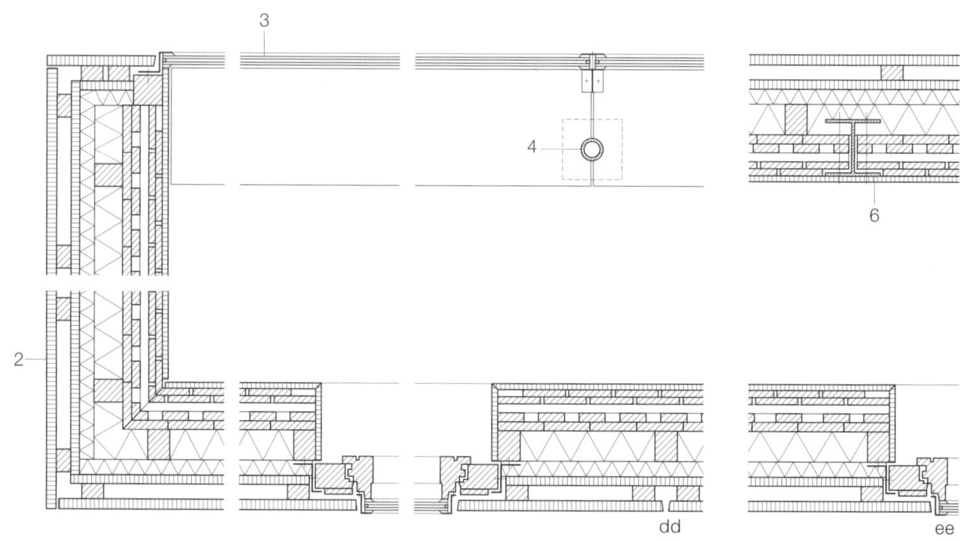

1 Gutter, 140 x 50 x 7 mm channel
2 Facade construction:
27 mm laminated veneer lumber, fir
40 x 60 mm battens/ventilation cavity
24 mm bitumen-impregnated wood fibre insulating board
40 x 60 mm battens, with thermal insulation in between
80 x 60 mm battens with thermal insulation between
110 mm glulam block panel wall element
3-ply core plywood, fir/spruce, opaque white paint finish
3 Double glazing:
8 mm toughened safety glass + 12 mm cavity + 8 mm toughened safety glass
4 Steel circular hollow section column, 60.3 mm dia. x 5 mm
5 Steel beam, IPE 240, web with cut-outs
6 Steel beam, HEB 160
7 Beam, 240 x 120 mm
8 Concrete plinth

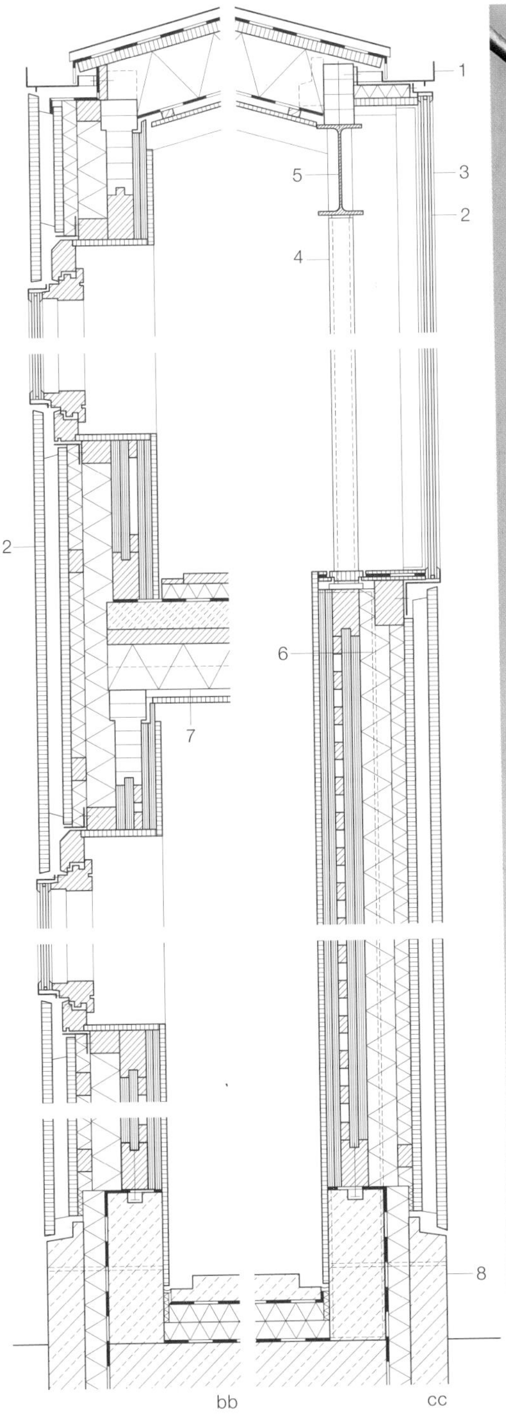

"GucklHupf"

Innerschwand am Mondsee, A, 1993

Architect:
Hans Peter Wörndl, Vienna

Architectural Record 04/1999
A+U 05, 1998
Techniques + architecture 441, 1999
Bahamon, Alejandro: PreFab. Barcelona 2002

- Lightweight construction for occasional use
- Facades can be varied from totally open to totally closed
- Changing relationship between inside and outside
- Okoume plywood, red, waterproof glue, 3 coats of extra-clear boat varnish

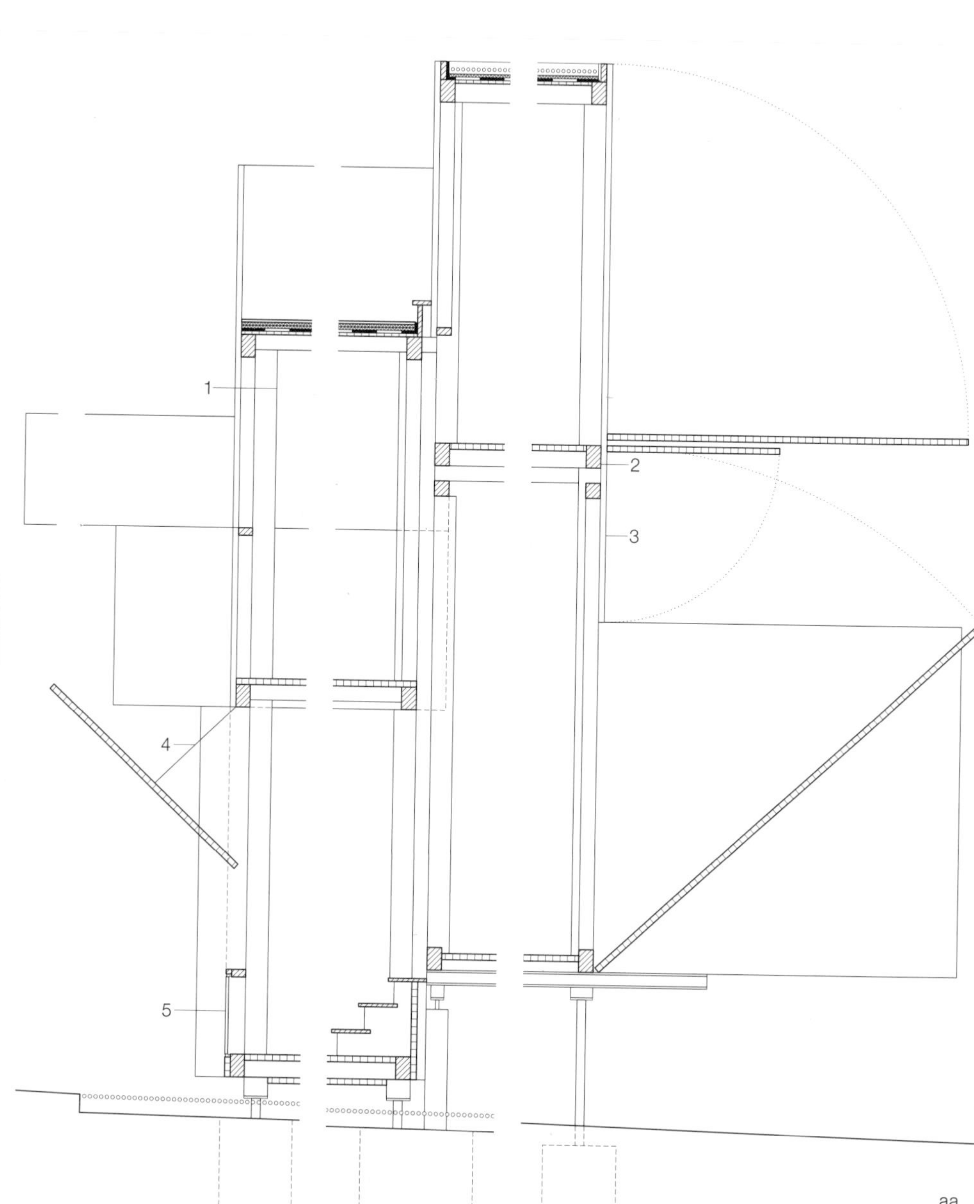

Plans
scale 1:200
Vertical section
scale 1:50

1 Column, 120 x 120 mm, spruce
2 Beam, 60 x 120 mm, spruce
3 External wall panel, 35mm:
6 mm okoume plywood, red, waterproof glue, 3 coats of extra-clear boat varnish, panel size 1200 x 2500 mm
8 mm plywood
building paper/airtight barrier
20 x 30 mm spruce battens, with 20 mm insulation in between
8 mm okoume plywood, red, 2 coats of extra-clear boat lacquer, panel size 1200 x 2500 mm
4 Cable with winch for variability by means of turning, folding, pivoting, pulling; silver anodized aluminium
5 Glazing, with foil backing

Student accommodation

Coimbra, P, 1999

Architects:
Aires Mateus e Associados, Lisbon

Architectural Review 12/2000
AV Monografías/Monographs 83, 2000
Casabella 691, 2001
Detail 07–08/2003

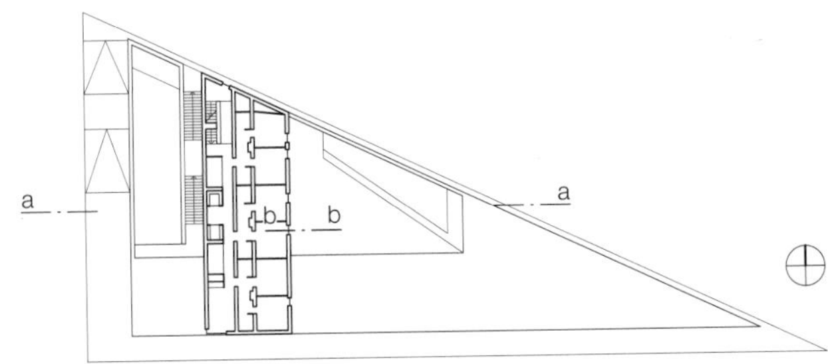

- Smooth timber panels, 800 mm wide, in three different heights
- Every apartment has windows of medium panel height and double panel width with pair of folding wooden shutters
- Timber facade elevation changes constantly with the coming and going of the occupants

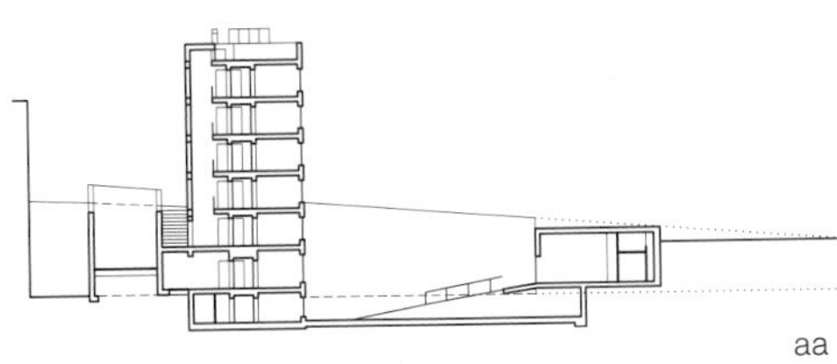

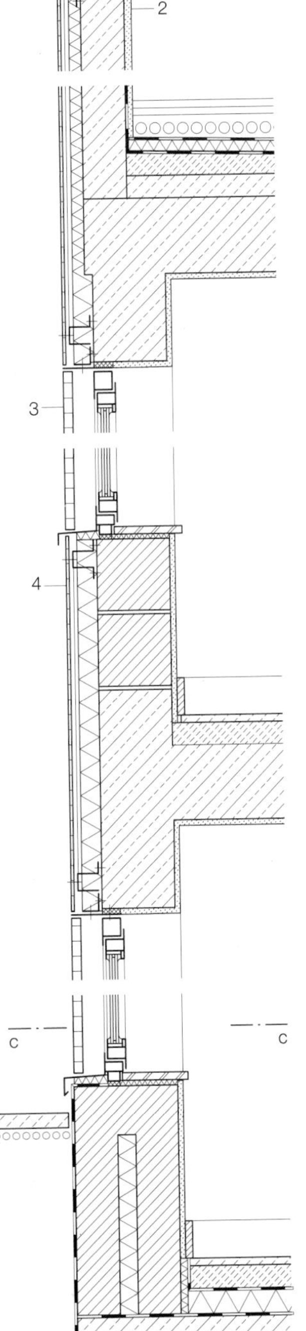

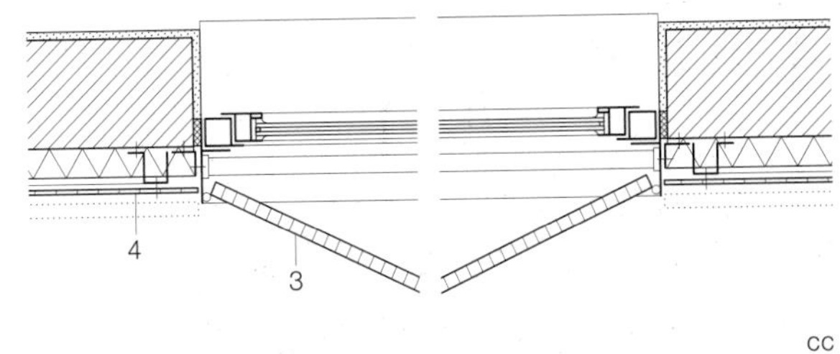

Plan of standard floor • Section
scale 1:1000
Vertical section • Horizontal section
scale 1:20

1 Concrete coping, 50 mm
2 Glass fibre-reinforced render
3 Plywood shutter, 20 mm
4 Facade construction:
8 mm phenolic resin-bonded plywood
20 mm ventilated air cavity
50 mm insulation
200 mm masonry
15 mm smooth-finish plaster

B 1.5 Metal

In many places the advance of human civilisation has gone hand in hand with the development of metalworking technologies. The discoveries of bronze (c. 2500 B.C.) and iron (c. 750 B.C.) are regarded as revolutionary. These new materials, which initially represented improvements primarily to tools and weapons, promoted an overall cultural revolution on a wider scale.
Besides casting, the archetypal way of forming metal, only a few other shaping techniques, e.g., forging, bending and beating, were available in those early days. Gradually, this repertoire was extended and refined through newly discovered metals and alloys which enlarged the range of applications.
This technological progress in metalworking is very much evident in the example of armour, because besides its protective function it also had to satisfy requirements of prestige and representation (figs B 1.5.2 and 1.5.3). These disparate needs led to numerous different forms of metalworking.
The building trade made use of metal from very early times. Lead, bronze and copper have been employed since ancient times, primarily for roofing. The Greeks also used large quantities of bronze and iron for the cramps to fix the stones of their temples and great walls, and also molten lead to subsequently seal the joints. Many of these structures were destroyed later in order to reclaim the coveted metals – particularly in times of war.
Many Gothic buildings would be unstable without the use of (mainly concealed) iron anchors and ties.

For a long time, the use of metals was restricted to these tasks plus local coverings and edgings to projections, canopies, etc. It was only with the appearance of large glass windows that metals started to appear in the facade in a new form and to a larger extent. From this point on, the development and spread of metal and glass are inseparable because the increasing resolution of the solid wall into individual members was only possible with the help of the loadbearing properties of metals (tensile and compressive strengths).

Wrought iron, cast iron and steel
After about 1720 the use of coke and coal instead of charcoal made possible the mass production of pig iron. By the mid-eigtheenth century the first iron sheets were being produced in England.
The use of metals in the construction of facades coincided with the development of rails for the emerging railway industry (from 1830) and the introduction of steel (from 1855). The very first I-beams of wrought iron were produced in France in 1854. In that same year James Bogardus built a five-storey street facade of prefabricated cast iron elements for the Harper & Brothers publishing company. Generally, the exposed steel and iron elements on the facades from this period formed part of

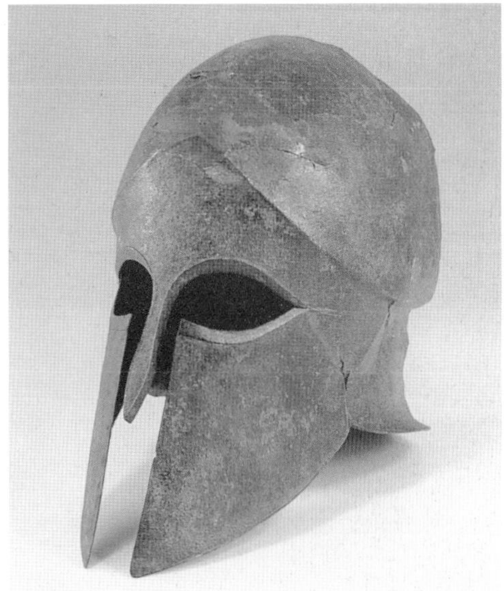
B 1.5.2

B 1.5.3

B 1.5.2 Greek bronze helmet
B 1.5.3 Squire's breastplate, c. 1480, ringmail, 16th c.
B 1.5.4 Cast iron balconies and arbours, London (GB), 19th c.

B 1.5.4

B 1.5.1 Distribution centre, Chippenham (GB), 1982, Nicholas Grimshaw & Partners

B 1.5.5 Prototype of a "Dymaxion Deployment Unit", 1929/45, Buckminster Fuller
B 1.5.6 Facade to the offices of the "Maison du Peuple", Clichy (F), 1939, Jean Prouvé
B 1.5.7 Demonstration of the stability and lightness of an aluminium caravan body
B 1.5.8 Streamlined aluminium panels on a railway carriage
B 1.5.9 Aluminium panels, Financial Times printing centre, London (GB), 1988, Nicholas Grimshaw & Partners
B 1.5.10 Smooth sheet aluminium, terrace houses, London (GB), 1988, Nicholas Grimshaw & Partners

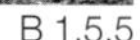
B 1.5.5

B 1.5.6

the loadbearing structure. Examples are found in conjunction with glass at the Sayner Works by C. L. Althans (1828–30), and in conjunction with clay bricks at Jules Saulnier's Menier chocolate factory in Noisel-sur-Marne (1871–72). Other typical applications for cast iron in the nineteenth century were prefabricated balustrading, spandrel panels and complete systems for balconies or arbours (fig. B 1.5.4), some of which still dominate the streetscape of New Orleans. Owing to its high strength, this material made delicate and open designs feasible. The method of production (casting) meant that the material was also very economical and so during this period cast iron was sometimes produced and stocked in large quantities. One early example of an almost completely opaque metal facade is the office building by Georges Chédanne in the Rue Réaumur in Paris (1905). Here, the infill panels to the exposed steel frame are made from riveted metal sheets. Steel dominates the appearance of this facade.

Prefabrication and systems

The use of metal enables a high degree of prefabrication and also great precision. Helped by parallel developments in the automotive and railway industries, these advantages led to the opportunity to employ systems and ideas for building units which could be mass produced. The steel house in Dessau by Georg Muches (1926), Buckminster Fuller's "Dymaxion Deployment Unit" (1929/45) (fig. B 1.5.5) or the series of system houses by Stahlhaus AG (from 1928) bear witness to these trends. However, none of these experiments led to larger series; indeed, the majority did not pass the prototype stage. But the situation was different with the attempt to produce not the entire facade but rather just parts of the external wall according to the aforementioned criteria. Based on the principle of the rail, whose cross-sectional shape is designed for a particular purpose and a defined position, "standard steel sections", optimised for certain loading cases, began to appear at the start of the twentieth century.
Similar ideas from architects like Ludwig Mies van der Rohe led to the development of special facade sections and elements. In the end this resulted in a completely new type of facade carrying only its own weight, which consisted of elements hung in front of the loadbearing structure; this became known as a "curtain wall". This new load-carrying principle in the facade, rendered possible by the very much smaller cross-sectional sizes in conjunction with improved glazing techniques, signified a great step forward on the way to the fully glazed facade, which had been appearing in the visionary sketches of Ludwig Mies van der Rohe, Bruno Taut and others since the beginning of the twentieth century.
This type of construction is important for the development of metal facades because it is often necessary to match the material of the plain panels beneath windows and across ends of floors between the exposed metal loadbearing sections. This leads to facades whose appearance, apart from the extensive use of glass, is determined primarily by the unifying effect of one type of metal. Besides coated steel, other metals such as stainless steel, aluminium, bronze or weathering steel are also feasible.
The Lake Shore Drive apartments (Mies van der Rohe, 1949/50) and the Chicago Inland Steel Building by SOM (1954/55, fig. B 1.5.11) are regarded as noteworthy examples of the use of stainless steel; the Seagram Building in New York by Mies van der Rohe (1955-57, fig. B 1.5.12) for the use of bronze; the Alcoa Building in Pittsburgh by the architects Harrison & Abramo (1950-53) for aluminium; and the Chicago Civic Center (Charles F. Murphy with SOM, 1963-66, fig. B 1.5.13) for weathering steel. In Europe proper curtain wall facades started to appear after about 1955.

The contribution of Jean Prouvé

Jean Prouvé (1901–84), who trained as an architectural metalworker, is regarded as a prominent designer of metal facades. The main

B 1.5.7

B 1.5.8

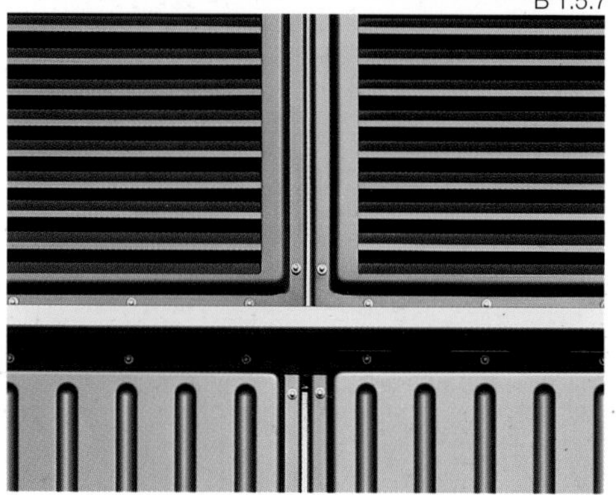
B 1.5.9

B 1.5.10

B 1.5.11 Stainless steel facade, Inland Steel Building, Chicago (USA), 1955, SOM
B 1.5.12 Bronze facade, Seagram Building, New York (USA), 1957, Ludwig Mies van der Rohe
B 1.5.13 Facade of weathering steel, Civic Center, Chicago (USA), 1966, Murphy and SOM

B 1.5.11

B 1.5.12

semi-finished products like metal sheet and sections were already available at the start of his career. However, the potential uses of these products on the facade had hardly been examined. Prouvé was interested in working metal with machinery and he focused on industrial practices, which advanced the shaping of sheet metal in particular.

By planning, experimenting and manufacturing in his own workshop, and thus having the main operations under his control, he went againsts the general trend towards more and more specialisation and division of labour. At the same time, he explored the options given by new manufacturing techniques, such as gas or electric-arc welding.

As a designer and manufacturer he worked with the leading architects of his time and was one of the first to investigate thermally isolated construction. He developed the first curtain wall made completely of sheet metal for the Maison du Peuple in Clichy (1955–39, fig. B 1.5.6). Other important works were the apartment block at Mozart Square in Paris (1954, see p. 258) with vertical sliding and fold-out sunshade elements and the Citroën showrooms in Lyon (1930/31). The facade with the large display windows is characterised by diamond-shaped sections, produced from bent sheet metal, surrounding the structural members [1].
Prouvé exerted a considerable influence on later architects who, from the 1960s onwards, promoted the use of metal for prefabricated panels and sandwich construction by improving the manufacturing and jointing techniques employed. Those architects include Fritz Haller (see p. 170), Norman Foster (see p. 172) and Nicholas Grimshaw (figs B 1.5.9 and 1.5.10).

Visual effect

Owing to its durability, metal is the favoured material for the outside skins of aircraft, vehicles, trains, ships, etc. This fact has led not only to engineering accomplishments; the aesthetics derived from the transport sector have also played a special role in architecture. The use of metal for the "outside skins" of buildings is ideal for conveying a "hightech" image (figs. B 1.5.9 and 1.5.10). Current developments in the field of metal facades are often primarily concerned with the possibility of using these materials to create a cladding over very freely designed building forms. This is possible due to the use of computer-assisted design and metalworking techniques, and the use of very thin sheet metal on highly elaborate supporting frameworks.

The special surface qualities of metals often enhance the highly sculptural effects of these structures. Examples that fall into this category are the smooth, iridescent titanium skin of the Guggenheim Museum in Bilbao (Frank Gehry, 1997, fig. B 1.5.16), the sheet zinc cladding on the Jewish Museum in Berlin (Daniel Libeskind, 1998), the structures of the Thames Barrier flood control facility (consulting engineers: Rendel Palmer & Tritton, 1982, fig. B 1.5.14), and the coarse lead outer envelope of the Auditorio Romano in Rome (Renzo Piano Building Workshop, 2003, fig. B 1.5.15).

New developments, new metals

The ongoing development of metal alloys continues to produce ever more accurately adjusted material properties to suit the most diverse applications. In addition, there are many new techniques which lead to different material structures, e.g., three-dimensional metal foams (fig. B 1.5.17). Their main potential lies in the construction of lightweight load-bearing structures, which is why these materials are currently the subject of experimentation in vehicle design especially.
New developments are expected in the use of composite materials, which amalgamate the specific properties of their individual constituent materials into an efficient combination.

B 1.5.13

B 1.5.14

B 1.5.15

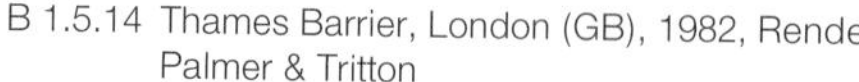

B 1.5.14 Thames Barrier, London (GB), 1982, Rendel Palmer & Tritton
B 1.5.15 Lead facade, Auditorio Romano, Rome (I), 2003, Renzo Piano Building Workshop
B 1.5.16 Titanium facade, Guggenheim Museum, Bilbao (E), 1997, Frank O. Gehry

Developments in coating technology are also seen as important for the visual appearance of facades. Extremely thin metal coatings designed to reflect radiation are being applied to more and more substrates these days, e.g. glass, plastics – including membranes, foils and films.

Material properties

Most of the metals used in facades are not employed in their pure form, but rather as alloys. Fig. B 1.5.19 lists the relevant properties of the most common metals used in facades, classified according to their atomic number. Metals with a density of max. 4.5 g/cm³ are classed as "light metals", and although titanium has a density of 4.51 g/cm³ it is normally counted among these. All metals are gas-tight and hence also vapour-tight.

The thermal expansion behaviour is particularly important in construction because this leads to movements which have to be accommodated by appropriate forms of jointing and assembly. Besides the temperature of the air, radiation is the main factor governing the temperature rise in the material. This is determined by the colour plus the reflection and absorption properties of the respective metal. Figs B 1.5.18 and 1.5.24 indicate the relationships with regard to metal surfaces. The majority of these materials react to environmental influences and thereby alter their appearance. In some metals these corrosive processes are a great problem with respect to their suitability for construction purposes. Steel, for example, can change its volume by a factor of seven when it corrodes. The leaching products of other metals either alter the colour (copper, weathering steel) or are in some circumstances highly toxic, even in small quantities (lead). Besides corrosion in the form of pitting, the phenomenon of galvanic corrosion can also occur when disparate metals are combined directly, or moisture – e.g., rainwater – forms a bridge for transporting ions (electrolysis) from the surface of one metal to that of another. To avoid this problem, refer to the electrochemical series, which divides metals into base metals (low electric potential, easy to oxidise) and noble metals (high electric potential, difficult to oxidise). The potential difference of the chemical form actually effective (often an oxide) determines the risk of corrosion. A neutral intermediate pad or isolating layer may be necessary [2].
As shown in fig. B 1.5.18, certain metals do not corrode, while others form a regenerative anti-corrosive layer (patina, see fig. B 1.5.24), either independently or with outside help. A third group (iron and steel) requires special treatment in order to resist environmental influences. The corrosion protection measures and, if necessary, other surface treatments must be compatible.

B 1.5.16

Manufacturing technologies and semi-finished products

The principal processes in metalworking are:

- forming and shaping
- separating
- jointing
- coating
- changing the material properties

These principal processes can be further subdivided into countless other techniques which are still undergoing development. Fig. B 1.5.20 shows the relationships between manufacturing methods and individual product groups related to uses in the facade.

Forming and shaping processes are employed in continuous or cyclic operations to produce semi-finished products which undergo subsequent processing in the form of folding, drilling, drawing or pressing to create ever more elaborate products. Besides the need to achieve certain visual effects, the metalworking processes generally serve to optimise certain

B 1.5.17 Metal foam (scale approx. 1:1)
B 1.5.18 Metallwerkstoffe und ihre Oberflächen
[1] Metals used in facades, content of primary metal > 90%
[2] Alloys commonly used in facades
[3] Basis: normal external environment
B 1.5.19 Properties of metals (selection), arranged according to atomic number of primary metal
• yes
○ no

B 1.5.17

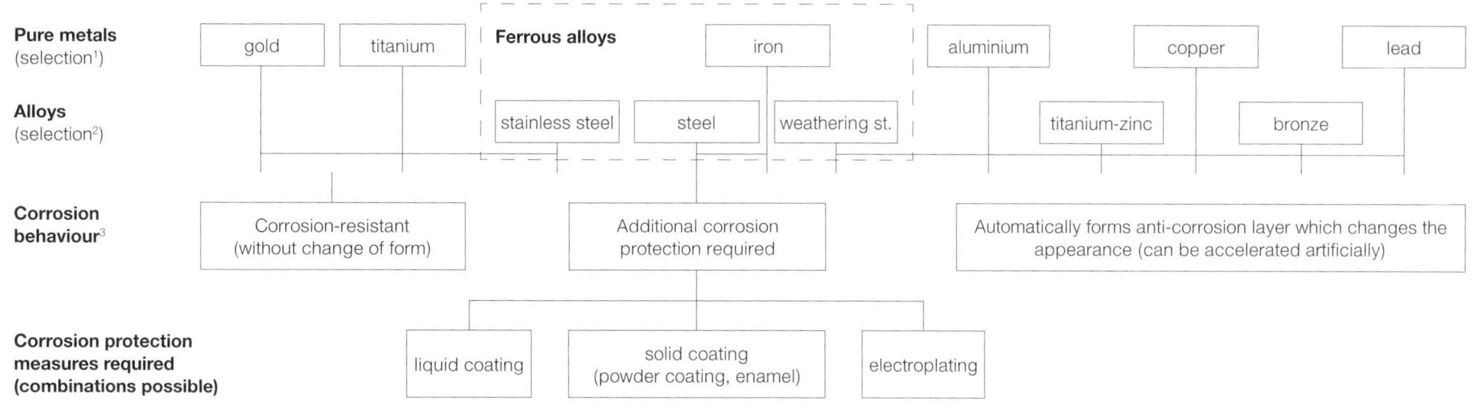

Surface finishes (combinations possible)

Mechanical surface treatment
- sandblasting
- shot-peening
- brushing
- grinding
- polishing
- water-jetting
- embossing
- cambering

Chemical surface treatment without coating formation
- cleaning
- chemical deburring
- acid etching
- pickling
- burnishing
- metal spraying

Chemical surface treatment with coating formation
- build-up welding
- hot dipping
- enamelling
- plating
- anodising
- oxidising
- electro-galvanising
- painting
- laminating
- printing

B 1.5.18

Metal		**aluminium**	**titanium**	**iron**				**copper**			**zinc**		**tin**	**gold**	**lead**
Chemical symbol (Z)		**Al (13)**	**Ti (22)**	**Fe (26)**				**Cu (29)**			**Zn (30)**		**Sn (50)**	**Au (79)**	**Pb (82)**
Alloy					**steel**	**stainless steel**	**weathering steel**		**bldg-grade bronze**	**tombac**		**titanium-zinc**			
density	[g/cm³]	2.7	4.51	7.87	7.8	7.98		8.92	8.73	8.5	7.2	7.2	7.29	19.32	11.34
modulus of elasticity[1]	[kN/mm²]	65	110	210	210	200	200	132	100	85	90	80	50	75	15
specific heat capacity	[J/(kg K)]	900	530	460	400			390	380	380	390	398	230	130	130
coef. of thermal exp.[2]	[10^{-6}m/(mK)]	23.8	10	12.1	11.7	17.3	11.7	16.8	18.5	19	36	20	20.5	14.2	28.3
thermal conductivity	[W/(mK)]	160	22	80.4	65	15		305	67	50	116	109	35.3	317	34
standard potential	[V]	−1.69		−0.44				+0.35			−0.76		−0.16	+1.38	−0.13
electrical conductance	[m/mm²Ohm]	35	1.25	10.3	10.2			60	9	16	16.9	17	8.7	45.7	4.82
[1] 1 kN/mm² = 1 GPa [2] at 20°C															
Corrosion behaviour															
forms protective oxide layer		•	•	○	○	○	•	•	•	•	•	•	•	○	•
additional corrosion protection required		○	○	•	•	○	○	○	○	○	○	○	○	○	○
coloured further development		○	○	•	•	○	•	•	•	•	•	•	•	○	•
discolours run-off water		○	○	•	•	○	•	•	•	•	○	○	○	○	○
Sheet metal (as semi-finished product) for facades															
standard thicknesses	mm		0.3–1			0.35–3	0.5–3	>3	0.6–0.8			0.7–1.5			2.25–3.0
recommended min. bending r. (inside)	t = sheet thickness		2 t			1–2 t	2 t					1.75 t			

B 1.5.19

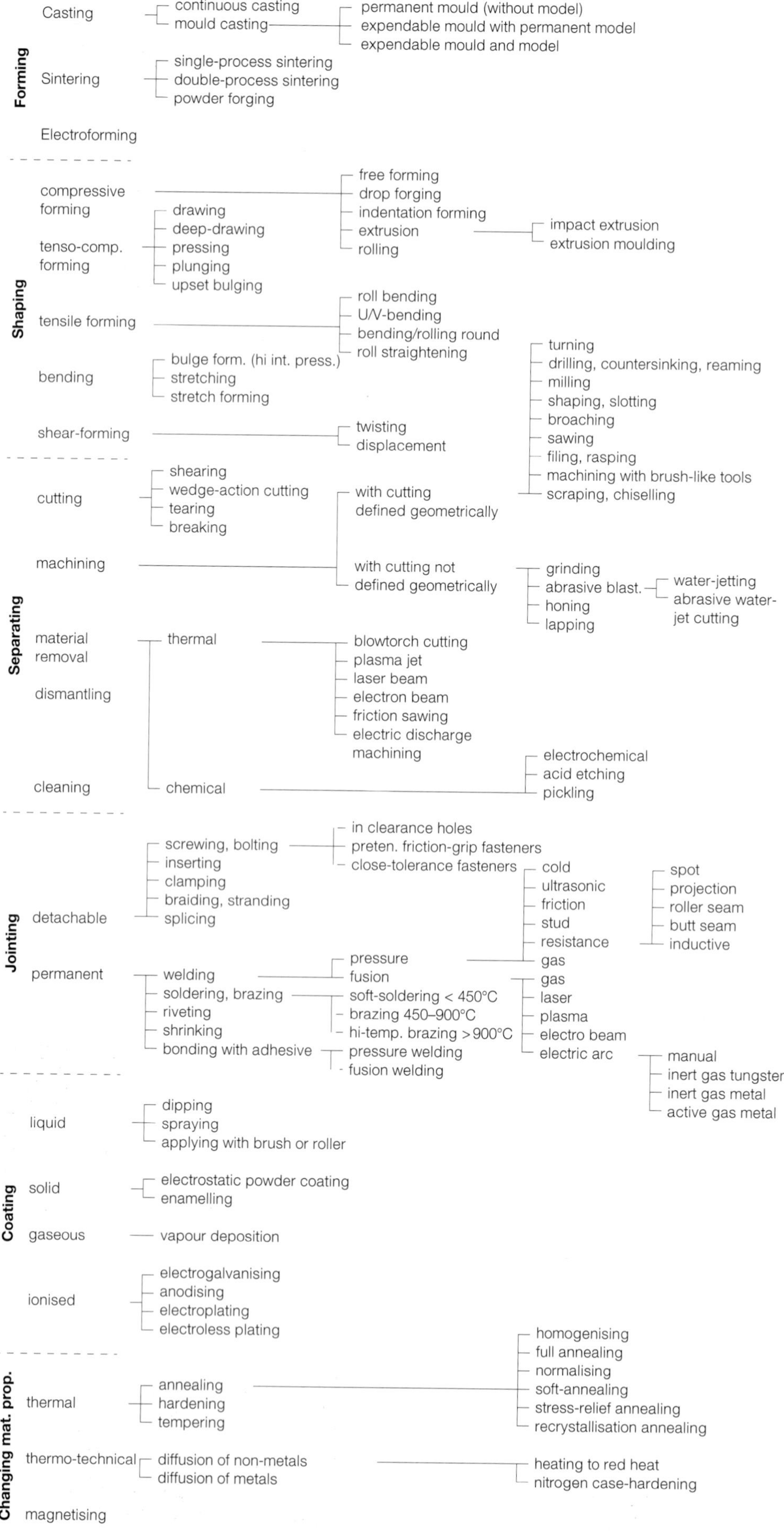

B 1.5.20

properties, e.g., to improve stability. Metals are often combined with other materials to create composite materials. These include, for example, the multi-layer sheets shown in fig. B 1.5.22. Steel and aluminium are the metals employed most frequently, but copper is also used.

Sheet metal sandwich elements

Owing to their properties (e.g., high strength and good formability), metals are particularly suitable for creating complete composite components – the sandwich elements (see fig. B 1.5.21 and chapter A 2.1). This usually involves bonding sheet metal to both sides of a shear-resistant core of insulating material to form a rigid, constructional unit (resembling the cross-section of a bone). This effect can be enhanced by pre-shaping the sheet metal, which increases the structural capacity in one direction (e.g., by folding) or in two directions (e.g., by deep-drawing).

Such elements, the sizes of which are essentially limited only by transport requirements and the widths in which sheet metal can be supplied, enable high stiffnesses and long spans – and fast erection times – to be achieved with a relatively thin, low-weight component. The principle of the sandwich element, owing to the insulation properties in conjunction with zero vapour transmission and mechanical stability, is frequently a sensible decision for the flat areas of a facade. However, careful detailing of joints and edges is necessary. Aluminium, steel and stainless steel are the metals normally employed for the outer layers of a sandwich element.

B 1.5.20 Methods of manufacture
B 1.5.21 Diverse sandwich panels, sheet steel
B 1.5.22 Diverse multi-layer panels, aluminium
B 1.5.23 Relationship between methods of manufacture and products
B 1.5.24 Surfaces of metals with natural and artificial corrosion and patination forms (selection)

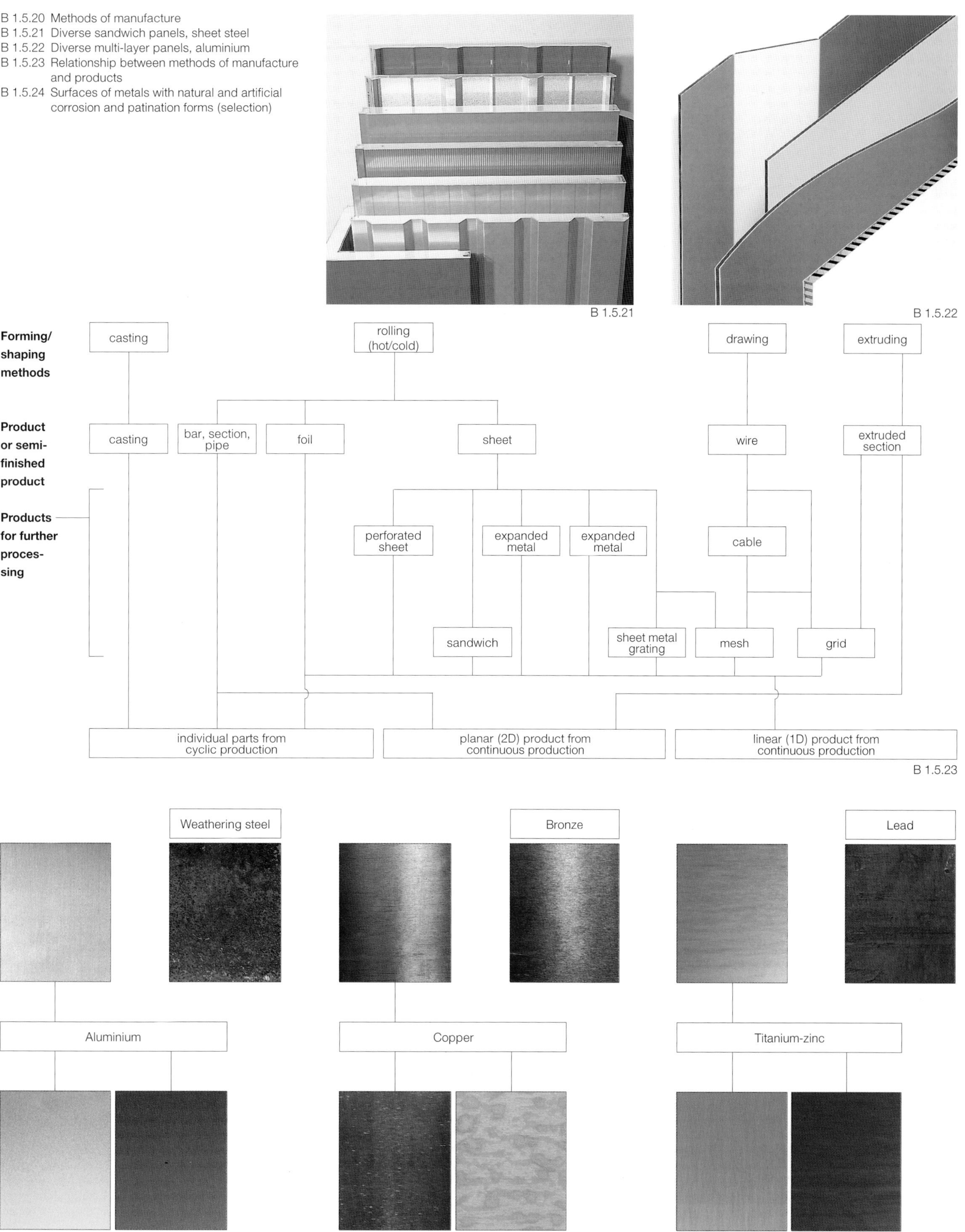

B 1.5.25
Mixed commercial and residential block, London (GB), 1991, Michael Hopkins & Partners

Metal facades – basic forms

Based on the systematic presentation of basic facade assemblies (see "Joining facade components", chapter A 2.1, pp. 32–34), fig. B 1.5.26 shows a selection of the most common basic forms for metal facades.
Besides the material-related treatment of the construction joints, the type of fixing to the supporting structure is important. In metal facades the fixings are usually detachable (e.g., screws, bolts) and can either be positioned within the joints or elsewhere. Certain soft metals, such as lead, are also suitable for nailing. Self-tapping screws are used to join metal sheets together or to the supporting construction. Exposed riveted connections, which for a long time were a characteristic feature of metalwork in building, are hardly ever employed these days.

Furthermore, we distinguish between ventilated and non-ventilated designs, with and without additional elements concealing or closing the joints.

Another distinguishing criterion is the way in which the facade elements are stabilised. This can be achieved by choosing a suitable format, depending on material and material thickness, but also through additional measures to stabilise the form, such as folding, corrugating, deep-drawing, or through combination to form sandwich elements. Certain shaping processes, such as extruding, can create intrinsically stable elements.

Unlike most other materials, metals are suitable for designing facades with a very high degree of prefabrication. The sizes of metal sheets and the shaping techniques available, plus the relatively low weight coupled with great robustness with respect to the weather (especially aluminium), enable the production of large-format elements. In the form of, for example, panels, pans or louvres, these represent highly economical solutions. However, traditional, manual metalworking methods, which require folding or bending plant on site, are still used.

The design of metal facades calls for considerable temperature-related movements to be accommodated in addition to erection and manufacturing tolerances. Restraint stresses which could damage the construction (and which can often be heard!) must be avoided at all costs. Therefore, the joints must be of adequate size and the connections capable of sliding.

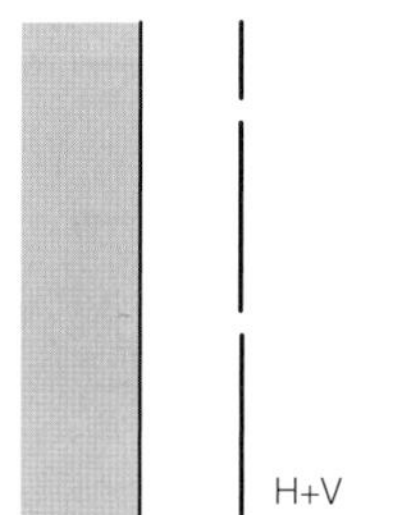

Panels with open (drained) joints

- fixings exposed or concealed
- second, water run-off layer required

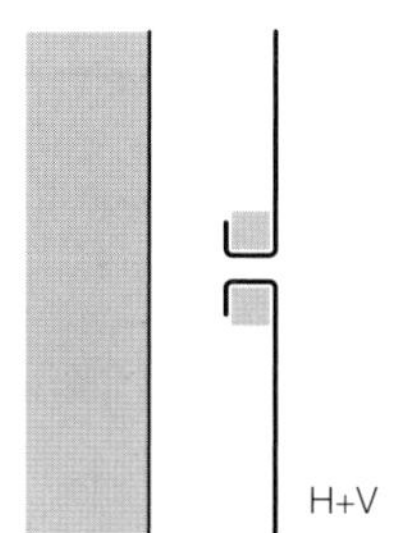

with sheet metal folded around frame

- difficult to maintain tension
- second, water run-off layer required

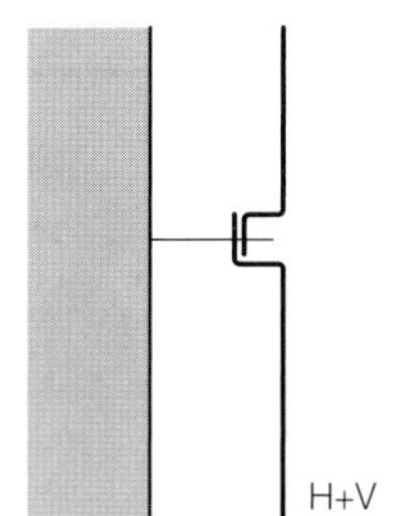

Pans

- peripheral folded edges stabilise the form
- detachable joints

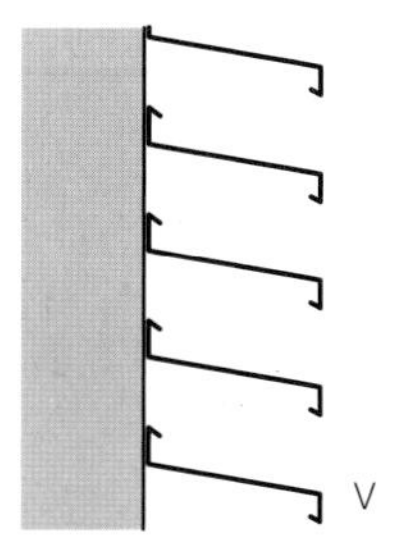

Louvres

- linear folds or extruded elements
- louvres spaced so that no water can penetrate
- butt joints require backing pieces

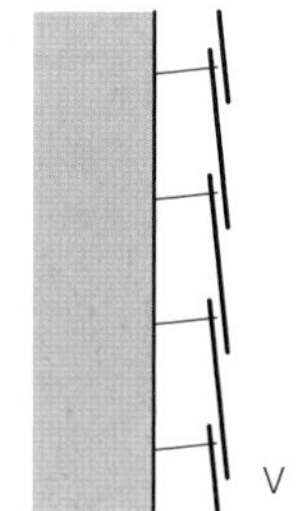

Plain overlapping elements (sheet metal)

- fixings concealed within overlap
- size of elements is limited
- risk of corrosion with steel owing to penetration
- typical materials: lead, zinc, copper

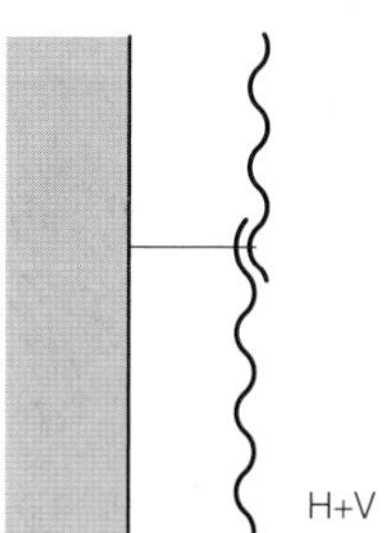

Overlapping planar elements with stable form

- formats depend on production
- stability not identical in both directions

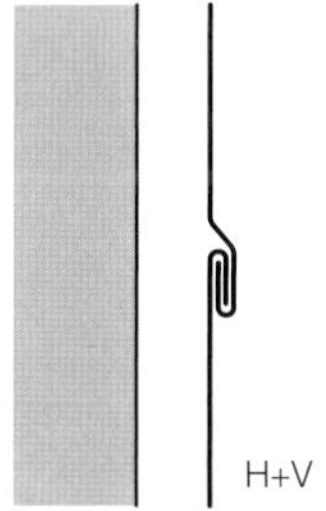

Welted seam

- can also be combined with fixing
- welt formed *in situ*
- detachable

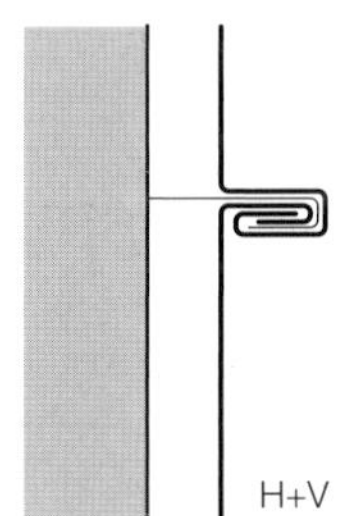

Standing seam

- also possible without fixing
- seam formed *in situ*
- forms very distinct pattern on surface
- detachable

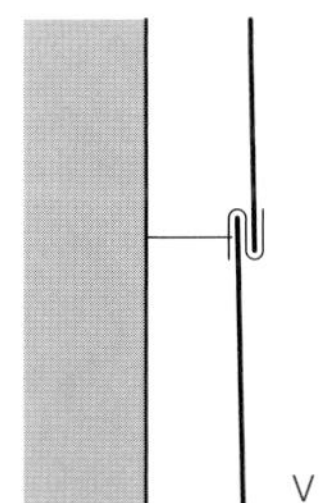

Overlapping panels with additional local fixing element

- risk of galvanic corrosion due to unsuitable combination of materials
- additional fixing element visible externally

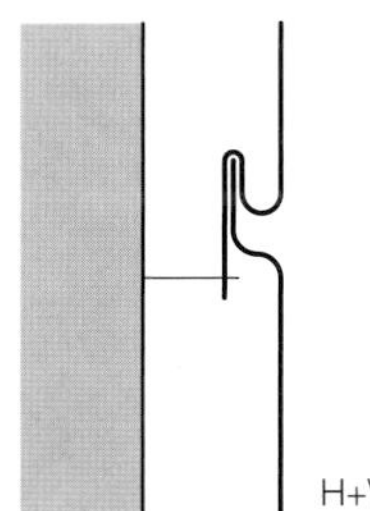

Concealed fixing

- elements cannot be replaced individually

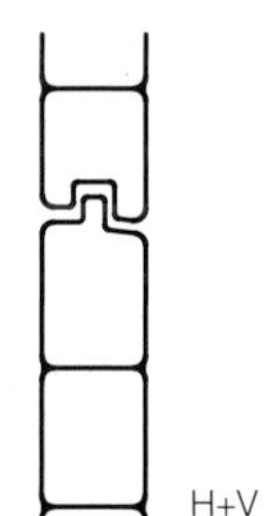

Special extruded forms

- ventilation of the cells must be ensured
- it is not possible to achieve an adequate joint in the other direction
- supporting construction only required perpendicular to webs
- elements cannot be replaced individually

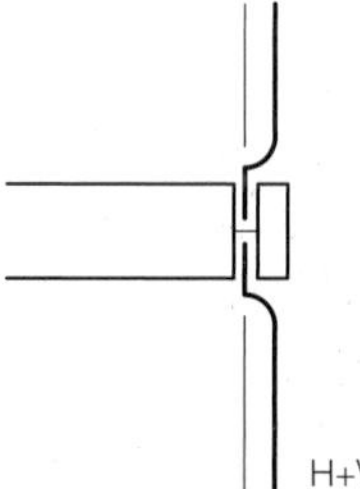

Stable individual elements fixed with third element concealing butt joint

- stable form, e.g., by means of tenso-compressive forming (deep-drawing)
- elements can be replaced individually

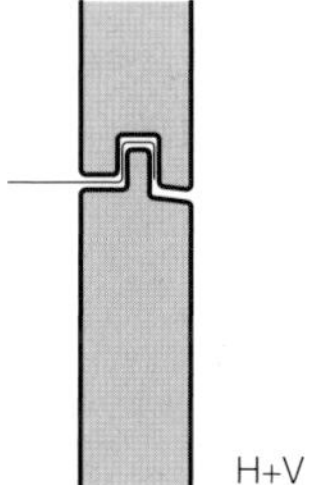

Sandwich elements with fixings in joints

- concealed fixings
- supporting construction only required in one direction
- elements cannot be replaced individually (erection sequence)

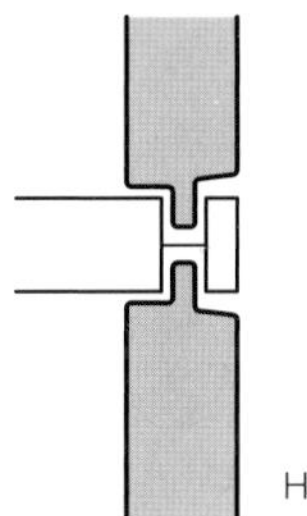

Sandwich elements fixed with third element concealing butt joint

- integrated into post-and-rail construction
- elements can be replaced individually

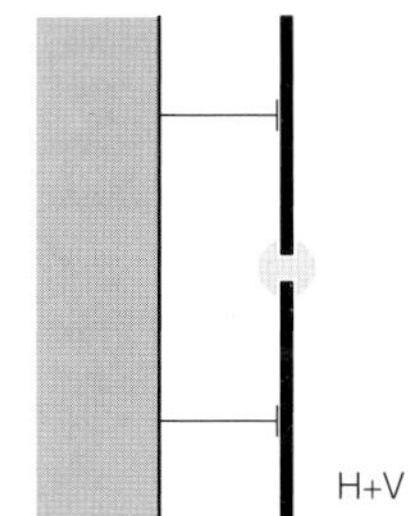

Panels with jointing by way of additional sealing element

- elements can be replaced individually if sealing element can be opened

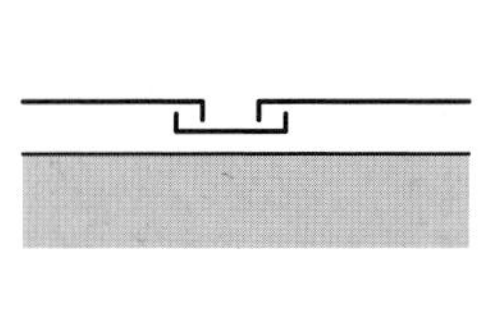

Sheet metal backing piece

- supplementary peripheral folds stabilise the form
- elements can be replaced individually

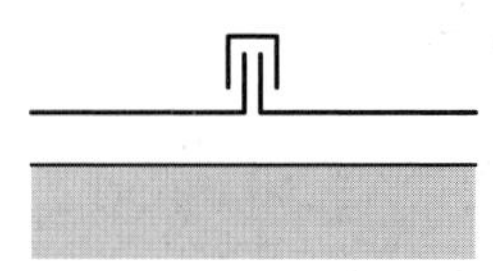

Cover strip over vertical butt joint

- cannot be used for horizontal joints because water run-off is interrupted
- elements can be replaced individually

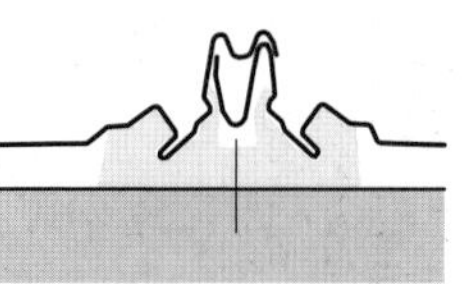

Multi-part clamped joint over local fixing element

- folds in element stabilise the form
- elements can be replaced individually

H = horizontal section, V = vertical section

B 1.5.26 Metal facades – basic designs (selection)

B 1.5.27 B 1.5.28 B 1.5.29 B 1.5.30 B 1.5.31 B 1.5.32 B 1.5.33 B 1.5.34

B 1.5.27 Stainless steel sheet embossed with pyramid-form studs
B 1.5.28 Sheet metal with studded surface
B 1.5.29 Sheet metal with lozenge pattern
B 1.5.30 Sheet metal with multiple lozenge pattern
B 1.5.31 Sheet metal, herringbone pattern
B 1.5.32 Sheet metal, fine rib pattern
B 1.5.33 Trapezoidal profile sheet
B 1.5.34 Asymmetric corrugated sheet

B 1.5.35 Square perforations, stainless steel
B 1.5.36 Triangular perforations, stainless steel
B 1.5.37 Offset slits, stainless steel
B 1.5.38 Expanded metal, aluminium
B 1.5.39 Expanded metal, aluminium
B 1.5.40 Circular perforations, folded, tombac
B 1.5.41 Lanced perforations, stainless steel
B 1.5.42 Grating (bars/wires), stainless steel

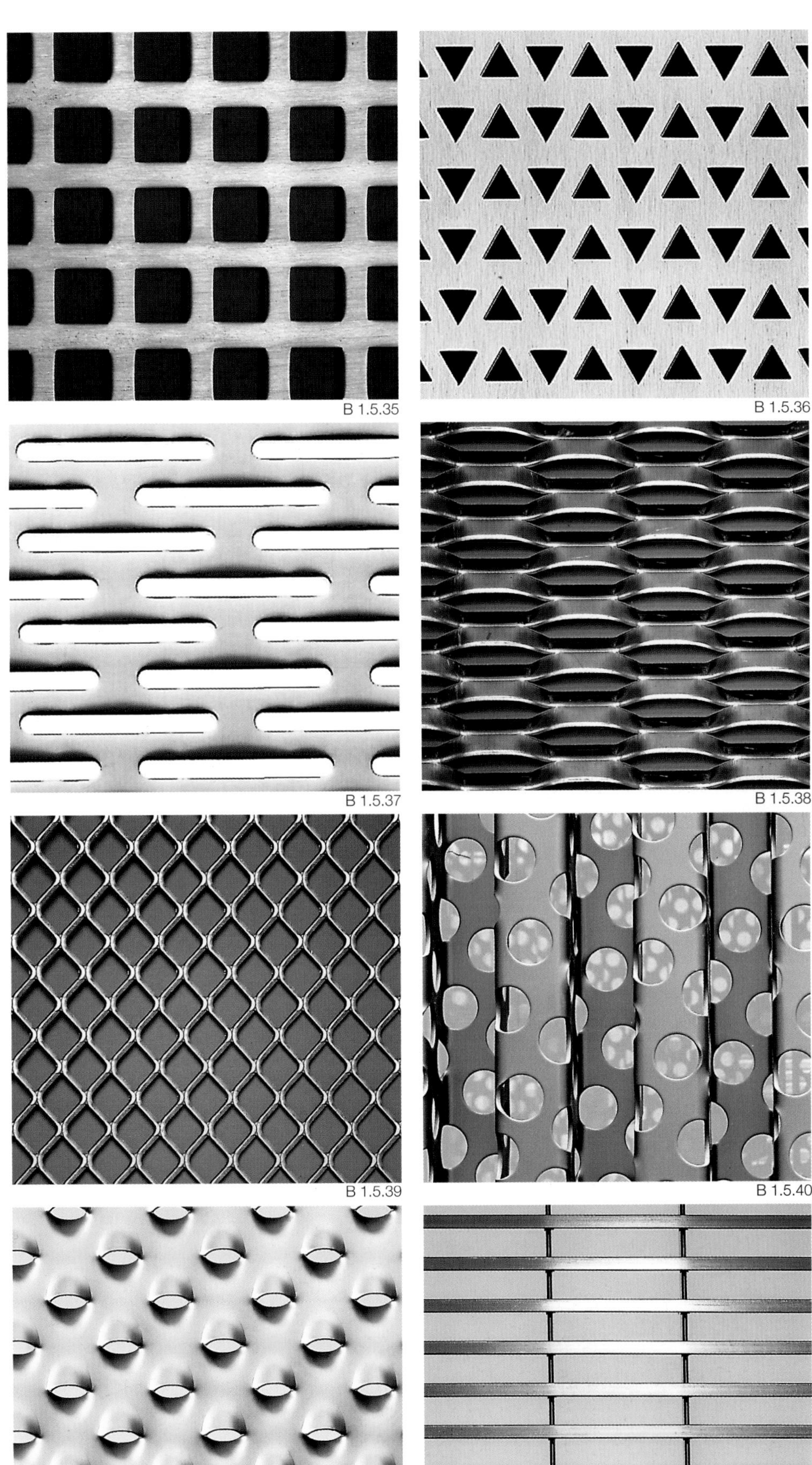

B 1.5.35 B 1.5.36 B 1.5.37 B 1.5.38 B 1.5.39 B 1.5.40 B 1.5.41 B 1.5.42

Metal fabrics (meshes)

The origin of diaphanous (translucent) metal fabrics can be found in industrial applications (e.g., filters and foodstuffs technology). Like perforated sheet metal, these offer the chance to construct permeable building envelopes. The effects possible depend heavily on the viewing angle and are essentially determined by the reflective properties of the material used, the aperture of the mesh/weave and the thickness and texture of the material. If required, metal fabrics can also fulfil functional requirements (e.g., sunshading, wind and weather protection, privacy, light redirection, lightning and radar protection, security) in addition to visual effects. The "translucency" can be adjusted by altering the size of the aperture and the depth of relief [3].
In a similar fashion to membrane materials, metal fabrics can be installed with a prestress (i.e. stabilised in the plane of the fabric). Springs are often used to maintain the prestress despite changing temperatures (fig. B 1.5.45). However, retensioning of connections is also feasible. Some products can be custom-fabricated in almost unlimited sizes – in both directions – in order to avoid having connections and seams visible in the final condition. The maximum width in which fabrics can be supplied is generally limited to 8 m.

Figs B 1.5.46a-h illustrate a selection of possible weaves for metal fabrics, which are similar to those used for traditional textiles. In addition, there are also knitted metal fabrics and metal nets [4]. Metal fabrics can be produced from various metals and also in combination with plastics. Examples of special fabrics available are:

- fabrics with integral optical fibres woven into the fabric for illumination purposes
- fabrics with graphics and texts (e.g., company logos) woven into the fabric
- fabrics with varying "translucency" (stepped or graded)

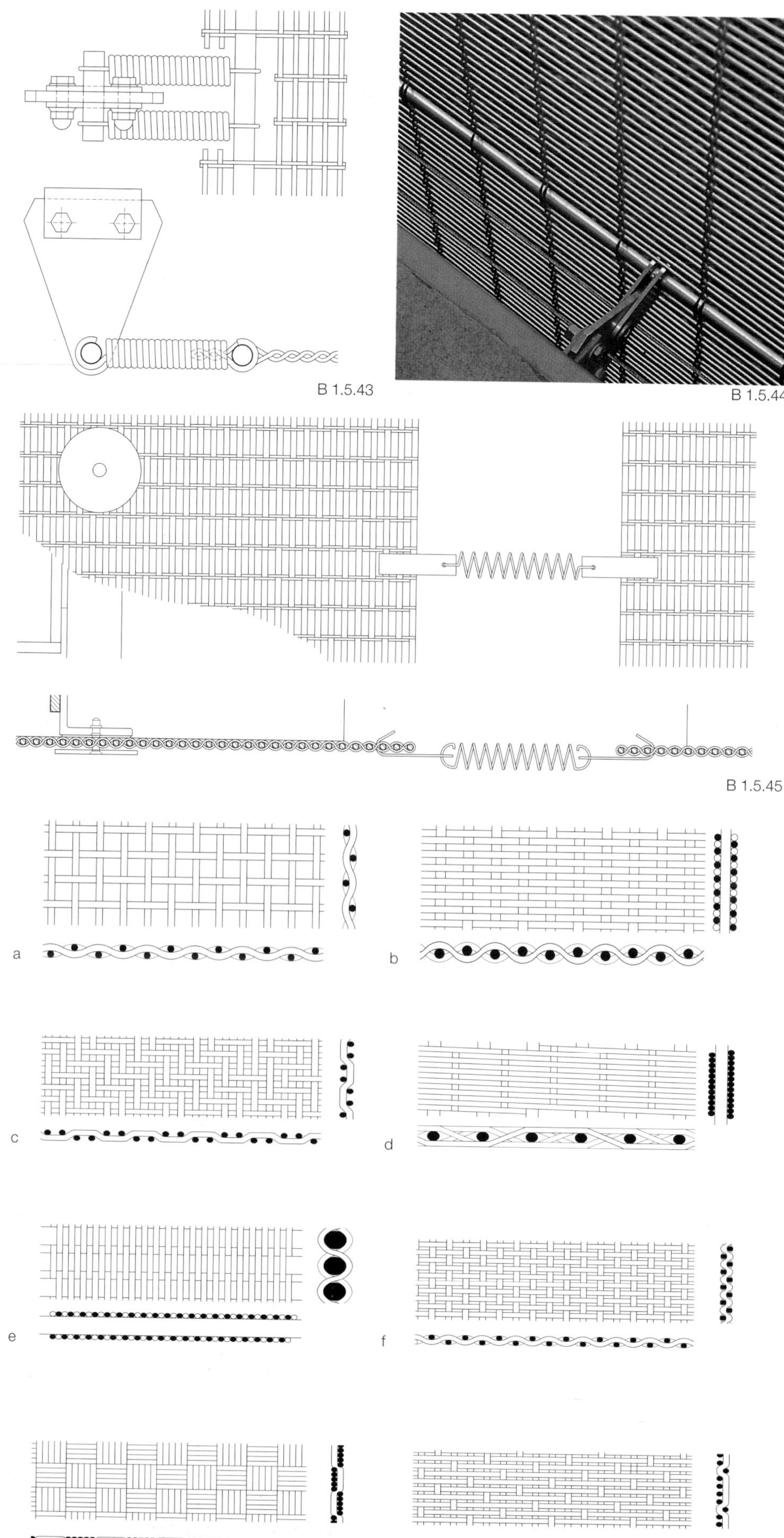

B 1.5.43

B 1.5.44

B 1.5.45

B 1.5.46

Notes:

[1] Among the works on Jean Prouvé, the publications by Peter Sulzer are definitive, e.g.: Sulzer, Peter: *Jean Prouvé, Oeuvre complète*, vol. 1: 1917-33, Berlin, 1995; vol. 2: 1934–44, Basel/Berlin/Boston, 1999; Highlights 1917–44, Basel, 2002.
[2] Karl Träumer & Söhne GmbH (pub.): *Dachdecker- und Spenglerarbeiten*. Munich, 1993, p. 95.
[3] A double-layer application of metal fabrics (e.g. on the altar wall of the Herz Jesu Church in Munich (D) by Allmann Sattler Wappner, 2000) is rare.
[4] Schäfer, Stefan: *Fassadenoberflächen aus metallischen Werkstoffen*; in: Detail 01-02/2003, pp. 90f.

B 1.5.43 Example of fixing with tensioning springs (vertical and horizontal sections)
B 1.5.44 Multi-storey car park, Cologne-Bonn Airport (D), 2000, Murphy/Jahn
B 1.5.45 Example of fixing
B 1.5.46 Types of weave for metal fabrics
a plain weave
b plain Dutch weave
c twilled weave
d twilled Dutch weave
e reverse plain Dutch weave
f long-mesh weave
g plain multiplex weave
h five-heddle twilled weave
B 1.5.47 Net of round, stranded cables, stainless steel, pressed sleeves of tinned copper
B 1.5.48 Stainless steel knitted fabric, round wire
B 1.5.49 Plain Dutch weave (warp and weft), stainless steel
B 1.5.50 Plain weave (warp and weft), stainless steel
B 1.5.51 Long-mesh weave with double wires, stainless steel
B 1.5.52 Plain weave with strands and bars, bronze
B 1.5.53 Plain weave with strands and bars, stainless steel
B 1.5.54 Spiral woven fabric of flat strips and round bars, stainless steel

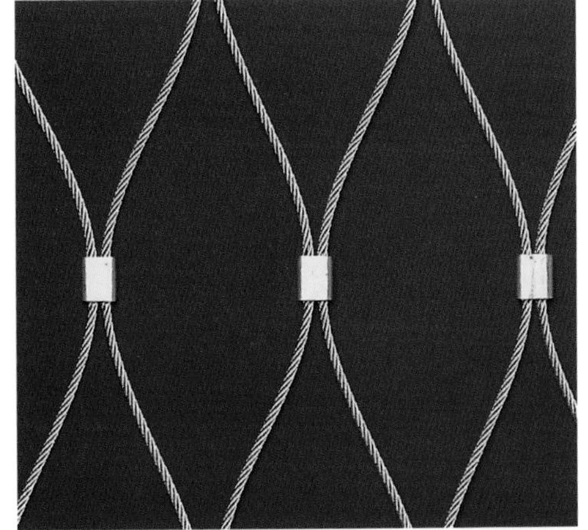

B 1.5.47

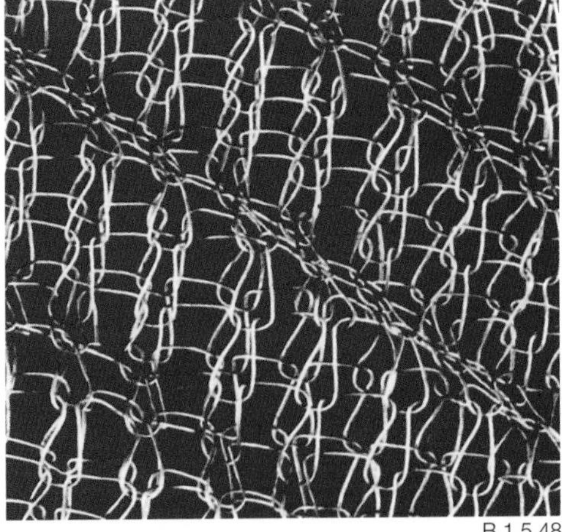
B 1.5.48

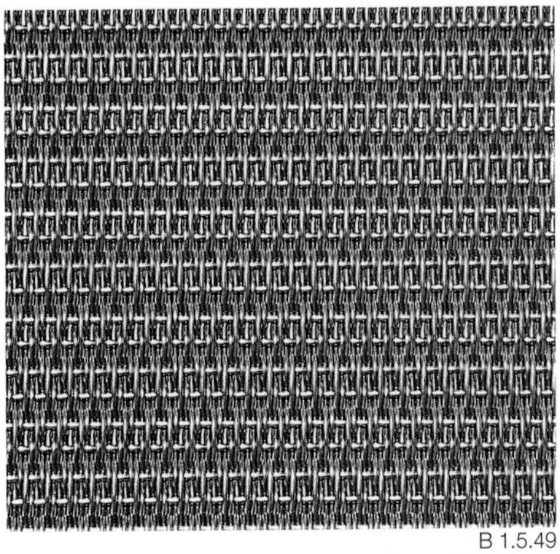

B 1.5.49

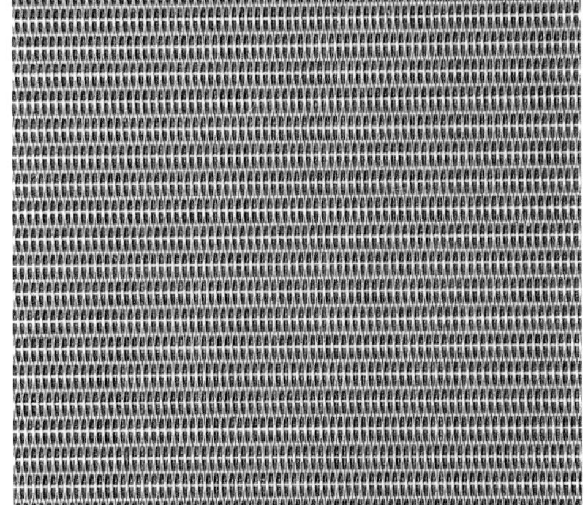
B 1.5.50

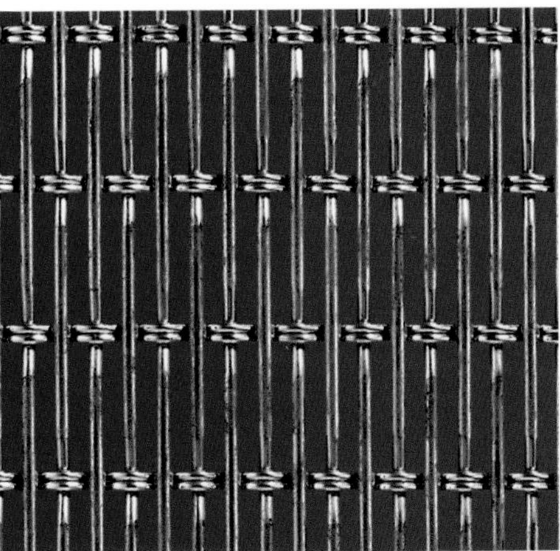
B 1.5.51

B 1.5.52

B 1.5.53

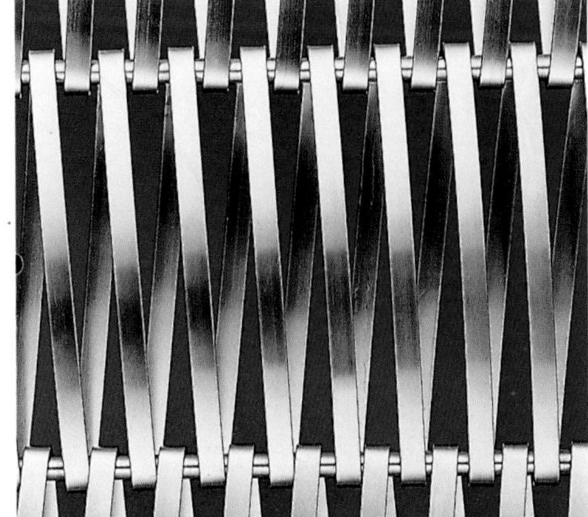
B 1.5.54

Semiconductor assembly plant

Wasserburg am Inn, D, 1968

Architect:
Von Seidlein, Munich
Peter C. von Seidlein, Horst Fischer
Responsible for facade:
Thomas Herzog

db 01/2002
Grube, Oswald W.: Industriebauten international. Stuttgart 1971
Von Seidlein, Peter C.: Zehn Bauten 1957–97, exhibition catalogue, Munich, 1997

- 1.50 m system grid throughout
- Use of heavy rolled sections instead of system resolved into individual members
- Use of a directional system resulting in different column connections
- Columns and wind bracing between external, non-insulated and internal, insulated sheet metal leaves
- Bent, vertical sheet aluminium with concealed fixings
- Auxiliary building with cooling hood slits in aluminium facade for cross-ventilation

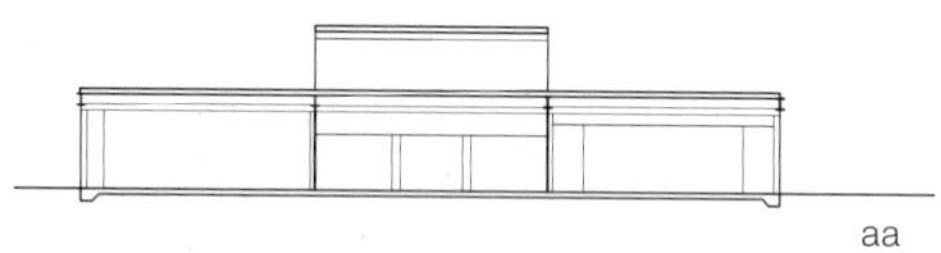

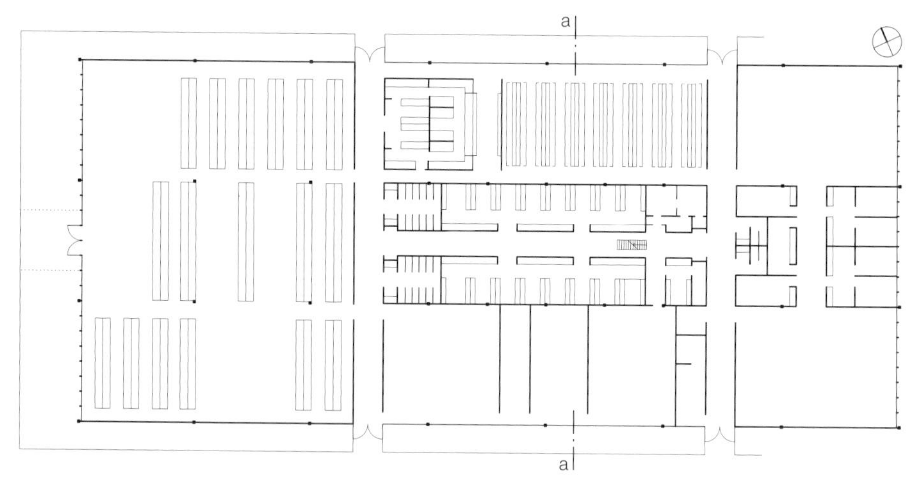

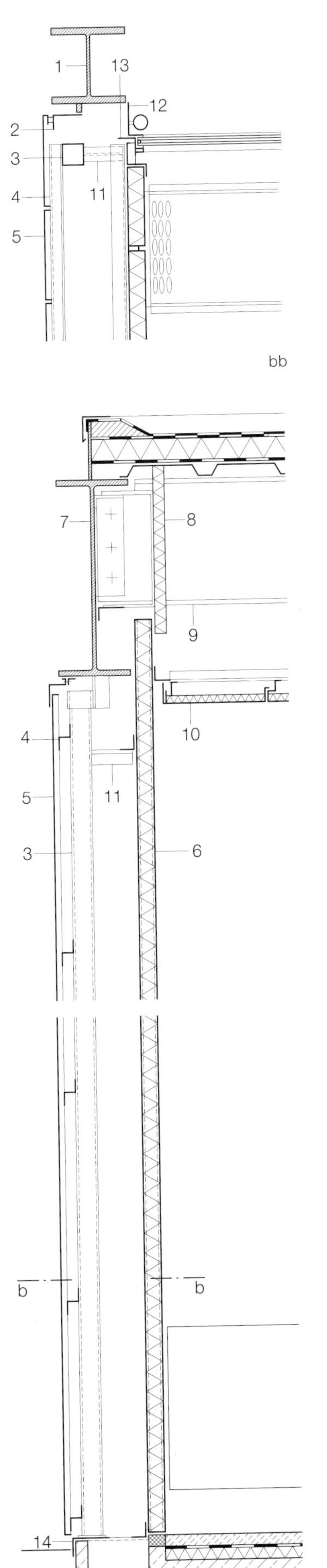

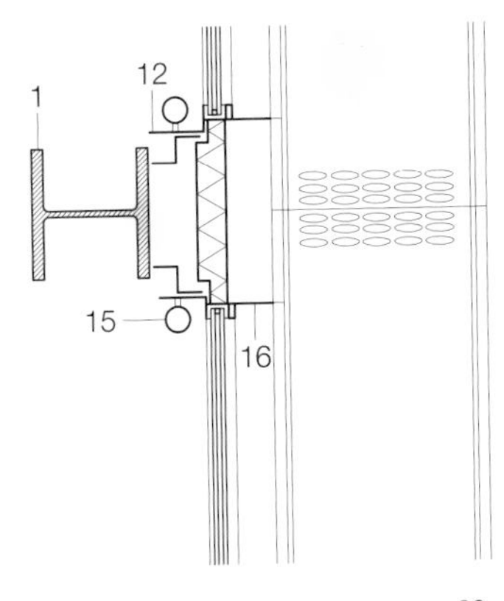

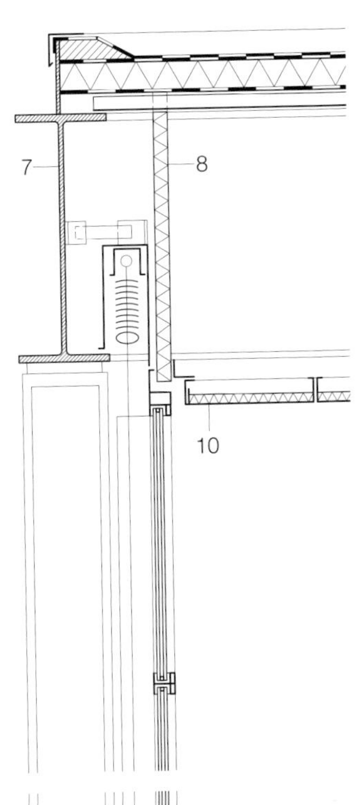

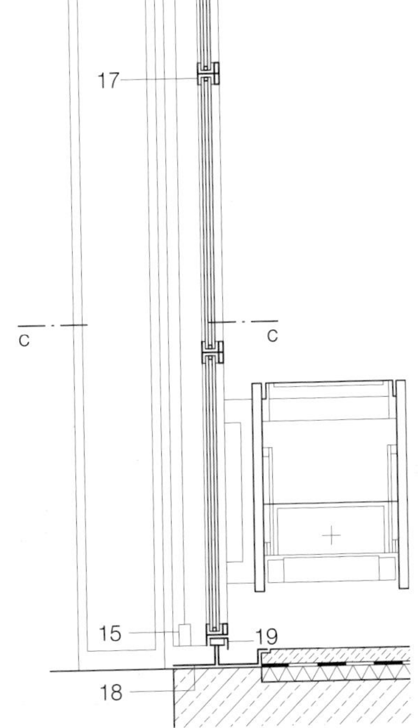

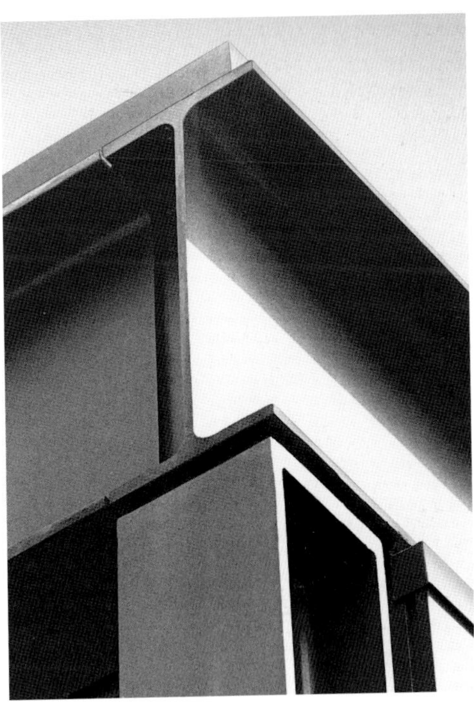

Plan • Section, scale 1:750
Horizontal sections • Vertical sections
scale 1:20

A Side elevation, plain facade areas
B End elevation with windows

1 Steel column, IPB 200
2 Steel angle, 80 x 40 x 5 mm
3 Steel square hollow section, 60 x 60 x 5 mm
4 Steel Z-section, Z 30
5 Sheet aluminium cladding, anodised, 250 mm
6 Thermally insulated panel, 25 x 500 mm
7 Steel secondary beam, IPE 550
8 Insulating panel
9 Steel beam, IPE 360
10 Suspended ceiling, insulated
11 Steel T-section, T 50
12 Steel angle, 100 x 50 x 5 mm
13 Steel angle, 40 x 40 x 4 mm
14 Steel angle, 100 x 50 x 5 mm
15 Tensioning pot for louvre blind, steel tube, 30 dia. x 2 mm
16 Steel flat or T-section for fixing radiatorr
17 Frame for fixed glazing, 50 x 50 x 5 mm T-section
18 Steel angle, 100 x 50 x 6 mm
19 Steel angle, 40 x 20 x 5 mm

Technical College

Brugg-Windisch, CH, 1966

Architect:
Fritz Haller, Solothurn
Facade design:
Hans Diehl, Neuenhof Baden

Bauen + Wohnen 08/1968
Detail 01/1969
Wichmann Hans (ed.): System Design Fritz Haller. Bauten – Möbel – Forschung. Basel 1989

- Early example and prototype for ideal industrial shaping of sheet metal
- Facade elements newly designed for this building
- No significant ageing phenomena due to solution chosen and use of stainless steel

Plan scale 1:1500
Vertical section • Horizontal section
scale 1:20
Details, scale 1:5

1 Sheet metal capping
2 Sheet metal edging to floor/roof slab
3 Plain facade element:
deep-drawn stainless steel sheet
thermal insulation
flat sheet steel
4 Louvre blind
5 Double glazing
6 Stainless steel transom
7 Column, steel circular hollow section, 318 mm dia., clad for fire protection
8 Condensation water drain, 8 mm dia.
9 Air-conditioning unit, sprayed cladding
10 Primary air duct
11 Sheet metal ceiling termination, sprayed
12 Steel angle, 70 x 70 x 6 mm
13 Stainless steel mullion
14 Supporting construction for curtain wall
15 Glazing bead
16 Glazing bead

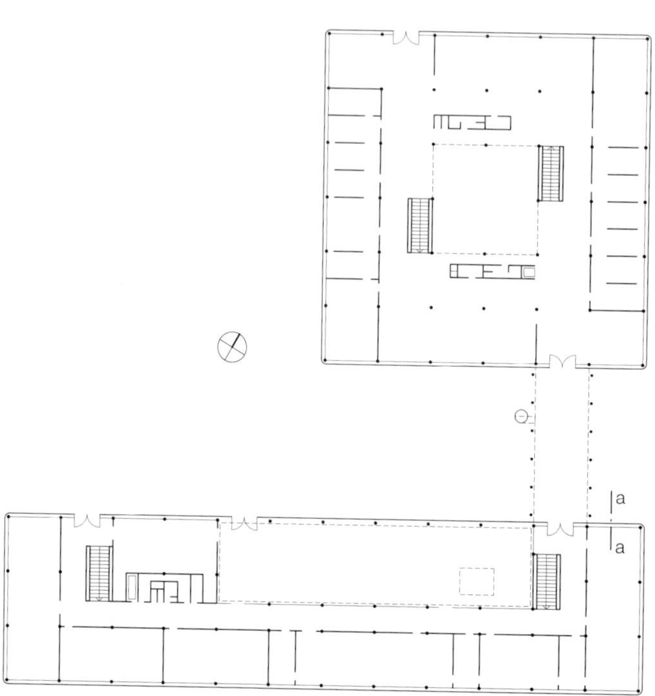

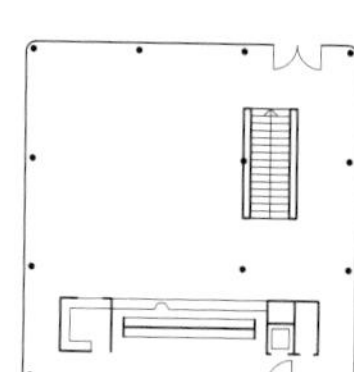

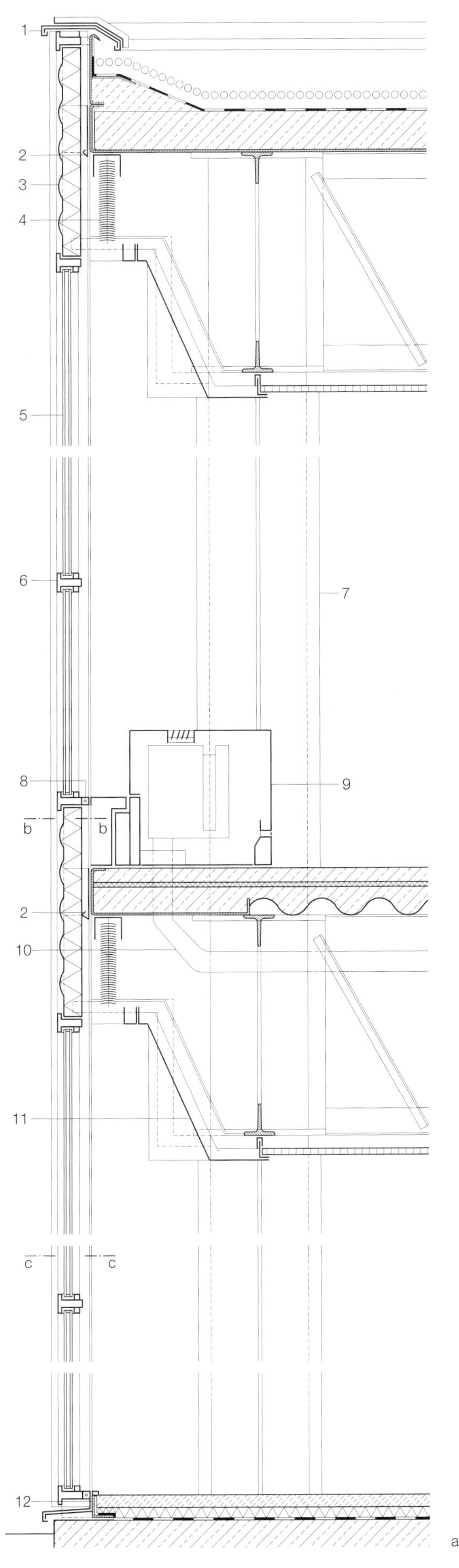
1
2
3
4
5
6
7
8
9
b
b
2
10
11
c
c
12
aa

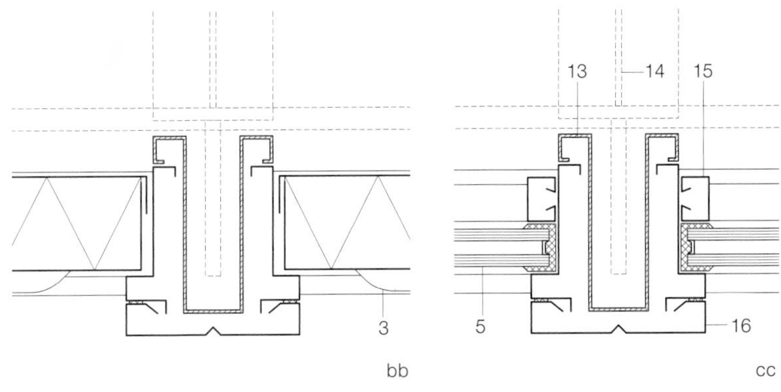
13
14
15
3
5
16
bb
cc

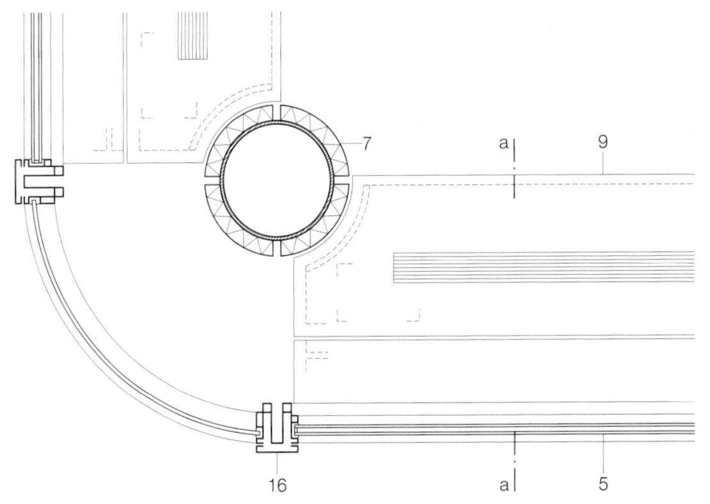
7
a
9
16
a
5

Sainsbury Centre for Visual Arts

Norwich, GB, 1978

Architects:
Norman Foster & Associates, London
Structural engineers:
Anthony Hunt Associates, Cirencester

l'architecture d'aujourd'hui 09/1991
Foster, Norman: Buildings and Projects of Foster Associates, vol. 2. Hongkong 1989
Busse, Hans-Busso et al.: Atlas Flache Dächer. Nutzbare Flächen. Munich/Basel, 1992

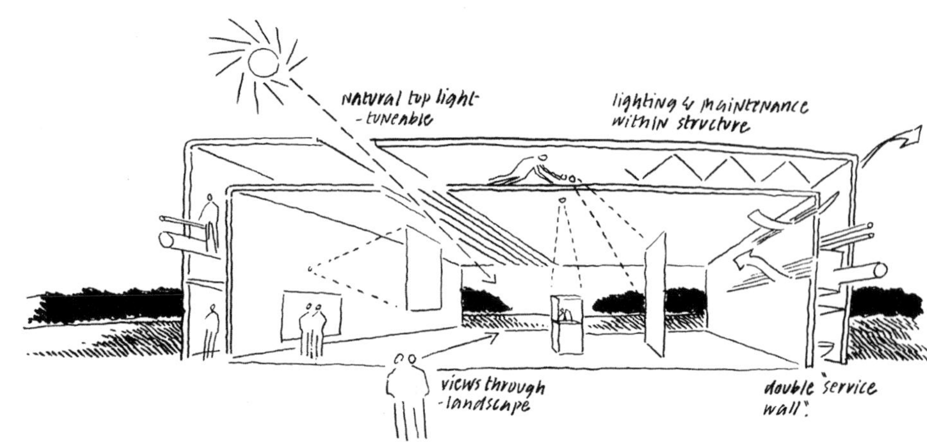

- Exhibition areas, restaurant, offices and common areas in one large "neutral" space
- Roof and facade built with similar elements
- Services incorporated in 2.40 m wide peripheral zone

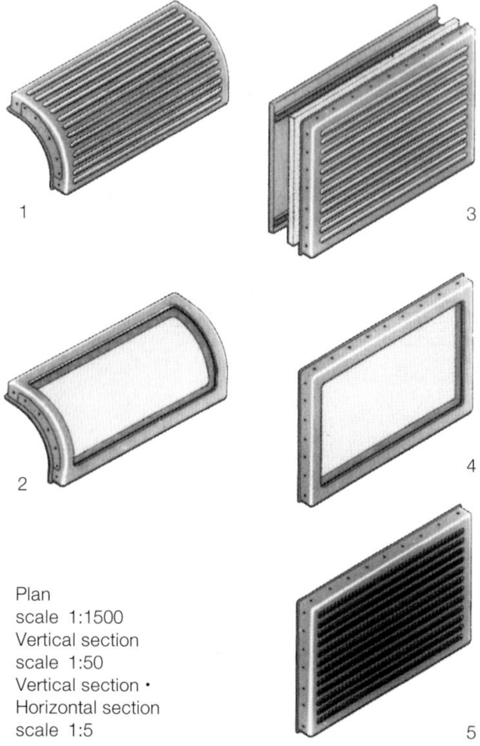

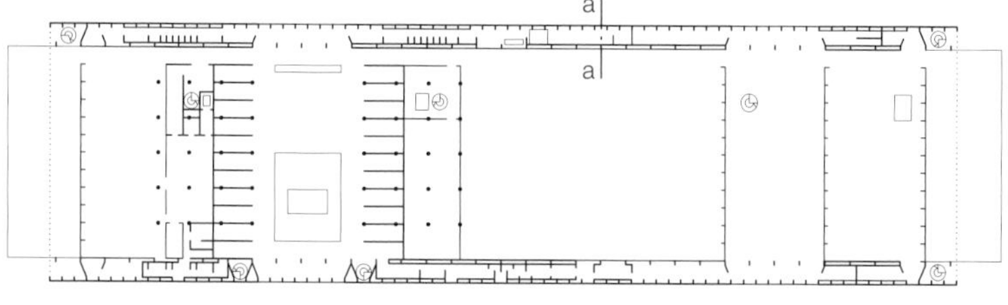

Plan
scale 1:1500
Vertical section
scale 1:50
Vertical section • Horizontal section
scale 1:5

1 Plain roof edge panel
2 Glazed roof edge panel
3 Plain panel:
ribbed aluminium sheet
75 mm thermal insulation
flat aluminium sheet
4 Glazed panel:
body-tinted laminated glass with UV filter
5 Panel with ventilation louvres
6 Steel circular hollow section, 120 mm dia., braced by diagonals,
7 Steel flat, 180 x 45 x 12.5 mm, welded to No. 6 and No. 8
8 Steel flat, 180 x 100 x 3 mm
9 Supporting construction of aluminium sections
10 EPDM sea
11 Bolted connection
12 Steel channel section

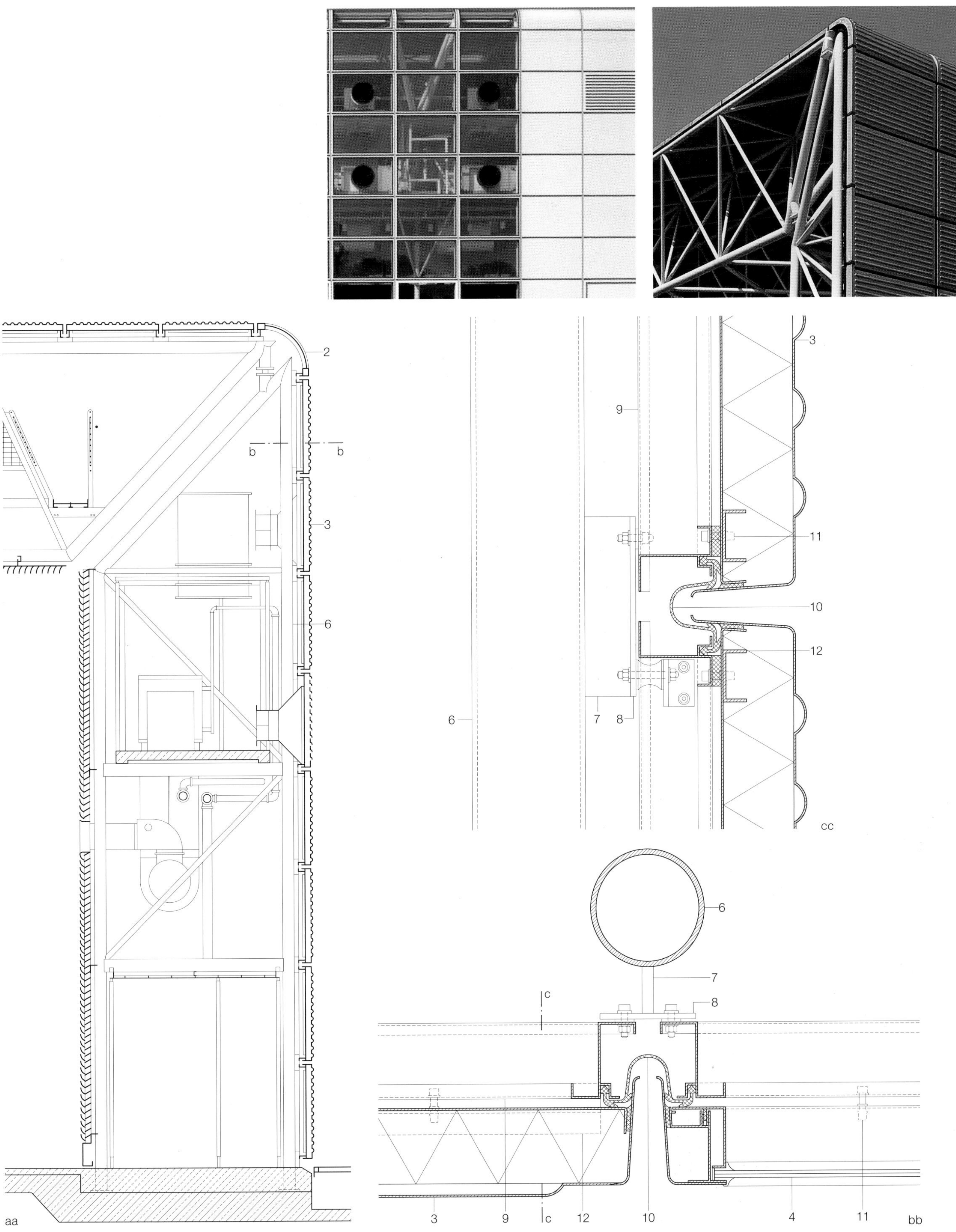
2
b
b
3
6
aa
3
9
11
10
12
6
7
8
cc
6
7
8
c
c
3
9
12
10
4
11
bb

Private house

Sottrum-Fährhof, D, 1995

Architects:
Schulitz + Partner, Braunschweig

Bauzeitung 04/2001
DBZ 12/1997
Schulitz, Helmut C. et al.:
Steel Construction Manual,
Munich/Basel, 1999

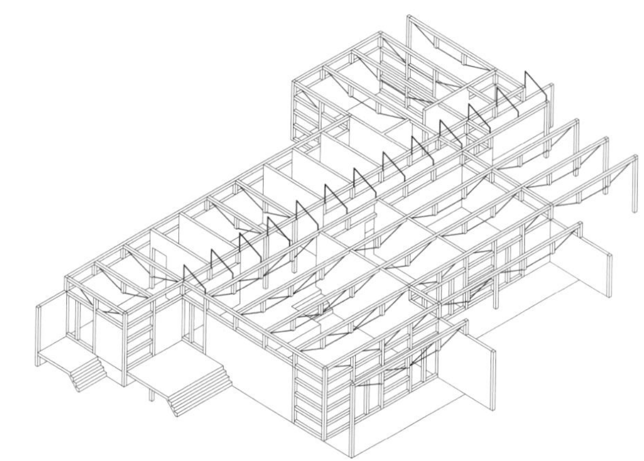

- Steel frame system building based on strict grid
- External wall in form of post-and-rail construction
- System grid: 1.80 x 1.80 m

Isometric view • Detail isometric view
(not to scale)
Vertical section • Horizontal section
scale 1:20
Horizontal section below rooflight (skylight)
scale 1:5

1 Light metal glazing cap, 45 x 26 mm
2 Double glazing, 24 mm
3 Gluelam post-and-rail construction
4 Facade construction:
18 x 76 mm light metal corrugated sheets
45 mm battens/ventilated cavity
airtight and rainproof membrane
19 mm chipboard
battens, with:-
50 mm mineral fibre insulation in between
70 mm mineral fibre insulation
vapour barrier
2 No. 12.5 mm plasterboard
5 Steel rectangular hollow section, 50 x 100 x 2.9 mm
6 Steel square hollow section, 80 x 80 x 2.9 mm
7 Squared timber section
8 Sheet steel bent to form spacer for ventilated cavity
9 EPDM seal

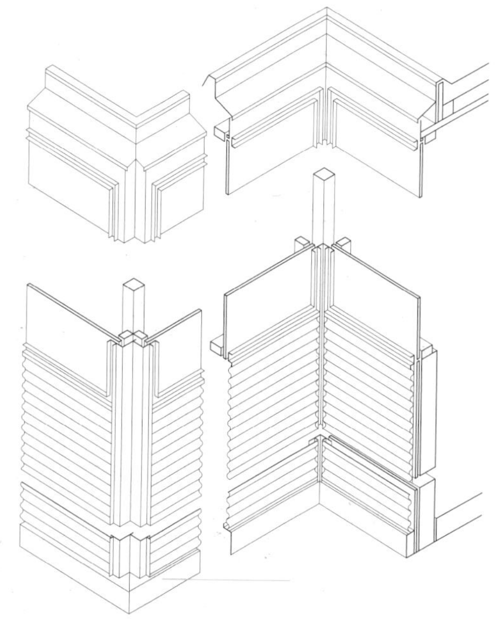

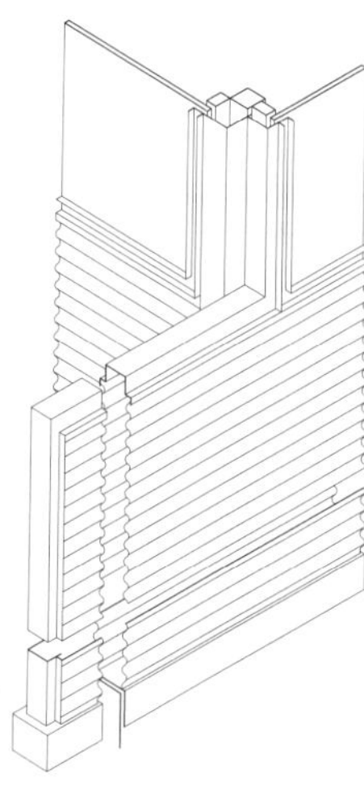

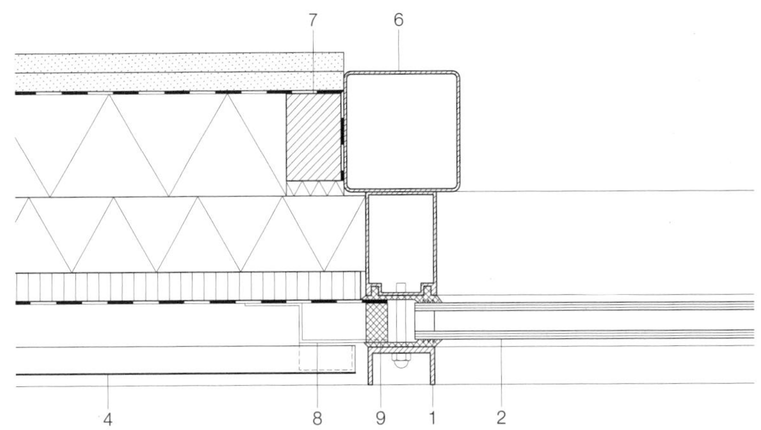
7
6
4
8
9
1
2

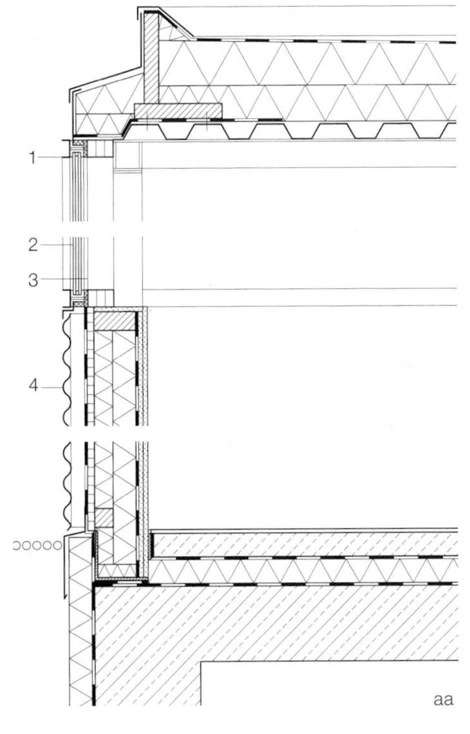
1
2
3
4
aa

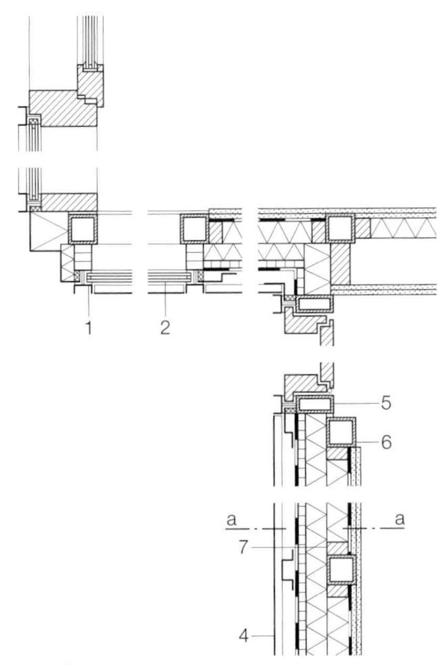
1
2
5
6
a
a
7
4

Pavilion

Amsterdam, NL, 2000

Architects:
Steven Holl, New York
Rappange & Partners, Amsterdam

Architectural Record 10/2000
Baumeister 09/2000
DBZ Sonderheft Büro + Architektur, 2001
domus 830, 2000
Schittich, Christian (ed.):
Gebäudehüllen, Munich/Basel, 2001

- Perforated copper sheet employed inside and outside
- Different levels of permeability achieved by overlapping three layers with cut-outs
- Very deep facade – approx. 1200 mm
- Large day/night contrast
- Partial use of fluorescent paint on the inside of the wall, with indirect redirection of light

1 Steel T-section, 60 x 60 x 5 mm
2 Steel flat, 100 x 6 mm
3 patinated copper sheet, perforated, panel size 1000 x 2100 x 4 mm, fixed with stainless steel screws to powder-coated supporting construction
glass-fibre panel with synthetic resin coating
80 mm rigid foam thermal insulation
150 mm calcium silicate masonry
60 x 55 mm timber section
16 mm medium density fibreboard, perforated, with birch veneer
4 Double glazing, transparent
5 Double glazing, translucent
6 Lighting

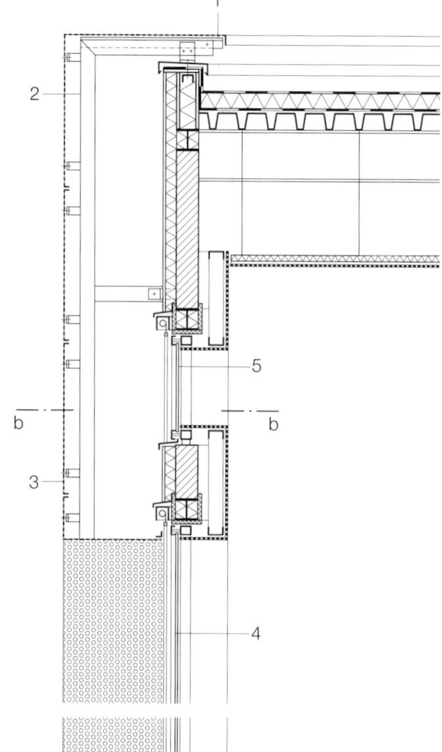

Plan, scale 1:750
Vertical section, scale 1:50
Horizontal section, scale 1:20

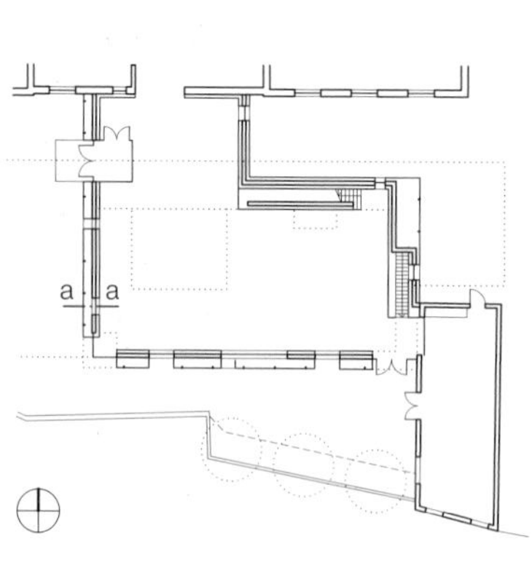

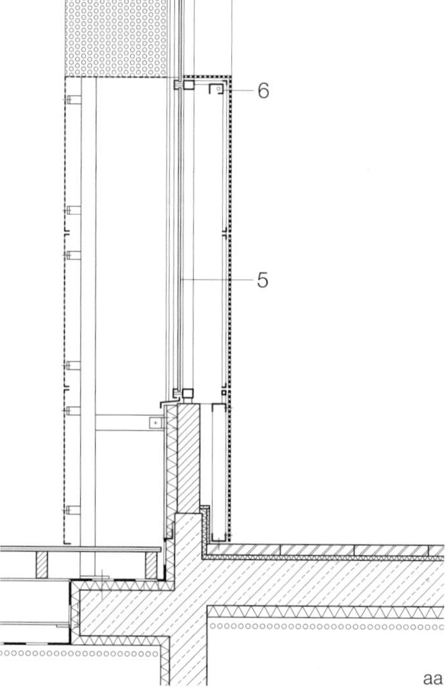

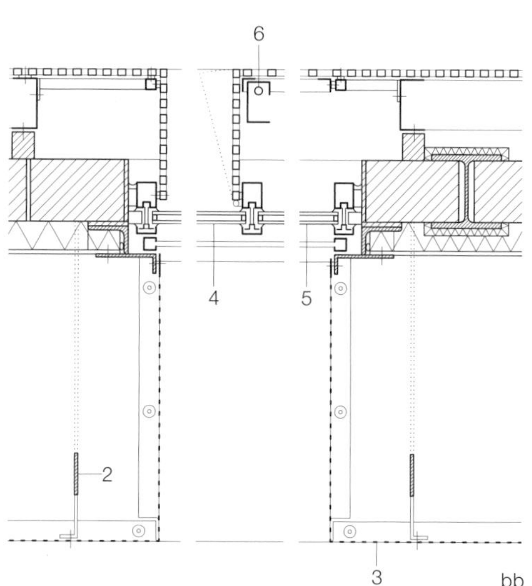

Scandinavian embassies

Berlin, D, 1999

Architects:
Berger + Parkkinen, Vienna
Pysall Ruge, Berlin
Structural engineers:
IGH, Berlin
Facade system (copper strip):
DEWI, Vienna

AIT 12/1999
l'architecture d'aujourd'hui 07–08/2000
A+U 384, 2002
domus 07–08/2000

- Permeable, separate wall of copper louvres at various angles as a curtain wall linking the six embassy buildings of the Scandinavian countries
- References to the buildings behind the facade achieved by the different openings
- Light, air and views controlled by the extent to which the louvres are "open"
- Total length of copper strip: 226 m
- Total number of copper louvres: 3926

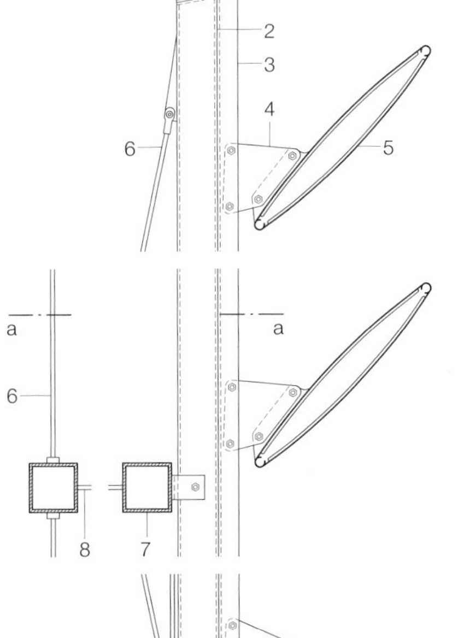

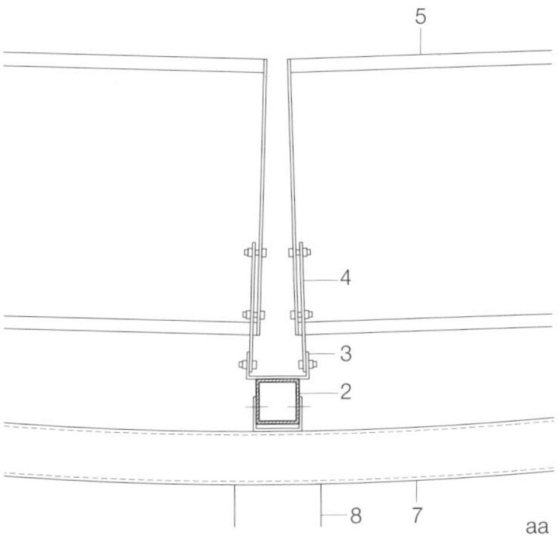

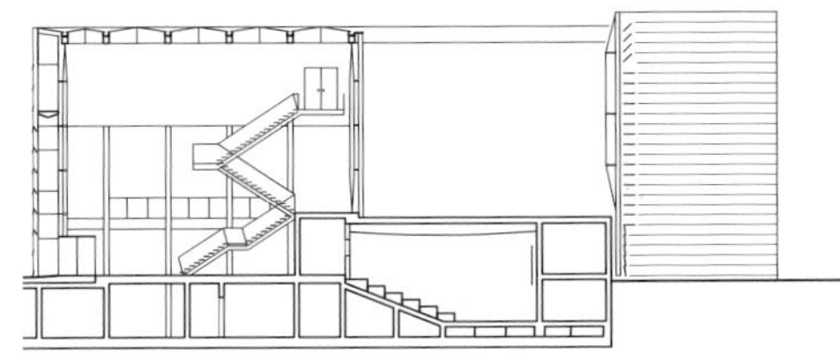

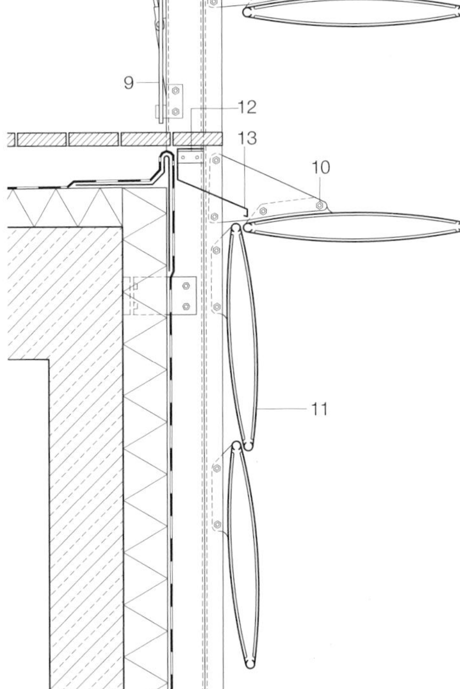

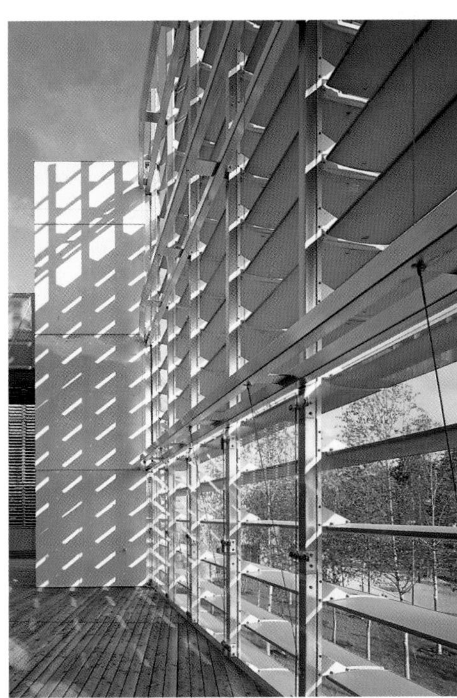

Part-section, scale 1:750
Vertical section through facade
Horizontal section through louvres
scale 1:20

1 Anti-pigeon wire
2 Stainless steel square hollow section, 100 x 100 mm
3 Stainless steel channel, 120 x 50 x 3 mm
4 Stainless steel sheet, 4 mm
5 Louvre, sheet copper, pre-weathered
6 Stainless steel guying cable
7 Stainless steel square hollow section, 120 x 120 mm, ground, screwed to No. 2 via channel section
8 Stainless steel web plate, 10 mm, ground
9 Glazing with discrete fixings as wind barrier at bottom of facade
10 Stainless steel screw fixings, with sleeves on one side to allow movement
11 Facade construction:
louvre, sheet copper, pre-weathered
100 mm air cavity
waterproofing
120 mm mineral wool insulation, laminated with black fleece
200 mm reinforced concrete
12 Stainless steel angle
13 Sheet copper flashing

Kalkriese Museum

Bramsche, D, 2002

Architects:
Gigon & Guyer, Zürich
with Volker Mencke

Architecture 09/2002
A+U 10/2000
Casabella 706–707, 2002/2003
DBZ 06/2002
Detail 01–02/2003
El Croquis 102, 2000

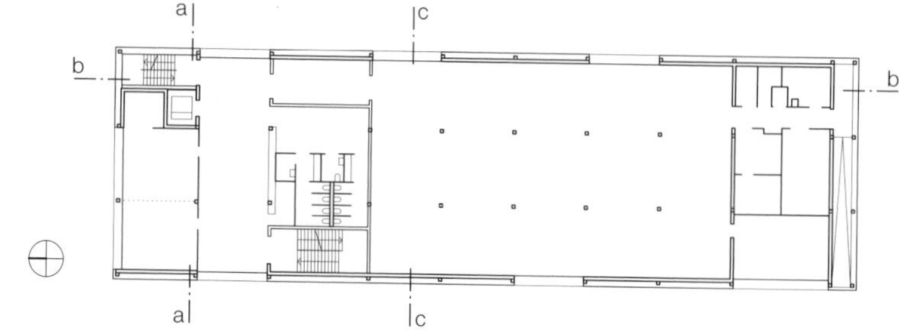

- Situated on the alleged site of the "Battle of Teutoburg Forest", 9 A.D.
- Weathering steel used throughout, also in the associated landscaped park
- Patinated material was chosen to symbolise the passage of time

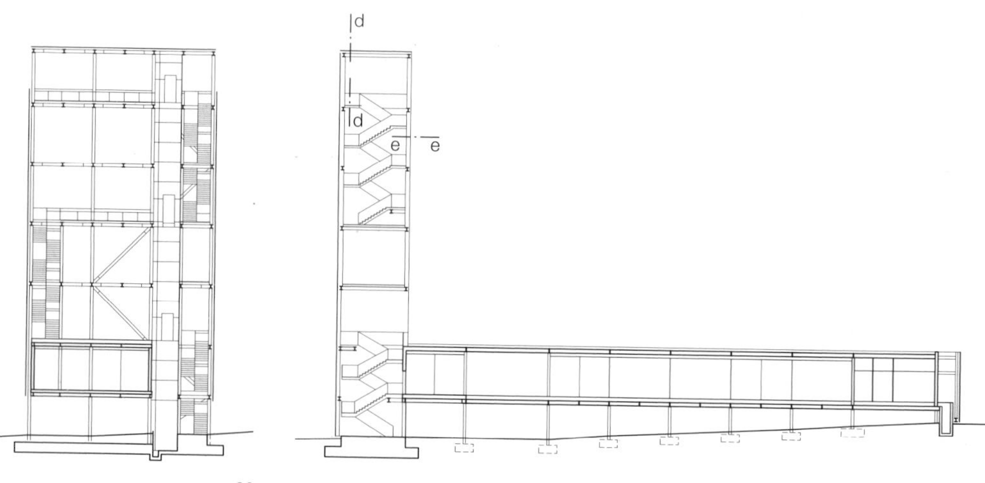

Plan • Sections, scale 1:750
Vertical section through facade, scale 1:20
Vertical section • Horizontal section through staircase tower, scale 1:20

1 Air outlet, perforated sheet metal
2 Steel beam, IPE 300
3 Steel beam, HEB 300
4 Fixed glazing:
15 mm toughened safety glass
steel angle frame, 90 x 60 x 8 mm
plus steel flat, 90 x 5 mm
5 Wall construction
double glazing: 2 No. 5 mm laminated safety glass + cavity + 8 mm float glass
65 mm steel frame, welded to sheet steel with foam core
6 Facade construction:
weatherproof steel sheet, 5900 x 3100 x 15 mm, sandblasted surface finish, horizontal edges with 10° chamfer, 20 mm joints
100 mm mineral fibre insulation
vapour check
175 mm precast aerated concrete units
sheet steel, hot rolled or pickled (depending on area of building), clear lacquer finish,
400 x 120 x 3 mm, 4 mm joints, positioned 100 mm in front of wall
7 Handrail, steel tube, 37 mm dia.
8 Landing element, sheet steel, 10 mm
9 Facade plate, weatherproof steel sheet, 15 mm, sandblasted surface finish
10 Horizontal fixing, steel angle, 6 No. per plate, connected to loadbearing structure and facade plates via bolts welded on
11 Vertical fixing with two adjusting screws per plate

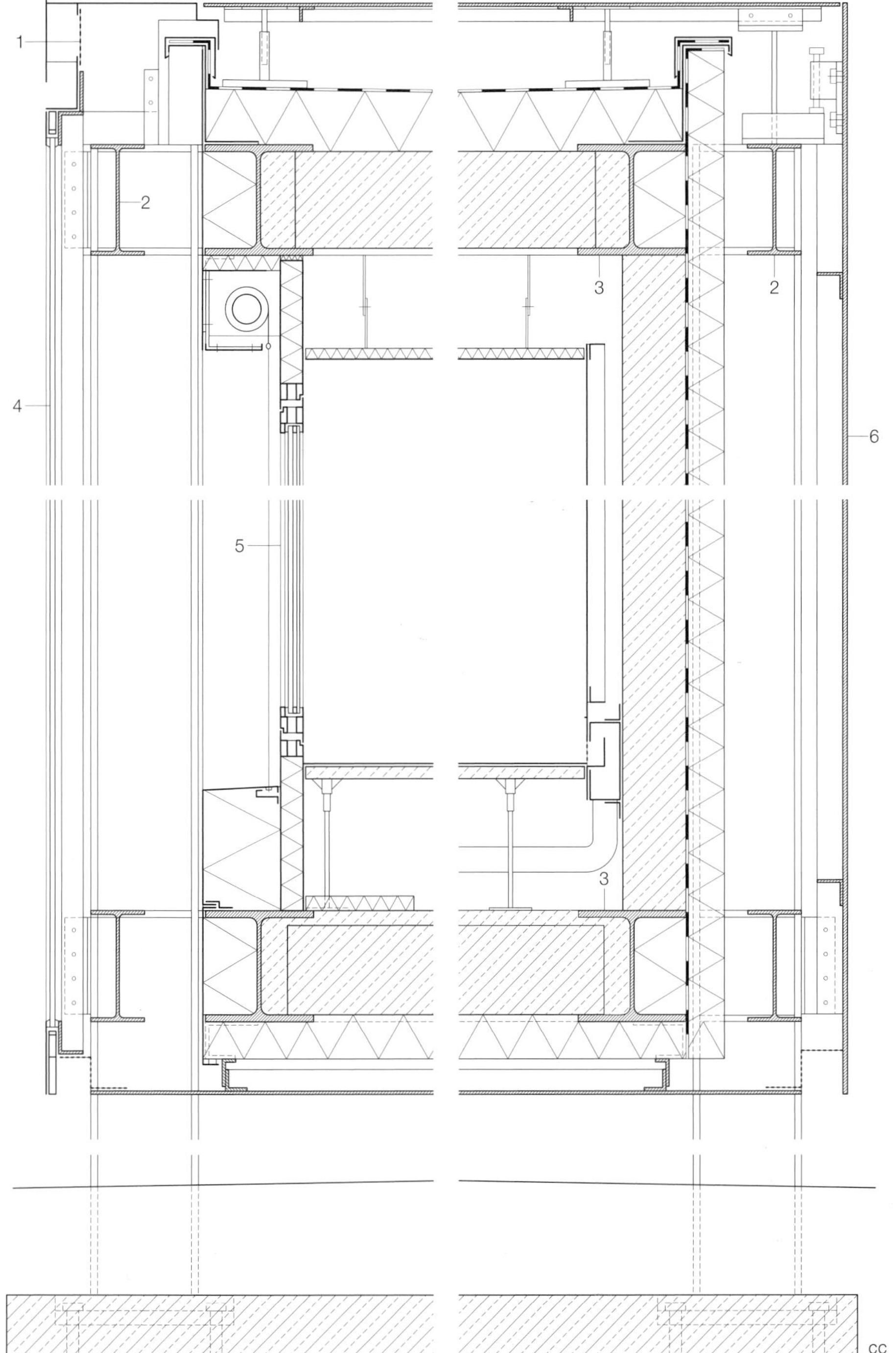

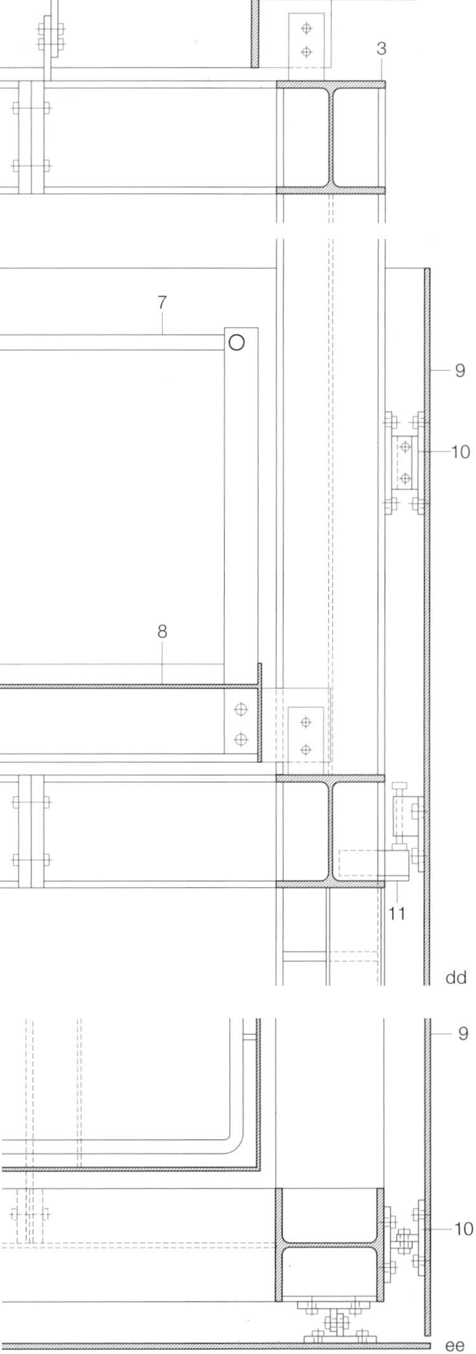

Museum of Contemporary Art

Chicago, USA, 1996

Architect:
Josef Paul Kleihues, Berlin/Dülmen
Local architect:
A. Epstein and Sons, Chicago

Architectural Record 08/1996
DBZ 03/1997
Mesecke, Andrea; Scheer, Thorsten (ed.): Museum of Contemporary Art Chicago. Josef Paul Kleihues. Berlin 1996

- The square forms the basis for the proportions determining the design; integrated into the facade in a grid of broad bands.
- Facade made from light, pyramid-shaped, iron filings-blasted cast aluminium plates, fixed with stainless steel screws to form a curtain wall
- The irregular shading (patination) of the facade is due to the corrosion of the small iron particles that remain embedded in the soft aluminium surface after the blasting process

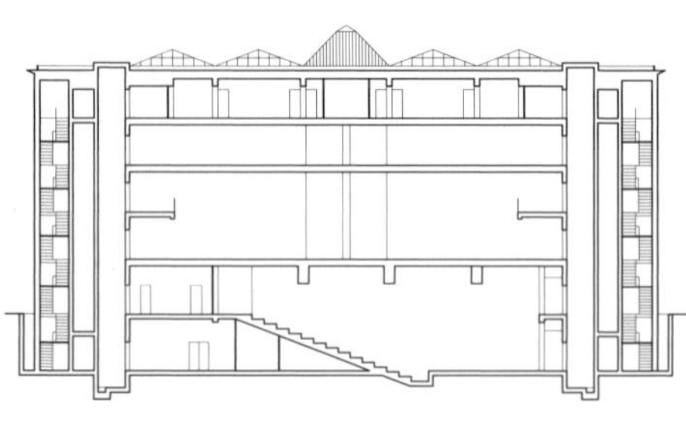

aa

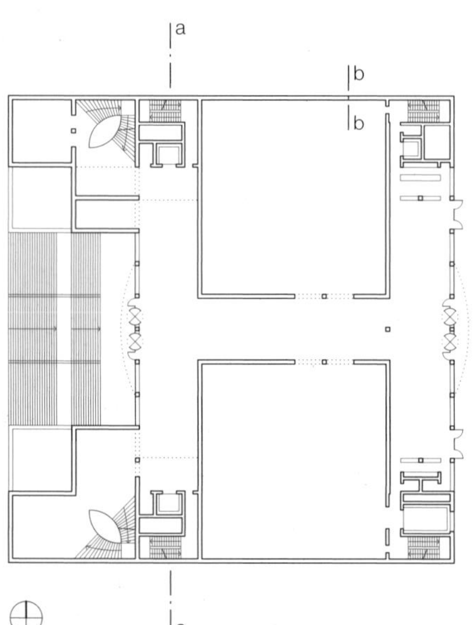

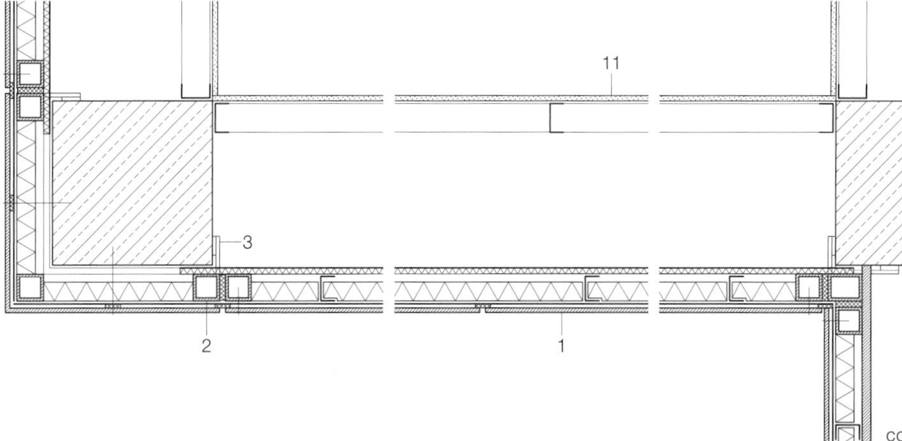

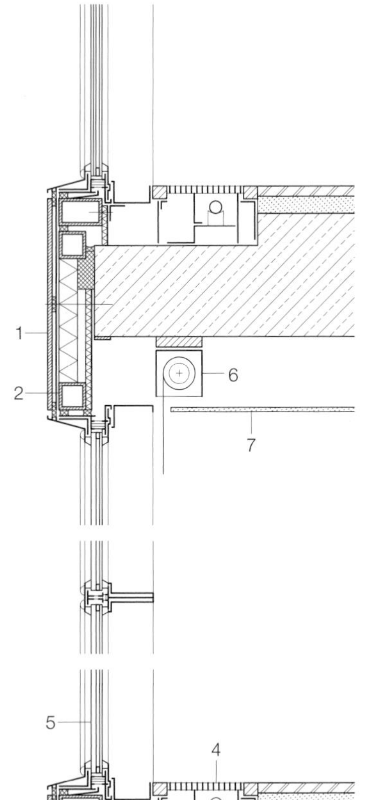

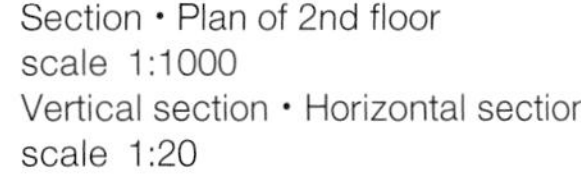

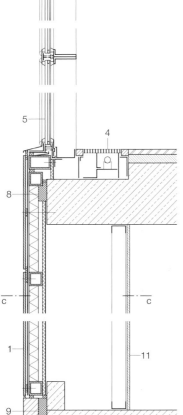

Section • Plan of 2nd floor
scale 1:1000
Vertical section • Horizontal section
scale 1:20

1 Facade panel:
square cast aluminium panels with textured surface finish, fixed with specially fabricated stainless steel screws, left exposed externally
air cavity
galvanised steel sheet
extruded rigid polystyrene thermal insulation
mineral fibre thermal insulation
sheet steel
2 Supporting framework of steel square hollow sections, 65 x 65 mm
3 Steel flat for fixing the steel supporting framework to the primary loadbearing structure
4 Grating over heating pipes, anodised aluminium in wooden frame
5 Aluminium window with double glazing, 16 mm toughened safety glass + 12 mm cavity + 6 mm toughened safety glass
6 Electrically operated roller blind to combat glare
7 Plasterboard suspended ceiling
8 Fire stop
9 EPDM sealing strip in open (drained) joint
10 Plinth area:
calcium silicate panel, fixed with specially fabricated stainless steel screws, left exposed externally
11 Plasterboard partition on metal framing

B 1.6 Glass

Glass is among the oldest building materials. Evidence for the use of natural glass from a volcanic origin for making knives and arrowheads, as well as for the production of opaque glass, has been discovered in finds dating back to 5000 B.C. One significant development on the road to the glass we use today was the invention by the Syrians of the blowing iron in the second century B.C. This invention made it possible to produce hollow vessels for the first time. Later, a casting method enabled the Romans to produce the first flat, albeit hardly transparent, glass.

From natural glass to the universal building material of the twenty-first century

Sheet glass

By the first century A.D., improved formulations for glass and the development of the blown cylinder sheet glass process made possible the production of the first flat, transparent glass. The invention of the crown glass method in the fourth century A.D. resulted in transparent panes with very smooth surfaces. These two methods, invented by the Syrians, underwent further development over the course of time and remained the dominant methods of glass production right up until the end of the nineteenth century [1].

The next major step in the evolution of sheet glass production methods was the introduction of drawn glass processes after 1905. These involved drawing the hot, still viscous, glass mass over rollers or through a nozzle of fired clay and subsequently cooling it. For the first time, it was possible to produce high-quality glass economically in large quantities.

The most important step on the road to producing inexpensive, high-quality glass was Alastair Pilkington's invention of the float glass process in 1959. Here, the viscous glass melt is passed over a bath of molten tin at approx. 1000°C. Owing to its lower specific weight, the glass floats on the level surface and spreads out absolutely evenly. At the exit from the bath, the plane-parallel and almost solid glass mass is conveyed on rollers, carefully cooled and finally cut to size. This method, now used worldwide, makes possible the economic production of high-quality glass in all manner of variations [2].

Glass bricks and blocks

One intriguing invention of the late nineteenth century is Gustave Falconnier's hand-blown glass brick (1886). These bricks were later used by well-known architects such as Guimard, Perret and Le Corbusier. The "reinforced concrete glass block", a solid design used from 1907 onwards, incorporated lateral grooves for interlocking with reinforced concrete cames. For the first time, it was possible to create large, loadbearing and light-permeable panels. There were also hollow glass blocks in shell

B 1.6.1 Bauhaus, Dessau (D), 1926/1976, Walter Gropius

B 1.6.2

B 1.6.3

B 1.6.4

B 1.6.2 Sainte Chapelle, Paris (F), 1244
B 1.6.3 Crystal Palace, London (GB), 1851, Joseph Paxton
B 1.6.4 Fagus Works, Alfeld (D), 1911, Walter Gropius
B 1.6.5 National Library, Paris (F), 1997, D. Perrault

B 1.6.5

B 1.6.6

Silicon dioxide	(SiO_2)	69–74%
Calcium oxide	(CaO)	5–12%
Sodium oxide	(Na_2O)	12–16%
Magnesium oxide	(MgO)	0–6%
Aluminium oxide	(Al_2O_3)	0–3%

This composition has been standardised for Europe in EN 572, part 1.

B 1.6.7

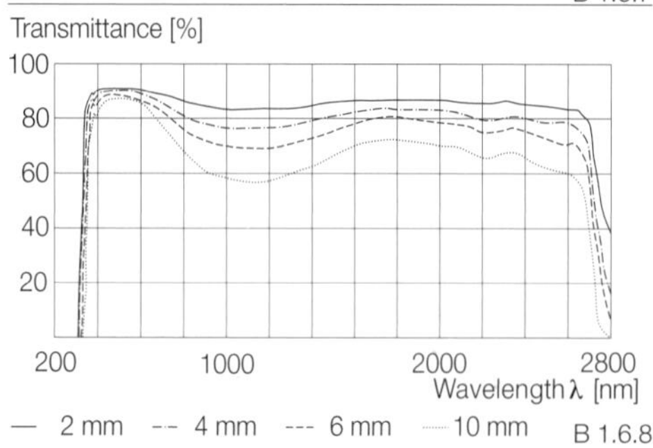

B 1.6.8

Property	Symbol	Value and unit
Density at 18 °C	r	2500 kg/m³
Hardness		6 units on the Mohs scale
Modulus of elasticity	E	7 x 1010 Pa
Poisson's ratio	m	0.2
Specific heat capacity	c	0.72×10^3 J/(kg x K)
Average coefficient of thermal expansion	a	9×10^{-6} K^{-1}
Thermal conductivity	l	1 W/(m x K)
Average refractive index for the visible range of wavelengths 380–780 nm	n	1.5

B 1.6.9

B 1.6.6 Sainsbury Centre, Norwich (GB), 1978, Norman Foster & Associates
B 1.6.7 Composition of glass
B 1.6.8 Spectral transmission curves for various thicknesses of float glass having an average Fe_2O_3 content of 0.10% Fe_2O_3
B 1.6.9 General physical properties of glass
B 1.6.10 Manufacture of glass products for facades

form which could be built in with the opening facing either inwards or downwards. These found applications in the glass-roofed arcades of Prague, Budapest and other European cities. The glass block familiar to us today first appeared around 1930, when two shell-like glass bricks were pressed together using heat and pressure – a method that is still used today.

Material properties

Composition

Glass essentially consists of sand, soda and lime, plus other additives that are melted down for production at temperatures exceeding 1000°C (fig. B 1.6.7). The melt gradually solidifies without crystallising at temperatures below about 680°C (float glass), and the transition from the liquid to the solid state remains reversible. The high transparency of glass is due to the lack of a crystalline molecular structure, which allows light to penetrate the material without scattering. Owing to its molecular structure, glass is an amorphous, isotropic material, that is, its physical properties do not depend on direction [3].

Optical properties

The spectral transmittance of glass with respect to solar radiation covers the wavelengths from approx. 300 to approx. 2500 nm. Glass is opaque to the long-wave infrared part of the spectrum above 2500 nm and ultraviolet radiation below 315 nm (fig. B 1.6.8). However, the majority of the short-wave solar radiation passes through the glass and heats up any objects or surfaces behind the glass. These in turn reflect long-wave heat radiation which can no longer penetrate the glass. This results in a temperature rise in the interior, which we call the "greenhouse effect".

Thermal properties

In building we use primarily soda-lime-silica glass, whose thermal expansion is roughly equivalent to that of steel. However, it lies far below the coefficient of thermal expansion of aluminium, which is something that needs to be considered in facade constructions especially. The high thermal conductivity of glass results in a thermal transmittance (U-value) of 5.75 W/m²K for a 4 mm thick pane of float glass.

Mechanical properties and strength

The siliceous content of glass provides hardness and strength, but at the same time a distinct brittleness. In contrast to elastically deformable materials like metal, glass fractures immediately once the limit of its elastic deformability has been exceeded. At 1000 N/mm², the compressive strength of glass is comparable to that of steel. However, the ultimate bending strength of conventional float glass is only about 30 to (max.) 60 N/mm².

Chemical properties

Owing to its siliceous composition, glass is highly resistant to aggressive substances, except for hydrofluoric acid, hot alkaline solutions and water. The latter becomes particularly problematic if allowed to remain standing on a glass surface for a long time, e.g., on horizontal panes.

Behaviour in fire

Glass is an incombustible material, but starts to soften at about 700°C and, owing to its low thermal stability, is unlikely to withstand a temperature difference exceeding 60 K. Almost 100% of the heat radiation that occurs during a fire can pass through the glass.

Acoustic properties

Owing to its low mass, glass is a better conductor of sound than other building materials; however, this effect can be countered by using more than one pane. The cavity between the panes achieves an acoustic decoupling between inside and outside, which hampers sound transmission. Double and triple glazing is available in various thicknesses and the cavities can also be filled with a heavy gas.

Types of glass for facades

Float glass

Float glass is a high-quality, clear sheet glass with flat, plane-parallel surfaces. Today, it serves as the basic material for the majority of single and double glazing systems used for facades. The maximum pane size is 3210 x 6000 mm, although longer lengths are available at higher prices. It is available in thicknesses between 2 and 19 mm [4].

Sheet glass

Sheet glass is an industrially manufactured transparent drawn glass with a quality slightly inferior to that of float glass, which can be attributed to the method of drawing glass. Characteristic of this type of glass are the corrugations perpendicular to the drawing direction, which are apparent when looking through the glass and also in the reflections.

Antique glass

Antique glass is a hand-blown type of glass in which a hollow cylinder is slit open and bent into a flat pane (blown cylinder sheet glass process). The dimensions possible are limited, but small quantities of coloured glass can be made in this way.

Rolled glass

Rolled glass (patterned glass) is produced by means of a continuous rolling process. Profiled rollers are employed to produce panes of glass with all manner of textured, profiled surfaces. This type of glass is used, for example, when privacy is important, or to achieve an even

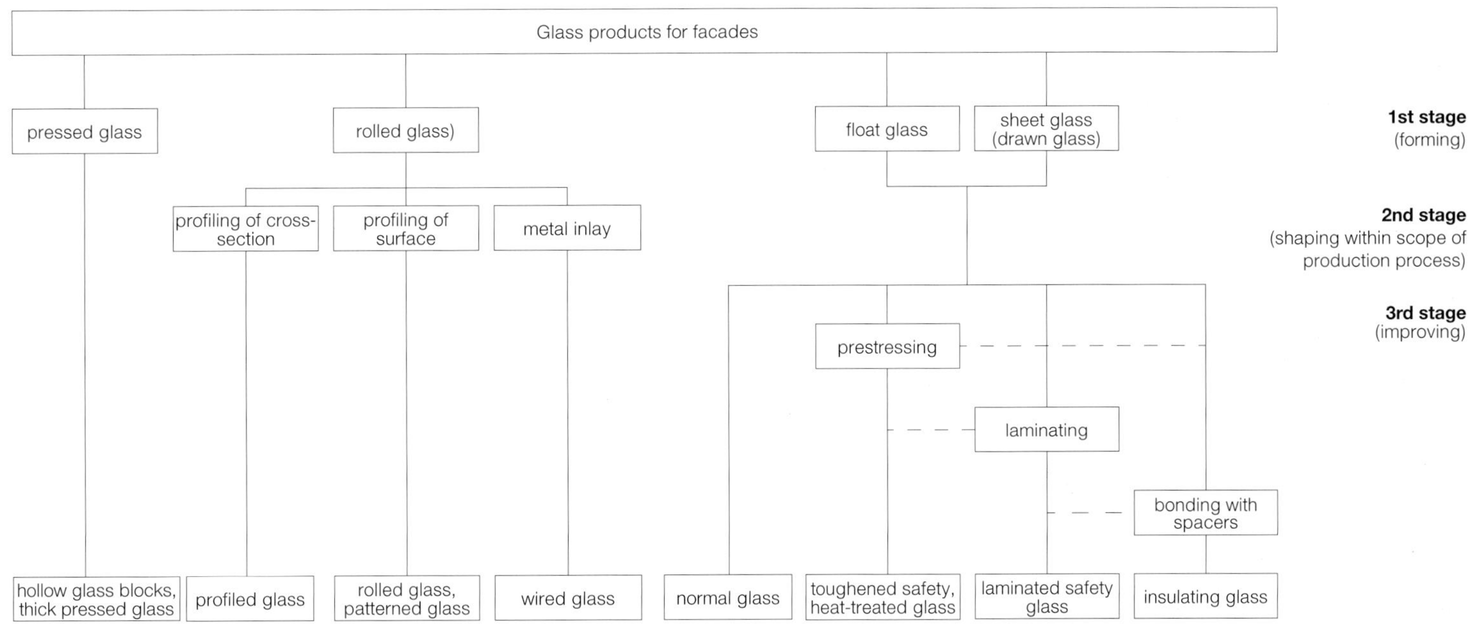

B 1.6.10

diffusion of daylight. The maximum sizes possible depend on the particular manufacturer.

Wired glass
Like rolled glass, wired glass is produced in a continuous rolling process. The difference here is that a fine wire mesh is incorporated in the glass in order to improve security and/or fire resistance.

Wired glass can be polished to achieve plane-parallel surfaces and to improve its optical properties. The maximum width available is 1980 mm, the maximum length 3820 mm.
The wire does carry a risk of rusting along the edges. These must be specially protected in order to prevent discoloration and breakage of the glass owing to the in-crease in volume associated with the corrosion of the steel wires.

Profiled glass
Profiled glass (channel glass) is produced in a second rolling process where the ribbon of glass, still hot, is bent into a "U" shape. This cross-sectional form gives the glass a high load-carrying capacity which renders possible the design of facade areas devoid of glazing bars. Profiled glass is available in widths of 220, 250, 320 and 500 mm, and a maximum length of 6000 mm.

Hollow glass blocks
These are produced by pressing two half-blocks together while still hot, whereupon the two halves fuse together. The cooling of the air in the sealed void leads to a partial vacuum, which improves the thermal insulation properties and prevents the formation of condensation water. Nevertheless, owing to the many thermal bridges, the thermal insulation values are considerably lower than those of modern double glazing units. Hollow glass blocks cannot accommodate vertical loads and may only be used for non-loadbearing applications. The standard sizes are 150 x 150 mm and 300 x 300 mm, with thicknesses of 80–100 mm.

Thick pressed glass
Thick pressed glass is a solid glass block produced in a pressing process and, in contrast to the hollow glass block, can also be used as a loadbearing component. Thick pressed glass is available in square, rectangular and circular forms. However, the number of applications is limited owing to the low thermal insulation value.

Influencing the properties of glass

The many options for influencing the properties of glass can be exploited to suit the particular application. This can be done by altering the composition of the glass, through thermal or chemical treatment, by modifying the surface of the glass, or by manufacturing laminated and insulating glass.

Modifying the composition of float glass
Minor impurities (e.g., iron oxide) in float glass give the glass a slight greenish tinge which can become noticeable with thicker panes and multi-pane or multi-layer glazing in particular. Absolute colourless glass, e.g., extra-clear glass, and also glass with certain physical properties, can be produced by modifying the chemical composition of the glass melt.

Special metal oxide additives add a hint of colour to the glass – from green or blue to bronze and grey. These body-tinted glasses cut down the transmission of radiation and hence reduce the heat gains in the interior; they also help to reduce glare. Furthermore, the addition of certain metal compounds to the glass melt enables virtually any colour to be achieved. Besides straightforward body-tinted glass, the casing or flashing method allows the application of a layer of glass over a layer of contrasting colour.

Thermal treatment
Flat or bent glass can be thermally prestressed to improve its ultimate bending strength and thermal stability. The heat treatment results in a higher characteristic strength.

Toughened safety glass
Toughened safety glass is manufactured by heating the glass to at least 640°C and then immediately quenching it in blasts of cold air. This causes the surfaces to cool and solidify quicker than the inner core of the glass, which is still hot and soft. As the core cools and solidifies, a compressive stress builds up at the surfaces which increases the ultimate bending strength (approx. 90–120 N/mm^2) and the thermal stability (float glass 40 K, toughened safety glass 200 K). Thermally toughened glass cannot be mechanically worked after the heat treatment. When it fails, toughened safety glass breaks into small, blunt fragments, which considerably decreases any risk of injury.

Heat-treated glass
The production of heat-treated (or heat-strengthened) glass likewise requires the glass to be heated to at least 640°C. However, the subsequent blasts of cold air are less intensive and so the compressive stress at the surfaces is lower. The ultimate bending strength (approx. 40–75 N/mm^2) and thermal stability (float glass 40 K, heat-treated glass 100 K, toughened safety glass 200 K) figures are therefore not as impressive as those of toughened safety glass.

The fracture patterned upon failure is different, with larger fragments (fig. B 1.6.15). When used in the form of laminated safety glass, this leads to a better residual load-carrying capacity upon failure, which is particularly beneficial for facades and overhead glazing. Just like toughened safety glass, it is not possible to work heat-treated glass mechanically after treatment.

Bent glass
Bent glass (curved glass) is produced by subjecting float glass to subsequent thermal treatment in a tunnel or muffle furnace. Cylinders or spherical forms, also with two panes on top of one another, are also possible. The possible bending radius depends on the thickness of the pane.

Fire-polished glass
Fire-polished glass is produced by heating glass to 500–700°C in a furnace. The plastic, viscous glass surface contracts as a result of the surface tension and is thus refined.

Mechanical treatments

Various methods are available for cutting and finishing the edges, and these are outlined briefly below.

Scoring
The glass can be scored with a material harder than glass, such as diamond, and subsequently broken along the score line. Various fluids (petrol, oil) can be used to reduce the risk of splintering along the score line.

Cutting
Thick panes and security glasses are generally cut with cutting discs (e.g., diamond) or lasers.

Abrasive water-jet cutting
Glass can be cut or – by reducing the pressure – notched with a very high-pressure water jet. In contrast to scoring, any shapes are possible in glass panes up to 70 mm thick. Laminated glass can also be cut with this method. The cut can be started in the solid material and the cut itself is narrower than that of other methods.

Edge work
The edge of the glass can be finished in such a way that any risk of injury is reduced. If after being installed there is no risk of the edge of the glass causing injuries, the edges can be left untreated. Grinding and polishing are employed to improve the edges, and there are several quality standards. For instance, an edge cut exactly to size may exhibit blank spots (i.e. imperfections), but a polished tabletop or the edge of a mirror should have perfect edges.

Surface treatments

Besides varying the chemical composition, the properties of the glass can be modified by surface treatments.

Obscuring process
Chemical and mechanical methods are available to decrease the transparency of a pane of glass. In acid-etching the glass is treated with hydrofluoric acid, or its vapours, which enables a fine graduation of the effect. The result is a smooth, matt surface finish with a very consistent appearance. Patterns and motifs can be included by applying wax prior to the acid treatment. This treatment does not make it more difficult to keep the glass clean afterwards.

Alternatively, the glass surface can be sandblasted. The degree to which the surface is roughened or removed depends on the size of the particles used. In contrast to acid-etched glass, the surface texture is relatively coarse, which can create traps for grease or cleaning agents, which in turn can permanently impair the appearance of the glass.

Chemical strengthening
Another surface treatment involves the chemical prestressing of the glass. To do this, the pane of glass is immersed in a bath of hot molten salt. The exchange of ions increases the compressive stress at the surface, thus improving the resistance to thermal and mechanical loads. In contrast to thermally treated glass, chemically strengthened glass can still be cut afterwards.

Surface coatings applied during manufacture
Such coatings can be applied either directly during the production of the glass as "online coatings" or, as with magnetron spattering, as "offline coatings". These coatings include those with metal oxides, which lead to a re-duction in the radiation transmission. Depending on the design of the coating, heat gains or heat losses can be reduced. Reflective coatings cut down the reflection of radiation at the surface of the glass; the reflectance of, for example, a single pane, can be reduced from 8% to 1%. Dichroic coatings, on the other hand, break up the incident light into its respective spectrum colours. Transmission or reflection in the respective different colours depends on the angle of incidence. The play of colours results in interesting artistic effects.

Online coatings
In the online method the coating material is applied in the form of a liquid, vapour or powder and forms a permanent bond with the surface of the glass in a chemical reaction. The panes treated with this "hard coating" are resistant to abrasion and chemicals, and can be used in single glazing.

Offline coatings
In the offline method the glass is coated in an immersion or vacuum process. Whereas the immersion process coats both sides of the pane of glass, the vacuum process can be used to treat just one side. The application of different individual coatings in several operations means it is possible to control the radiation properties of the glass very specifically. The coatings applied in the offline process are usually softer and less resistant than the hard coatings of the online process. Glass treated in this way is therefore used exclusively for insulating and laminated glass.

B 1.6.11

Other surface coatings
Treatments such as enamelling, silk-screen printing or painting can be used to change not only the appearance but also the radiation properties of glass.

Enamelling
The application of an enamel frit (ground glass plus additives and pigments) to the pane of glass, followed by re-firing, produces a ceramic, corrosion-resistant coating. Only glass toughened by thermal means is suitable for enamelling, because only this type of glass is able to withstand the higher thermal stresses caused by the pigments.

Silk-screen printing
Panes of glass printed with motifs and patterns can be produced in the silk-screen printing process with the help of mesh stencils. The pigments are printed on to the glass and then baked in an oven. Multi-colour graphics and photographs can be produced in this way. The size of the pane is determined by the size of the stencils and is usually max. 2.0 x 3.5 m.

Staining
Finely ground coloured glass is dissolved in a liquid (e.g., turpentine), applied to the glass and subsequently baked on at 550°C.

Multi-layer laminated glass

Multi-layer laminated glass consists of two or more panes of glass bonded together by means of a viscoplastic interlayer or a casting resin. Depending on the type of glass used, such a multi-layer assembly renders possible numerous combination options to allow the properties of the glazing to be matched to specific requirements. Laminated glass can be manufactured in flat or curved forms.

Laminated safety glass
When broken, splinters of glass from a multilayer pane of glass, bonded together by means of a viscoplastic interlayer, remain attached to this interlayer and thus substantially reduce the risk of injury. Laminated safety glass can be manufactured from conventional float glass, toughened safety glass or heat-treated glass.

Bonding with printed or coloured films
One – inexpensive – way of producing coloured laminated glass is to use a coloured interlayer; printed films can also be used. Laminated glass incorporating holographic optical elements (HOE) can be used to redirect the incoming light in various ways – similar to the use of prisms and lenses. The high luminous intensity of holograms is due to the directional emission of light within a certain angular range, which is governed by the diffraction grating used during the creation of the hologram.

Fire-resistant glass
Interlayers containing aqueous gels are employed to produce fire-resistant glass. The protective effect is based on the heat of evaporation of the water. When the pane on the side exposed to the fire breaks, the gel remains bonded to the side not exposed to the fire and gradually releases water vapour in the direction of the fire. A large proportion of the radiated energy is thus consumed. When combined with safety glass, such panes can achieve long fire-resistance times (fig. B 1.6.16). G-glasses prevent the passage of flames and smoke for much longer than the specified fire-resistance time but do not prevent the transmission of heat radiation, for which F-glasses are available.

B 1.6.12

B 1.6.13

B 1.6.11 Peckham Library, London (GB), 1999, Alsop + Störmer
B 1.6.12 Laminated safety glass with ceramic printing, new trade fair grounds, Leipzig (D), 1996, von Gerkan Marg and Partner
B 1.6.13 Profiled glass facade, institutional building, Paris (F), 1998, Brunet & Saulnier
B 1.6.14 Art Gallery, Bregenz (A), 1997, Peter Zumthor
B 1.6.15 Fracture patterns (not to scale) of:
a normal glass
b heat-treated glass
c toughened safety glass

B 1.6.14

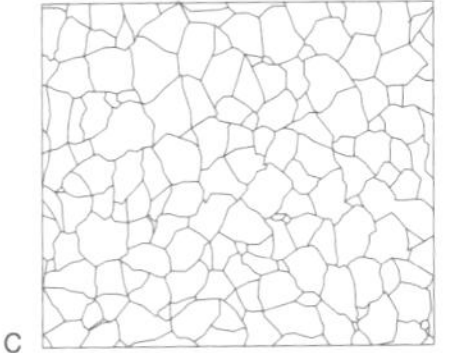

B 1.6.15

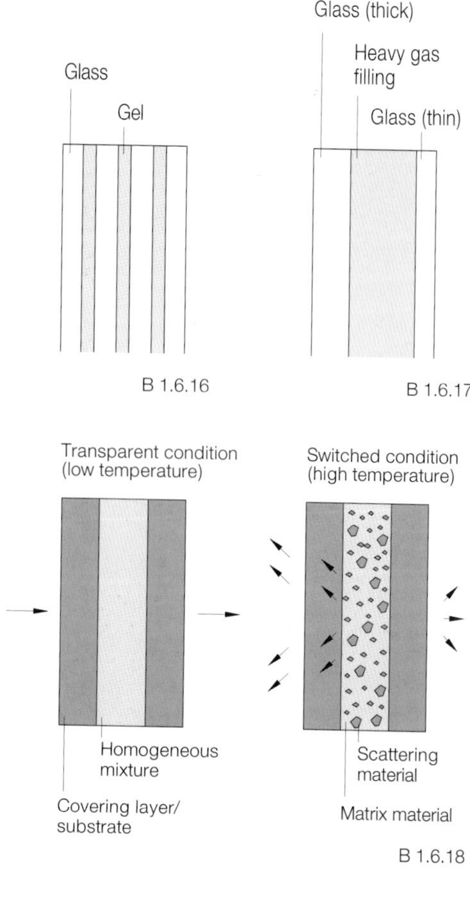

B 1.6.16

B 1.6.17

B 1.6.18

	g-value	Light trans-mittance	Appea-rance
Thermotropic solar-control window	0.18-0.55	0.21–0.73	white to clear
Electrochromic window	0.12-0.36	0.20–0.64	blue to neutral
Gasochromic solar-control window	0.15-0.53	0.15–0.64	blue to neutral

The figures quoted here may alter considerably in the course of further developments.

B 1.6.19

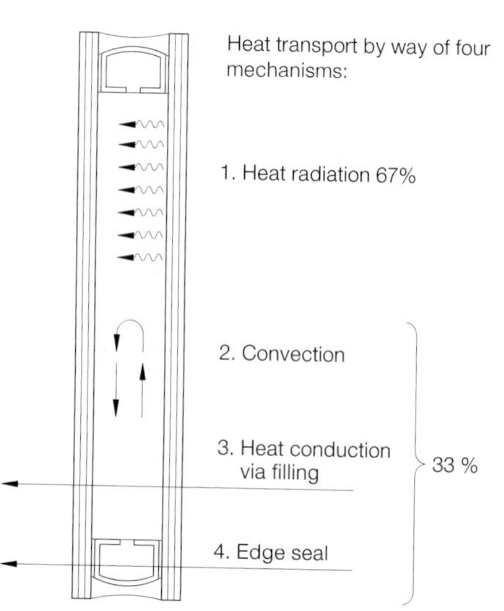

B 1.6.20

Glass for sound insulation

The use of laminated glass with different pane thicknesses plus a heavy gas filling in the cavity between the panes considerably improves the sound reduction index of insulating glazing. Whereas conventional double glazing (4 mm + 16 mm cavity + 4 mm) achieves a sound reduction index R_w of 30 dB, an asymmetric arrangement (4 mm + 16 mm cavity + 8 mm) plus a suitable gas filling can achieve 35 dB [5]. An increase in the sound reduction index of up to 47 dB is possible by using multi-layer laminated glass and a wider cavity. However, the heavier glass requires special frames and fittings (fig. B 1.6.17). A certain frequency range of the external noise is filtered out depending on the thickness of the individual panes. This means that the acoustic performance properties of the glazing can be matched exactly to the requirements.

Thermotropic glass

A liquid comprising two components, such as water and a gas (hydrogel), is fixed between two panes of glass or glass films. Up to a certain temperature this mixture remains homogeneous and the "captive" layer remains transparent. However, once this limiting temperature is exceeded, the two components segregate. The layer turns a cloudy white colour and reflects the majority of the incoming light in a diffuse fashion, which leads to a reduction in the amount of transmitted radiation (fig. B 1.6.18).

Electrochromic glass

Here, the cavity between the panes contains liquid crystals whose state can be altered as required by applying an electric current. When switched off, the liquid crystal layers have a milky white appearance and scatter the incoming light; when the voltage is applied, they are almost transparent. Once switched on, the light transmission rises from 40% to 70% (fig. B 1.6.19).

Laminated glass with photovoltaic modules

As they are very thin, solar cells can be fixed between panes of glass by means of an interlayer or casting resin, which guarantees optimum protection for the cells and the wiring. Depending on the arrangement and the type of cells used, also the spacing between them, it is possible to create transparent, translucent or opaque modules (see chapter B 2.3 "Solar energy").

Insulating glass

Insulating glass consists of two or more panes of glass with a cavity of 8–24 mm sealed against ingress of air (fig. B 1.6.20). There are numerous ways of improving the U-value (thermal transmittance) and g-value (total energy transmittance) by using special fillings or additional films in the cavity or by coating the inner surfaces of the panes.

Filling the cavity with gas

The cavity can be filled with dried air or, to improve the thermal performance, with an inert gas. The use of argon, krypton or xenon can decrease the U-value of the glazing unit, because these gases exhibit a lower heat conduction and convection tendency than air. For economic reasons, argon is preferred to the more expensive krypton and xenon, although these latter two offer the best thermal performance. To prevent water condensing on the inner surfaces of the panes, the hermetic edge seal contains a desiccant.

Evacuating the cavity

A vacuum has a low thermal conductivity and so evacuated insulating glazing can achieve a U-value of approx. 0.6 W/m²K for a total thickness of 8 mm. However, the high vacuum in the cavity calls for spacers at regular intervals to prevent contact between the inside and outside panes as they collapse inwards under atmospheric pressure. In addition, a hermetic edge seal must be guaranteed (fig. B 1.6.23).

Transparent thermal insulation in the cavity

Incorporating a transparent thermal insulation material in the cavity prevents convection and improves the thermal performance. Transparent and translucent materials such as glass, acrylic glass, polycarbonate and quartz foam in various forms (fig. B 1.6.24) are available (see chapter B 2.3 "Solar energy", pp. 287-93).

Sunshades in the cavity

A whole range of elements for controlling the incoming sunlight, preventing glare and redirecting daylight can be incorporated in the cavity fully protected from the weather. These include electrically operated sunshades like louvre or roller blinds, but also fixed systems like sunshading grids, mirrors or prisms (fig. B 1.6.25).

Glass in facades

The majority of glass facades erected today consist of flat glass in the form of insulating, toughened safety or laminated safety glazing in the most diverse variations and combinations. The weight of the glass acting in the plane of the glass plus the wind and impact loads acting perpendicular to the plane of the glass must be taken into account in order to ensure that the loads are transferred to the structure and that the panes of glass are fixed adequately [6].

Transfer of forces

Transferring the loads acting on the panes of glass is achieved in three different ways depending on the nature and magnitude of the loads.

B 1.6.16 Build-up of layers in fire-resistant glazing
B 1.6.17 Build-up of layers in sound insulation glazing
B 1.6.18 Build-up of layers in thermotropic glazing
B 1.6.19 Properties of switchable functional glasses
B 1.6.20 Heat transport in insulating glass
B 1.6 21 Coloured glass, Zumtobel Staff exhibition room, Berlin (D), 1999, Sauerbruch Hutton
B 1.6 22 Glass products and their main applications and functions

Main applications / functions

	improved breakage behaviour (security/safety)	better breakage resist. (load-carrying capacity)	fire resistance	sound insulation	thermal insulation	permanent sunshading	controllable sunshading	self-regulating sunshading	protection against electromagnetic radiation	light filtration	light scattering	light redirection	light transmission	view in	view out
Sheet glass															
Basic methods of production															
float glass													•	•	•
plate glass													•	•	•
antique glass													•	•	•
Modified methods of production															
cased glass											•		•		
metal-coated glass (online)					•	•				•			•		•
UV-permeable glass													•	•	•
radiation protection glass									•				•	•	•
phototropic glass						•		•					•		
sheet glass with low expansion			•										•	•	•
coloured glass						•				•	•		•	•	•
opalescent glass						•				•	•		•		
wired glass	•		•										•		
Profilglas		•											•		
patterned glass											•		•		
1st processing stage															
toughened safety glass, heat-treated glass	•	•											•	•	•
chemically strengthened glass	•	•											•	•	•
acid-etched glass											•		•		
sandblasted glass											•		•		
2nd processing stage															
enamelled glass						•							•		
metal-coated glass (offline)					•	•				•			•		•
broadband anti-reflection glass													•	•	•
Cavity between planes filled with:															
gas				•	•								•	•	•
vacuum				•	•								•	•	•
hydrogel			•	•							•		•	•	•
thermotropic layer								•			•		•		
electrochromic materials							•			•	•		•		
Laminated safety glass produced with:															
plastic interlayer	•				•								•	•	•
printed and coloured interlayers	•				•	•				•	•		•		
casting resin		•											•	•	•
Moulding resin															
glass blocks	•	•	•	•									•		
thick pressed glass	•	•											•		
Glass fibres															
membranes						•					•		•		
glass wool				•	•										
Cellular glass															
aerogel				•	•								•		
foamed glass				•	•										

B 1.6.22

B 1.6.21

Contact
Only compressive forces acting perpendicular to the contact faces may be transmitted in this way. The contact faces must therefore be of such a size that adequate distribution of the stress is guaranteed. This fact is especially important with small contact faces (e.g., "patch" fittings). Hard bearings, such as glass on steel, must be avoided by incorporating intermediate resilient pads (EPDM or plastic). Panes can be fixed in the plane of the glass but also perpendicular to the glass, either continuously or at individual points (fig. B 1.6.26). The forces are transferred by means of patent glazing wings/caps, clamping plates, screws/bolts and/or setting blocks.

Friction
In this case the forces are transferred by a mechanical interlocking of the microscopic surface imperfections of the contact faces and by adhesion. To avoid local stress peaks glass should not be directly in contact with other hard materials such as steel; the elasticity and fatigue strength of an intermediate pad is absolutely crucial to the durability of a friction connection. Soft metals such as pure, annealed aluminium, fibre-reinforced plastics or natural materials (e.g., cork, leather, cardboard) are suitable as intermediate pads.

Adhesives (material bond)
Adhesive joints are very common in modern glass engineering, provided the forces to be transferred are relatively small. In addition, great care must be taken to ensure sufficiently large mating faces which can be permanently joined with an elastic adhesive. Besides the magnitude of the forces to be transferred, temperature and duration of loading are important influencing factors. In the event of a fire, the excessive heat normally leads to failure of the adhesive joints. In Germany adhesive connections in facades are permissible above a height of 8 m only when additional mechanical fixings are provided to prevent glass components from becoming detached.

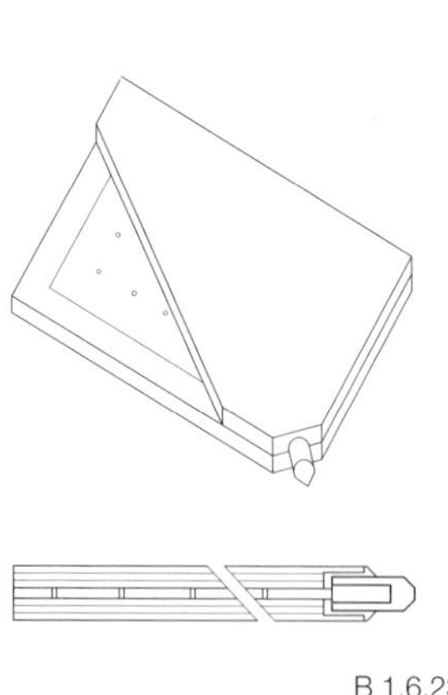

B 1.6.23

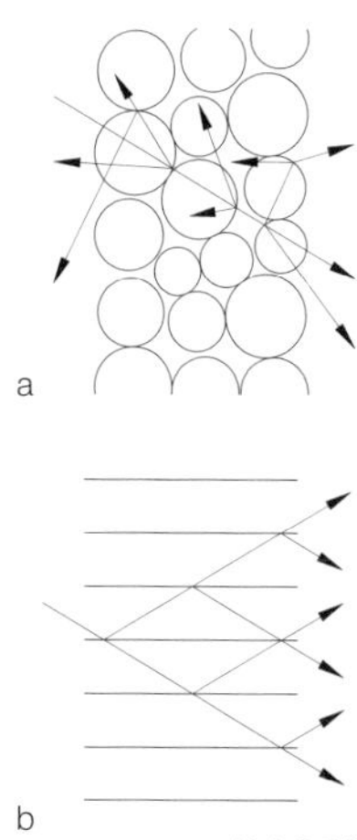

B 1.6.24

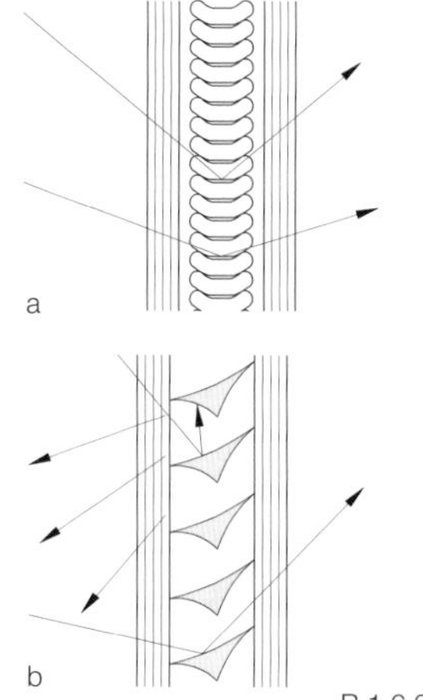

B 1.6.25

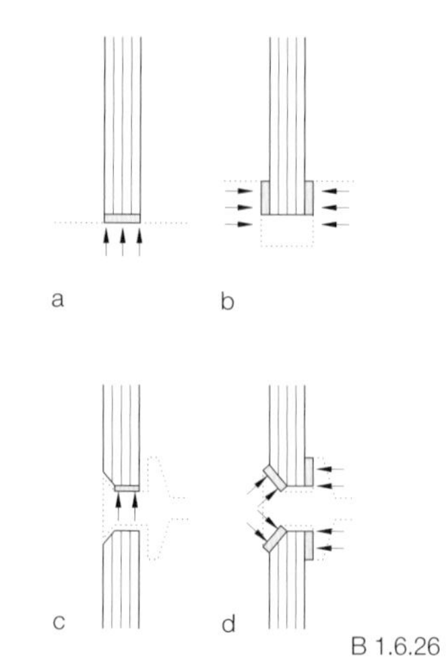

B 1.6.26

Jointing

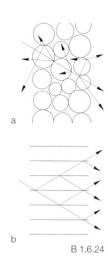

The necessary joints between the individual glass elements must be designed to allow mechanical movement (e.g., due to thermal expansion) and to exclude rain and wind permanently.

Contact seals

The seal between the surface of the glass and its supporting member can be achieved by way of a permanently elastic preformed sealing profile or gasket, which can be designed as a solid block or with sealing lips. To guarantee a sealed connection, adequate contact pressure and clean glass surfaces are essential (fig. B 1.6.27a).

Putty

This traditional method of sealing glass is less important these days, because once the putty cures, the joint is very rigid, and this excludes the possibility of movement or other deformations. Any cracks that do occur usually lead to saturation of the joint (fig. B 1.6.28b).

Adhesive seals

Permanently elastic "putty" (silicone) enables the formation of elastic connections which, owing to the adhesion, are stable even in the presence of tensile forces to a certain extent. The width of the joint and the type of adhesive used determine the resilience of the joint (fig. B 1.6.27b).

Glazing

Window and facade constructions consist of the following functional elements:

- glazing elements (e.g., panes of glass)
- supporting construction (e.g., posts, rails, frames)
- fixings (e.g., glazing bead)
- joints (e.g., EPDM, silicone)

Depending on the jointing and the way the forces are transferred, various combination options are possible. Whereas in a traditional patent glazing arrangement the various functional elements are combined, in designs with individual fixings the jointing and load transfer functions are separate, and require separate installation operations. In other types of glazing, e.g., structural sealant glazing, the "joint" and "fixing" functions are amalgamated.

Leaded glazing

Leaded glazing represents the oldest technique for producing larger areas of glass. Small individual pieces of glass are inserted into H-section lead cames and tapped into place. Besides the complete enclosure of the edge of the glass, subsequent filling with putty creates an additional bond (fig. B 1.6.28a).

Rebate with putty fillet

In this traditional form of glazing the pane of glass is fitted in an open rebate and held in position with a fillet of putty. The rebate itself is formed directly in the masonry or timber frame, or forms an integral part of a cast or rolled metal section (fig. B 1.6.28b). Owing to the bond between the glass and the supporting member, this simple form of glazing has enabled the construction of many delicate forms over the centuries, as the glasshouses of the nineteenth century show so well.

Rebate with glazing bead

Curtain wall facade systems require the use of special loadbearing constructions onto which the panes of glass are fixed by means of glazing wings/caps. These allow two adjoining panes to be fixed with one section, which leads to simple erection and slim frames (fig. B 1.6.28d). Preformed, permanently elastic gaskets seal the glazing bars both inside and outside. When using insulating glass where enhanced thermal performance is desired, the provision of a thermal break between glazing wing/cap and supporting construction is particularly important.

Linear support without glazing bar

In this variation of patent glazing only two opposite sides of the pane of glass are held in place by glazing wings/caps. The other two sides are finished flush with an adhesive silicone joint. When using laminated safety glass it is particularly important to ensure that the jointing material is compatible with the interlayer. It is advisable to work with two sealing levels (inside and outside) when using insulating glass and to drain the void between the two levels separately.

Individual supports at the edge

This form of glazing combines the advantages of patent glazing with those of individual fixings. Small fittings that disturb the appearance only minimally carry the loads in the plane of the glass pane via narrow setting blocks, while clamping plates take on this task in the case of forces acting perpendicular to the plane of the glass. Owing to the high mechanical loads at the fixings, heat-treated glass is used, which increases the cost of such glazing. However, no drilling is required with this type of glazing. The unsupported edges merely have to provide sealing functions.

Individual drilled fixings

In this case the supporting and sealing functions are completely separate. The loads are carried in the plane of the glass via the fixings fitted into the drilled holes, while the sealing function is provided at the unsupported edges. Besides its delicate appearance, this approach results in great design freedom. However, great attention must be given to ensuring high-quality seals around the edges and at the individual fixings (especially with insulating glass) (fig. B 1.6.28e).

Adhesive fixing (structural sealant glazing)

In this type of glazing the adhesive bond between glass and supporting construction serves the load-carrying and sealing functions. However, the adhesive is only suitable for transferring short-term loads, e.g., wind. The self-weight of the panes of glass is carried by mechanical fixings (fig. B 1.6.28f).

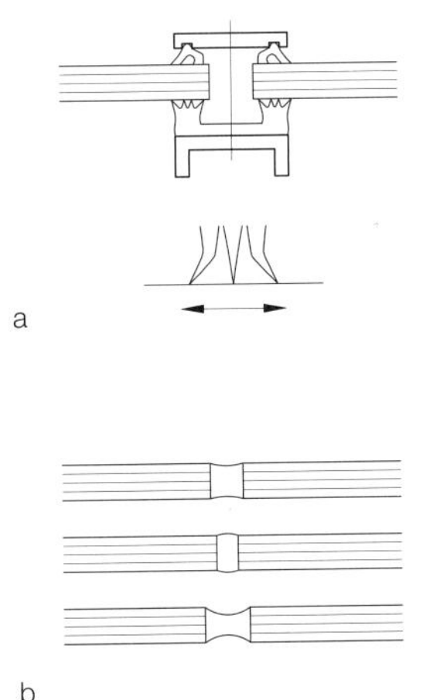

B 1.6.27

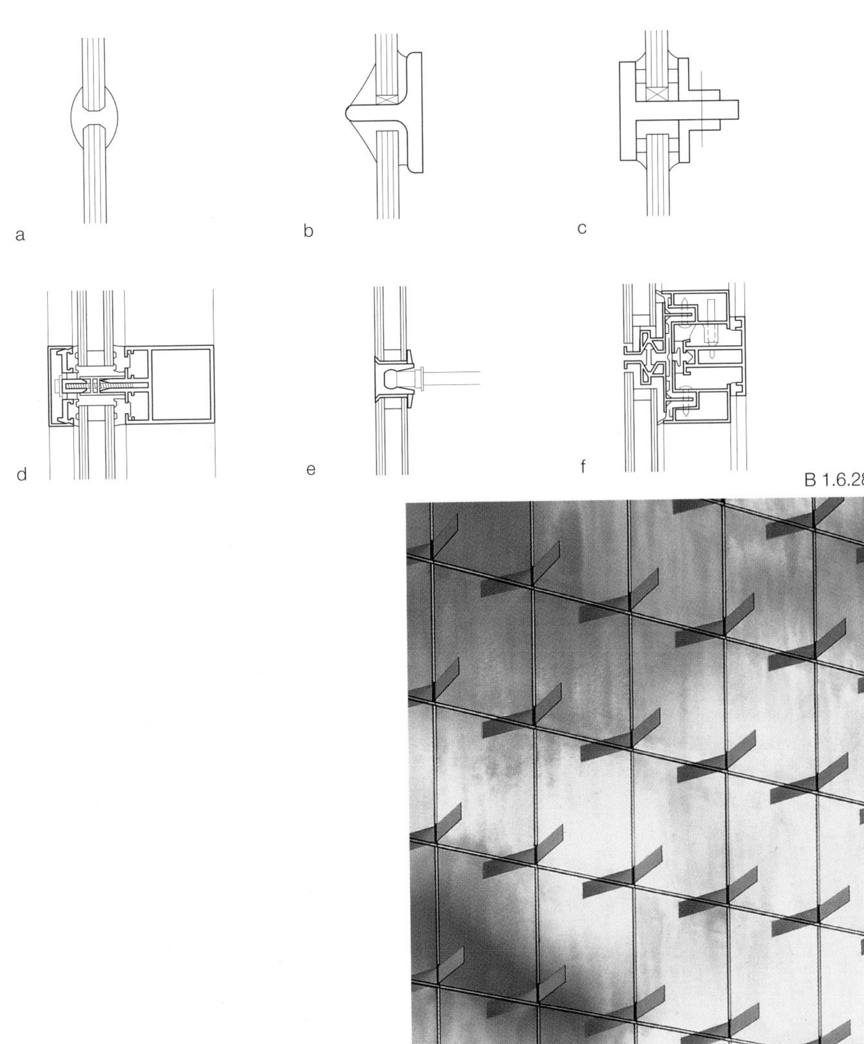

B 1.6.28

B 1.6.29

Notes regarding insulating glass
Spacers and panes of glass are normally bonded with "Thiokol", which guarantees a gastight edge seal of very high quality. However, as Thiokol is not resistant to ultraviolet light, the edge seal must be totally en-closed. This can be achieved by using glazing wings/caps or UV-opaque printing. Alternatively, it is possible to create a hermetic edge seal with silicone. In that case, however, the gastightness cannot be guaranteed and so it is not advisable to use noble gases with this type of glazing [7].

Notes:

[1] Staib, Gerald: From the origins to classical modernism; in: Schittich, Christian et al.: *Glass Construction Manual*, Munich/Basel, 1998, pp. 9–33.
[2] Button, David: *Glass in Building. A Guide to Modern Architectural Glass Performance*; Pye, Brian (ed.), Oxford, 1993.
[3] Herzog, Thomas: Seminar report "Sonderthemen Baukonstruktion – Materialspezifische Technologie und Konstruktion – Gläser, Häute und Membranen", Teil 1 Grundlagen; Munich Technical University, Chair of Building Technology, Munich, 1998 (unpublished manuscript).
[4] Compagno, Andrea: Glass as a building material – Developments and trends; in: Kaltenbach, Frank (ed.): *Translucent Materials*, Basel/Munich, 2004, pp. 10–25.
[5] Saint Gobain Glass (pub.): *Memento Glas Handbuch*, 2000 ed., p. 32.
[6] Sobek, Werner et al.: Designing with glass – strength and loadbearing behaviour; in: Schittich, Christian et al.: *Glass Construction Manual*, Munich/Basel, 1998, pp. 91–92.
[7] A detailed overview of different types of glazing together with detailed drawings can be found in: Schittich, Christian et al.: *Glass Construction Manual*, Munich/Basel, 1998, pp. 152–677.

B 1.6.23 Schematic diagram of vacuum glazing
B 1.6.24 Principles for transparent thermal insulation:
a cell-like structure
b tube-like structure
B 1.6.25 Insulating glazing with integral elements for controlling incoming daylight:
a acrylic glass elements for redirecting the light
b louvres with a highly reflective coating
B 1.6.26 Methods of supporting panes of glass and transferring forces:
a supported in the plane of the glass, forces transferred via edge
b supported perpendicular to the plane of the glass, forces transferred via edge
c supported in the plane of the glass, forces transferred via surface of drilled hole
d supported perpendicular to the plane of the glass, forces transferred around edge of hole
B 1.6.27 Jointing options for glass:
a contact seal, movement possible through sliding and deformation of lips
b adhesive joint with permanently elastic "putty" (silicone), movement
B 1.6.28 a Leaded glass
b Rebate with putty fillet
c Rebate with glazing bead
d Patent glazing
e Individual fixing in drilled hole
f Adhesive (structural seal and glazing)
B 1.6.29 Dichroic glass (glass coated with coloured crystals, the colour of which changes depending on the angle of incidence and the direction of oscillation of the light), New York (USA), 1999, James Carpenter

Office building for Willis Faber & Dumas

Ipswich, GB, 1975

Architects:
Foster Associates, London
Structural engineers:
Anthony Hunt Associates, Cirencester
Facade consultant:
Martin Francis, with Jean Prouvé

Architectural Review 09/1975
A+U 02/1974
Bauen + Wohnen 02–03/1976
Wiggington, Michael: Glass in Architecture. Stuttgart 1996

- One of the first examples of a single-glazed curtain wall facade
- Panes of glass fixed with "patch" fittings
- Facade braced with glass fins
- Minimal junction with floors achieved with EPDM preformed seal between edge of floor and glass
- Use of grey body-tinted solar-control glass

Isometric view (not to scale)
Plan, scale 1:2000
Isometric view of facade, scale 1:50
Detail of glass suspension, scale 1:20
Ilsometric detail of "patch" fitting (not to scale)

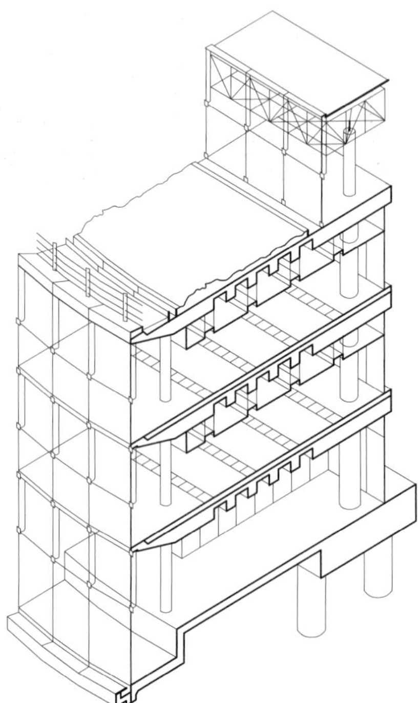

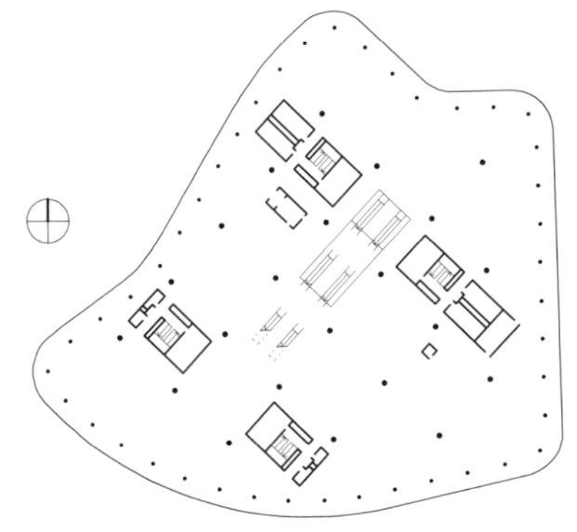

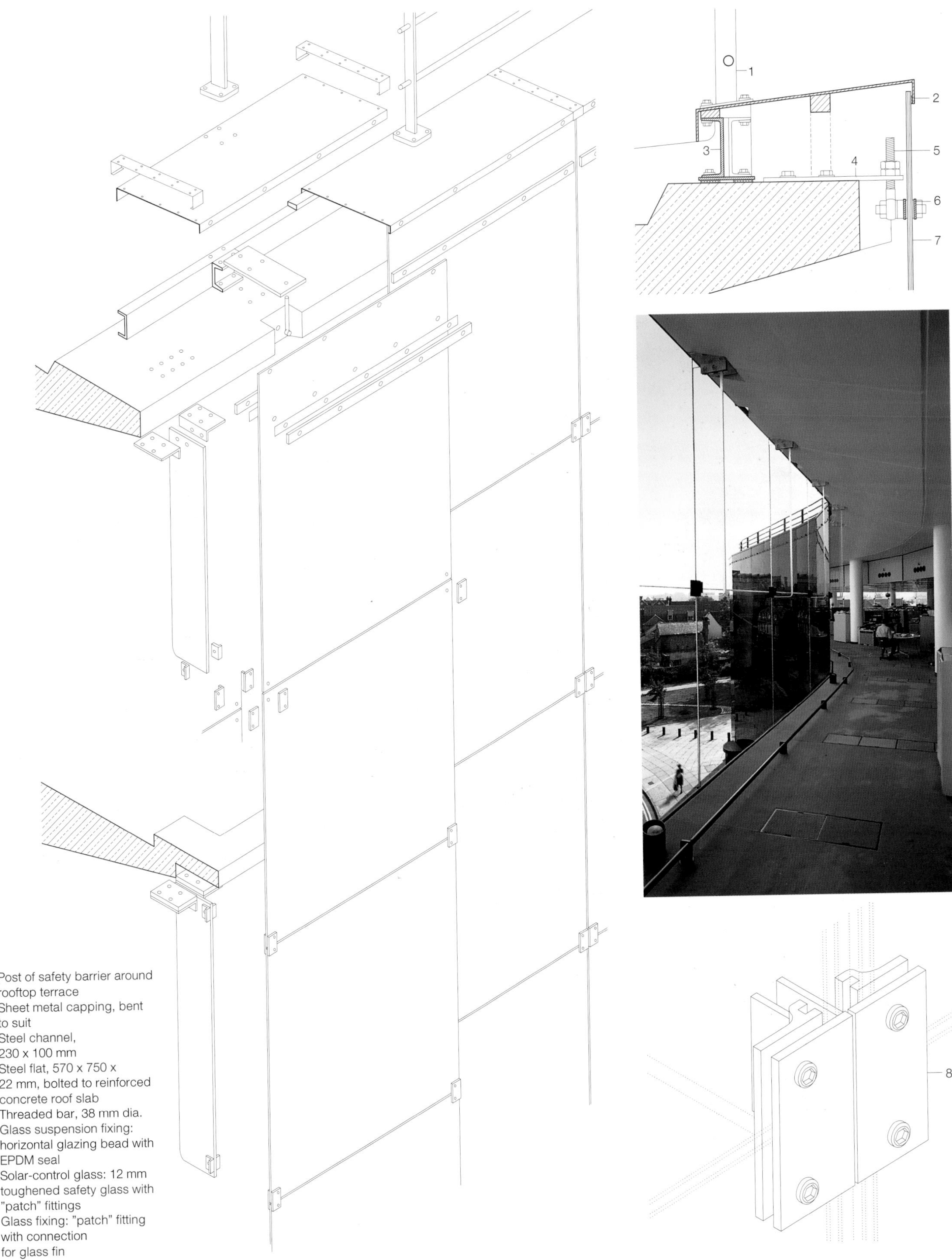

1 Post of safety barrier around rooftop terrace
2 Sheet metal capping, bent to suit
3 Steel channel, 230 x 100 mm
4 Steel flat, 570 x 750 x 22 mm, bolted to reinforced concrete roof slab
5 Threaded bar, 38 mm dia.
6 Glass suspension fixing: horizontal glazing bead with EPDM seal
7 Solar-control glass: 12 mm toughened safety glass with "patch" fittings
8 Glass fixing: "patch" fitting with connection for glass fin

Herz Jesu Church

Munich, D, 2000

Architects:
Allmann Sattler Wappner, Munich
Facade design:
R+R Fuchs, Munich

Bauwelt 47/2000
DBZ 03/2001
Detail 02/2001
GLAS 02/2001

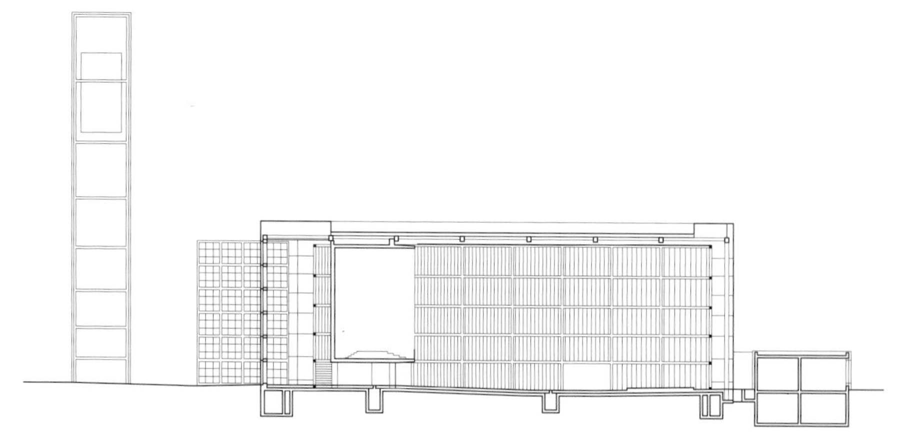

- External leaf in the form of a post-and-rail framework with double glazing, outer pane set back to allow glazing bars to be fitted flush with glass
- Wind bracing to glass facade by means of horizontal and vertical glass fins
- Panes of glass printed in different intensities, graded from transparent at the entrance to opaque/translucent around the alta
- Entrance portal formed by 14 m high printed glass leaves, designed by Alexander Beleschenko
- Internal leaf made from vertical, light-coloured maple louvres
- Structural steelwork of hollow sections

Section • Plan
scale 1:750
Vertical section, scale 1:50
Horizontal section through corner/door
scale 1:20

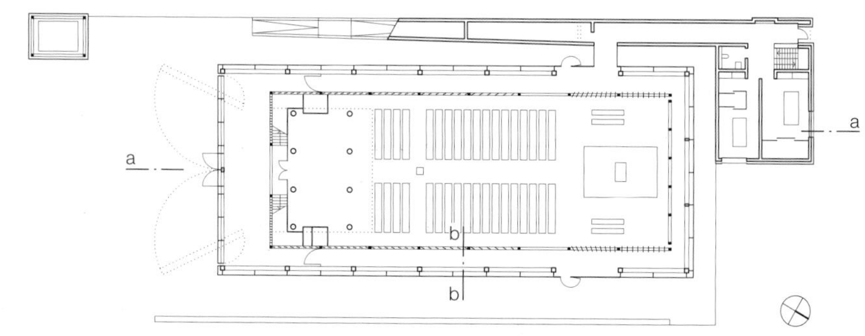

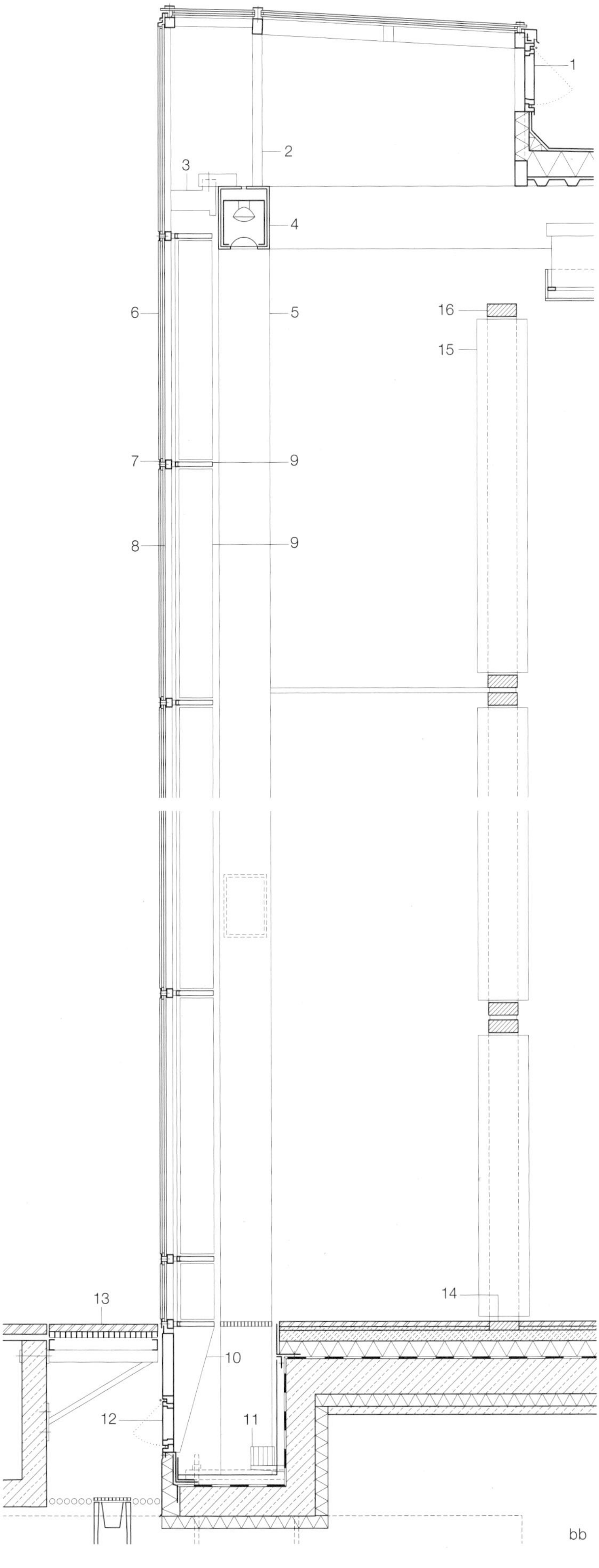

1 Hinged panel, aluminium, exhaust-air outlet/smoke vent
2 Steel square hollow section
3 Height-adjustable facade suspension fitting
4 Edge beam, steel rectangular hollow section, 420 x 500 x 20 mm, welded, with lighting units integrated
5 Twin box column, 2 No. 170 x 60 mm + 2 No. 300 x 30 mm steel flats
6 Double glazing, 8 mm laminated safety glass + 16 mm cavity + 10 mm toughened safety glass, printed, outer pane set back on all sides, printed
7 Glazing cap fitted flush
8 Steel rectangular hollow section 50 x 70 x 5 mm
9 Stiffening glass fin, 36 x 300 mm
10 Steel bracket to support glass fin
11 Convector
12 Hinged panel, aluminium, air inlet
13 Sandstone paving slab, 60 mm, with ventilation slits, 60 mm grating, steel bracket
14 Sole plate, solid maple, 240 x 50 mm
15 Maple veneer louvres, fixed to frame
16 Solid maple frame, 240 x 120 mm
17 Steel angle, 170 x 90 x 10 mm
18 Glazing for door: 5 mm toughened safety glass + 20 mm cavity + 5 mm toughened safety glass, printed both sides pane size: 755 x 767 mm
19 Aluminium glazing cap
20 Secondary framing: steel rectangular hollow sections 100 x 60 x 4 mm
21 Primary framing: steel rectangular hollow sections 280 x 150 mm

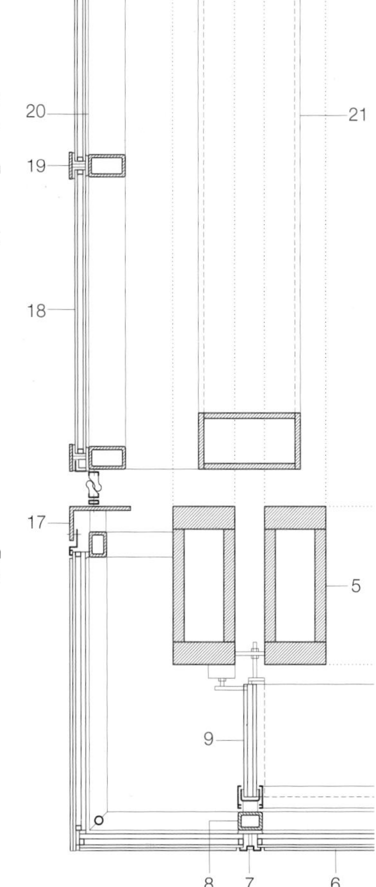

bb

Faculty of Law

Cambridge, GB, 1995

Architects:
Sir Norman Foster and Partners, London
Structural engineers:
Anthony Hunt Associates, Cirencester
Facade design:
Emmer Pfenninger Partner, Munichstein

Architectural Review 03/1993
Bauwelt 35/1995
Foster Catalogue 2001. Munich/London/New York 2001

- Barrel-shaped north elevation with silicone structural glazing construction to maximise utilisation of daylight on all floors
- Supporting construction comprises a double layer of curving steel circular hollow sections in a triangulated arrangement to provide the necessary stiffness
- Panes of glass can be adjusted in all three directions at the corners; pane side lengths: 2800 or 3800 mm
- Thermal movement accommodated via sliding bearings
- Insulating glass with solar-control and low-e coatings

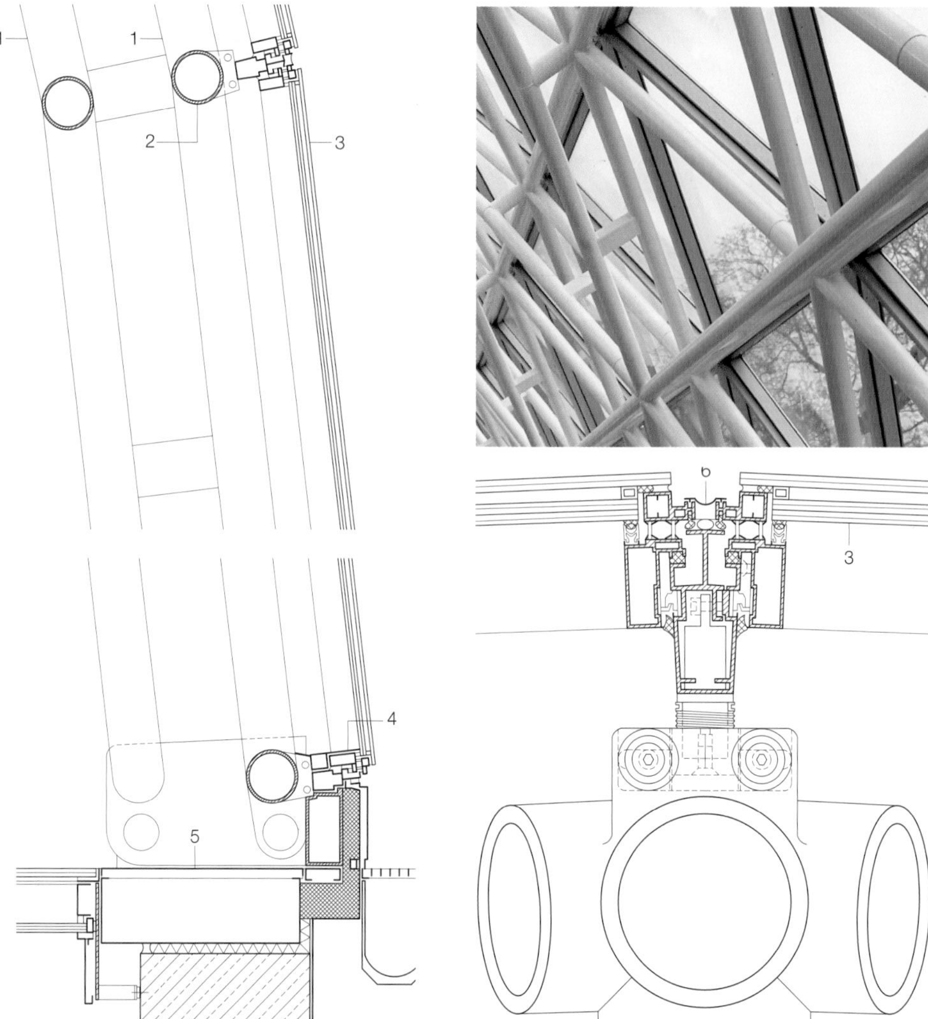

Section, scale 1:500
Vertical section, scale 1:20
Detail, scale 1:5

1 Primary loadbearing framework of steel circular hollow sections, 140 mm dia., welded nodes, painted white
2 Horizontal bracing, steel circular hollow sections, 140 mm dia., painted white
3 Stepped insulating glass, 10 mm toughened safety glass + cavity + 2 No. 8 mm laminated safety glass, projecting outer pane bonded to aluminium frame
4 Specially formed framing member
5 Inspection opening for district heating duct
6 Silicone seal

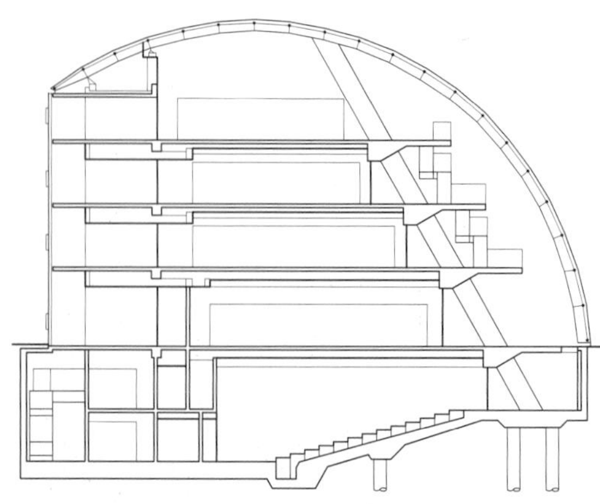

Warehouse

Marktheidenfeld, D, 1999

Architects:
schneider+schumacher, Frankfurt am Main

Archithese 04/1999
Baumeister 04/2000
GLAS 05/2000

- Main facade to longitudinal elevation comprises two layers of translucent profiled glass, 0.24 x 7.30 m
- Continuous rooflight in flat roof runs parallel to glass facade and allows facade to appear bright and lightweight
- Primary structure of structural steelwork on reinforced concrete plinth

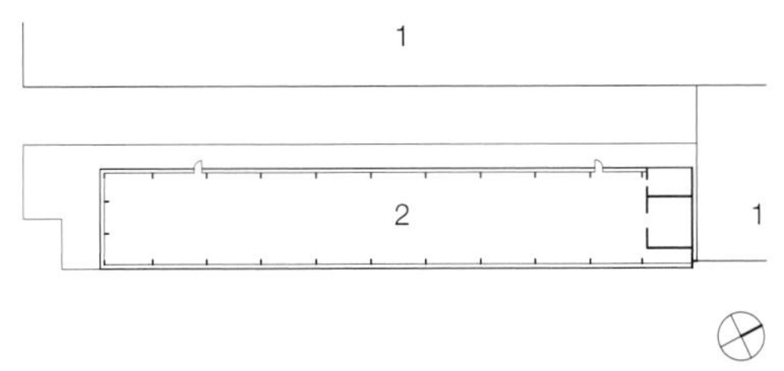

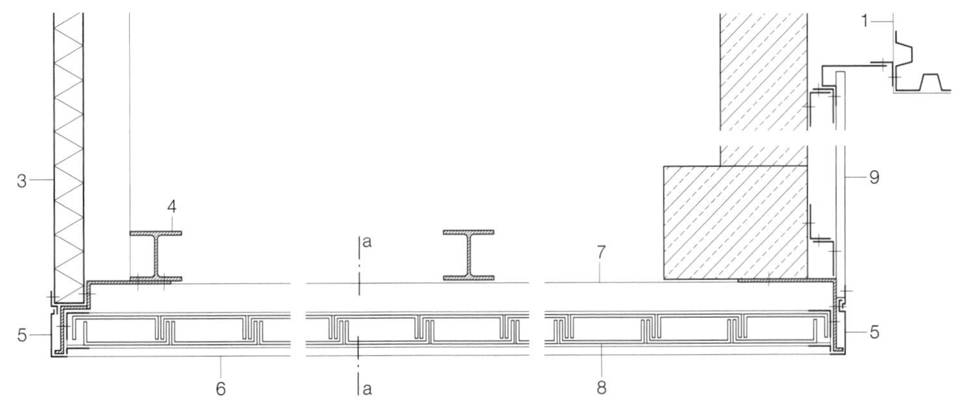

Plan, scale 1:1500
Horizontal section • Vertical section
through profiled glass facade, scale 1:20

1 Existing building
2 New warehouse
3 Thermally insulated panel, 80 mm, concealed screws
4 Column, HEA 140
5 Corner trim, aluminium-zinc
6 Flashing, aluminium-zinc
7 Reinforced concrete plinth
8 Profiled glass
9 Flat panels, 250 mm, laid horizontally
10 Sheet metal capping plate, bent to suit and fixed to clips
11 Retaining bracket for top of profiled glass, bent sheet metal, d = 3mm

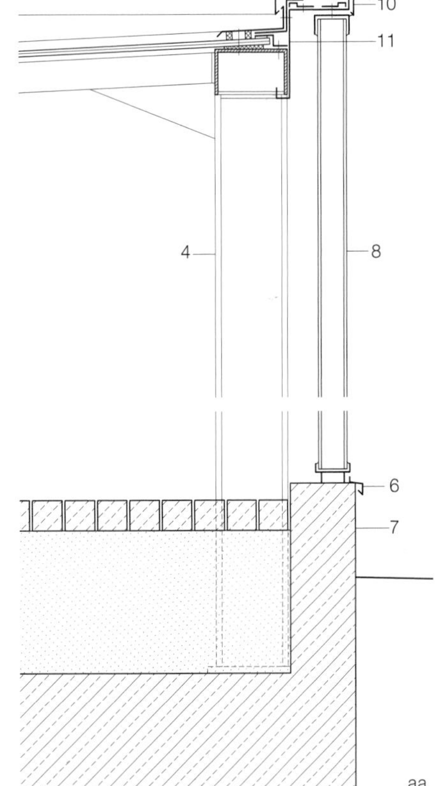

Rodin Museum

Seoul, ROK, 1997

Architects:
Kohn Pedersen Fox Associates, London/ New York

Architecture 11/1998
Kennon, Kevin et al.: The Rodin Museum, Seoul. New York 2001

- Prime location in the centre of Seoul
- Double-leaf glass envelope for facade and roof made from translucent laminated safety glass with varying cavity width, countersunk screw fixings
- Rustproof steel loadbearing construction
- Cavity ventilated with preheated incoming air in winter and precooled incoming air in summer

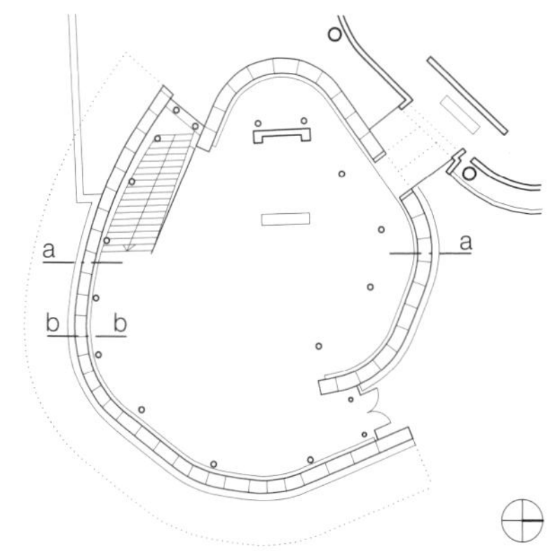

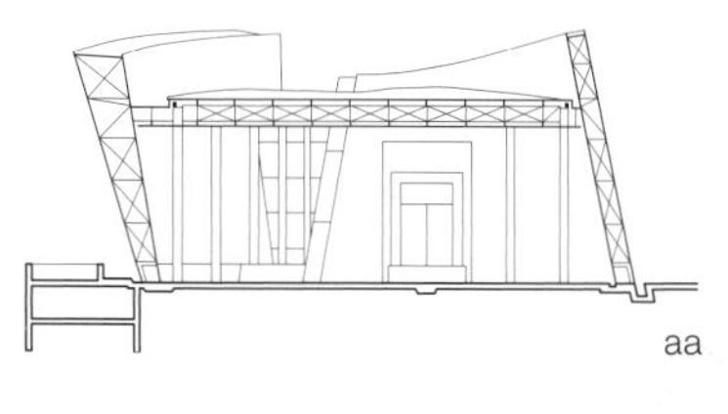

Section • Plan, scale 1:750
Vertical section • Horizontal section
scale 1:50
Details, scale 1:5

1 Laminated safety glass, 2 No. 10 mm, coated
2 Glass fixing, stainless steel, rigid
3 Glass fixing, stainless steel, non-rigid
4 Silicone seal
5 Stainless steel rectangular hollow section, 100 x 40 x 4 mm
6 Steel rectangular hollow section, 60 x 40 x 4 mm
7 Bracing, round steel bar, 20 mm dia.
8 Steel rectangular hollow section, 100 x 60 x 4 mm
9 Stainless steel sheet, 2 mm, bent to suit
thermal insulation
powder-coated aluminium sheet
10 Aluminium capping plate, 3 mm
11 Steel rectangular hollow section, 450 x 250 mm
12 Double glazing, 8 mm toughened safety glass + 12 mm cavity + 2 No. 7 mm laminated safety glass
13 Steel rectangular hollow section, 300 x 150 mm
14 Suspended glass ceiling, 2 No. 8 mm laminated safety glass
15 Stainless steel panel, thermally insulated, 83 mm
16 Hinged stainless steel grating
17 Lighting unit
18 Stainless steel sheet, removable
19 Cover to ventilation duct, stainless steel grating, 35 x 35 mm

1
16
17
cc

2
3
5
1
4

5
7
6
4
1

1
7
1
10
11
12
9
5
6
8
13
14
c
c
16
17
15
18
19
bb

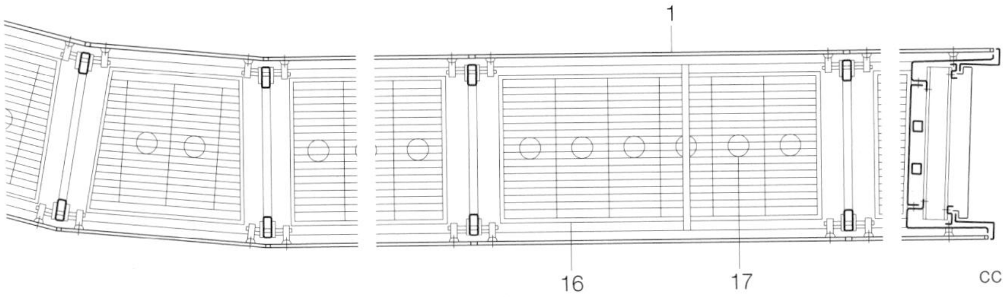

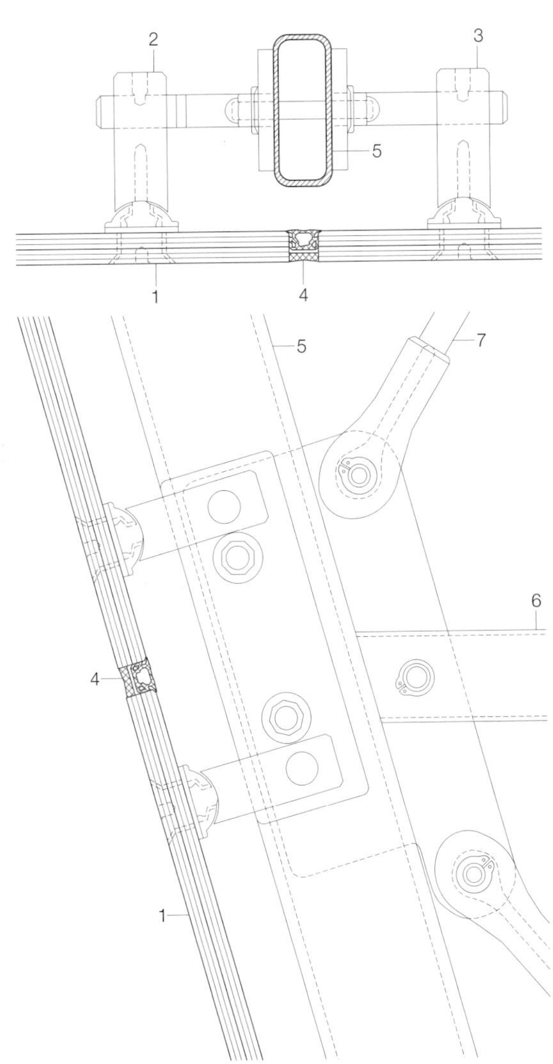

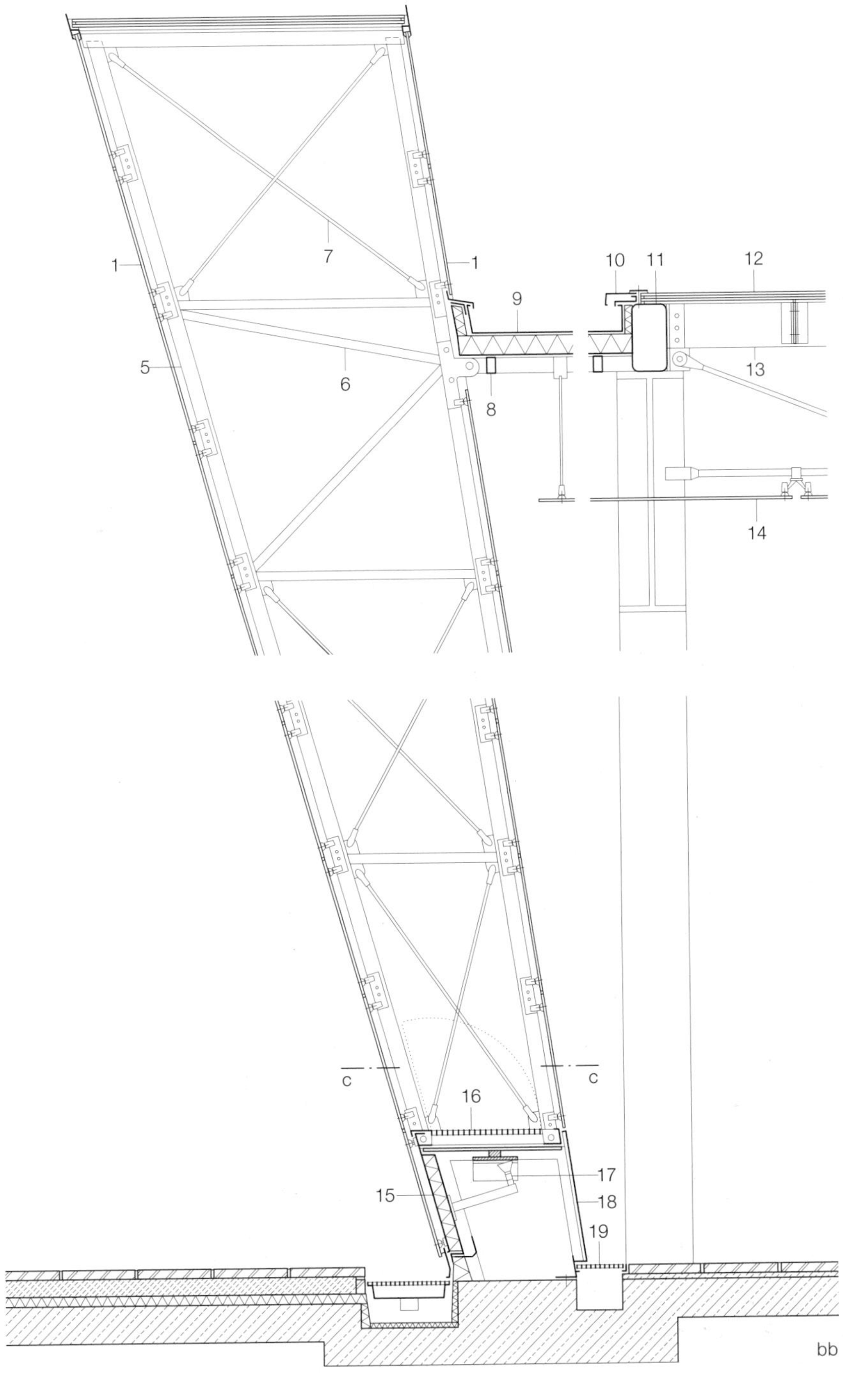

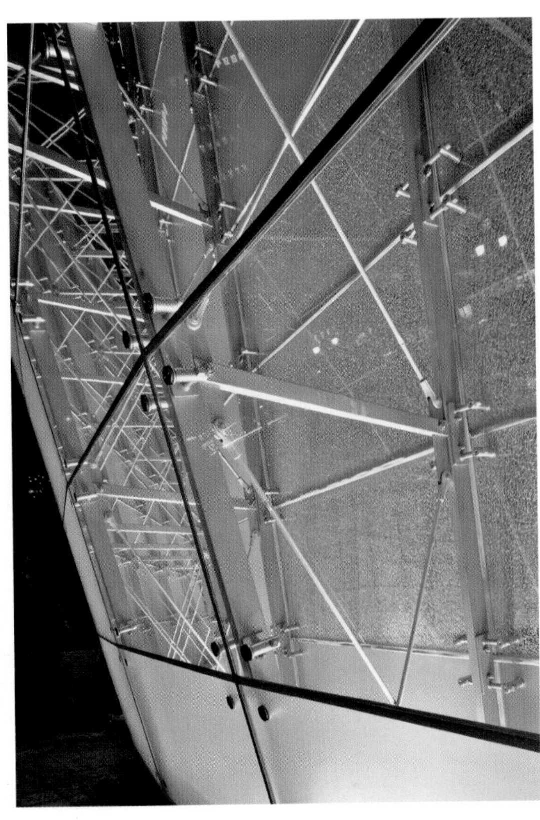

Hermès department store

Tokyo, J, 2001

Architects:
Renzo Piano Building Workshop, Paris
Structural engineers:
Ove Arup & Partners, London/Tokyo

Detail 07/2001
Fassade/Façade 03/2002
GLAS 02/2002

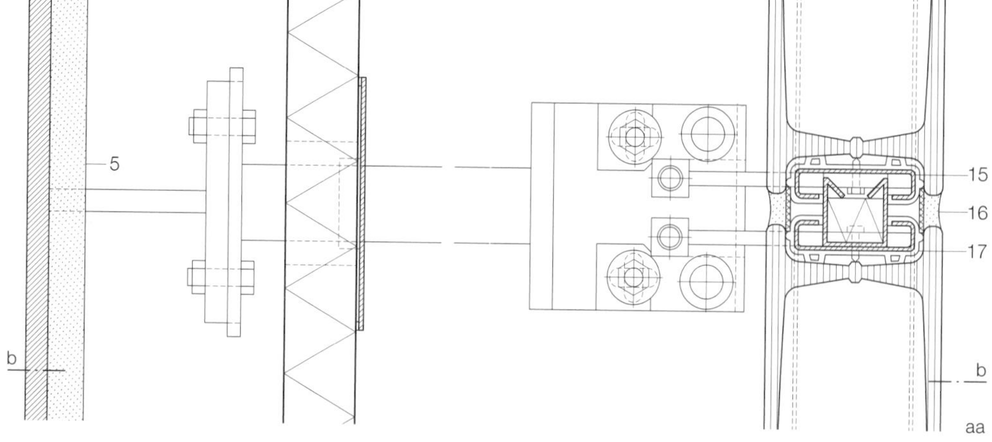

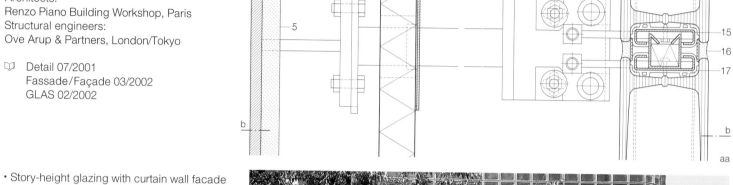

- Story-height glazing with curtain wall facade made from specially designed glass blocks, 430 x 430 mm
- Glass blocks fixed with steel sections fitted into the joints to carry vertical and horizontal loads and improve the seismic resistance
- Casing for steel sections made from EPDM with lip seals forming junction with glass blocks – creates a resilient bearing for the blocks and can accommodate up to 4 mm movement

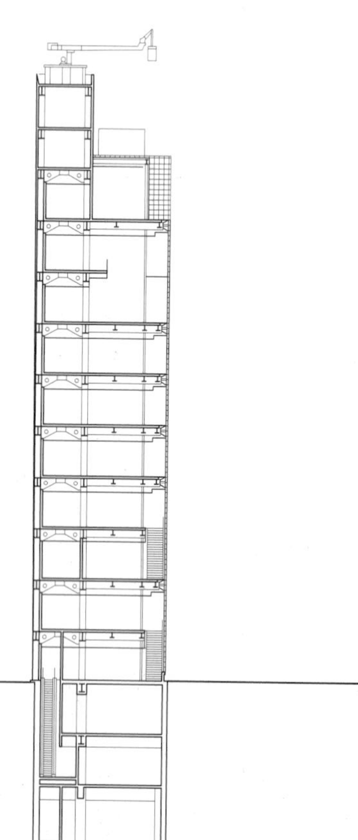

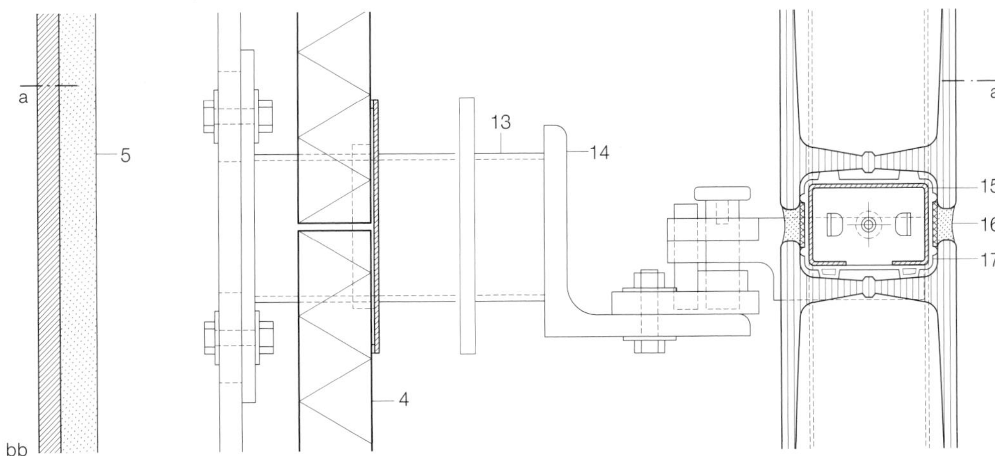

Section, scale 1:600
Horizontal section, scale 1:5
Vertical section showing special glass edge block scale 1:5
Vertical section, scale 1:20

1 Glass block, 430 x 430 x 4 mm
2 Cavity floor with parquet flooring finish
3 Reinforced concrete floor slab on permanent formwork of trapezoidal profile sheets, 150 mm
4 Sheet metal panel, insulated, 50 mm
5 Steel beam, IPE 375 x 300 mm, with 25 mm fire-resistant coating
6 Steel section, HEA 200
7 Round steel bar with threaded ends, 16 mm dia., with intumescent paint finish
8 Inspection opening
9 Steel beam, IPE 250 x125 mm, with 25 mm fire-resistant coating
10 Articulated support with spherical head, steel, 140 mm dia., fire-resistant design
11 Suspended ceiling, 12.5 mm plasterboard
12 Column, steel circular hollow section, 180 dia. x 40 mm, with 10 mm fire-resistant coating
13 Steel rectangular hollow section, 100 x 50 x 5 mm
14 Steel angle, 140 x 140 x 15 mm
15 Steel section, 80 x 53 x 3 mm
16 Permanently elastic silicone seal
17 EPDM preformed seal
18 Special glass edge block

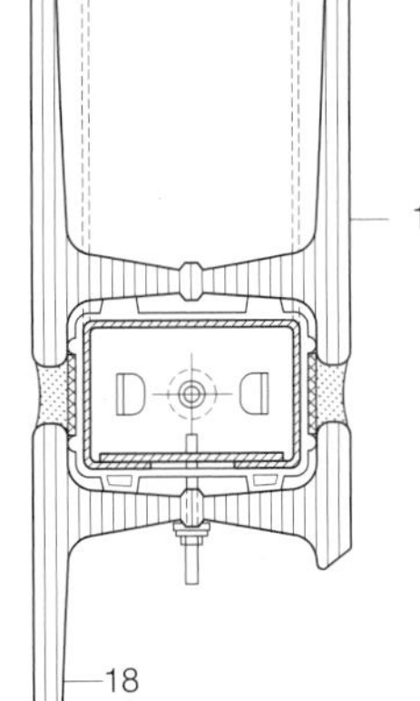

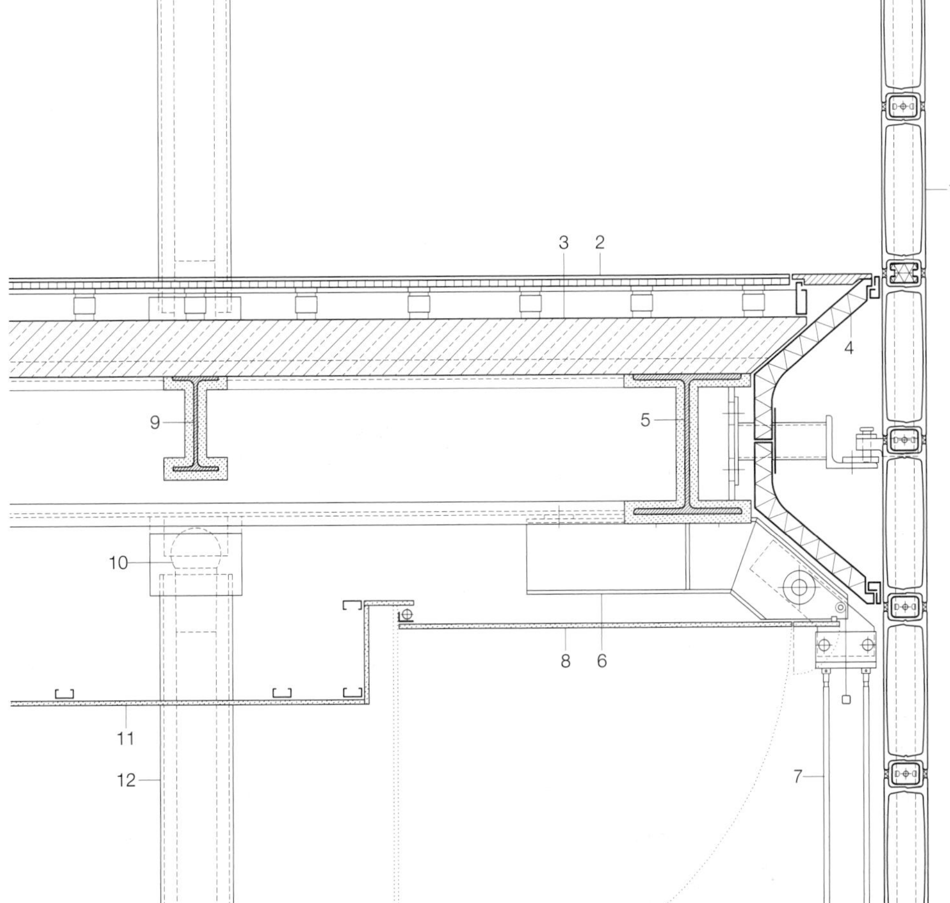

Mediathèque

Sendai, J, 2001

Architect:
Toyo Ito, Tokyo

Detail 07/2001
El Croquis 98/99, 1999
Witte, Ron: CASE: Toyo Ito – Sendai Mediathèque. Munich/Berlin/London/New York 2002

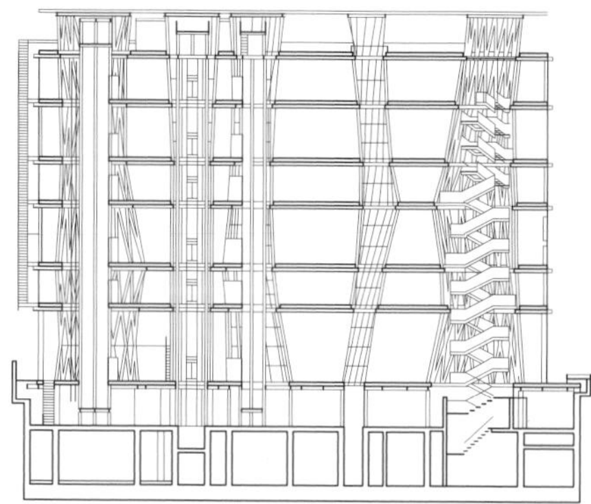

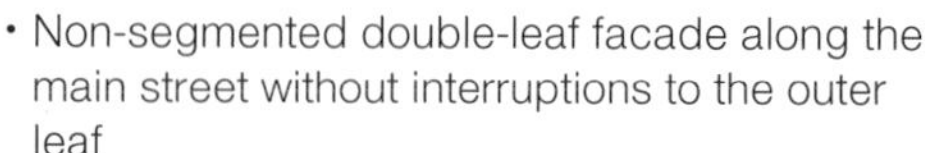

- Non-segmented double-leaf facade along the main street without interruptions to the outer leaf
- 1000 mm facade cavity
- Facade acts as thermal buffer in winter, heat gains in summer dissipated by means of ventilation
- Inner and outer glazing stiffened by glass fins
- Panes of glass with silk-screen printing to imitate spandrel panels
- Glass facades to east, west and north sides all in different designs
- All exposed steel parts protected with intumescent paint

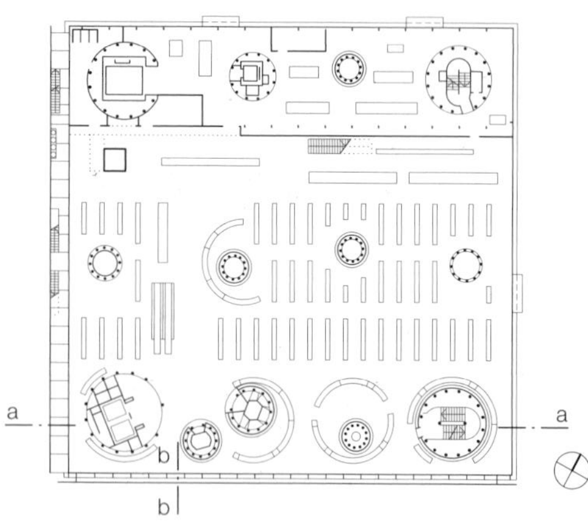

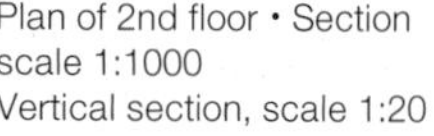

Plan of 2nd floor • Section
scale 1:1000
Vertical section, scale 1:20

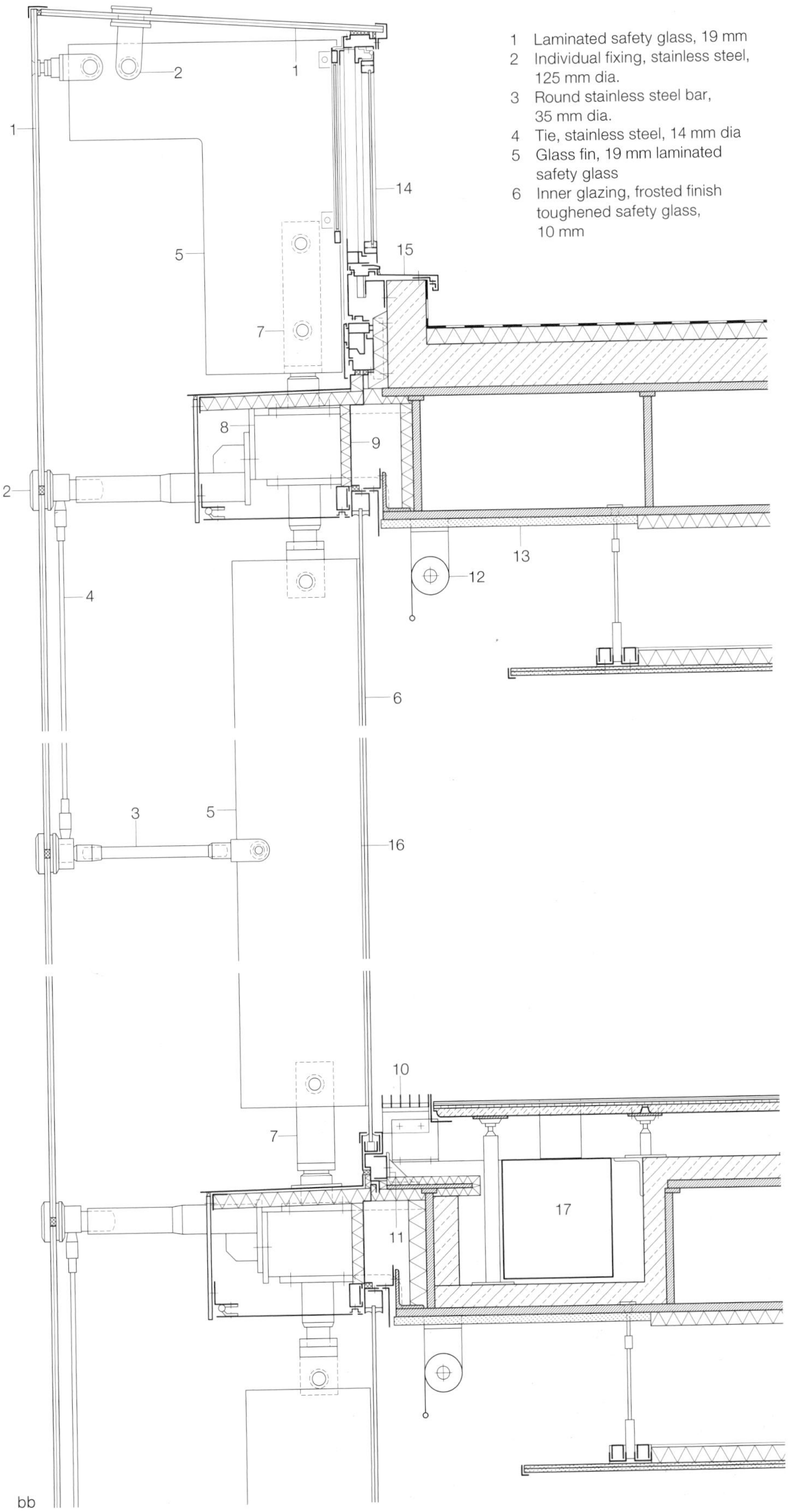

1 Laminated safety glass, 19 mm
2 Individual fixing, stainless steel, 125 mm dia.
3 Round stainless steel bar, 35 mm dia.
4 Tie, stainless steel, 14 mm dia
5 Glass fin, 19 mm laminated safety glass
6 Inner glazing, frosted finish toughened safety glass, 10 mm
7 Stainless steel glass fixing
8 Steel fin
9 Sheet steel, 1.6 mm
10 Lüftungsgitter Stahl verzinkt
11 Steel angle, 110 x 110 x 10 mm
12 Adjustable anti-glare blind
13 Fire-resistant lining
14 Opening light for ventilationl
15 Aluminium capping plate
16 Silicone joint
17 Heating/ventilation duct

bb

Lift towers, Reina Sofia Museum

Madrid, E, 1990

Architects:
Ian Ritchie, London
with José Luiz Iñiguez & Antonio Vazquez
Structural engineers:
Ove Arup & Partners, London

📖 l'ARCA 11/1991
Architectural Design 11–12/1991
Architectural Review 12/1991
Baumeister 09/1991
DBZ 10/1992
Progressive architecture 02/1994

- Glazed cable lattice wall with a total height of 36 m
- Panes of glass fixed with individual fasteners attached to specially designed "dolphin" arms
- Spring-type compensation elements at the base of the cable lattice accommodate the temperature-related changes in length of the glass panes
- Wind forces carried at the corners via bracing elements which transfer the loads to the primary loadbearing structure

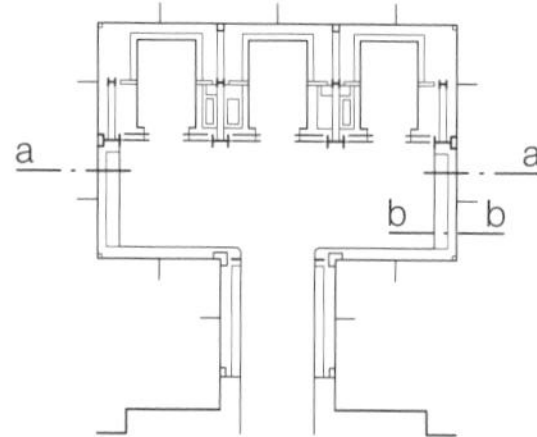

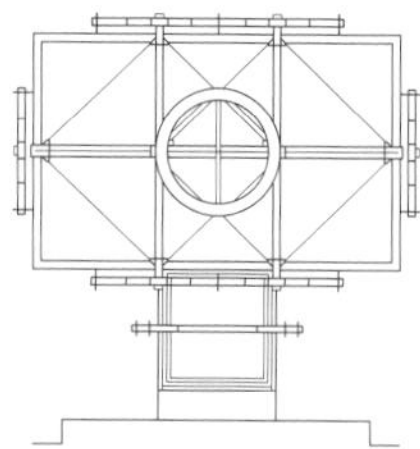

Plan of standard floor •
Roof plan • Section
scale 1:500
Isometric views (not to scale)
Vertical section, scale 1:20

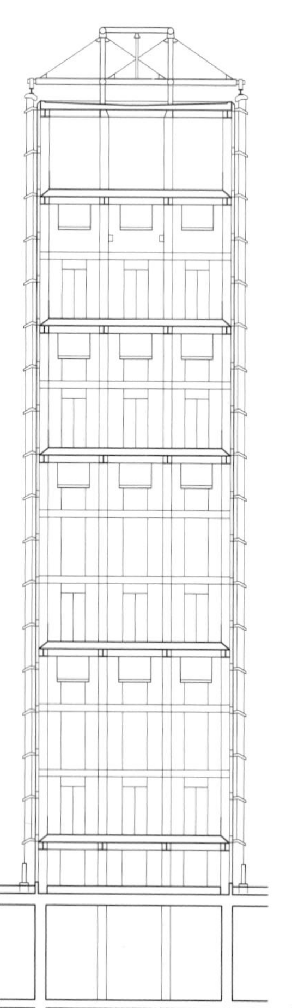

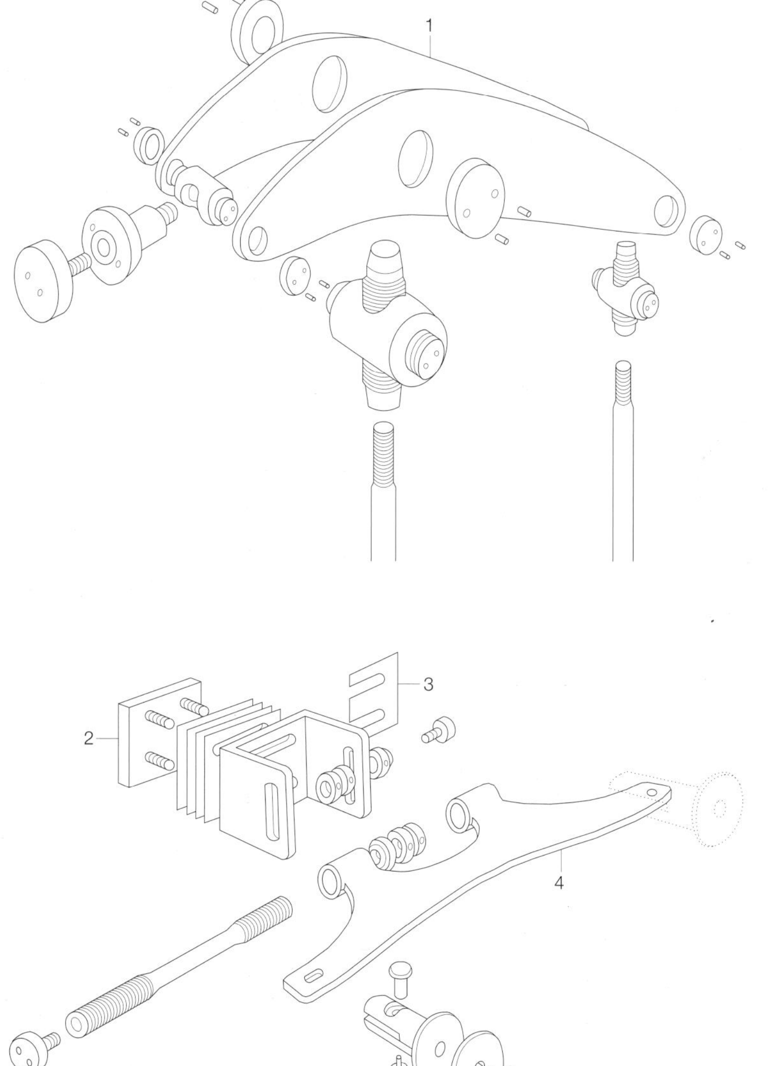

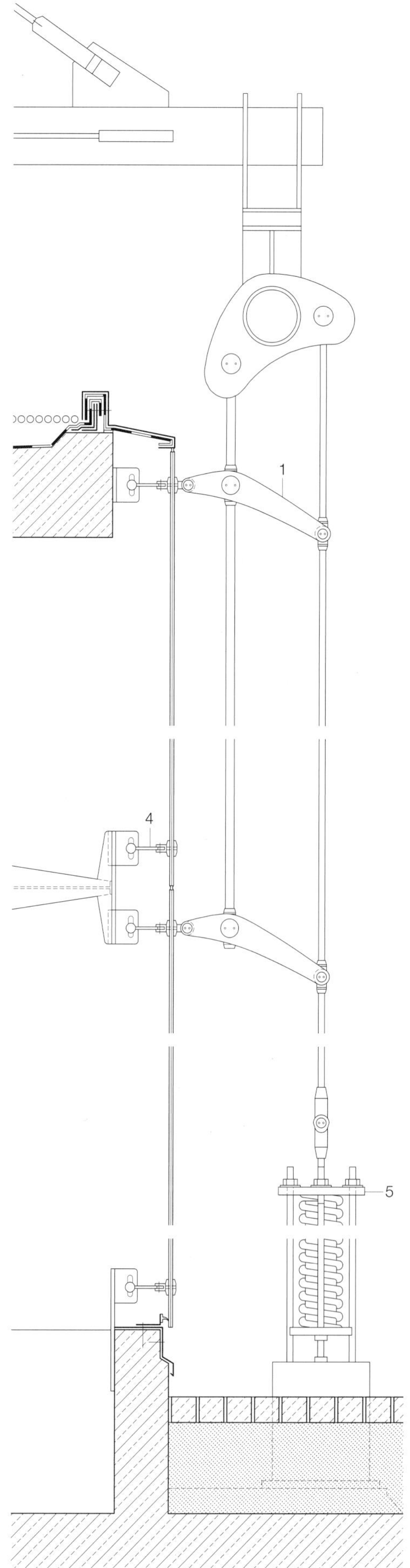

1 "Dolphin" arm, stainless steel
2 Fixing plate
3 Special shims
4 Stainless steel support bracket, 6 mm
5 Circular stainless steel plate for securing the spring

Entrance lobby, Hotel Kempinski

Munich, D, 1994

Architects:
Murphy/Jahn, Chicago
Structural engineers for cable lattice facade:
Schlaich Bergermann und Partner, Stuttgart

📖 Arch+ 124–125, 1994
A+U Extra Edition: Hotel Kempinski. Tokyo 1995
Knaack, Ulrich: Konstruktiver Glasbau, Cologne, 1998

- Suspended cable lattice construction, 45 m wide x 25 m high
- Wind loads carried via horizontal prestressed stainless steel cables, 22 mm dia.
- Vertical loads carried via prestressed stainless steel cables, 22 mm dia.
- Glazing comprises laminated safety glass panes, 1.5 x 1.5 m, 10 mm thick
- 45° chamfer at all corners to accommodate the screws of the cable clamps
- Panes mounted in "floating" arrangement in cable clamps to compensate for facade movements of up to 900 mm

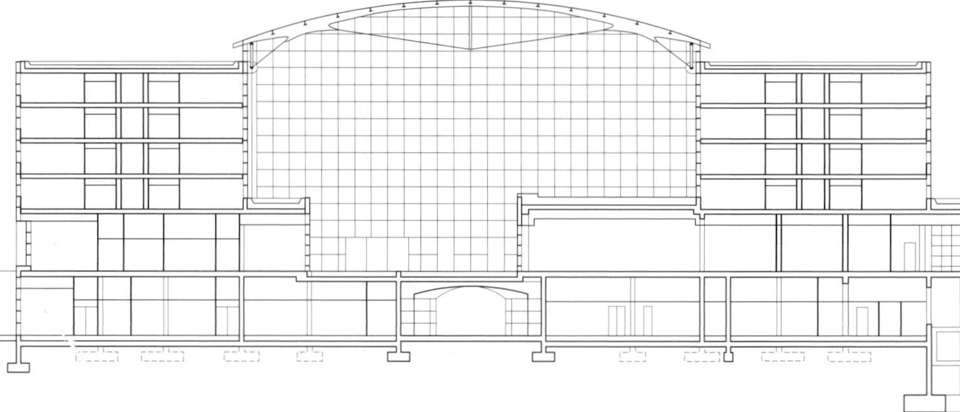

Section, scale 1:750
Vertical sections • Horizontal section scale 1:20
Detail of cable clamp, scale 1:5

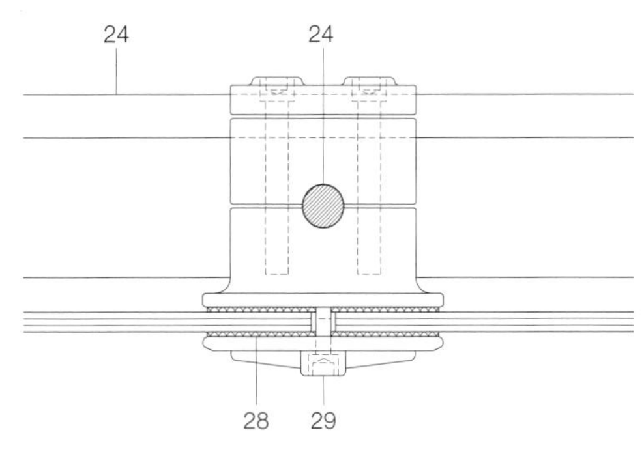

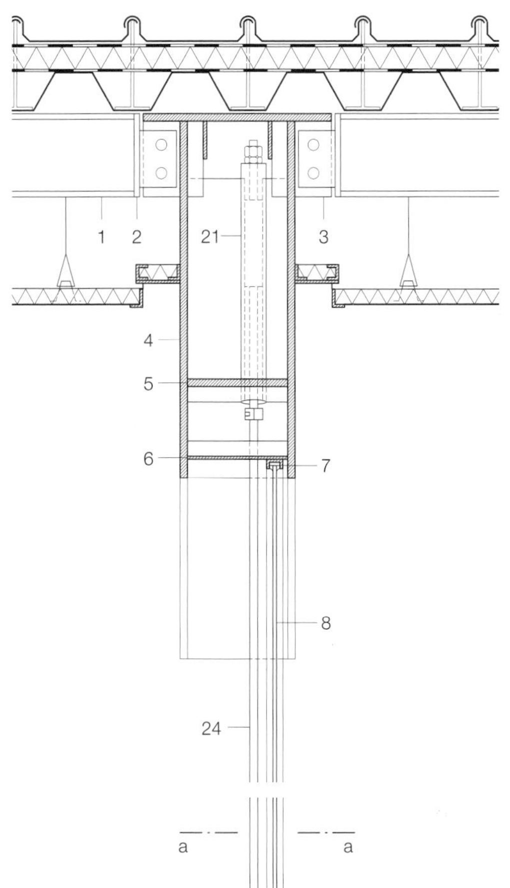

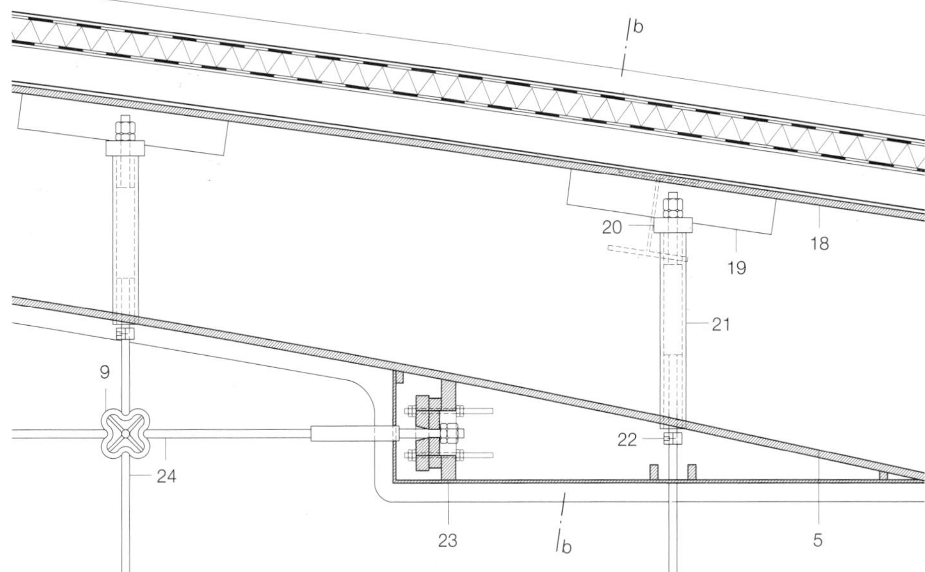

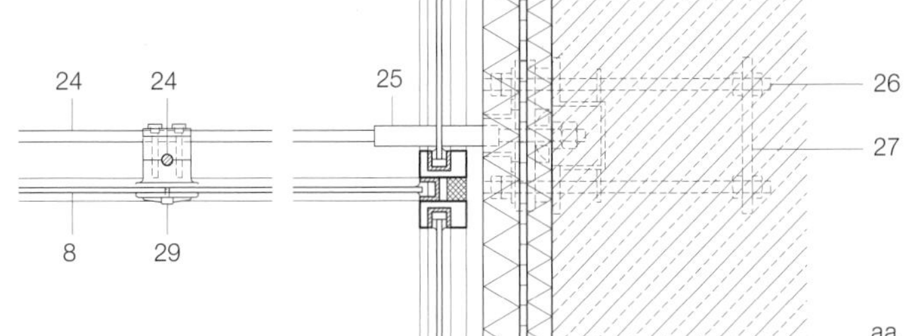

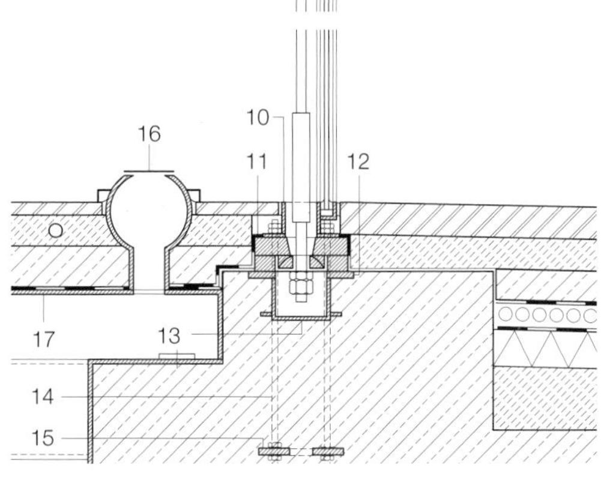

1 Steel section, HEB 220
2 End plate, 220 x 220 x 15 mm, 100 x 100 x 100 mm angle welded on both sides
3 Steel flat, 175 x 200 x 20 mm
4 Steel flat, welded to steel circular hollow section and web plates
5 Bottom flange, 265 x 20 mm
6 Sheet metal capping, 265 x 5 mm
7 Steel channel
8 Laminated safety glass, 10 mm
9 Cable clamp
10 Steel tube, 101.6 dia. x 71 x 2.6 mm, with plates (160 x 160 x 4 mm) for screw fixings welded on, screwed to tensioning plate with 4 No. M6 x 15 after prestressing
11 Tensioning plate, 210 x 190 x 40 mm, with 65 dia. x 52 mm countersunk hole and 106 x 2 mm recess
12 Steel frame, 240 x 220 x 15 mm, 120 mm dia. hole, 127 dia. x 3.2 mm steel tube welded on
13 Steel insert, 240 x 220 x 15 mm, with 120 mm dia. hole in centre, 127 dia. x 3.2 mm steel tube welded on, 4 No. M18 holding-down bolts
14 Threaded bars, 4 No. M16
15 Sheet metal frame, 70 x 15 mm, 4 No. 17 mm dia. holes
16 Air nozzles, brushed stainless steel
17 Ventilation duct
18 Top flange, 500 x 200 mm
19 Steel flat, 560 x 100 x 10 mm
20 Bearing plate, 100 x 80 x 40 mm, with 40 mm dia. hole in centre
21 Steel circular hollow section, 70 dia. x 10 mm, with EPDM ring inside
22 Clamping ring, 2-part
23 Steel transverse member, 260 x 40 mm, 120 mm dia. hole
24 Stainless steel cable, 22 mm dia.
25 M36 threaded fitting, M36 locknut
26 M22 threaded bar
27 Steel anchor plate, 300 x 70 x 20 mm, 2 No. 23 mm dia. holes, M22 nut both sides
28 EPDM bearing
29 M10 x 20 screw

Extension to glass museum

Kingswinford, GB, 1994

Architects:
Design Antenna, Richmond
Structural engineers:
Dewhust Macfarlane & Partners, London

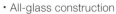

Detail 01/1995
Knaack, Ulrich et al.: Konstruktiver Glasbau 2. Cologne, 2000

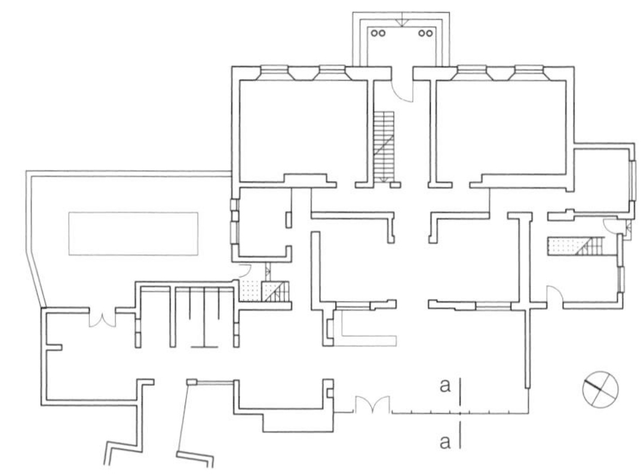

- All-glass construction
- Demonstration of the possibilities offered by glass technology
- Loadbearing structure made from triple-glazed, casting resin-bonded, prestressed glass
- Spacing of beams: 1.10 m, span: 5.70 m
- Bridle joint connection between column and roof beam
- Use of insulating glazing made from solar-control glass
- Ceramic coating to underside of roof glazing acts as sunshading
- Roof construction accessible for cleaning purposes

Plan, scale 1:500
Axonometric view (not to scale)
Detail of door lintel, scale 1:5
Vertical section • Horizontal section
scale 1:5

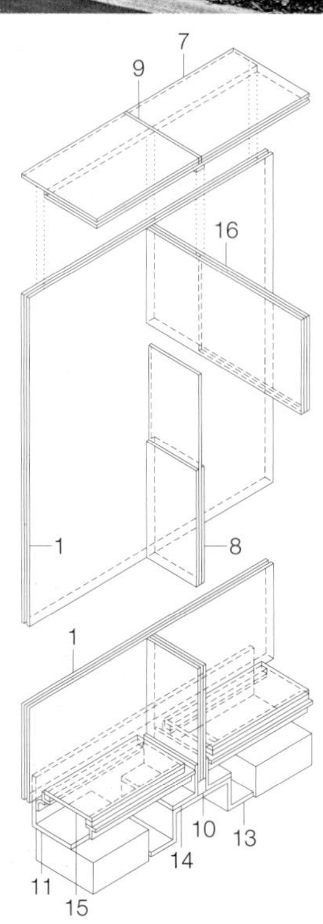

1 Insulating glazing, 8 mm solar-control toughened safety glass + 10 mm cavity + 8 mm toughened safety glass
2 Preformed silicone sea
3 Hole for door pivot
4 Acrylic glass corner trim
5 Stainless steel door fitting
6 Glass door, 15 mm
7 Insulating glazing, 8 mm solar-control toughened safety glass + 10 mm cavity + 2 No. 6 mm laminated safety glass.
8 Glass column, 32 x 200 mm laminated glass
9 Silicone joint
10 Preformed silicone seal
11 Stainless steel angle, 150 x 150 x 10 mm
12 Cork board, 5 mm
13 Steel support bracketr
14 Acrylic glass mounting
15 Toughened safety glass, 10 mm
16 Glass beam, 300 x 32 mm laminated glass

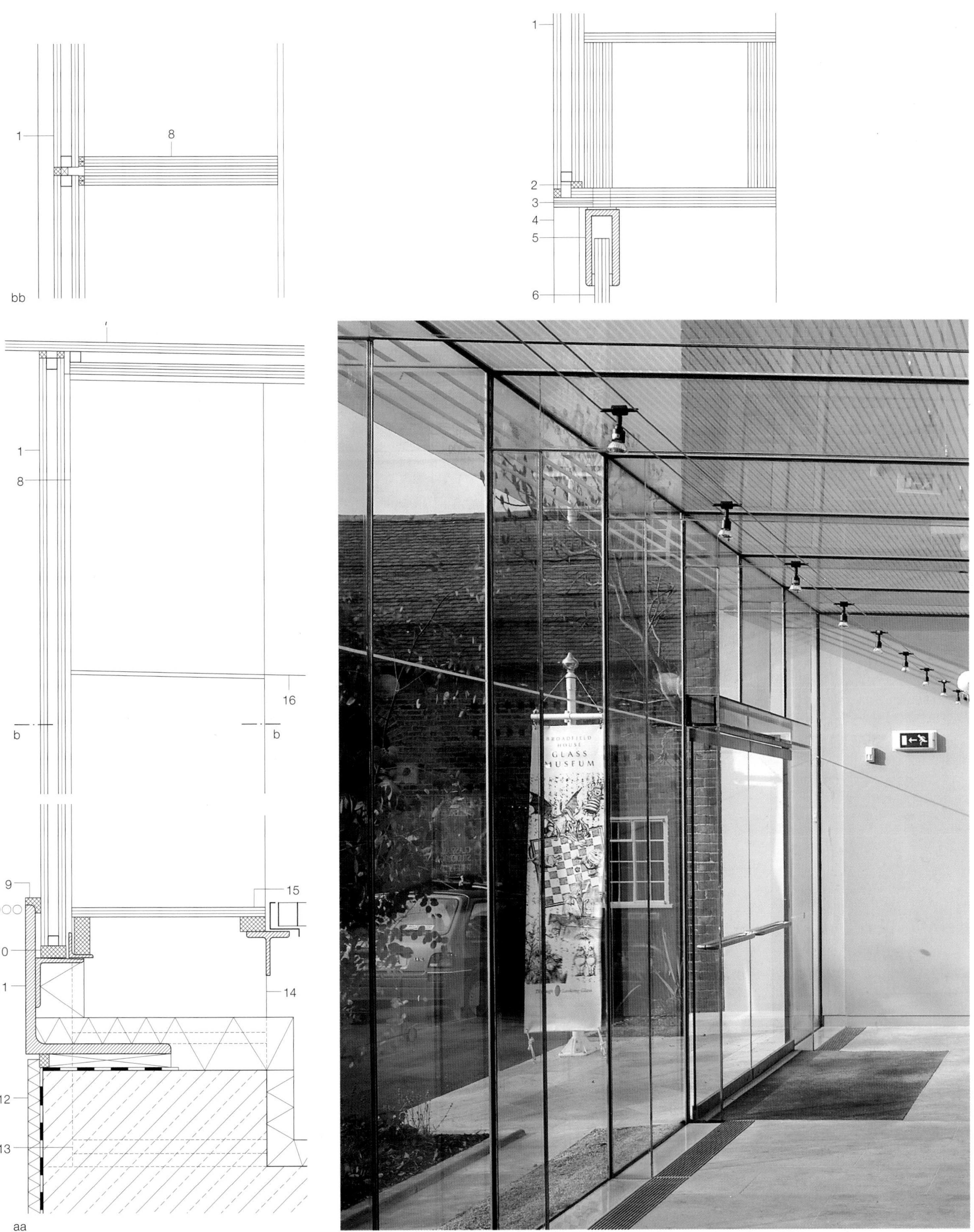
1
8
bb
1
2
3
4
5
6
1
8
16
b
b
9
15
10
11
14
12
13
aa
BROADFIELD HOUSE
GLASS
MUSEUM

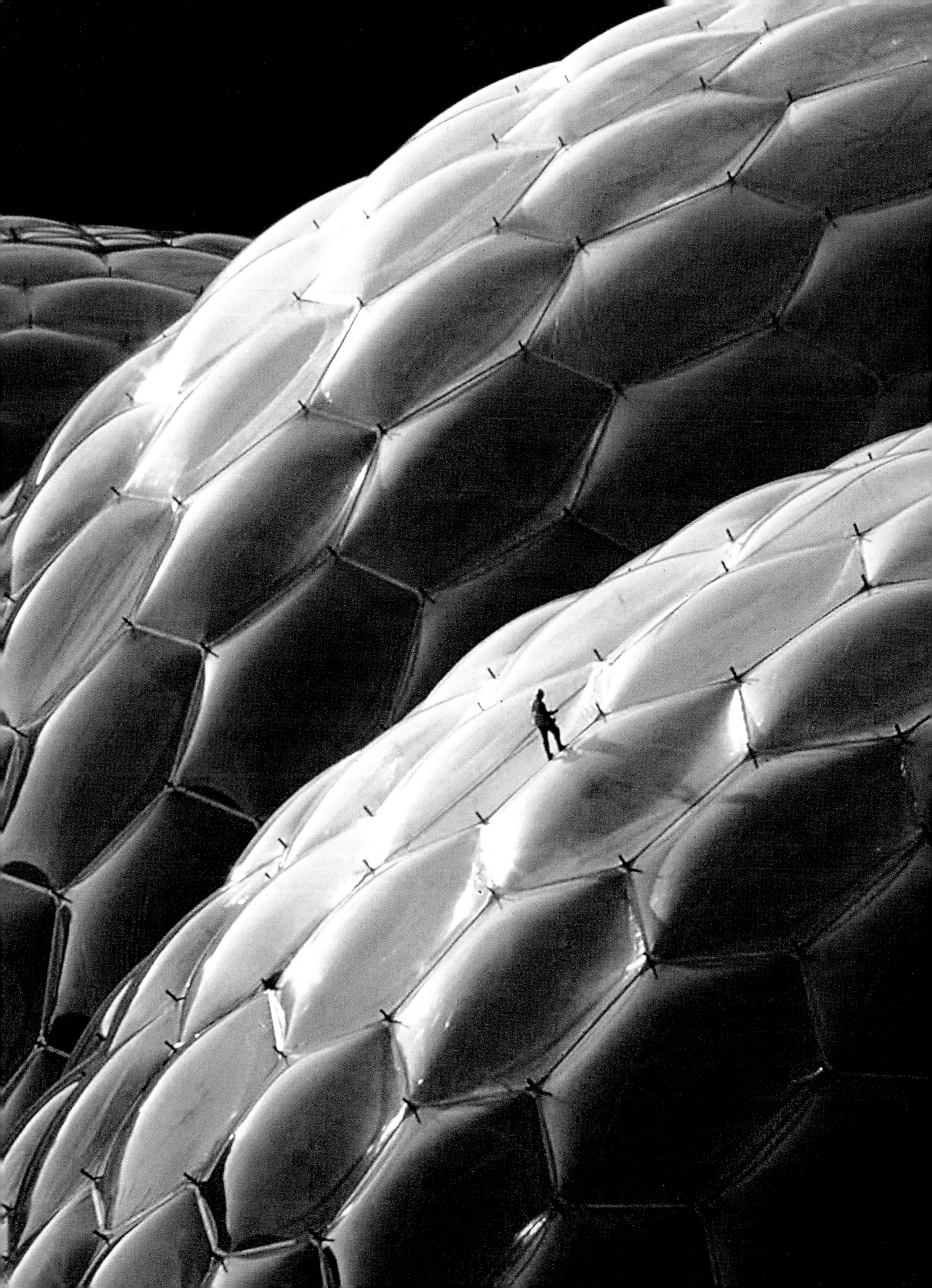

B 1.7 Plastics

The majority of the plastics that play an important role in the modern building industry were invented and developed ready for large-scale production between 1931 and 1938. PVC, for instance, was already marketable as a material for pipes and fittings by 1935. Initially, plastics were used only for internal fitting-out and for producing furniture (figure B 1.7.2). But the late 1950s saw the start of intensive development work surrounding the production of complete buildings made from these synthetic materials [1]. This was made possible by the use of new processes such as laminating and by the production of glass fibre-reinforced moulded parts.

The development of the use of plastics for external walls

Sheets and panels

In their "House of the Future" (1957) Richard Hamilton and Marvin Goody, together with the Monsanto company, managed to implement the new manufacturing techniques in an influential design (figure B 1.7.3).

In 1968 Matti Suuronen designed the "Futuro House", built using self-supporting GFRP sandwich elements with polycarbonate facings and polyurethane foam as the core insulation (figures B 1.7.4 and 1.7.5).

One outstanding example from this period is the Olivetti Training Centre in Haslemere, southern England, (1969-73) designed by James Stirling. The various material properties of the plastics were systematically combined here in order to create sheet-like, self-supporting components. There is a seamless transition between external wall and roof, and the sheets used have thermal insulation properties (fig. B 1.7.8).

When we turn to transparent building envelopes, the roof to the Olympic Stadium in Munich (1972) is regarded as exemplary. This was the first project in the world to make use of transparent, oriented acrylic sheets (3 x 3 m) on a larger scale (figure B 1.7.10). Other milestones along this road are the temporary IBM pavilion (1984) and the BMW exhibition pavilion (1999) (figure B 1.7.12), both by Renzo Piano.

Tension structures

Besides the development of panels and sheets, synthetic fabrics and films were used from a relatively early stage of the history of plastics. These helped to create lightweight, often light-permeable tension envelopes.

Pneumatic structures

Walter Bird designed the first pneumatic, air-supported structure for protecting sensitive radar installations in 1948. This was used as the basis for the further development of air-inflated and air-supported structures for non-

B 1.7.2

B 1.7.2 Stacking chair, 1960, Verner Panton
B 1.7.3 House of the Future, demonstration building as part of "Tomorrowland", Disneyland, California (USA), 1957, Richard Hamilton and Marvin Goody
B 1.7.4–5 Futuro Haus, external and internal views, 1968, Matti Suuronen

B 1.7.3

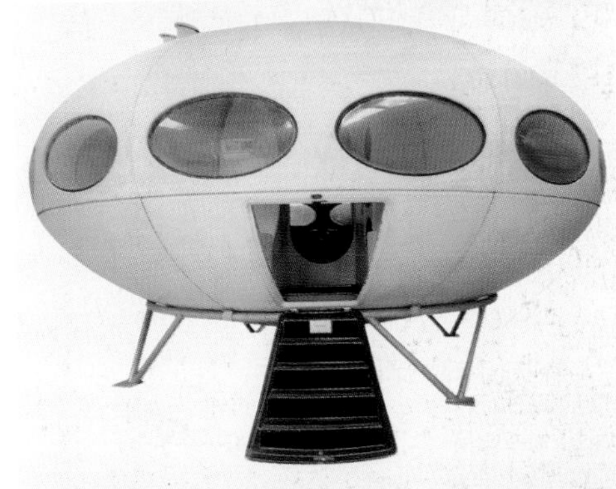
B 1.7.4

B 1.7.5

B 1.7.1 Air-inflated cushion structure, Eden Project, St Austell, (GB) 2001, Nicholas Grimshaw & Partners

B 1.7.6

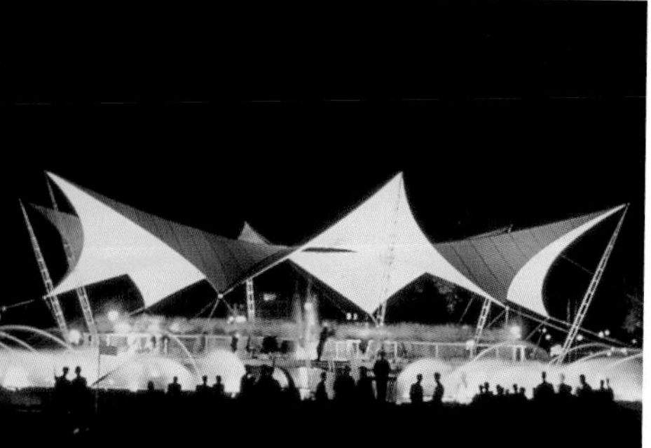

B 1.7.7

B 1.7.6 Giant dome over Manhattan (USA), 1960, Richard Buckminster Fuller
B 1.7.7 Tanzbrunnen Pavilion, Cologne (D), 1957, Frei Otto
B 1.7.8 Olivetti Training Centre, Haslemere (GB), 1973, James Stirling

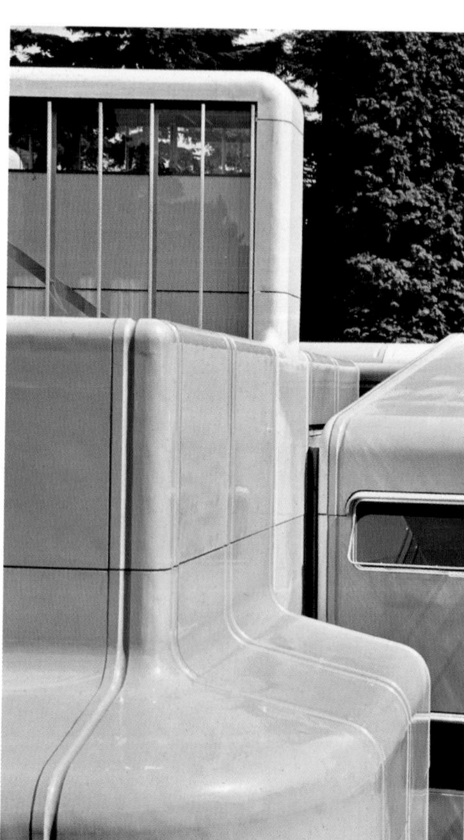

B 1.7.8

military applications, e.g., coverings for swimming pools and tennis courts.

Buckminster Fuller caused an uproar in 1950 with his proposal to erect a giant dome over Manhattan [2] (fig. B 1.7.6). From 1959 onwards Frei Otto collaborated with Kenzo Tange on concepts for roofing over residential developments in the Arctic [3]. One highlight of this development work was the 1970 World Exposition in Osaka, Japan, which could be said to have been a demonstration of the possibilities of pneumatic structures at that time [4].

Tent structures
Frei Otto started intensive development work on the applications of tension structures in the early 1950s; up until then they had been built using only natural materials (fig. B 1.7.7). He therefore created the foundations for the use of plastics in this field. Owing to their outstanding material properties, synthetic fabrics and films are today among the most popular materials used for tent structures.

Manufacture of plastics

Plastics are made from materials that do not occur in nature in their final form. They are mostly obtained from petroleum derivatives and their main characteristic is their macromolecular structure. Plastics are manufactured by means of controlled chemical reactions in which the hydrocarbon molecules are split and then recombined in the form of long macromolecules.

We distinguish between the following processes [5]:

- polymerisation
- polycondensation
- polyaddition

Classification of plastics

Irrespective of the method of manufacture, the macromolecules of plastics are present in the form of long molecular chains – in linear, branched or cross-linked forms. Depending on the degree of cross-linking, we distinguish between the following types of plastics (fig. B 1.7.9):

- thermoplastics (thermosoftening plastics)
- elastomers
- thermosets (thermosetting plastics)

Material properties

General properties
The modern building industry is the second-largest user of plastics after the packaging industry. More than 30 different types of plastics are in use, with PVC accounting for the largest share. Most of the others are in the form of polystyrene foams, polyethylene and polypropylene.

The great importance of plastics in building is due to their advantageous properties:

- adequate tensile and compressive strength, stiffness, hardness and abrasion resistance
- good transparency, dyeable in every gradation from "clear as glass" to black
- adequate to outstanding toughness
- high elasticity
- low density
- adequate thermal stability
- good electrical insulation properties and low thermal conductivity
- weather resistance
- low water absorption
- high resistance to chemicals
- ease of working and processing
- very good surface finish qualities
- paintable surfaces

The material properties can be considerably modified through the method of manufacture and the formulation. In this way it is possible for building materials with the same designation to be designed differently to suit specific applications.

In terms of their resistance to ageing, it should be remembered that many plastic products are considerably younger than the life expectancy of buildings. This aspect should be given careful consideration, especially for building components at risk, such as facade elements and roof waterproofing systems.

Behaviour in fire
This aspect needs very careful consideration when using plastics in the building envelope. The most important criteria here are:

- flammability
- ignition temperature
- decomposition temperature
- release of smoke and fumes
- toxicity of decomposition products
- corrosion caused by decomposition products

Besides the release of – sometimes – highly toxic gases, smoke given off during a fire can severely impair vision for persons trying to escape or fire-fighters. The choice of a suitable plastic therefore depends to a large extent on the toxicity and smoke to be expected. Apart from that, the decomposition products in the smoke can also have very corrosive effects on other materials.
Flammability can be decreased by applying a flame retardant.

Semi-finished products for facades

A wide range of semi-finished products is available for use in facades. These allow the construction of rigid (resistance to mechanical loads) or non-rigid (stable under compression or stressed only in tension) designs depending on the type of loads to be expected.

The range of different physical properties is likewise very broad. This is because the specific combination of various materials or the modification of the material properties results in great diversity. Suitable subsequent processing of the raw materials enables the production of a huge variety of semi-finished products (fig. B 1.7.11).

Plain, corrugated and multiple-web sheets

Extruding, calendering and pressing are the methods of manufacture most commonly employed for producing plastic sheets. Plain, corrugated and multiple-web sheets can be manufactured in this way. The usual materials for producing plain, transparent sheets are polymethyl methacrylate (PMMA) and polycarbonate (PC), which, owing to their good transparency, weather resistance and impact resistance, are well suited to applications in facades. The conventional sheet format is 2050 x 3050 mm. At 4 mm thick, the light transmittance is approx. 90%. These materials fall into building materials class B 2 for fire classification purposes.

Plain sheets made from thermoplastic polyesters (PET, PETG) have a high breaking resistance and belong to building materials class B 1. The use of glass fibres to reinforce plastics (GFRP) enables the production of opaque sheets in any form. PMMA corrugated sheets, owing to their corrugated cross-section and the resulting increased stiffness, can be produced in sizes up to 10 450 x 4000 mm, PC corrugated sheets up to 10 970 x 7000 mm. Corrugated sheets made from GFRP can be produced in sizes up to 3000 x 20 000 mm.

Multiple-web sheets with a whole range of different cross-sections can be manufactured from PMMA, PC and GFRP (fig. B 1.7.16). The stiffening effect of the webs mean that PMMA sheets can be manufactured in lengths up to 7000 mm, PC sheets up to 11 000 mm. The formation of hollow cavities leads to a relatively low thermal transmittance (U-value) of approx. 2.5 W/m²K for twin-wall multiple-web sheets and as low as 1.6 W/m²K for the triple-wall varieties. This figure can be further im-proved by using mutli-wall multiple-web sheets and by filling the cavities with an insulating material.

The acoustic and optical properties can be controlled as required by applying protective coatings or by forming multi-layer cavities. Multiple-web sheets made from GFRP are particularly noteworthy in this respect and are

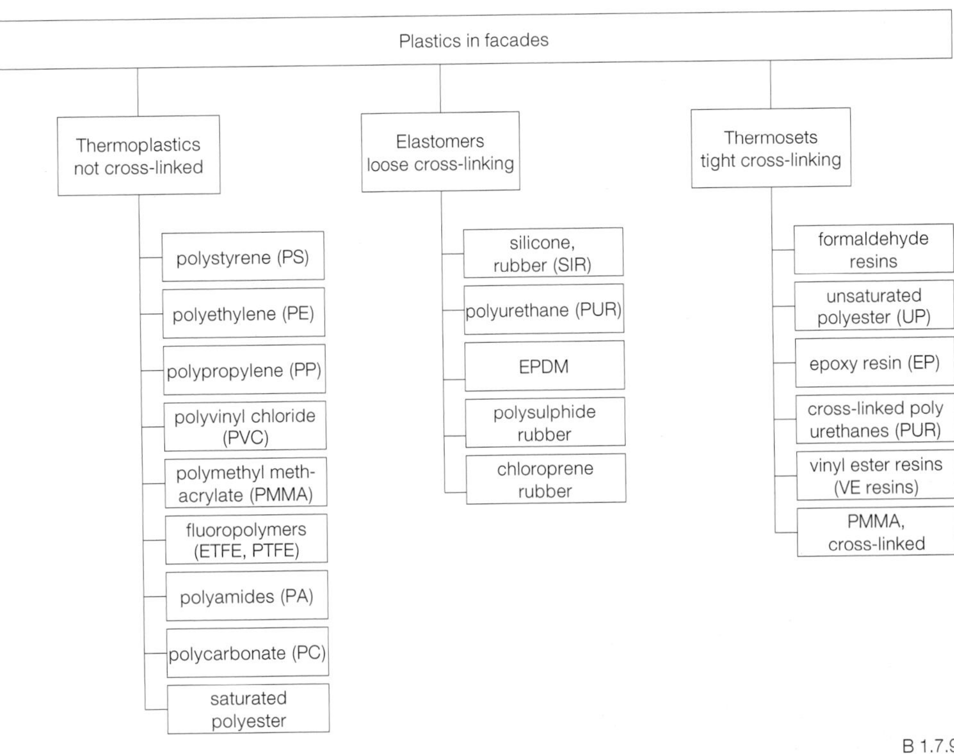

B 1.7.9

B 1.7.9 Classification of plastics according to the degree of cross-linking and the resulting material properties [6]

B 1.7.10 Tent roof of acrylic sheets, Olympic Stadium, Munich (D), 1972, Günter Behnisch + Partner, Frei Otto et al.

B 1.7.10

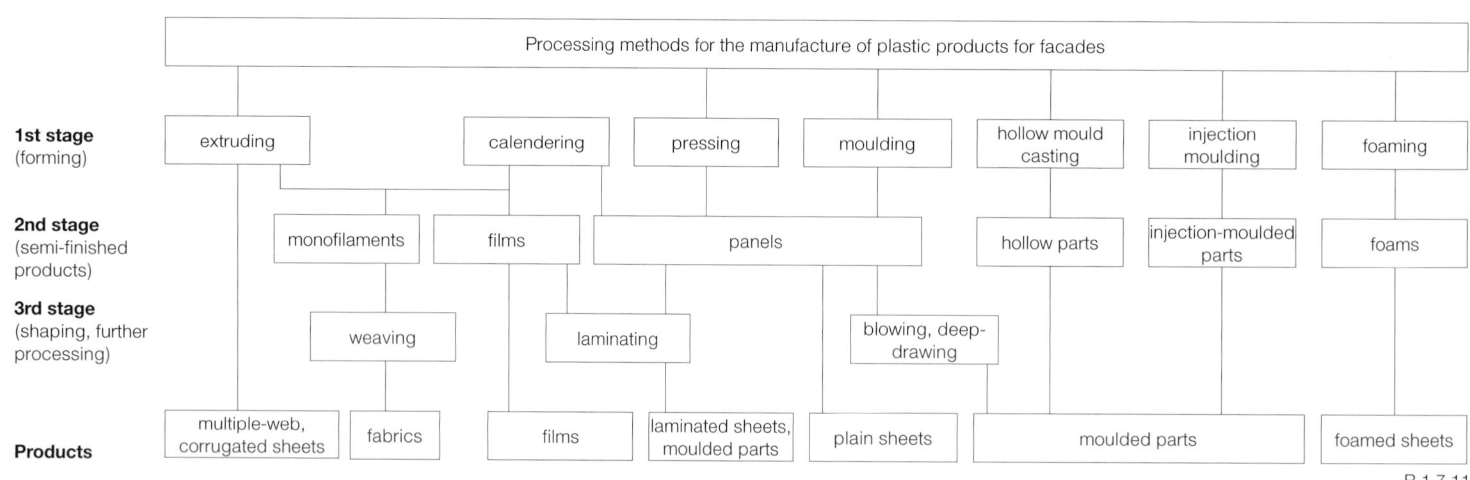

B 1.7.11

B 1.7.12

B 1.7.11 Processing methods for the manufacture of plastic products for facades
B 1.7.12 Temporary exhibition pavilion, 1999, Bernhard Franken with ABB
B 1.7.13 Private house, Tokyo (J), 1996, FOBA

B 1.7.13

available with many different cross-sections in lengths of up to 15 000 mm.

Moulded parts (casting and laminating)
Building components in virtually any shape or form and with almost any dimensions can be produced by means of casting and laminating methods. Besides large-scale industrial production, a considerable proportion of the work is carried out manually, although this results in high production costs.

One of the best-known products is the type of moulded part made from glass fibre-reinforced plastic (GFRP). The use of reinforcing materials (fibres, knitted or woven fabrics) substantially improves the mechanical properties of the plastics used to make building components.

Synthetic resins are employed for producing the parts. The components of the resin are mixed together and placed in or on the moulds, together with the aforementioned reinforcing materials (laminating materials) for curing.

The type of reinforcement (fibres, knitted or woven fabrics) and the material (glass, carbon, aramide) together determine the strength, elasticity and cost of the building components produced in this way. A mould is required for all laminating processes, and the component has to remain in the mould until the plastic is fully cured.

Facades made from sheets and moulded parts
Semi-finished products in the form of sheets and the plastics used for producing moulded parts all exhibit a high coefficient of thermal expansion. Therefore, all clamping and drilled fixings must include flexible seals and gaskets to ensure that the material is not restrained at these positions.

With the great diversity of materials and semi-finished products available, it is possible to adapt materials and cross-sections to suit specific requirements.

The selection criteria that apply here are:

- structural requirements
- thermal performance
- acoustic performance
- weather resistance
- optical properties
- fire classification
- mechanical loads
- resistance to chemicals
- temperature range
- working and fixing options
- recycling

Connections between moulded parts and the way they are fixed to the supporting construction usually require the respective interfaces to be individually designed. Depending on the materials used, the connections and fixings range from mechanical fasteners and butt joints to bonded and welded joints.

Sheet-type semi-finished products, on the other hand, can make use of the well-known jointing and connecting techniques familiar in timber, metal and glass facades [7].

Synthetic films and fabrics
More and more, the plastic films and fabrics made from PVC, polyester, PTFE and ETFE in general use are being combined to exploit the advantages of the individual materials. Figure B 1.7.18 provides an overview of such products.

Thanks to the development of new plastics, which are both transparent and loadbearing, and promise to be long-lasting (e.g., ethylene-tetrafluoroethylene, abbreviated to ETFE), it is now possible to construct single- or multi-leaf facades. Such membrane designs offer a series of advantages which result in long-span facades with a hitherto unknown degree of transparency.

The advantages of ETFE films are:

- low self-weight
- high light and UV permeability
- good transparency
- high resistance to chemicals
- long useful life
- high degree of recyclability

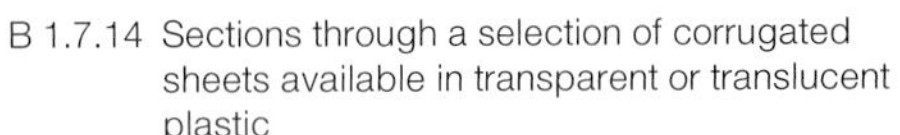

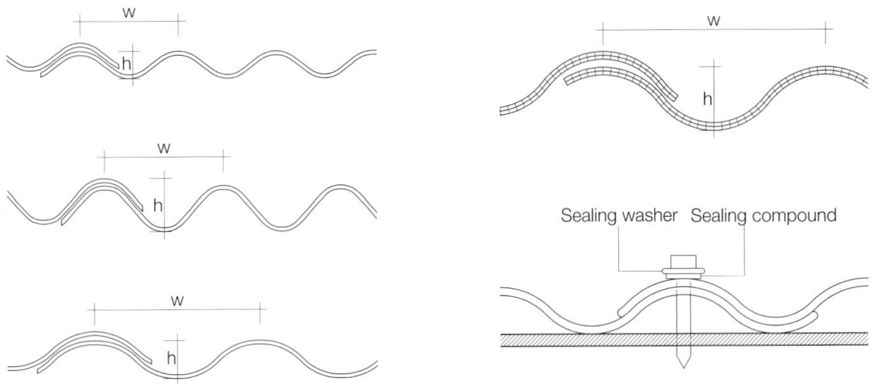

B 1.7.14 Sections through a selection of corrugated sheets available in transparent or translucent plastic
B 1.7.15 Methods of fixing multiple-web sheets by means of glazing bars (a), in rebates or with screws (b–f)
B 1.7.16 Sections through a selection of sheet-type semi-finished products made from transparent or translucent plastic

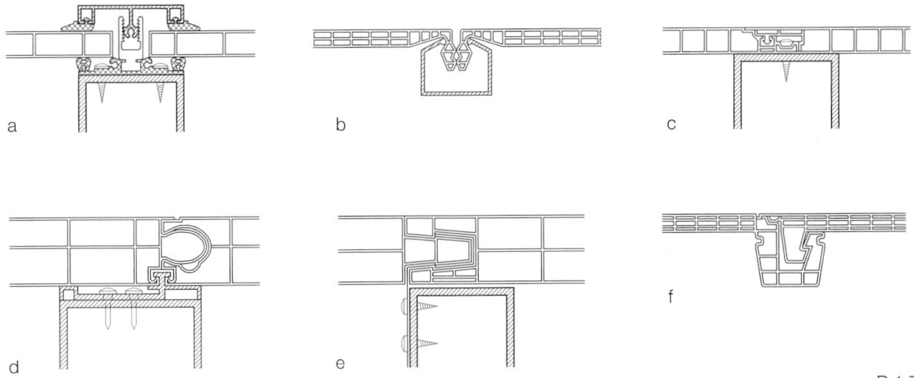

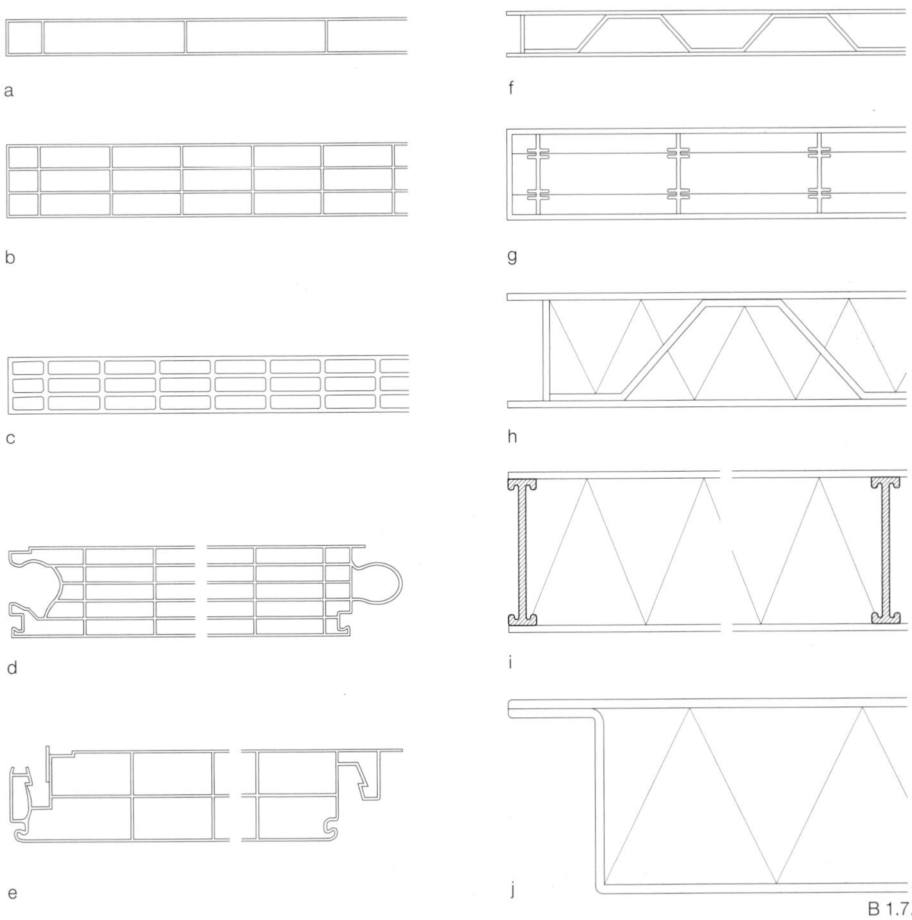

Surface coatings

In contrast to the films, the fabrics used for membranes very often comprise several layers. These make use of a "coating" of PVC-P, PTFE or silicone to protect the fabric itself against moisture, UV radiation, microbes and fungal growths. Besides achieving watertightness, the coating substantially decreases soiling and lengthens service life. Far better than PVC-P in terms of ageing behaviour are PTFE and silicone, where no changes have been observed over periods of 25–30 years. PVC-P has a much shorter useful life, owing to its sensitivity to ultraviolet light.

Tension facades

In membrane designs, external actions due to wind and snow loads are usually carried by creating an – as far as possible – even prestress within the film or fabric. As a rule, this is achieved by employing either a pneumatic structure or a mechanically prestressed construction [8].

Pneumatic structures

The development of gastight and flexible but also loadbearing synthetic materials has enabled the construction of air-supported and air-inflated envelopes. Large areas can be covered with a minimal amount of material, mainly thanks to the use of translucent, PVC-P-coated polyester fabrics.

One variation is the foil cushion system made from UV-permeable fluoropolymer films, which were developed primarily for greenhouses.

In order to be able to accommodate the mechanical stresses and strains that occur with longer spans, coated ETFE, polyester or aramide fibre fabrics are used. These possess many times the tensile strength of unreinforced materials.

Mechanically prestressed structures

The prestress required to stabilise the material against external actions is achieved by restraining the membrane or film material along a defined edge of high and low points. The curvature of the surface results in two effective

B 1.7.14

B 1.7.15

B 1.7.16

B 1.7.17

directions for every point on the surface of the membrane. Depending on the particular action, one direction assumes the loadbearing function, the other the tensioning function [9]. The smaller the curvature of the surface, the larger the resultant stresses in the membrane. When using materials with a lower load-carrying capacity, this aspect should be taken into account because the area which can be covered will be very much smaller. One principal advantage of tension structures is the small amount of material needed to cover a large area. However, this is also a major disadvantage because the thin, single-layer membrane material has thermal insulation and acoustic properties that are comparatively poor for an external wall. Multi-layer materials, which considerably improve the acoustic properties, represent one optimisation route. And multi-leaf materials or the use of fabrics with mineral-fibre fillings can achieve considerable improvements to the thermal insulation properties [10].

Woven fabric (excl. mesh weaves)	Designation	Weight per unit area [g/m²] to DIN 55352	Min. tensile strength of fabric [N/50 mm] warp/weft to DIN 53354	Elongation of fabric at tear [%] warp/weft to DIN 53354	Tear propagation strength of fabric [N] warp/weft to DIN 53363	Buckling stability	UV-resistance	Building materials class for fire classification to DIN 4102	Translucency [%]	Useful life expectancy (years)
cotton fabric		350 520	1700/1000 2500/2000	35/18 38/20	60 80	very good	adequate	B2	varies	<5
PTFE fabric		300 520 710	2390/2210 3290/3370 4470/4510	11/10 11/10 18/9	approx. 500/500	very good	very good	A2	up to approx. 37	>25
ETFE fabric, THV-coated		250	1200/1200			very good	very good	B1	up to approx. 90	>25
polyester fabric, PVC-P-coated	type I type II type III type IV type V	800 900 1050 1300 1450	3000/3000 4400/3950 5750/5100 7450/6400 9800/8300	15/20 15/20 15/25 15/30 20/30	350/310 to 1800/1600 580/520 800/950 1400/1100 1800/1600	very good	good	B1	up to approx. 4.0	>20
glass fibre fabric, PTFE-coated		800 1150 1550	3500/3500 5800/5800 7500/6500	7/10 to 2/17	300/300 to 500/500	adequate	very good	A2	up to approx. 13	>25
glass fibre fabric, silicone-coated		800 1270	3500/3000 6600/6000	7/10 to 2/17	300 570	adequate	very good	A2	up to approx. 25	>20
aramide fibre fabric, PVC-P-coated		900 2020	7000/9000 24500/24500	5/6	700 4450	good	adequate	B1	basically zero	>20
aramide fibre fabric, PTFE-coated			project-related, limited adjustability			good	adequate	A2	basically zero	>25

Film	Designation	Weight per unit area [g/m²] to DIN 55352	Tensile strength of film [N/mm²] to DIN 53455	Elongation of film at tear [%] to DIN 53455	Tear propagation strength of film [N/mm] to DIN 53363	Buckling stability	UV-resistance	Building materials class for fire classification to DIN 4102	Translucency [%] nach	Useful life expectancy (years)
ETFE-films	50 μm 80 μm 100 μm 150 μm 200 μm	87.5 140 175 262.5 350	64/56 58/54 58/57 58/57 52/52	450/500 500/600 550/600 600/650 600/600	450/450 450/450 430/440 450/430 430/430	adequate	very good	B1	up to approx. 95	>25
THV-film	500 μm	980	22/21	540/560	255/250	good	good	B1	up to approx. 95	>20
PVC-P-films						good	adequate	B1	up to approx. 95	<5

B 1.7.18

B 1.7.17 New football stadium, Munich (D), currently at the design stage, Herzog & de Meuron
B 1.7.18 Overview of the properties of materials used in modern membrane structures
B 1.7.19 Seams for membrane materials used in facades
a) welded seam, PVC-polyester fabric
b) welded seam, PTFE-glass fibre fabric with PTFE intermediate strip
c) sewn seam with cover strip, PVC-polyester fabric
d) butt joint with clamping strips
B 1.7.20 Membrane reinforcing details for facades
a) double layer of material for low stresses
b) double layer of material for high stresses
c) cable in pocket
d) strap in pocket
B 1.7.21 Edge details for membrane materials used in facades
a) edge strap sewn on
b) cable in pocket
c) edge cable and strap
d) tube in pocket
e) edge clamping strip without perforating the membrane, without tensioning option
f) edge clamping strip without perforating the membrane, with tensioning option
g) edge clamping strip with connecting straps
h) tied edge
B 1.7.22 "Airquarium", temporary structure, 2000, Axel Thallemer

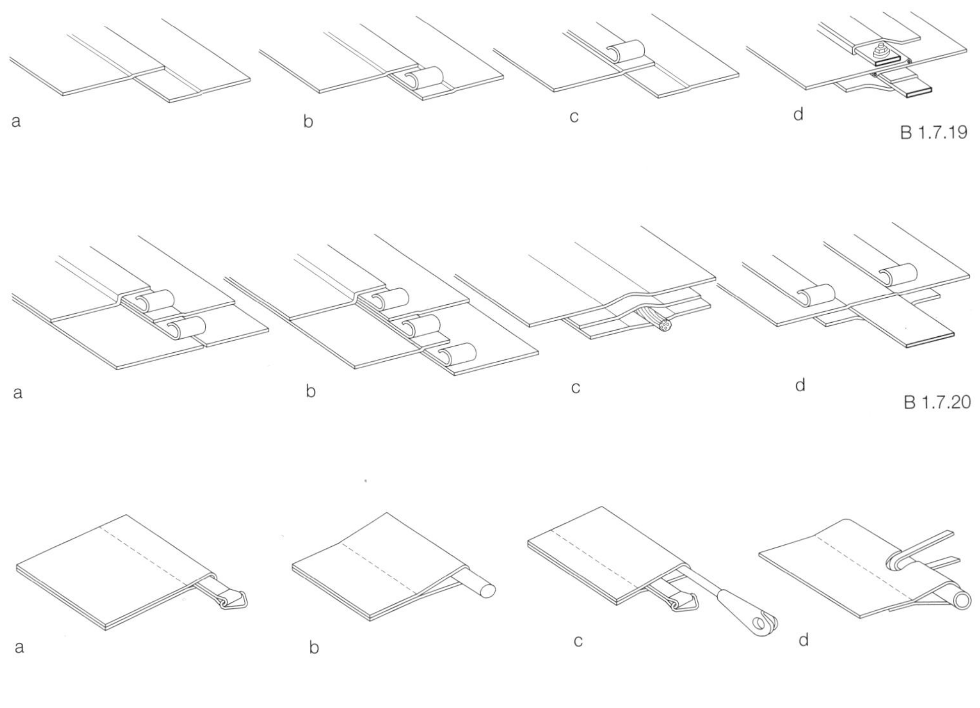

B 1.7.19

B 1.7.20

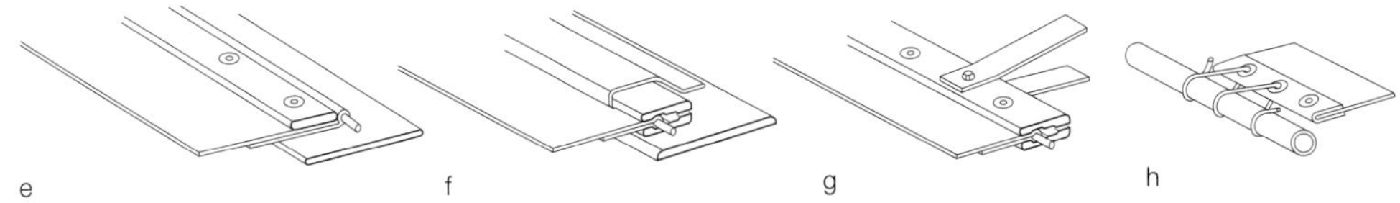

B 1.7.21

Notes

[1] Eisele, Jo; Schoeller, Walter: Kunststoffe in der Architektur; in: Detail 12/2000, pp. 1540–43.
[2] Fuller, Buckminster Richard: Your private Sky. Design als Kunst einer Wissenschaft. Museum für Gestaltung, catalogue pub. by Krausse, Joachim; Lichtenstein, Claude, Baden, 1999, pp. 436–37.
[3] Otto, Frei: *Das hängende Dach. Gestalt und Struktur*, Berlin, 1954.
[4] Herzog, Thomas: Pneumatic Structures. *Handbook of Inflatable Architecture*, New York, 1976.
[5] Weber, Anton: Kleine Werkstoffkunde der Baukunststoffe; in: *Bauen mit Kunststoffen*, pub. by Institut für das Bauen mit Kunststoffen e.V., Berlin, 2001, pp. 44–45.
[6] This overview was drawn up in cooperation with Rainer Letsch, scientific assistant at the Chair for Building Materials and Materials Testing, Munich Technical University.
[7] Kaltenbach, Frank: Plastic – Semi-finished sheet products; in: Kaltenbach, Frank (ed.): *Translucent Materials*, Munich, 2003, pp. 40–56.
[8] Koch, Klaus-Michael: *Bauen mit Membranen*, Munich, 2004.
[9] Rein, Alfred; Wilhelm, Viktor: Das Konstruieren mit Membranen; in: Detail 06/2000, pp. 1044–49.
[10] Moritz, Karsten; Barthel, Rainer: Building with ETFE sheeting; in: Kaltenbach, Frank (ed.): *Translucent Materials*. Munich, 2003, pp. 70–78.
[11] Moritz, Karsten: Membranwerkstoffe im Hochbau; in: Detail 06/2000, pp. 1050–58.

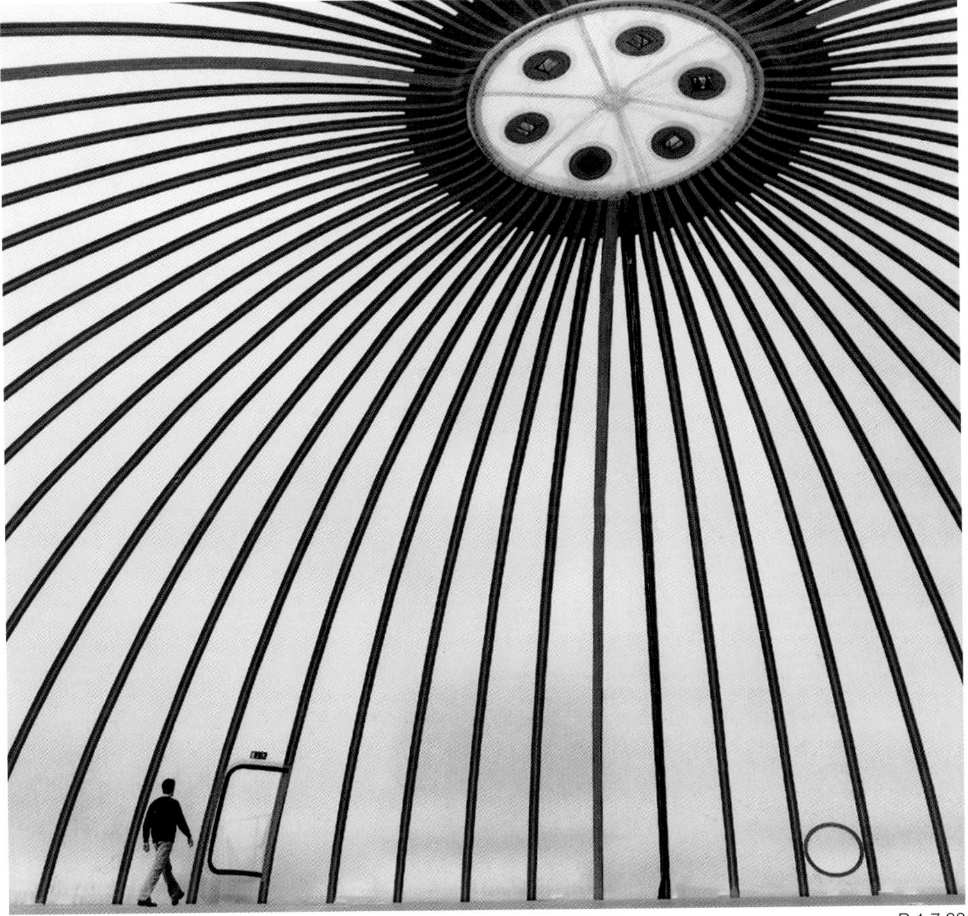
B 1.7.22

Information pavilion

Bologna, I, 2003

Architect:
Mario Cucinella, Bologna

Abitare 03/2003
l'ARCA 10/2003
Architectural Review 10/2003
THE PLAN 004/2003

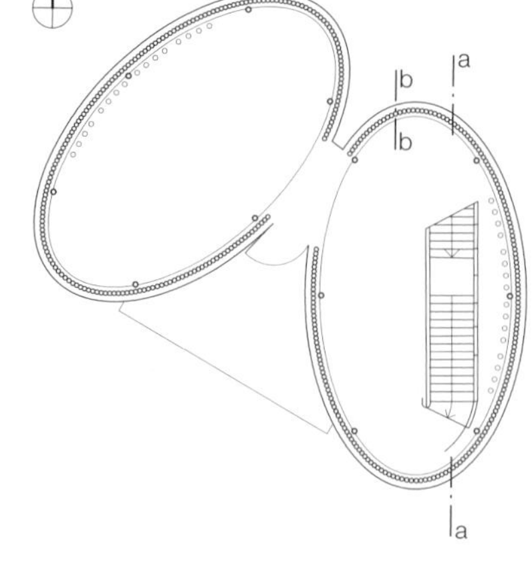

- Exhibition and access pavilion with associated underground exhibition rooms
- 4.50 m high, double-leaf glass/plastic facade
- External glass leaf made from curved panes of laminated safety glass
- Ventilated facade cavity
- Integral, coloured LED lights at the base of the acrylic tubes can be used to create different lighting effects
- Roof loads carried by separate steel stanchions

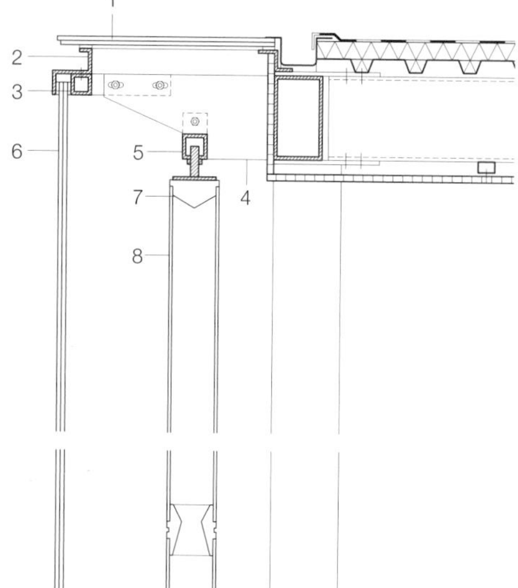

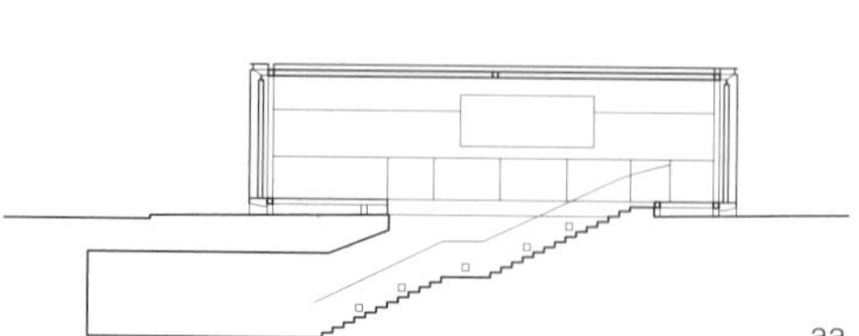

Plan • Section scale 1:400
Vertical section scale 1:20

1 Laminated safety glass, coated
2 Edge trim, 5 mm
3 Steel square hollow section, 50 x 50 x 5 mm
4 Steel flat, galvanised and painted, 10 mm
5 Steel square hollow section, 60 x 60 x 4 mm
6 Laminated safety glass, 2 No. 10 mm panes, curved
7 Acrylic capping piece, satin finish, 120 mm dia.
8 Acrylic tube, 120 mm dia., with acrylic splicing pieces, satin finish
9 Steel angle, 150 x 35 x 5 mm
10 LED lights, white and blue
11 Steel square hollow section, 60 x 60 x 4 mm
12 Sheet metal, 3 mm
13 Laminated safety glass, 2 No. 12 mm panes, anti-slip silk-screen printing
14 Steel rectangular hollow section, 80 x 40 x 4 mm
15 Steel flat, 10 mm

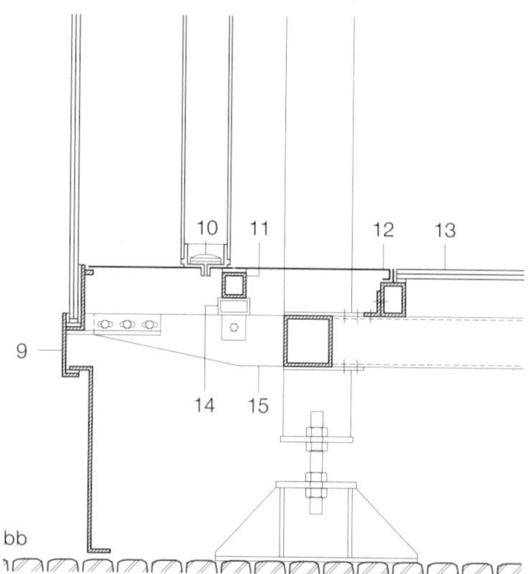

Leisure studio

Espoo, FIN, 1992

Architects:
Kaakko Laine Liimatainen Tirkkonen, Helsinki

Baumeister 03/1995
Detail 01/1993
UME 04/1997
Riley, Terence: Light Construction.
Barcelona 1996

- 5 m high facade made from hollow acrylic sheets, d = 16 mm
- Sheets attached to acrylic bars with transparent self-adhesive tape, bars screwed to supporting construction of timber battens
- Trees provide shade
- Doors and ventilation flaps in the roof to dissipate hot air
- Building used from spring to autumn
- Solid sanitary core within the building serves to store heat energy
- 450 mm high cavity floor provides storage space

Plan • Section, scale 1:200
Vertical section, scale 1:20

1 Transparent plastic film, fixed with adhesive (closes off hollow cavities)
2 Hollow acrylic sheets, d = 16 mm
3 Metal hinges
4 Acrylic bar, 40 x 8 mm
5 Timber batten, 45 x 45 mm
6 Rafter, 220 x 68 mm
7 Column, 145 x 68 mm

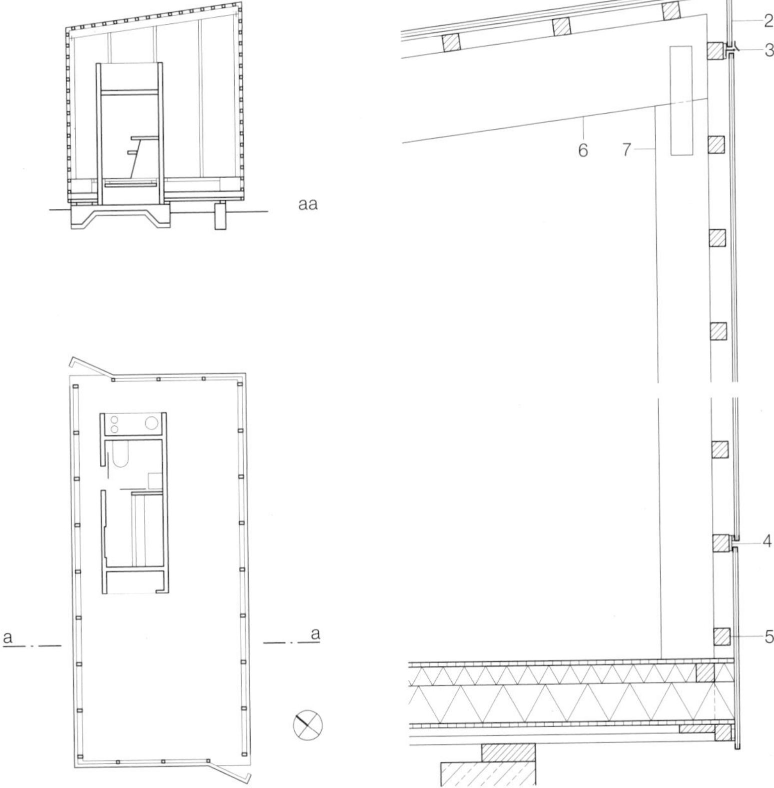

Institutional building

Grenoble, F, 1995/2001

Architects:
Anne Lacaton & Jean Philippe Vassal, Paris

Bauwelt 16/1996
Detail 12/2002
Architekturzentrum Wien (pub.):
Hintergrund 19. Vienna 2003

- North and south elevations have conservatory-type buffer facades with internal planting
- North and south elevations have conservatory-type buffer facades with internal planting
- Inner leaf of sliding aluminium windows
- Fair-face concrete components with insulation on the inside
- Facade cavity ventilated via strip of opening lights in outer leaf

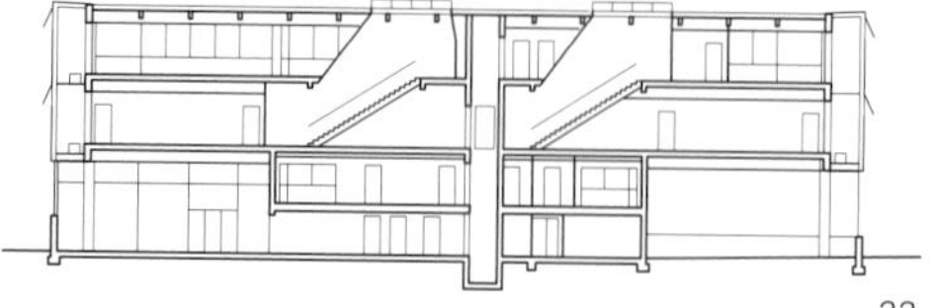

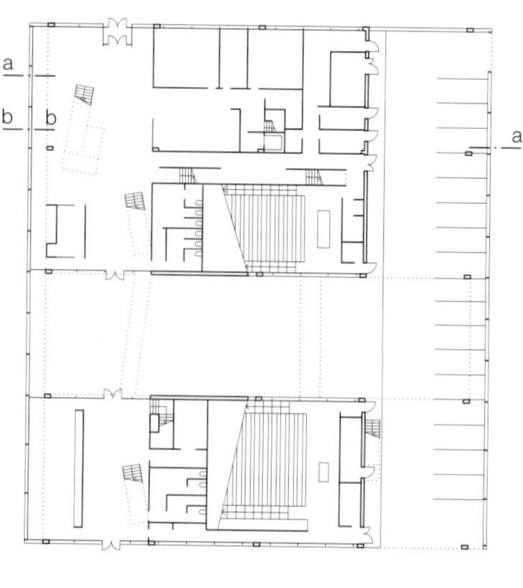

Section • Plan of ground floor scale 1:800
Horizontal section • Vertical section scale 1:20

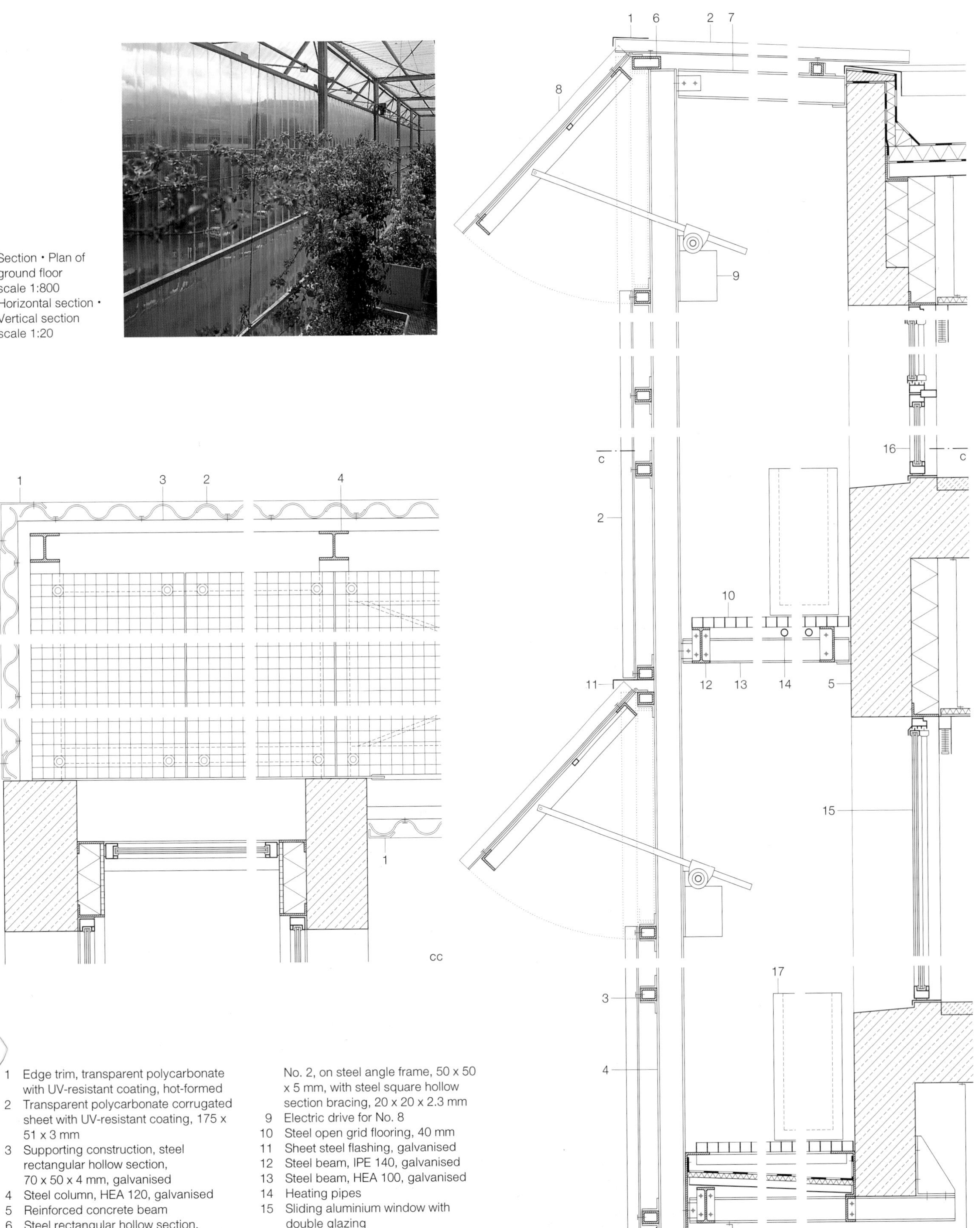

1 Edge trim, transparent polycarbonate with UV-resistant coating, hot-formed
2 Transparent polycarbonate corrugated sheet with UV-resistant coating, 175 x 51 x 3 mm
3 Supporting construction, steel rectangular hollow section, 70 x 50 x 4 mm, galvanised
4 Steel column, HEA 120, galvanised
5 Reinforced concrete beam
6 Steel rectangular hollow section, 100 x 50 mm
7 Steel beam, HEA 120, galvanised
8 Opening lights, corrugated sheet as in No. 2, on steel angle frame, 50 x 50 x 5 mm, with steel square hollow section bracing, 20 x 20 x 2.3 mm
9 Electric drive for No. 8
10 Steel open grid flooring, 40 mm
11 Sheet steel flashing, galvanised
12 Steel beam, IPE 140, galvanised
13 Steel beam, HEA 100, galvanised
14 Heating pipes
15 Sliding aluminium window with double glazing
16 Spandrel panel with double glazing
17 Container for plants

Factory building

Bobingen, D, 1999

Architect:
Florian Nagler, Munich

📖 db Sonderheft 06/2000
Detail 03/2001
GLAS 05/2000
Schwarz Ulrich: Neue Deutsche Architektur. Stuttgart 2002
Herzog, Thomas et al.: Timber Construction Manual, Basel/Boston/Berlin, 2004

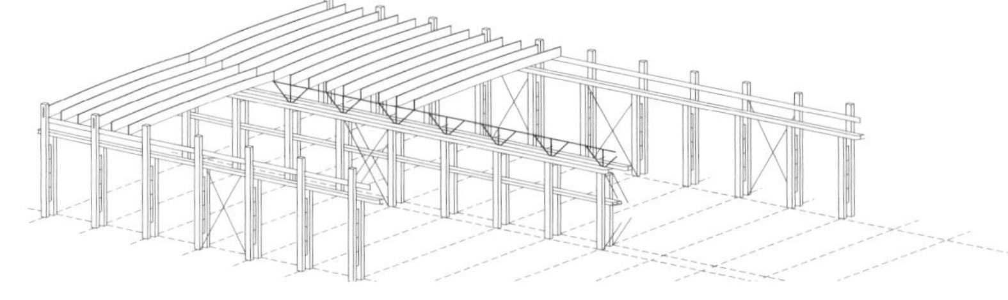

- Translucent external skin of polycarbonate double-layer sheets, 10 m high
- Length of sheet = height of building, fixed at the base but with sliding fixings at the eaves to accommodate temperature-related changes in length
- Sheets fixed to the suspended facade rails by means of concealed metal clips
- Building ventilated via large doors, emergency exits and smoke vents in the roof

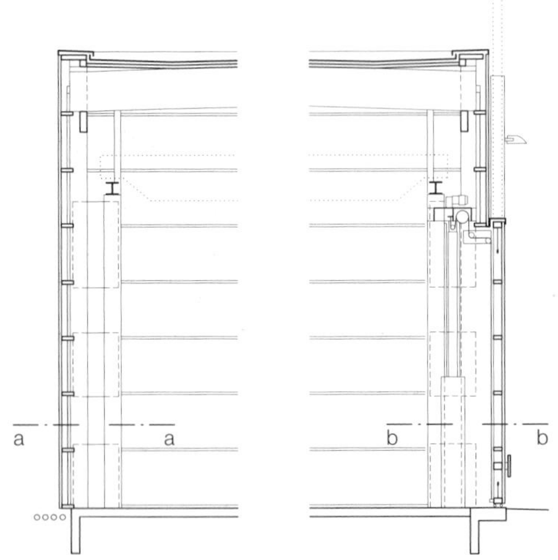

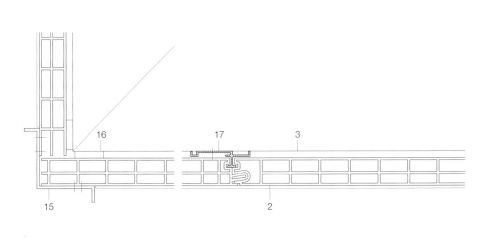

Axonometric view (not to scale)
Section scale 1:200
Detail of corner scale 1:5
Vertical section • Horizontal section
scale 1:50

1 Sheet aluminium capping, bent to suit, 2 mm, semi-machined finish
2 Polycarbonate multiple-web sheets, 40 x 500 mm, U-value = 1.65 W/m²K
3 Glulam rail, spruce, 60 x 280 mm
4 Round steel bar, 12 mm dia.
5 Column, 4 No. 120 x 400 mm glulam sections in pairs joined by 40 mm 3-ply core plywood webs
6 External lighting
7 Steel section, IPE 330
8 Motor for vertical lift door
9 Door driveshaft
10 Door drive gear
11 Counterweight
12 Solid polycarbonate sheet, 8 mm
13 Safety barrier, 27 mm timber boards on galvanised steel frame
14 Glulam facade post, 160 x 400 mm
15 Polycarbonate angle, 80 x 80 mm, riveted to No. 2
16 Spacer piece
17 Aluminium wind suction anchor, semi-machined finish

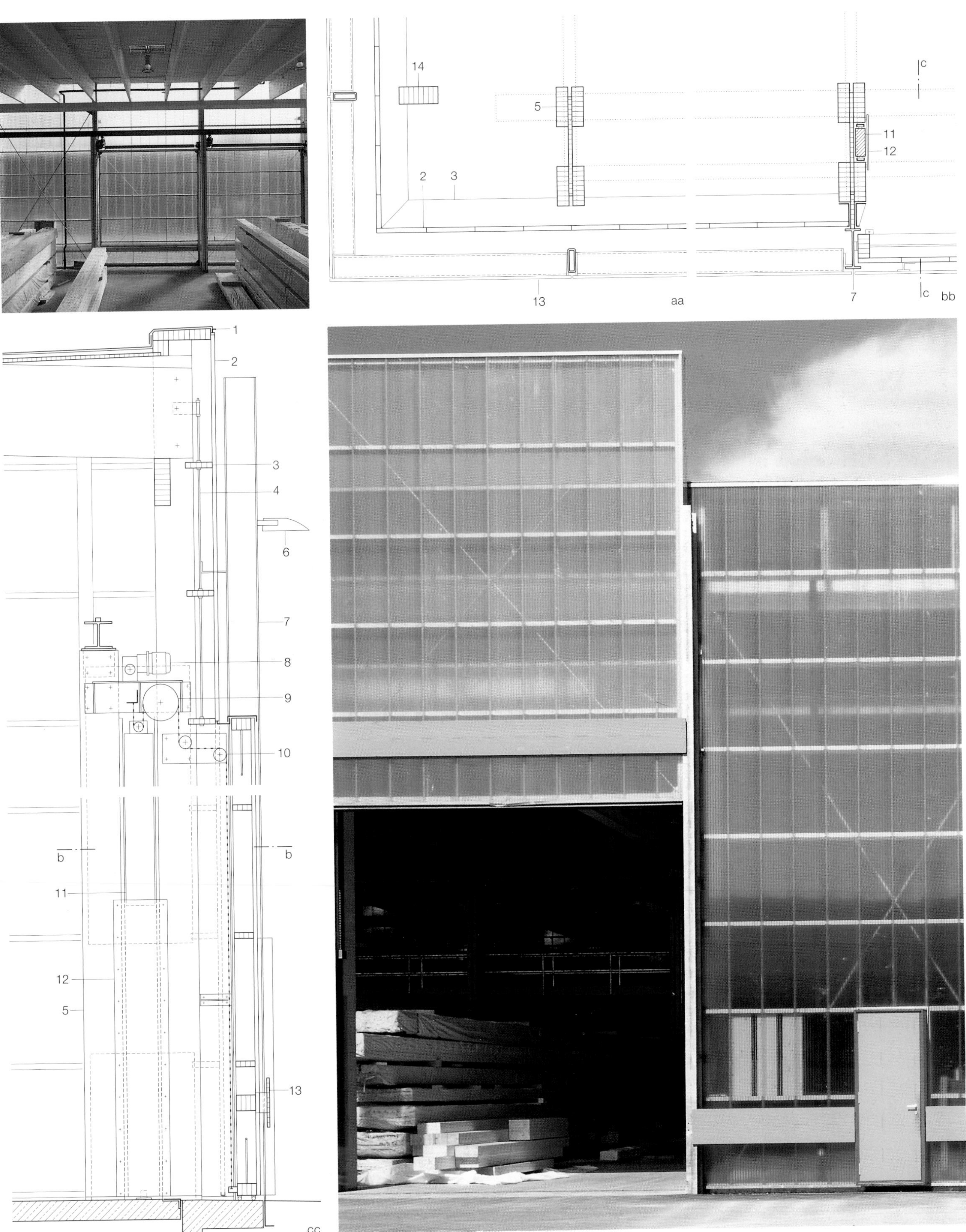
14
5
2
3
11
12
c
13
aa
7
c
bb
1
2
3
4
6
7
8
9
10
b
b
11
12
5
13
cc

Artist's studio

Madrid, E, 2002

Architects:
Abalos & Herreros, Madrid
Facade design:
Jesus Rodriquez

Arquitectura 331, 2003
El Croquis 118, 2003
Detail 12/2002

- Facade, rooflights and external doors made from double layer of translucent polycarbonate corrugated sheet
- Even distribution of daylight is important because building used as artist's studio
- Opaque facade areas made from insulated reinforced concrete walls clad with single layer of corrugated sheets
- Rear of building let into slope

1 Steel section, IPE 80
2 Steel rectangular hollow section, 100 x 50 mm
3 Double layer of polycarbonate corrugated sheets with UV-resistant coating, fixed with screws
4 Facade construction: polycarbonate corrugated sheets with UV-resistant coating, fixed with screws
steel rectangular hollow section, 100 x 50 mm
30 mm thermal insulation, rigid extruded polystyrene foam
waterproofing
250 mm reinforced concrete
46 mm ventilated cavity/ battens
2 No. 13 mm plasterboard
5 Steel square hollow section, 100 x 100 mm
6 Planting to wall: reinforcing mat
rigid foam, 50 mm
polyethylene
filter mat
40 mm thermal insulation, rigid extruded polystyrene foam
1.2 mm waterproofing
250 mm reinforced concrete
46 mm ventilated cavity/ battens
2 No. 13 mm plasterboard
7 Steel rectangular hollow section, 100 x 50 mm
8 Door opening: steel frame/double layer of polycarbonate corrugated sheets with UV-resistant coating

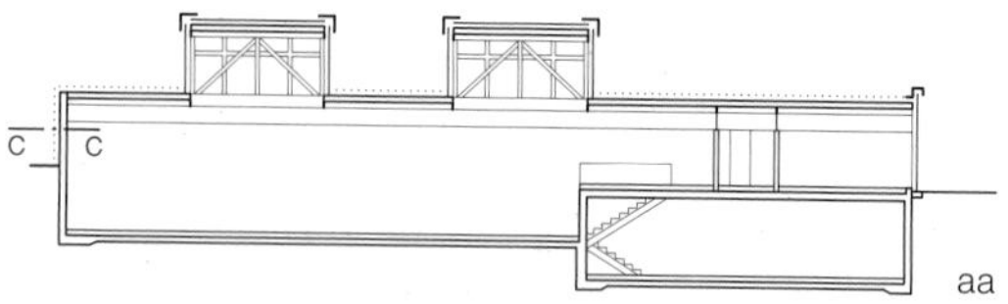

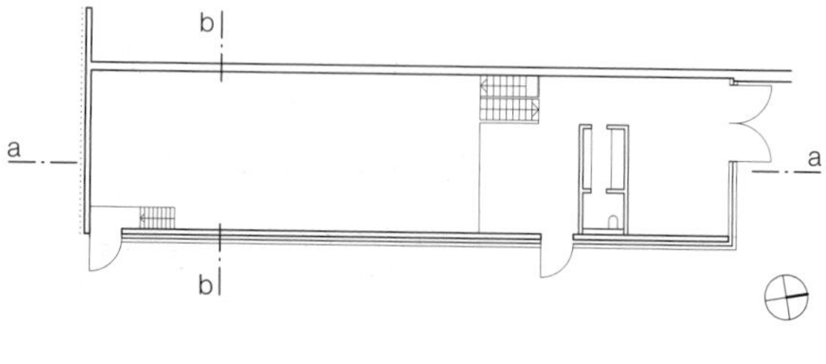

Section • Plan
scale 1:500
Vertical section •
Horizontal section
scale 1:20

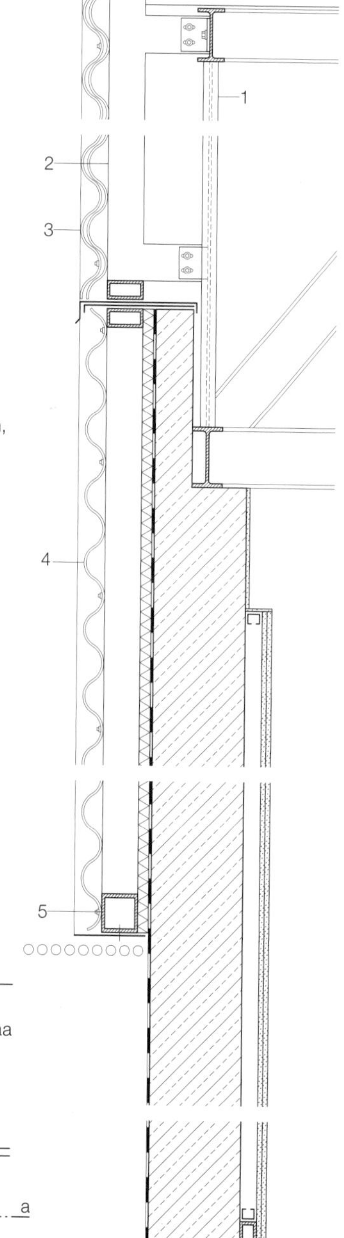

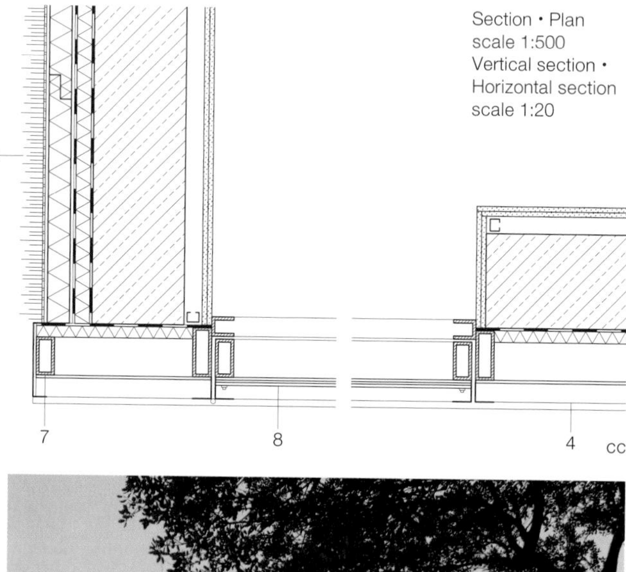

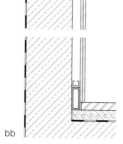

Artist's studio

Ottobrunn, D, 1993

Architect:
Peter Haimerl, Munich
Assistants:
Ralph Feldmeier, Maria Laurent

Blueprint 125, 1996
Newhouse, Victoria: Wege zu einem Neuen Museum. Ostfildern 1998

- 6 m high studio with walls and roof comprising double-layer membrane cushion of PTFE film
- Sunshading provided by aluminium louvre blinds fitted within membrane cushion

Isometric view (not to scale)
Plan • Section scale 1:250
Vertical section • Horizontal section
scale 1:20

1 Steel square hollow section, 90 x 90 x 8 mm, with steel flats welded on, 25 x 10 mm
2 Aluminium clamping rail
3 PTFE film as air cushion, double layer, permanently inflated
4 Aluminium louvre blind, natural colour, anodised, fitted between membranes
5 Air supply hose, flexible PVC
6 Aluminium angle, 150 x 50 x 8 mm
7 Glass louvre vent
8 Fluorescent tube lighting unit

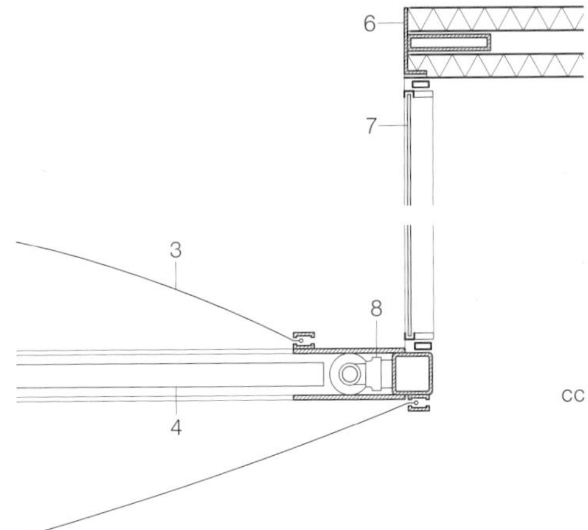

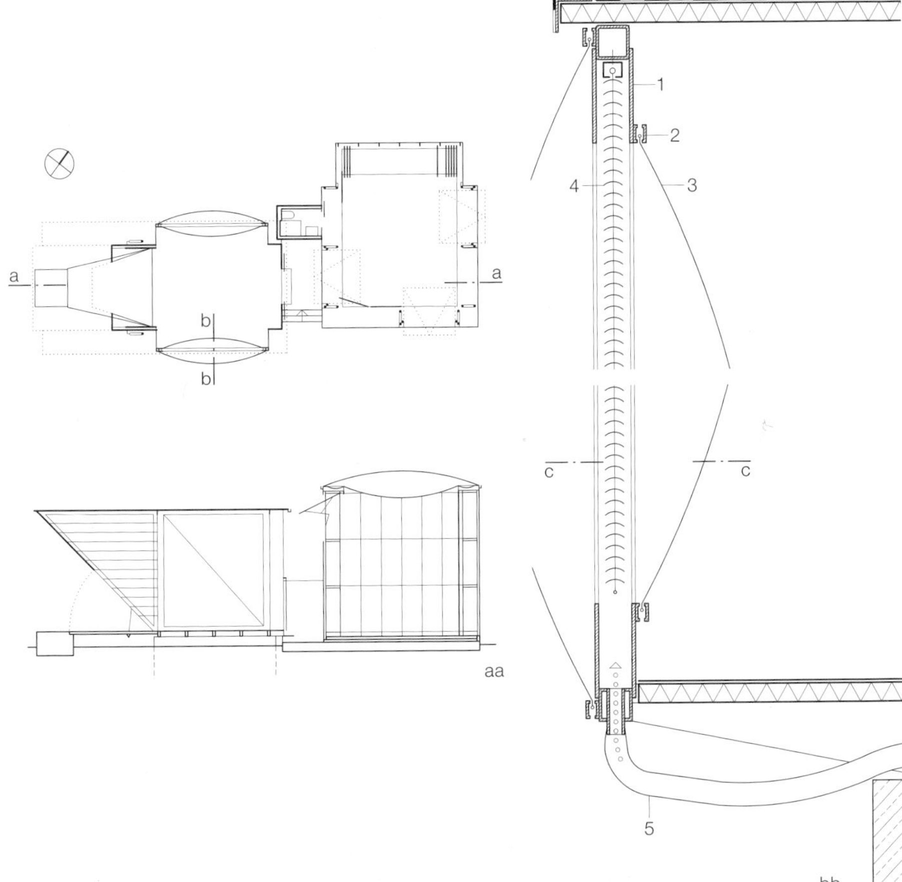

Gerontology technology centre (GTZ)

Bad Tölz, D, 2004

Architect:
Diethard Johannes Siegert, Bad Tölz

- "Climate-control envelope" comprising a single layer of ETFE film to protect the access area, facing the inner courtyard of the building, against the weather and to create a zone with an intermediate temperature
- Pattern of silver-colour dots printed on film to serve as a sunshade and to distribute the incoming daylight
- ETFE film fixed with the aid of special aluminium sections
- Access area ventilated naturally by means of controllable, glazed flaps at the top and bottom of the facade
- Main building consists of reinforced concrete frame plus prefabricated facade

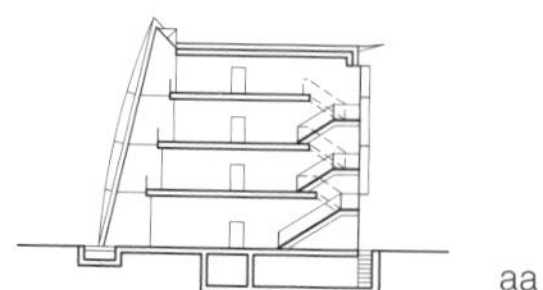

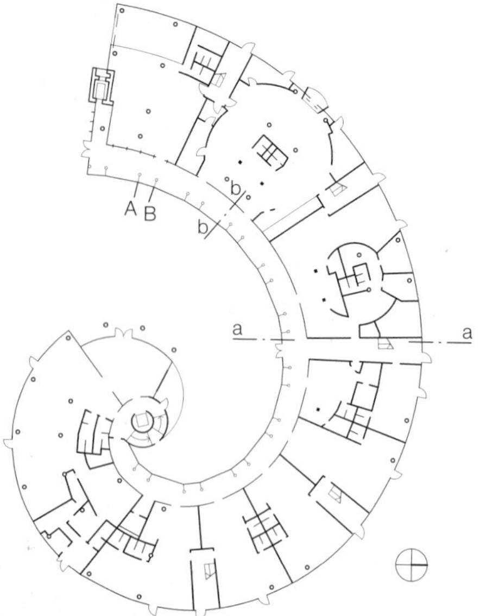

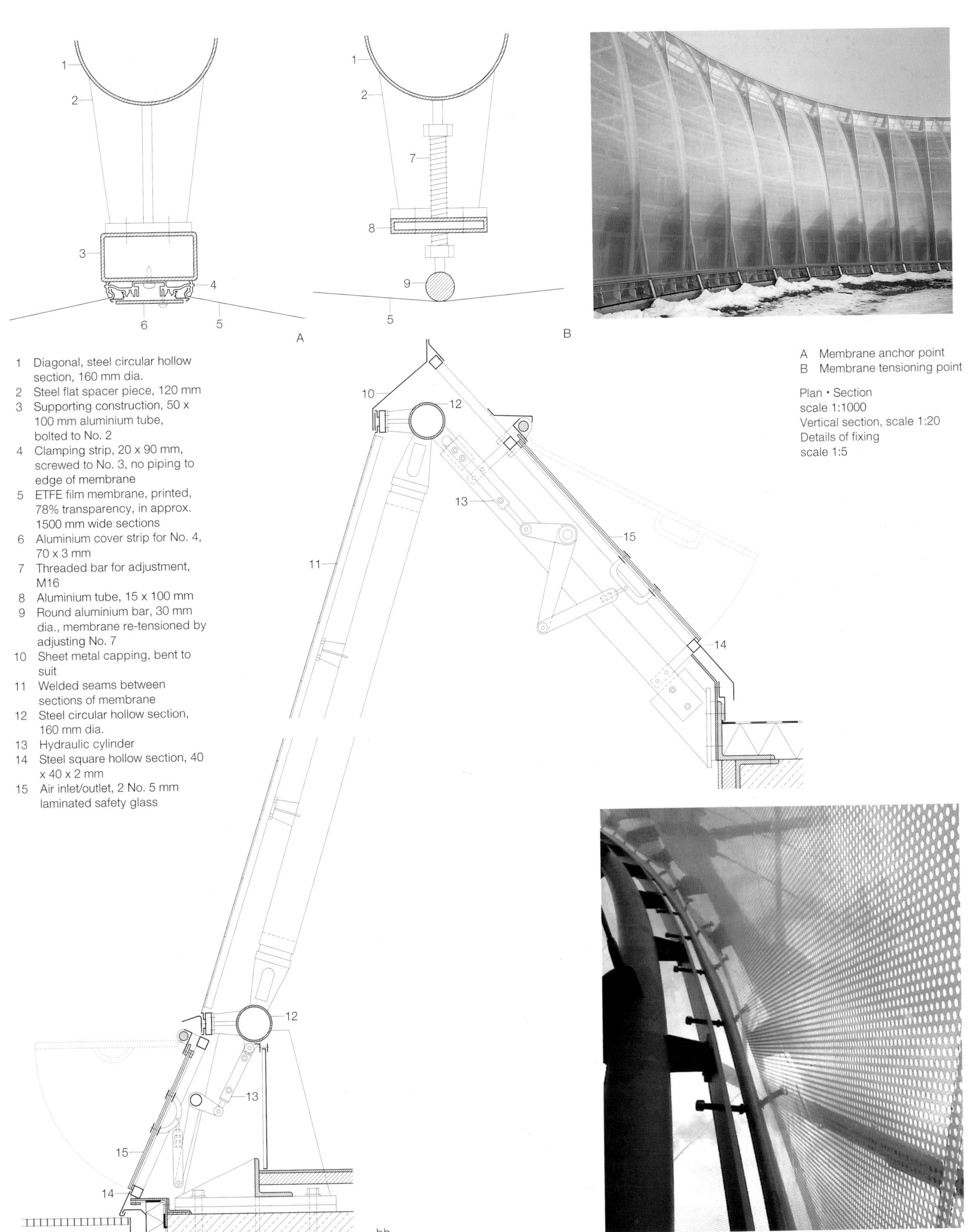

1 Diagonal, steel circular hollow section, 160 mm dia.
2 Steel flat spacer piece, 120 mm
3 Supporting construction, 50 x 100 mm aluminium tube, bolted to No. 2
4 Clamping strip, 20 x 90 mm, screwed to No. 3, no piping to edge of membrane
5 ETFE film membrane, printed, 78% transparency, in approx. 1500 mm wide sections
6 Aluminium cover strip for No. 4, 70 x 3 mm
7 Threaded bar for adjustment, M16
8 Aluminium tube, 15 x 100 mm
9 Round aluminium bar, 30 mm dia., membrane re-tensioned by adjusting No. 7
10 Sheet metal capping, bent to suit
11 Welded seams between sections of membrane
12 Steel circular hollow section, 160 mm dia.
13 Hydraulic cylinder
14 Steel square hollow section, 40 x 40 x 2 mm
15 Air inlet/outlet, 2 No. 5 mm laminated safety glass

A Membrane anchor point
B Membrane tensioning point

Plan • Section
scale 1:1000
Vertical section, scale 1:20
Details of fixing
scale 1:5

National Space Centre Museum

Leicester, GB, 2001

Architects:
Grimshaw, London
Facade consultants:
Montresor Partnership, Chippenham

Architectural Review 04/2000
Detail 12/2002

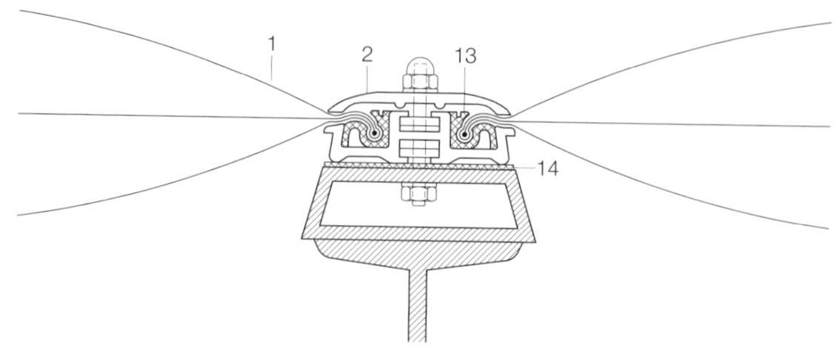

- 42 m high tower made from partially printed, partially transparent three-layer ETFE cushion
- Film cushion made from ethylene-tetrafluoroethylene (ETFE), self-cleaning, 60-year life expectancy
- Primary and secondary loadbearing construction comprises steel circular hollow sections, spacing = 3000 mm, curving in three dimensions
- Reinforced concrete tower housing stairs and lifts

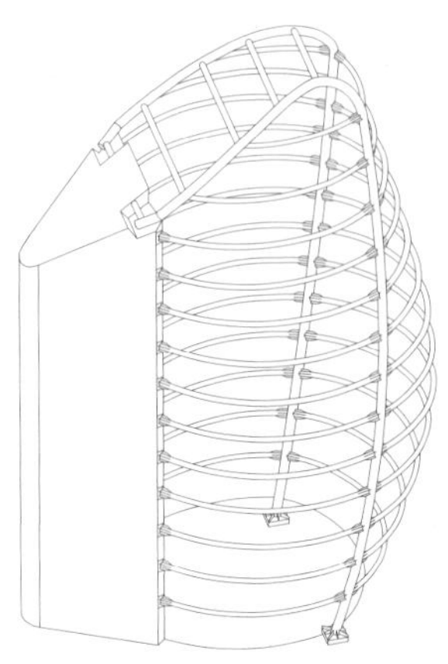

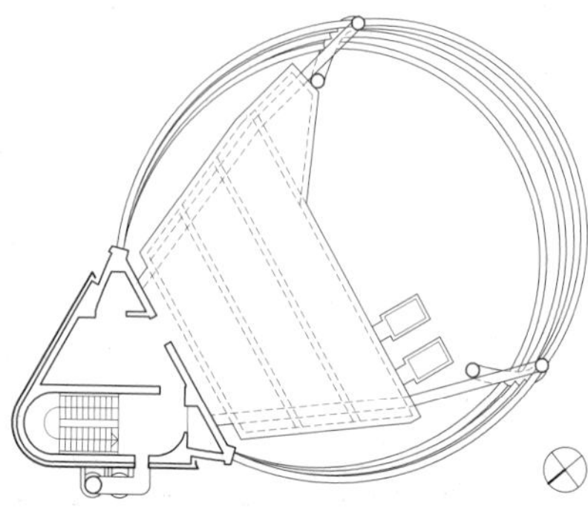

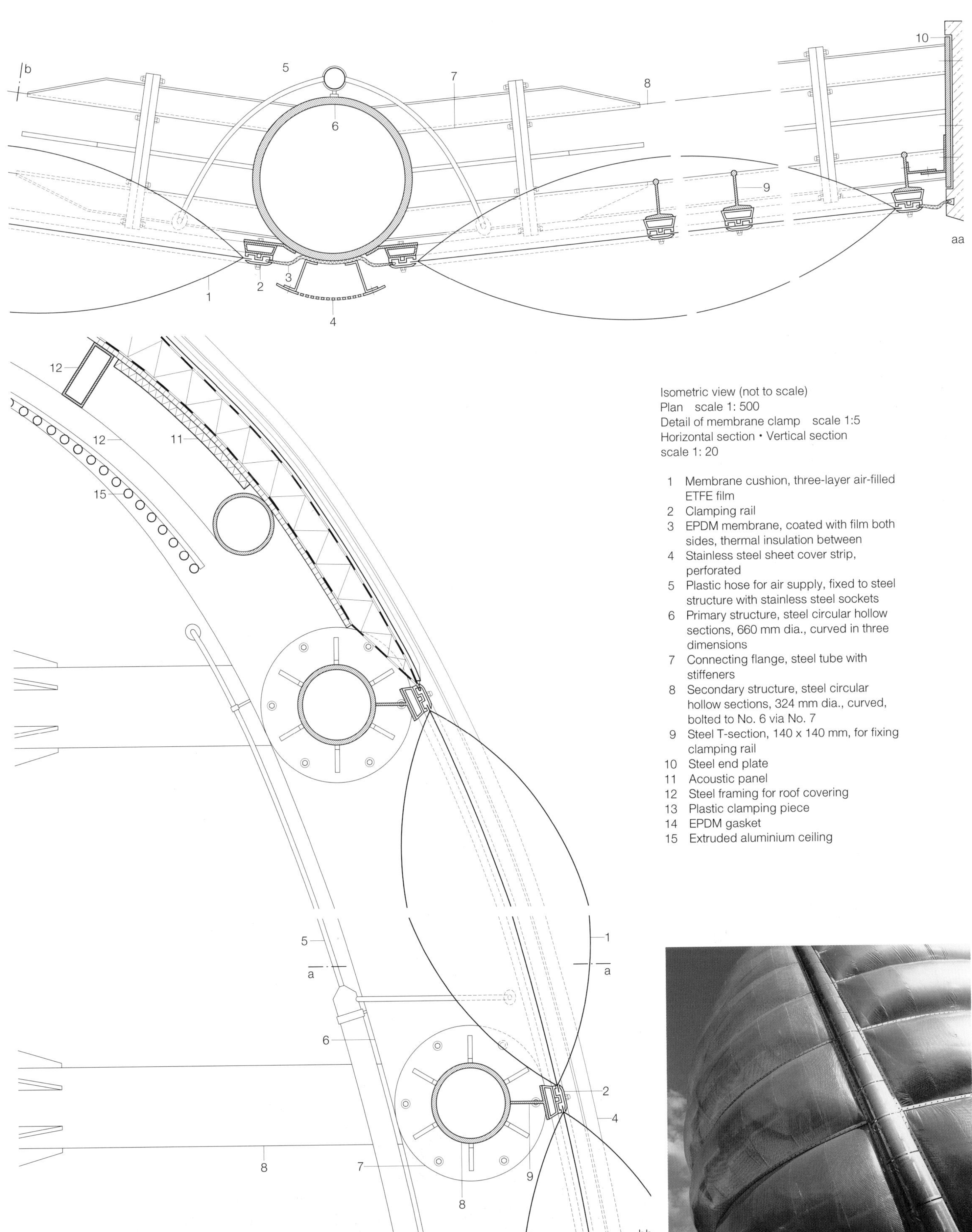

Isometric view (not to scale)
Plan scale 1: 500
Detail of membrane clamp scale 1:5
Horizontal section • Vertical section
scale 1: 20

1. Membrane cushion, three-layer air-filled ETFE film
2. Clamping rail
3. EPDM membrane, coated with film both sides, thermal insulation between
4. Stainless steel sheet cover strip, perforated
5. Plastic hose for air supply, fixed to steel structure with stainless steel sockets
6. Primary structure, steel circular hollow sections, 660 mm dia., curved in three dimensions
7. Connecting flange, steel tube with stiffeners
8. Secondary structure, steel circular hollow sections, 324 mm dia., curved, bolted to No. 6 via No. 7
9. Steel T-section, 140 x 140 mm, for fixing clamping rail
10. Steel end plate
11. Acoustic panel
12. Steel framing for roof covering
13. Plastic clamping piece
14. EPDM gasket
15. Extruded aluminium ceiling

Hotel Burj al Arab

Dubai, VAE, 1999

Architects:
Atkins, Epsom

Bauwelt 44/1998
DBZ 07/2000
Fassade/Façade 01/2001

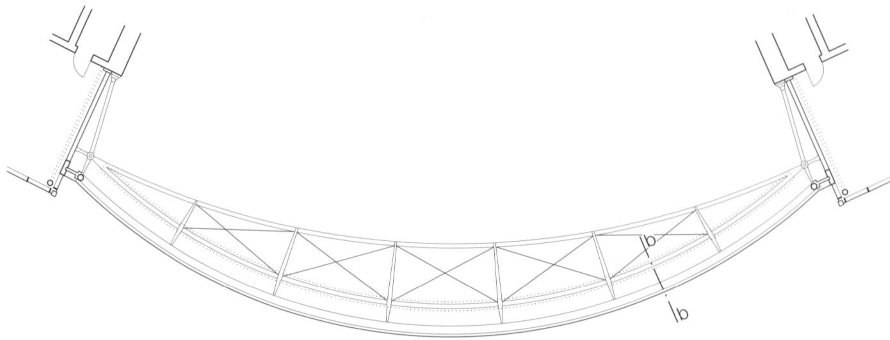

- PTFE-coated glass fibre fabric double-layer membrane construction, height = 200 m, area = 7500 m²
- Facade divided into 12 individual bays supported by structural steelwork
- 50% reduction in self-weight compared to a glass facade
- Translucent membrane facade with 10% light permeability
- Membrane surface used as projection screen at night

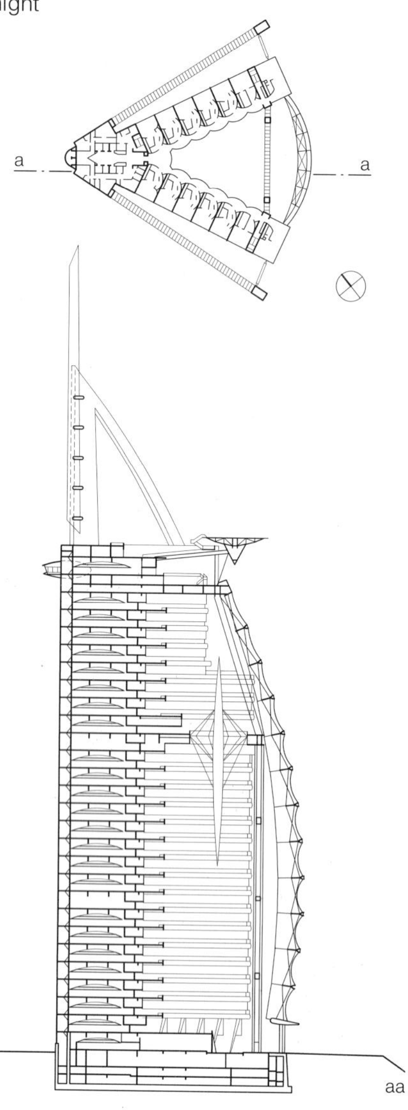

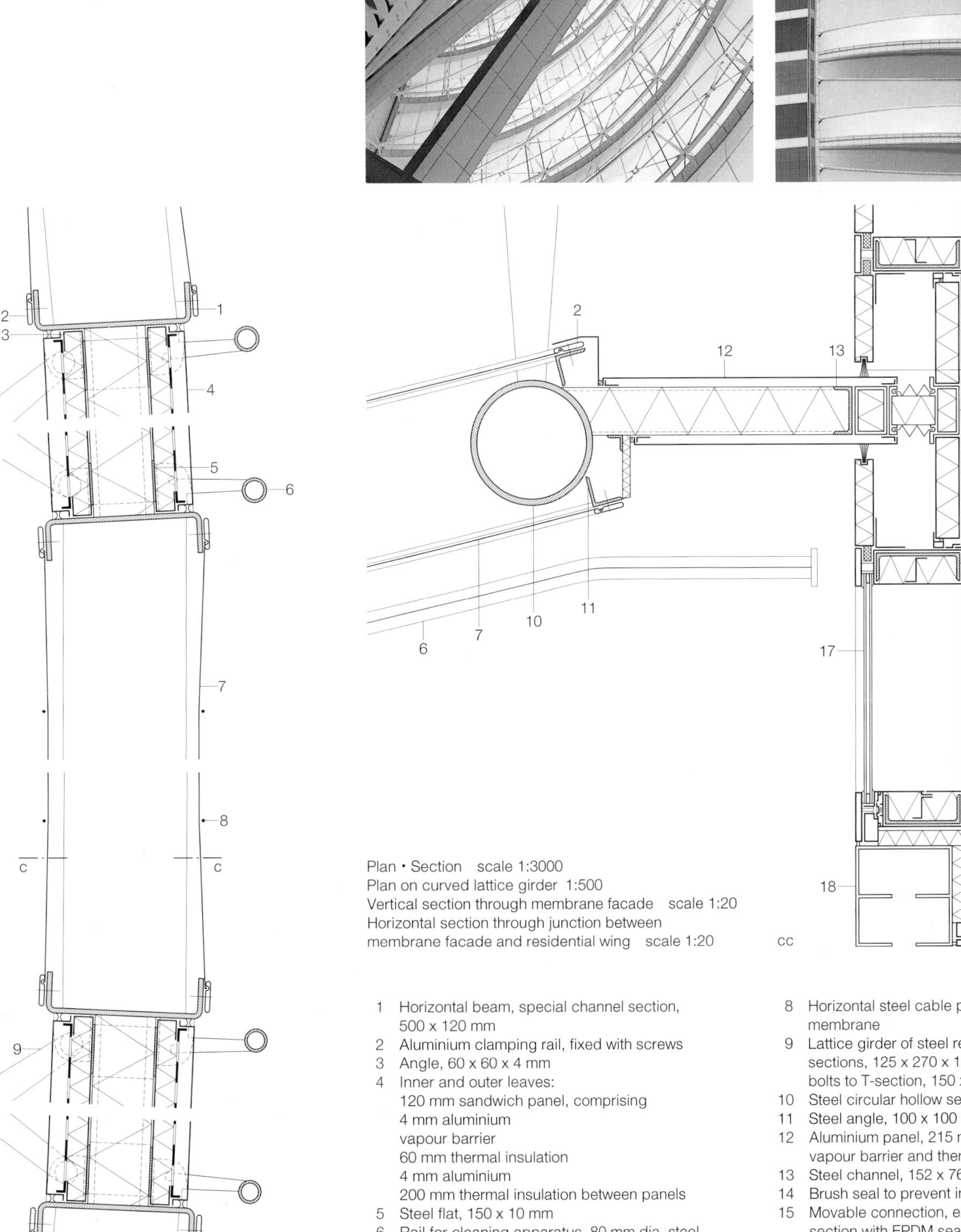

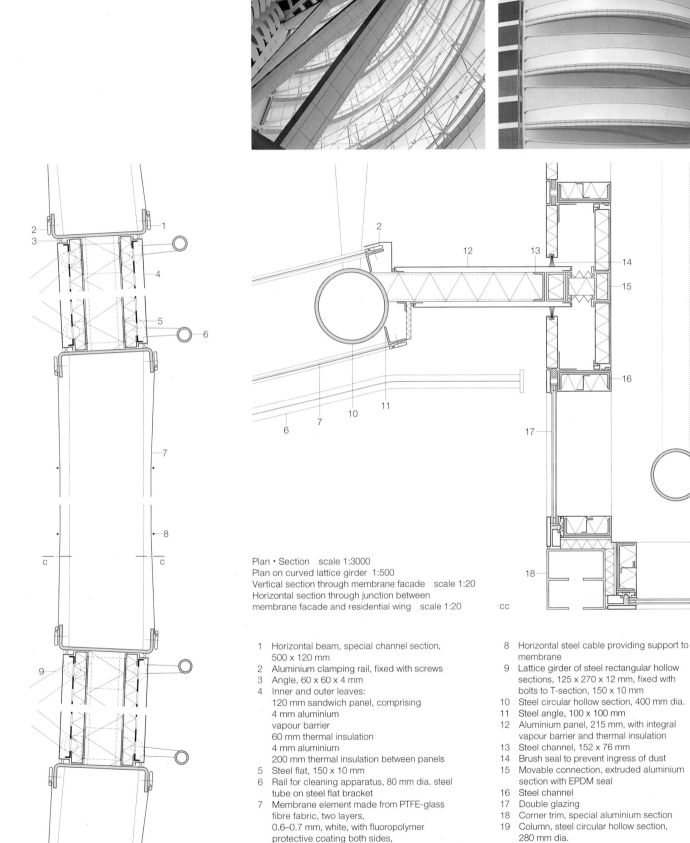

Plan • Section scale 1:3000
Plan on curved lattice girder 1:500
Vertical section through membrane facade scale 1:20
Horizontal section through junction between membrane facade and residential wing scale 1:20

1 Horizontal beam, special channel section, 500 x 120 mm
2 Aluminium clamping rail, fixed with screws
3 Angle, 60 x 60 x 4 mm
4 Inner and outer leaves:
120 mm sandwich panel, comprising
4 mm aluminium
vapour barrier
60 mm thermal insulation
4 mm aluminium
200 mm thermal insulation between panels
5 Steel flat, 150 x 10 mm
6 Rail for cleaning apparatus, 80 mm dia. steel tube on steel flat bracket
7 Membrane element made from PTFE-glass fibre fabric, two layers,
0.6–0.7 mm, white, with fluoropolymer protective coating both sides,
UV-resistant, 500 mm air cavity in between
8 Horizontal steel cable providing support to membrane
9 Lattice girder of steel rectangular hollow sections, 125 x 270 x 12 mm, fixed with bolts to T-section, 150 x 10 mm
10 Steel circular hollow section, 400 mm dia.
11 Steel angle, 100 x 100 mm
12 Aluminium panel, 215 mm, with integral vapour barrier and thermal insulation
13 Steel channel, 152 x 76 mm
14 Brush seal to prevent ingress of dust
15 Movable connection, extruded aluminium section with EPDM seal
16 Steel channel
17 Double glazing
18 Corner trim, special aluminium section
19 Column, steel circular hollow section, 280 mm dia.

B 2.1 The glass double facade

B 2.1.1 Posttower, Bonn (D) 2003, Murphy/Jahn

From storm window to glass double facade

The development of multi-leaf, transparent building envelopes is closely linked with the increasing use of glass in buildings during the seventeenth and eighteenth centuries. This was made possible by progress in the manufacture of glass, such as the invention of the cast glass process in 1687, which allowed flat panes of glass to be produced with a better quality but with reduced effort [1].

The first references to the use of multi-leaf window systems are found in a publication dating from the late seventeenth century [2]. Talking about improving the thermal performance of windows, the use of a "double window" is recommended; during the cold months this is closed in front of the actual window. The use of such "storm windows" has been common in Central Europe since the beginning of the eighteenth century and can still be found today in many rural buildings. During the eighteenth and nineteenth centuries, variations on this theme appeared – multi-leaf window constructions such as coupled windows and glazed loggias, some of which still remain in buildings dating from this period.

The possibility of improving the functional properties of single glazing by placing an additional glass skin in front was also used when glazing large facade openings. For example, in the "Iron House" in the Nymphenburg Palace Park in Munich, when it was rebuilt in 1867 after a fire, or in the "Galerias Galegas" in La Coruña on the north-west coast of Spain (mid-nineteenth century), where the entire external wall is clad with a second skin of glass. At roughly the same time, other variations appeared, such as oriels, glazed loggias, atria and lean-to conservatories – all with the same aim of improving the thermal insulation properties and also using the greenhouse effect to heat up the interior air. One of the best-known historic examples is the buffer facade to the Steiff soft toys factory in Giengen an der Brenz. This was enclosed in a second skin of glass in 1903 to improve the thermal insulation properties of the building envelope while still ensuring maximum use of daylight (see p. 241). The aforementioned window and facade systems remained the most common solutions up until the 1950s, when the development of secondary glazing and, a little later, insulating glazing, enabled the thermal performance of windows to be improved more conveniently.

In the mid-1970s dwindling resources and the ensuing discussion concerning the finiteness of the Earth's resources led to awareness of the chance to improve the thermal performance and utilise solar energy through the use of external, additional glass leaves. The ongoing development of existing systems such as lean-to conservatories and atria was thus spurred on, but this time with insulating glazing.

B 2.1.2

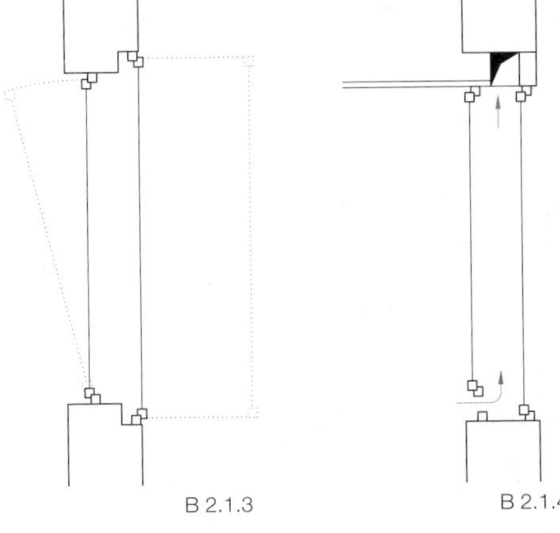

B 2.1.3 B 2.1.4

B 2.1.2 Storm window, Farmhouse, Flims (CH)
B 2.1.3 Storm window
B 2.1.4 Exhaust-air window
B 2.1.5 Double window, Krems (A)

B 2.1.5

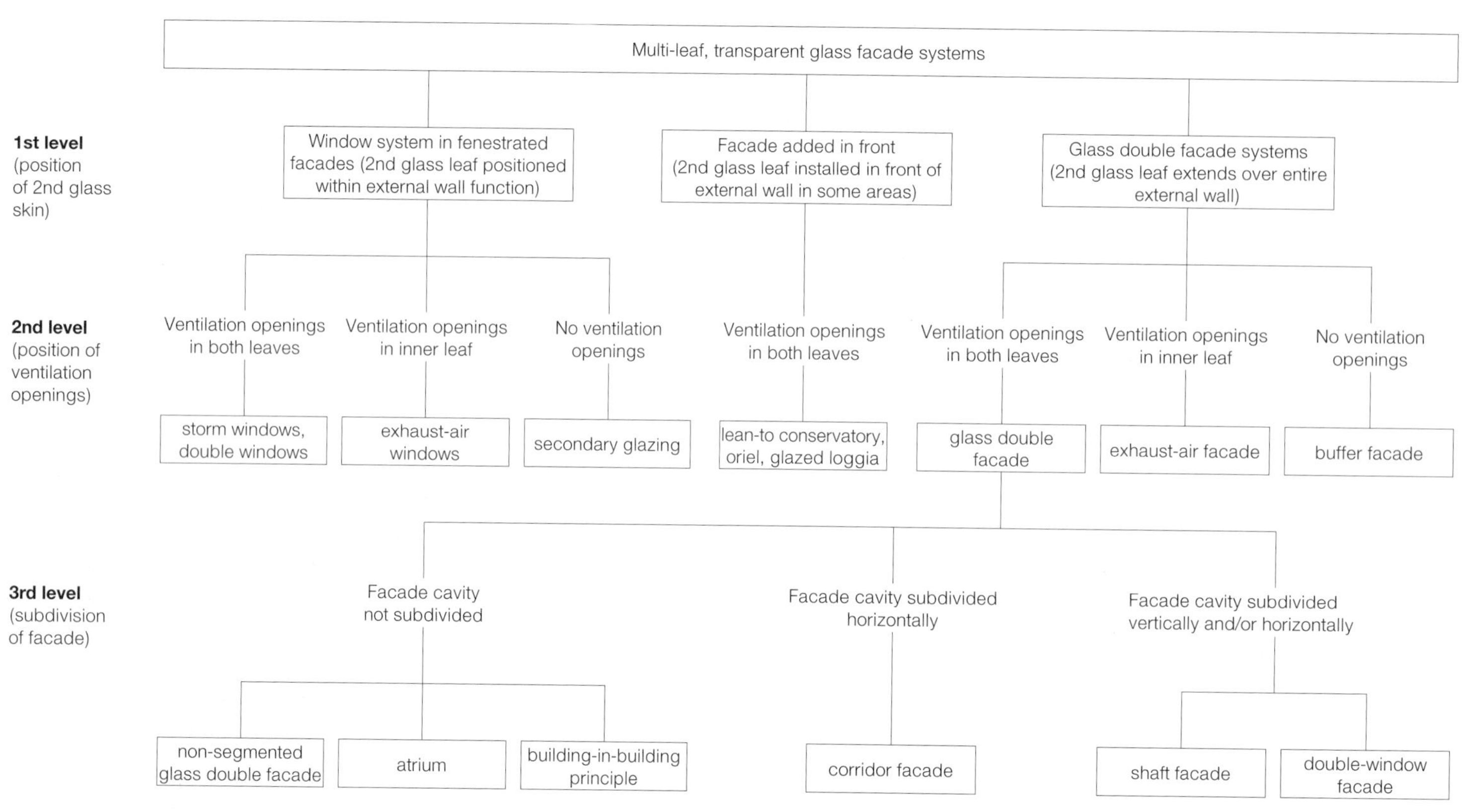

B 2.1.6

B 2.1.7

B 2.1.6 Design variations for multi-leaf, transparent glass facade systems
B 2.1.7 Exhaust-air window, Museum of Applied Arts, Frankfurt (D), 1984, Richard Meier
B 2.1.8 Atrium, private house, Regensburg (D), 1979, Thomas Herzog
B 2.1.9 Lean-to conservatory
B 2.1.10 Glazed loggia
B 2.1.11 Glazed loggias to apartment blocks, La Coruña (E), 2nd half of 19th century

B 2.1.8

New systems, such as exhaust-air windows and facades, in which the window or facade cavities are combined with the building's ventilation system to help control the interior climate, started to appear in the 1980s.

Advances in multi-leaf glass facade systems continued into the 1990s as the increasing demands of users and more stringent energy saving regulations favoured the development of "glass double facades". The potential for reducing energy consumption and optimising interior conditions makes the glass double facade one of the most interesting new developments in the facade sector. The feature common to all these systems is the ventilated glass leaf in front of the actual glazed facade, which enables a natural exchange of air even in heavily polluted, noisy or extremely windy locations. Important variations on this type of facade are the glass double facades with an uninterrupted or interrupted facade cavity, whose design and functional aspects will be explained later. Besides the increased energy and comfort awareness of the user, one primary reason for the rapid spread of such glass double facades may be the discussion surrounding the "sick building syndrome" [3] associated with fully air-conditioned buildings; this has helped to highlight the advantages of naturally ventilated, individually controllable multi-leaf glass facades.

Design features of multi-leaf glass facades and their importance for the function of the facade

The functional and architectural aspects of multi-leaf glass facades are primarily governed by a number of design features.

Starting with the breakdown illustrated in fig. B 2.1.6, the design features and their importance for the functional properties of various glass double facade systems [4] are outlined below:

Position of second glass skin in relation to the external wall

The positions of the two planes of glazing in relation to the external wall has a substantial effect on the function and, above all, the appearance of the facade.

Second glass leaf positioned within the external wall construction

This approach includes window systems in fenestrate facades with two panes of glass, one behind the other (double windows), in the openings. The loads are usually carried by a solid external wall; the layout and size of openings in the wall therefore depend on the structural requirements. Owing to the relatively small proportion of glazing, the design of fenestrate facades is very different from that of outer leaves or curtain walls made of glass or other materials. Double windows include storm windows, coupled windows, exhaust-air windows

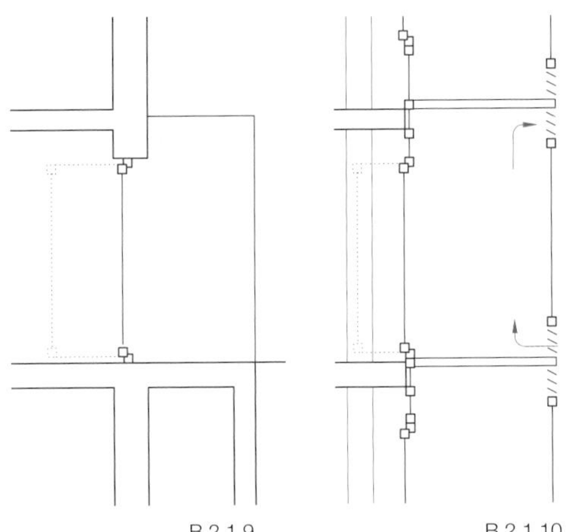

B 2.1.9 B 2.1.10

B 2.1.11

(figs. B 2.1.2 to 2.1.5) and secondary glazing. However, such window systems have played only a minor role in the modern building industry since the introduction of insulating glazing and the associated improvement in thermal insulation properties. Only a few buildings with air conditioning, e.g., museums, still make use of exhaust-air windows.

Second glass leaf positioned in front of parts of the external wall

One intrinsic feature of outer leaves is the positioning of a second glass envelope at a considerable distance from (in front of) the actual external wall. This creates a zone with an intermediate temperature suitable for temporary uses [5]. A glass envelope in front of just parts of the external wall has a considerable effect on the appearance of the facade, although the actual external wall is generally still recognisable as the true building envelope. Outer leaves of this kind include lean-to conservatories, oriels and glazed loggias. The opportunities mentioned above mean that these facade systems are still met with frequently today.

Second glass leaf extends over the entire external wall

In this facade variation there is a second plane of glazing either in front of or behind the actual glass facade. This multi-leaf glass facade construction is generally called a "glass double facade", but this term is not necessarily linked with any particular functional properties. These glass double facades include exhaust-air facades and buffer facades. Although the design of these different types of facade can be very similar, they differ quite distinctly with regard to the natural ventilation options.

Arrangement of ventilation openings in inner and outer leaves

The arrangement of the ventilation openings governs the type of system, as described below:

Window systems in fenestrate facades – ventilation openings in both leaves:

Storm windows
In this type of window a second single-glazed window is placed in front of an existing single-glazed window during the cold months in order to improve the thermal performance. This second window is usually fitted flush with the outer face of the external wall (fig. B 2.1.3).

Characteristic of these windows is the way the thermal insulation properties of the window can be adapted to suit the season. Daily temperature fluctuations are, however, not taken into account [6].

Double windows
These consist of two, usually single-glazed, opening lights, fitted one behind the other. The gap between the two opening lights is approx. 100–200 mm. In many instances separate opening lights within the frame or integral bottom-hung (hopper) lights at the top of the window provide night-time ventilation. Such an arrangement ensures that protection against rain or intruders is still maintained. The thermal and acoustic properties, as well as the air change rate, are easily adapted to changing external conditions over the course of the day.

Window systems in fenestrate facades – ventilation openings in the inner leaf:

Exhaust-air windows
In the exhaust-air window the conventional functions of the window are combined with the tasks of interior climate control by allowing the exhaust air from the interior to escape via the cavity of a double window arrangement. The outer glazing is generally made of insulating glass in order to avoid transmission heat losses. As it flows through the cavity, the warmer interior air causes the surface temperature of the inner pane to rise. This results in a distinct improvement in the level of comfort near the windows, especially during the colder months. At the same time, the cavity can be used to accommodate sunshading devices, which can even be lowered during strong winds because they are completely protected from the weather. Another major advantage of the exhaust-air window is the opportunity to minimise the radiation-related cooling loads in the interior because radiation heat gains are dissipated directly by the flow of exhaust air through the cavity. However, these two advantages can only be accomplished by using mechanical energy to force the exhaust air through the cavity, and this aspect usually leads to an increased capital outlay and higher operating costs than in buildings without a mechanical ventilation system (fig. B 2.1.7).

Window systems in fenestrate facades – no ventilation openings in the leaves:

Secondary glazing
This consists of two windows, one fitted directly behind the other, without any true cavity. As the two lights cannot be separated, except for cleaning purposes, in contrast to the double window there is no chance of adapting the window to changing external conditions over the course of a day or a year.

Facade with outer leaf – ventilation openings in both leaves:

Generally, facades with outer leaves have ventilation openings in both the inner leaf and the outer leaf, thereby guaranteeing a natural exchange of air. The following types of facade belong to this category:

Lean-to conservatory
A single-glazed glass leaf is built at a large distance from the actual external wall. This creates a zone with an intermediate temperature which can be used for a whole range of uses (fig. B 2.1.9). This zone can act as a thermal buffer, as an "airlock" for preheating the incoming air, but can also make use of the solar energy directly [7].

Oriel, bow or bay window, glazed loggia
Like the lean-to conservatory, these features create zones with an intermediate temperature. The window openings enable the external envelope to be harmonised with the facade

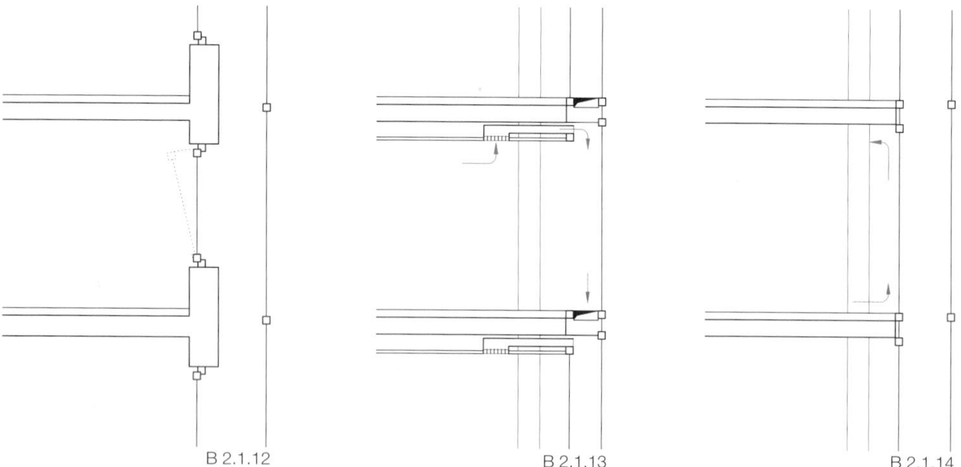

B 2.1.12 Glass double facade
B 2.1.13 Exhaust-air facade (air flows from top to bottom)
B 2.1.14 Buffer facade
B 2.1.15 Exhaust-air facade, Lloyds headquarters, London (GB), 1986, Richard Rogers Partnership
B 2.1.16 Atrium
B 2.1.17 Non-segmented glass double facade
B 2.1.18 Glass double facade, office building, Munich (D), 1997, Steidle + Partner
B 2.1.19 Glazed atrium, Museum of Hamburg History, Hamburg (D), 1989, von Gerkan Marg and Partner

functions, such as exchange of air, summer and winter thermal performance and sound insulation.

One important difference from the lean-to conservatory is, however, is the positioning with respect to the actual facade. As a rule, oriels, bow/bay windows and glazed loggias are positioned in a direct functional relationship with the adjoining room, forming a unit in terms of heat, ventilation and sound requirements.

The zone of intermediate temperature is therefore directly linked to the warmer interior air, which means that – in contrast to the lean-to conservatory – permanent (year-round) utilisation is possible (fig. B 2.1.11).

Curtain wall facade systems – ventilation openings in both leaves:

Glass double facade
The main feature of the glass double facade is the arrangement of the second leaf in front of the actual external wall in such a way that natural ventilation is still possible. Normally, the outer leaf is suspended in front of the wall as a non-loadbearing element (fig. B 2.1.12). Compared with single-leaf facades, this arrangement usually improves the sound and thermal insulation properties while also allowing natural ventilation even in locations with high wind speeds [8]. The glass double facade includes shaft, corridor and double-window facades, likewise facades in which the outer leaf is positioned at some considerable distance from the external wall (atrium, building-in-building principle).

B 2.1.15

Curtain wall facade systems – ventilation openings in the inner leaf:

Exhaust-air facade
Like the exhaust-air window, in the exhaust-air facade the warm interior air flows through the facade cavity and back to a central air-conditioning plant [9]. Here, the outer leaf ensures not only weather protection, thermal insulation and adequate daylight, but in its function as an air duct should also be considered a part of the air-conditioning system (fig. B 2.1.13).

To reduce transmission heat losses, the outer leaf usually makes use of insulating glazing. The advantages of this type of facade are very similar to the aforementioned benefits of the exhaust-air window.

Curtain wall facade systems – no ventilation openings in either leaf:

Buffer facade
A buffer facade in the form of a second glass skin provides additional screening of the interior against certain external conditions without impairing the daylight utilisation (fig. B 2.1.14).

Enhanced thermal performance is very often the primary reason for choosing such a facade, but other ambient conditions, such as traffic noise, also determine the actual application.

From the ventilation viewpoint, the facade cavity represents a discrete unit in which an exchange of air is not possible. The exchange of air takes place via separate, box-like window elements incorporated in the buffer facade.

Subdividing the facade cavity

In this respect it is the glass double facades that are addressed because only these exhibit a wide range of practical variations with very different functional properties. We distinguish between two main groups:

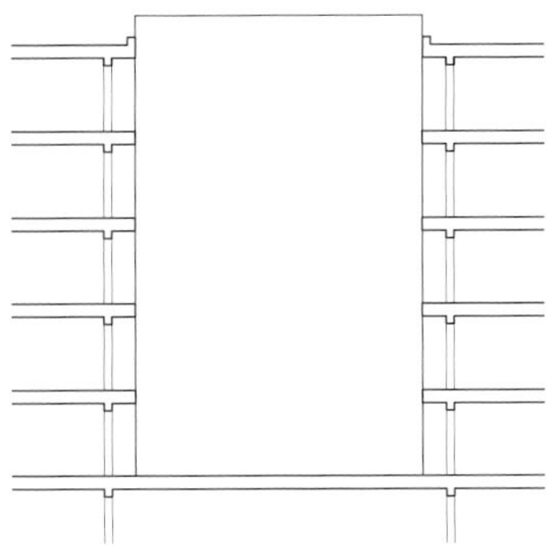

B 2.1.16

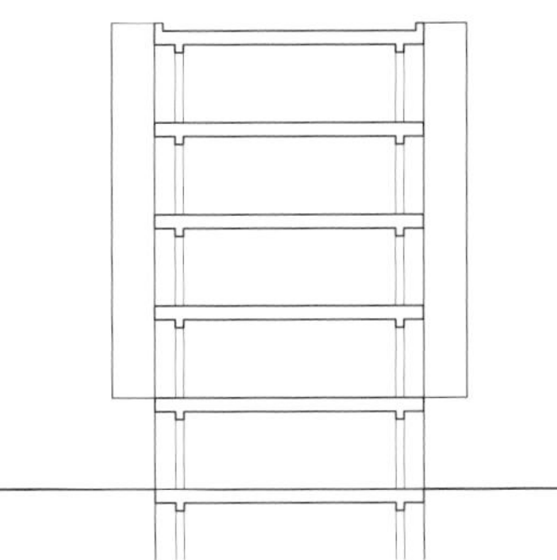

B 2.1.17

B 2.1.18

- uninterrupted glass double facades
- interrupted glass double facades

Glass double facades with an uninterrupted cavity:

The following types of facade belong to this group:

- non-segmented glass double facade
- atrium
- building-in-building principle

An uninterrupted cavity results in the following performance characteristics:

- Acoustic performance:
  Sound can propagate unhindered throughout the facade cavity, which can lead to problems in the adjoining rooms.
- Behaviour in fire:
  Flames and smoke can spread throughout the facade cavity unhindered.
- Thermal performance:
  A noticeable temperature gradient from the highest to the lowest point can ensue within the facade cavity. The magnitude of this gradient depends on the vertical dimension of the cavity, and the stack effect (thermal buoyancy) can be exploited in summer to improve the flow of air through the cavity. However, an inadequate flow of air very quickly leads to high air temperatures at the top of the cavity, which affects the adjoining rooms. The simple controllability of the air change rate and the buffer effect of the facade are, however, beneficial effects. The inclusion of ventilation flaps at the base and at the eaves is an adequate way of varying the cross-sectional area of the ventilation inlets and outlets.

B 2.1.19

In functional terms, these facade variations exhibit similar properties. However, the position of the curtain wall in relation to the actual fabric of the building leads to differences in terms of their construction, architecture and utilisation.

Non-segmented glass double facade
In this variation, a second glass skin is built in front of the actual external wall and the cavity is not subdivided for ventilation purposes. From the design point of view, this is one of the simplest forms. However, the uninterrupted facade cavity calls for more stringent requirements to prevent the spread of odours, sound, flames and smoke internally, especially in multi-storey facades.

Atrium
Frequently referred to as a climate buffer, this type of glass skin exhibits properties similar to those of the lean-to conservatory, but normally adjoins the building on at least two sides, and can even be surrounded by the building completely. This results in an intensive interaction between the glass atrium and the surrounding wall surfaces (figs B 2.1.16 and 2.1.19).

In terms of the interior climate of the building, the internal air temperature remains higher than that of the exterior air throughout the year due to the solar gains and transmission heat losses of the adjoining building facades. This broadens the utilisation options of the space and also significantly reduces the constructional requirements of the adjoining facade components.

Building-in-building principle
In this type of multi-leaf transparent building envelope, the glass skin surrounds the building completely (fig. B 2.1.20). Solar energy can be used directly to raise the ambient temperature of parts of the building within the envelope, whereas transmission and ventilation heat losses are reduced. Just like the lean-to conservatory and the atrium, the biggest advantage is the creation of a multi-purpose zone with an intermediate temperature. Air can circulate unhindered within the envelope, and this internal air circulation raises the

B 2.1.20

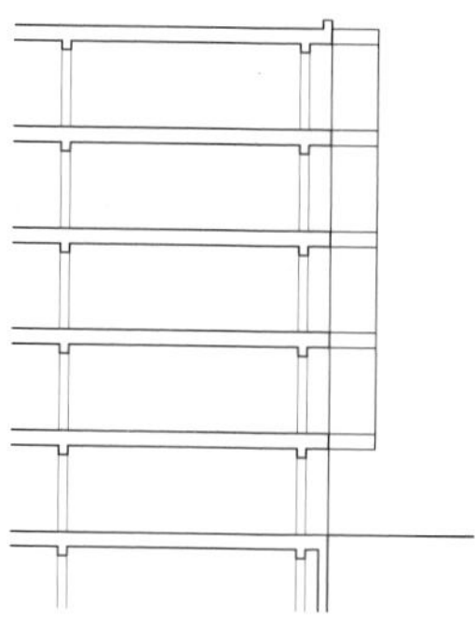

B 2.1.21

B 2.1.22

temperature of less sunny areas quite considerably compared to the external air [10].

Glass double facades with an interrupted cavity:

We distinguish between the following variations:

- corridor facade
- shaft facade
- double-window facade

The common feature of these facade systems is the way they influence the functional characteristics of the facade, although this depends on whether the facade cavity is subdivided horizontally, vertically or in both directions. The effects are most noticeable in terms of the behaviour in fire and the acoustic performance, but the nature and controllability of the flow of air through the cavity is also affected.

Corridor facade
Here, the facade cavity is subdivided by horizontal walkways alongside the floors of the adjoining rooms. These form accessible corridors within the facade in every storey (fig. B 2.1.21). The supply air generally flows along the bottom part of each storey, while the exhaust air leaves the facade corridor at the top. To minimise the risk of the outgoing exhaust air and the incoming supply air becoming mixed, the ventilation openings must be offset laterally or an adequate vertical spacing guaranteed (fig. B 2.1.24). The extra construction necessitated by this type of facade is minimal compared to the non-segmented glass double facade, but it does allow a series of building performance problems to be substantially reduced or even avoided altogether. Spread-of-fire requirements can be met by providing a suitably designed corridor floor or ceiling. The segmentation prevents a build-up of heat at the top of the facade. The problem of internal sound transmission has not been completely solved, but with a suitable design, sound transmission in the vertical direction is virtually eliminated. As with the uninterrupted glass double facades, the inclusion of controllable ventilation flaps allows the buffer effect of the facade to be regulated. In comparison to uninterrupted facades, however, more work is required because a large number of ventilation openings is necessary.

Shaft facade
In contrast to the corridor facade, the shaft facade employs vertical separating elements to subdivide the facade cavity. The inner leaf of the facade alternates between shaft-like facade zones with a closed inner leaf, and sections designed like double windows with opening lights. The construction therefore represents a combination of non-segmented glass double facade in the shaft zones and double-window facade in the areas with opening lights (fig. B 2.1.25). The temperature differential that ensues in the shafts and the resulting stack effect are exploited to improve the exchange of air between the cavity and the rooms behind.

External air enters at the bottom of the double window, which is also the location of the opening light to the adjoining room. The exhaust-air outlets are situated at the top of the elements closing off the shaft section which, due to the thermosiphon effect, is subjected to a marginal negative pressure. A suction draft ensues which removes the stale air and simultaneously replenishes with external air.

Again, however, this type of facade suffers from spread-of-fire problems, the risk of supply and exhaust air becoming mixed during unfavourable weather conditions, and internal sound transmission problems.

Double-window facade
In terms of its construction, this type of glass double facade is the most complicated because the cavity is separated horizontally, storey by storey, and also vertically, to form shaft-like sections. Like the conventional double window, each individual window element is a discrete unit, with no ventilation links to the neighbouring elements. Each of these facade units has its own supply-air and exhaust-air openings. The risk of short-circuiting the incoming and outgoing air flows between units in the vertical direction can be minimised by offsetting the ventilation openings diagonally, as with the corridor facade. The most important advantages of this type of facade are the avoidance of spread-of-fire and sound insulation problems because the vertical and horizontal separating elements are normally aligned with the interior layout. In contrast to non-segmented glass double facades and shaft facades, it is virtually impossible to exploit the stack effect to help move the air through the facade cavity. Adequately sized ventilation openings are therefore essential if overheating problems are to be avoided (fig. B 2.1.26).

Costs and benefits

Glass double facade systems, depending on their design and construction, exhibit a wide range of functional characteristics which would appear to favour their use over single-leaf facades. The merits of such facades are:

- natural ventilation options in very windy locations
- avoidance of sick building syndrome due to better opportunities to realise user-controlled natural ventilation
- enhanced comfort because of the higher surface temperature of the inside of the facade during cold weather
- improvements to the energy balance thanks to the option of night-time cooling of buildings with exposed thermal masses
- better use of and protection for sunshading devices
- better sound insulation in noisy locations
- the opportunity for storey-height glazing to optimise daylight utilisation

Owing to the risk of spread of fire in high-rise buildings, this last benefit can only be achieved by including horizontal fire barriers in the form of, for example, a cantilevered floor slab. Otherwise, fire-resistant spandrel panels must be incorporated, which not only limit the amount of incoming daylight but are often undesirable from the architectural viewpoint. When using insulating glazing for the outer leaf,

B 2.1.23

B 2.1.20 Building-in-building principle
B 2.1.21 Corridor facade
B 2.1.22 Building-in-building principle, training academy, Herne (D), 1999, Jourda et Perraudin, with Hegger Hegger Schleiff
B 2.1.23 Corridor facade, office building for Olivetti, Ivrea (I), 1940, Figini & Pollini
B 2.1.24 Double-window facade
B 2.1.25 Shaft facade
B 2.1.26 Double-window facade, office building for RWE AG, Essen (D), 1997, Ingenhoven Overdiek Kahlen & Partner

the fire barrier can be realised very inexpensively by means of a cantilevered floor slab, because the enhanced thermal insulation properties of the external glazing obviate the need for a complicated thermal break in the floor slab.

On the other hand, feasibility studies that point out the poor "profitability" of the second leaf are frequently cited. The studies mainly concentrate on the extra cost of the additional glass leaf, which can amount to between 30 and 100% of the cost of a single-leaf facade construction, depending on the size and nature of the project, and where the construction details are a major issue. However, such deliberations must be considered in the light of the simplification of the inner glass facade in terms of the construction, thermal, acoustic and fire aspects. In order to arrive at a costs-optimised design irrespective of the type of construction chosen, constructional issues must be matched exactly to the functional profile. Generalisations must therefore always be considered critically with regard to the respective individual solution.

The following aspects may turn out to be especially important when optimising the costs of a multi-leaf glass facade and should be investigated in each individual case to assess their relevance [11]:

- The use of standard components or near-standard components
- The inclusion of simple, manually operated opening elements
- Sizing the facade cavity with respect to the functional requirements in terms of interstitial sunshading devices and cleaning options
- Sizing the facade cavity with respect to the functional requirements in terms of interstitial sunshading devices and cleaning options
- Designing the floor of the facade cavity as a non-loadbearing element or for cleaning loads only
- The use of closed external wall components to prevent the spread of fire
- Keeping the number of different types of construction to a minimum
- The use of prefabricated facade components to achieve further cost savings on larger projects

One important factor in the cost-benefit analysis is the everyday running costs of such a facade. The cost of cleaning, operation, maintenance, inspection and upkeep are the main issues here. It should be realised that it is not necessary to clean the double facade any more often than a single-leaf design.

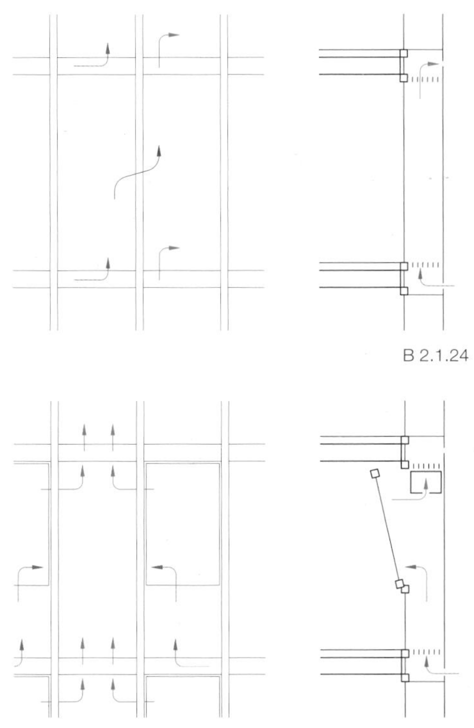
B 2.1.24

B 2.1.25

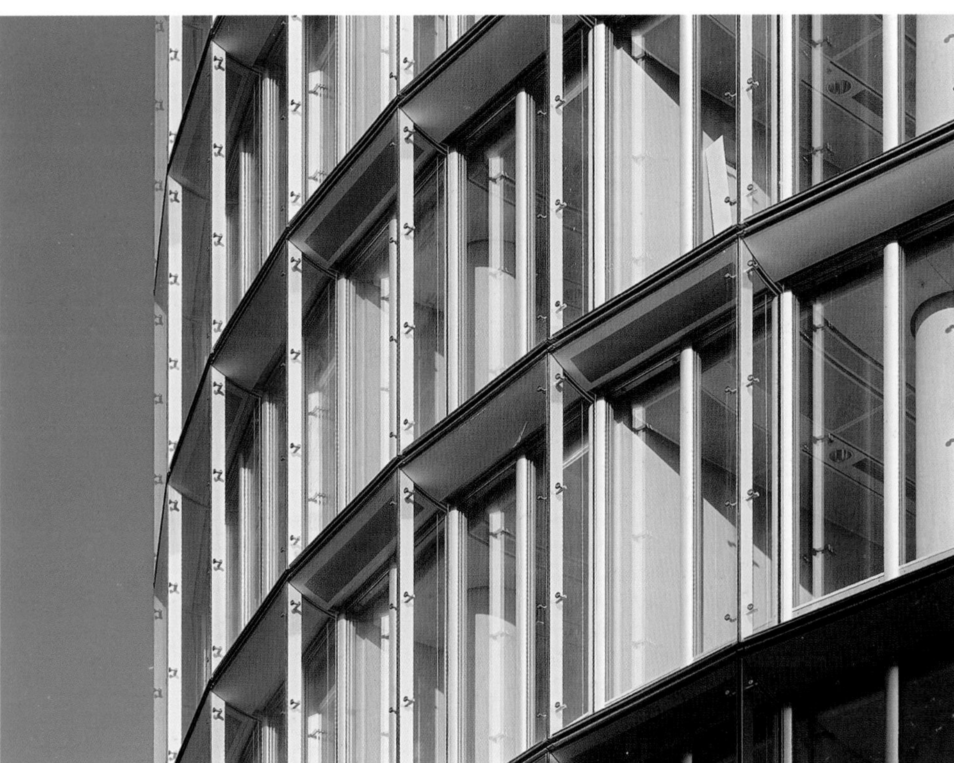
B 2.1.26

B 2.1.27 Deutsche Bank, Berlin (D), 1997, Benedict Tonon plus Nowotny Mähner and Associates

B 2.1.27

The increased cost of cleaning two additional glass surfaces is limited to about once a year, and the cost of cleaning sunshading devices within the cavity is considerably less because they are protected from the weather (unlike the single-leaf glass facade with sunshades fitted outside).

The annual cost of maintenance, inspection and upkeep varies between about 0.5 and 3% of the investment, with the exact cost varying greatly depending on the number and quality of the movable components [12].

Looking at such factors it transpires that multi-leaf glass facades can be economical when their use means that the building's air-conditioning systems can be reduced to a minimum. Besides correct design of the ventilation openings, depending on the type and proportion of glazing plus the sunshades, this factor is very frequently influenced by the behaviour of the building's occupants.

However, it should be remembered that profitability calculations should not be based on the level of investment and operating costs alone, but instead should include user satisfaction, staff productivity and potential savings in the (much shorter-lived) air-conditioning and heating systems in order to obtain a complete picture. The great variety of systems available calls for a precise definition of the requirements profile for the functions of the glass double facade in each case in order to guarantee satisfactory solutions for all areas when deciding on the design. The glass double facade is in a position to achieve notable improvements in the level of comfort and the energy balance in windy, noisy or polluted locations. Nevertheless, the success of a glass double facade depends to a not inconsiderable extent on the skills of the design team, which should not regard this facade system as just an isolated subsystem but rather as a part of the entire structure in terms of its interaction with users, building services, construction and energy balance.

Notes

[1] Staib, Gerald: From the origins to classical modernism; in: Schittich, Christian et al.: *Glass Construction Manual*, Munich/Basel, 1998, p. 10.

[2] The first references to this type of construction are found in a publication on building dating from 1691 (Davilers: Cours d'Architecture); see: Lietz, Sabine: *Das Fenster des Barock*, Munich/Berlin, 1982, p. 123.

[3] Kröling, Peter: Das Sick-Building-Syndrom in klimatisierten Gebäuden: Symptome, Ursachen und Prophylaxe; in: *Innenraum Belastungen: erkennen, bewerten, sanieren*, Wiesbaden/Berlin, 1993, pp. 22–37.

[4] Lang, Werner: Zur Typologie mehrschaliger Gebäudehüllen aus Glas; in: Detail 7/1998, pp. 1225–32.

[5] The intermediate temperature zone and building envelopes is dealt with at length in: Herzog, Thomas; Natterer, Julius (ed.): *Gebäudehüllen aus Glas und Holz. Maßnahmen zur energiebewussten Erweiterung von Wohnhäusern*, Lausanne, 1986.

[6] Zimmermann, Markus: Fenster im Fenster; in: Detail 4/1996, pp. 484–89.

[7] A detailed description of the main interactions between energy transmission and energy storage can be found in: ibid [5], pp. 4–11.

[8] Depending on location and building design, we can assume that in office buildings the operating times of mechanical ventilation systems can be reduced to 35% of the occupancy time of the building when natural ventilation is possible; see: Werkbericht 12. Gebäudetechnik für die Zukunft – "weniger ist mehr", pub. by HL-Technik AG, Munich, 1994, pp. 39–53. In terms of the operating times of an air-conditioning system, we can assume that it is not necessary to use the system at all on days with an outside air temperature between 5 and 22°C. In the case of Düsseldorf this accounts for 63% of the occupancy time of the building; during 29% of the year it is colder than 5°C when the building is occupied, and temperatures exceeding 22°C occur during only 8% of the occupancy time over the year; see: Thiel, Dieter: Doppelfasssaden – ein Bestandteil energetisch optimierter und emissionsarmer Bürogebäude; in: proceedings of the workshop "Lichtlenkende Bauteile" and the International Forum for Innovative Facade Technology, Institute of Light and Building Technology, Cologne Polytechnic (pub.), Cologne, 1995, p. 30.

[9] First references to the effects and potential applications of multi-leaf glass walls with internal air circulation can be found as early as 1929; in: Le Corbusier: *Precisions on the Present State of Architecture and City Planning* (English translation: E. Schrieber, 1991).

[10] An overview of the options for improving the air with internal planting can be found in: Daniels, Klaus: *Technology of Ecological Building*, Basel/Boston/ Berlin, 1997, pp. 194–97.

[11] An interesting discussion of the economic issues with respect to glass double facades can be found in: Oesterle, et al.: *Doppelschalige Fassaden: Ganzheitliche Planung*, Munich, 1999, pp. 178–98.

[12] ibid, p. 187.

Steiff soft toys factory

Giengen an der Brenz, D, 1903

Planning:
Eisenwerke München, Munich
Richard Steiff, Giengen an der Brenz

Bauen + Wohnen 07/1970
Baumeister 11/2003
Bauwelt 44/1992
Finke, Barbara: Der Ostbau der Steiff-Fabrik in Giengen/Brenz, dissertation, master's degree in art, Berlin Technical University, 1998

- First known storey-height, fully glazed buffer facade in Germany for improving thermal performance compared to a single-leaf glass facade
- Structural steelwork positioned within the facade cavity
- Facade members connected via specially shaped steel lugs
- Double windows integrated into the facade enable an exchange of air and views of the surroundings

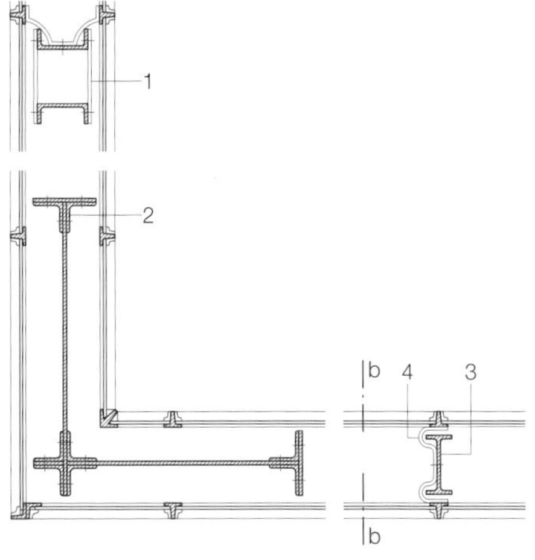

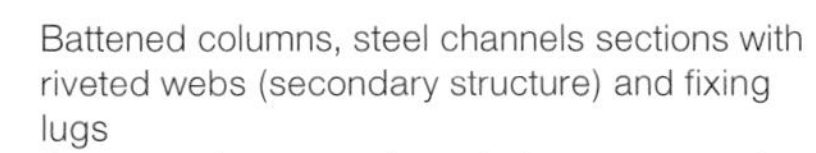

Plan of 1st floor • Section, scale 1:500
Horizontal section, scale 1:20
Details of facade post, scale 1:5

1. Battened columns, steel channels sections with riveted webs (secondary structure) and fixing lugs
2. Compound corner column (primary structure)
3. Facade post, IPE 130 x 80 mm steel section, with fixing lug riveted on
4. Fixing lug, riveted to No. 3
5. Facade rail, steel channel section, 60 x 140 mm
6. Transverse beam, IPE 140 x 70 mm steel section
7. Clay hollow pot floor with concrete topping and wooden floorboards on battens
8. Glazing bar, steel T-section, 20 x 30 or 25 x 30 mm
9. Putty
10. Outer glazing to curtain wall extends over all floors
11. Storey-height inner glazing, 3 mm obscured glass

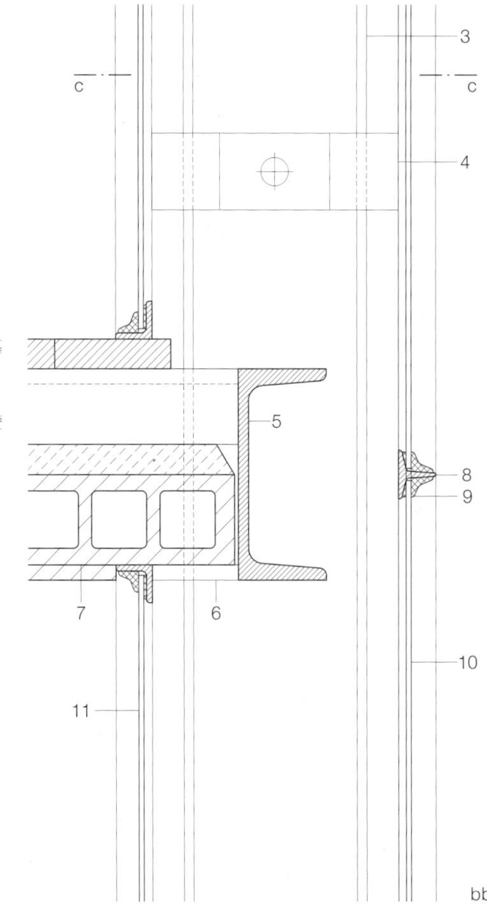

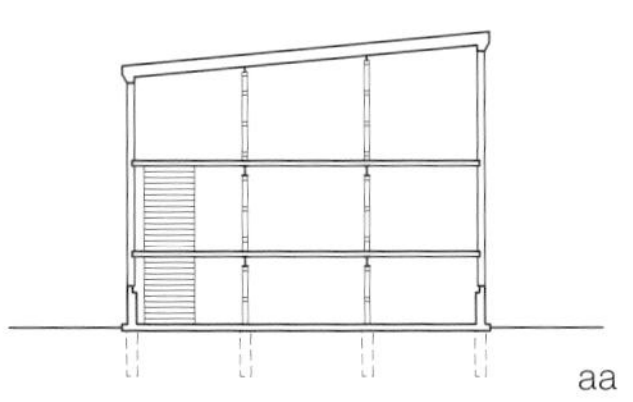

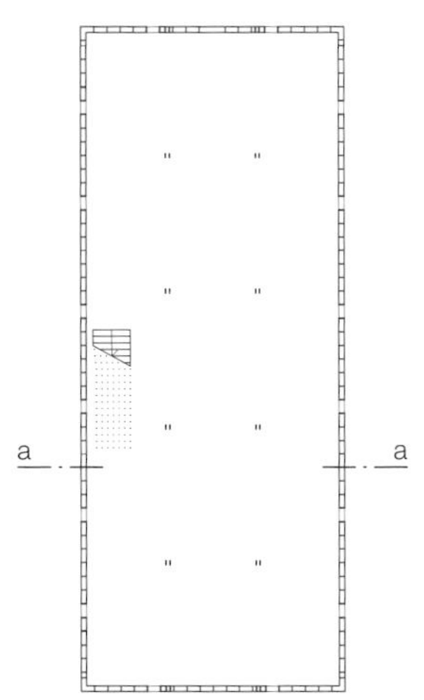

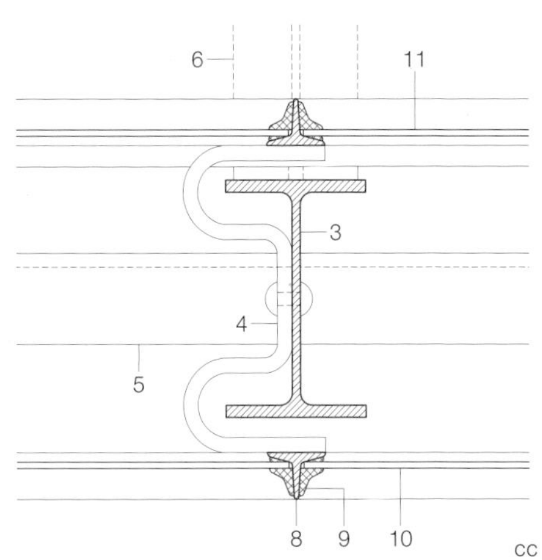

Events and Congress centre

San Sebastián, E, 1999

Architect:
Rafael Moneo, Madrid

Detail 03/2000
domus 722, 1990
El Croquis 98, 2000: special edition – Rafael Moneo, 1995–2000

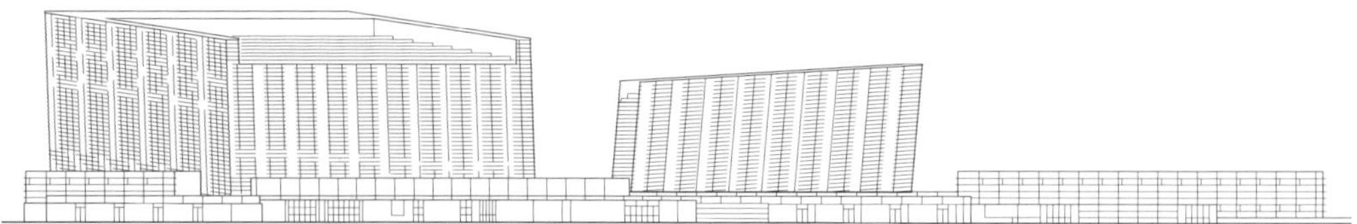

- Buffer facade with internal structural steelwork
- Width of cavity = 2500 mm
- Glass provides protection against the salt-laden air
- Window openings with insulating glazing incorporated in the buffer facade offer limited views of the surroundings

Elevation, scale 1:1500
Vertical section through facade
scale 1:20
Details, scale 1:5

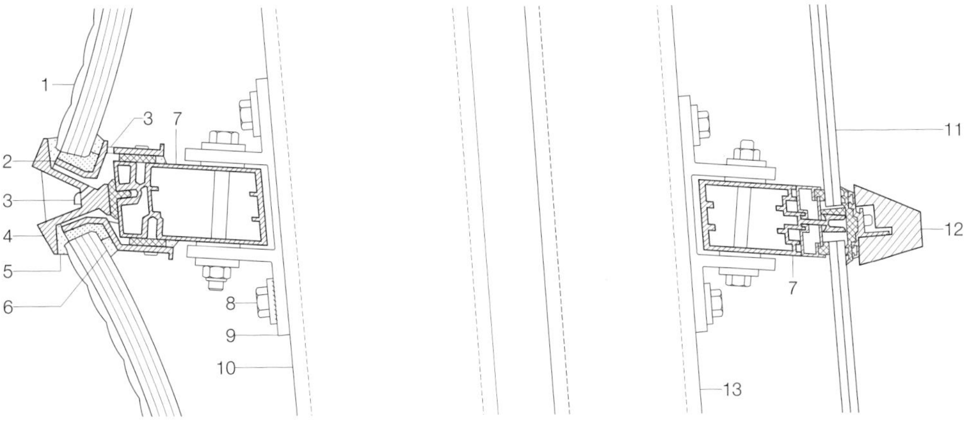

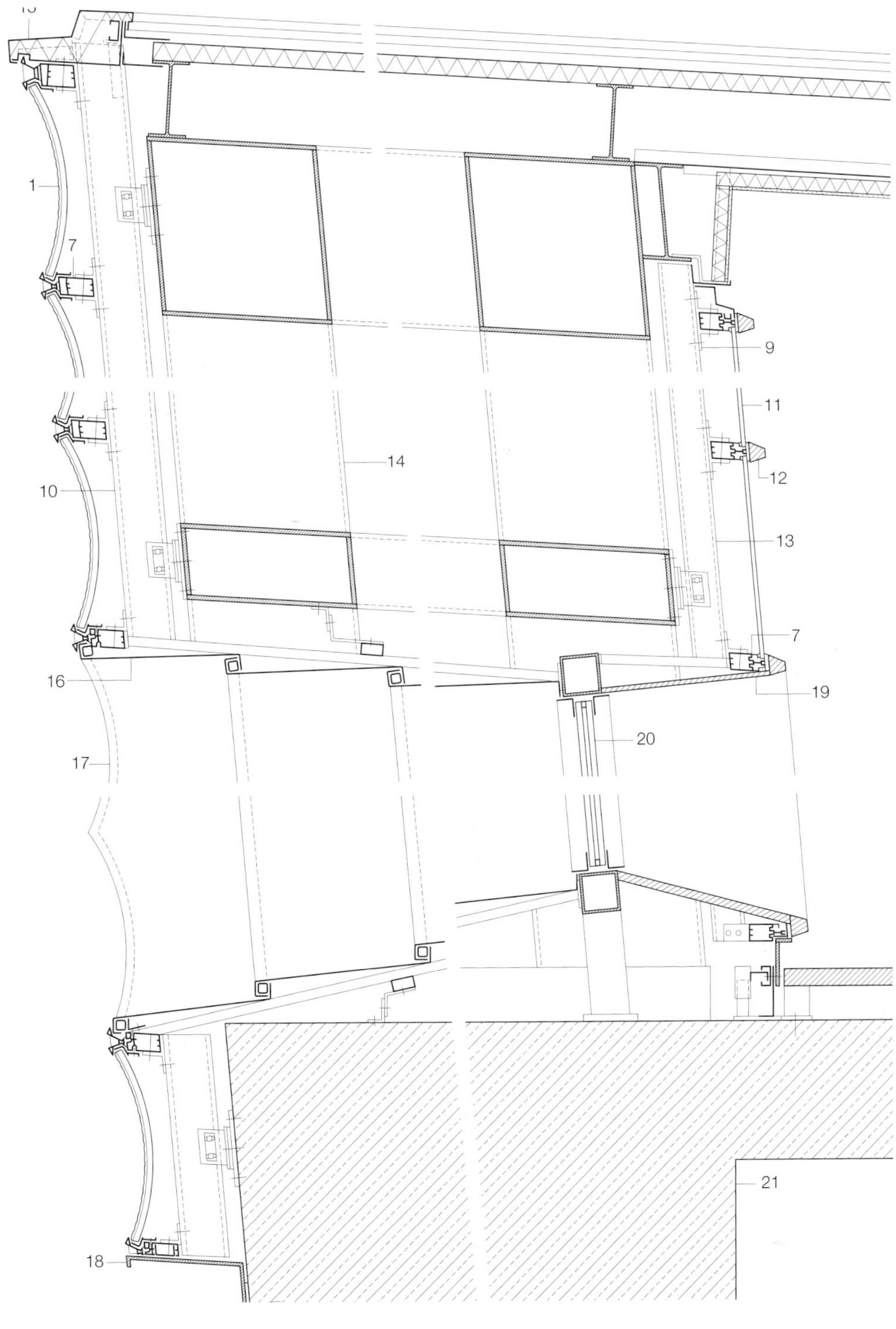

1 Curved glass panes, laminated safety glass, 2500 x 600 mm, made from translucent 4-5 mm rolled glass plus 19 mm sandblasted float glass, vertical joints sealed with silicone
2 Glazing bar, extruded aluminium, glass bonded with silicone adhesive
2 Hole for drainage and pressure equalisation, shielded from the wind on the outside
4 Cast aluminium section
5 Translucent silicone sealant
6 White silicone sealant
7 Extruded aluminium rail
8 Stainless steel bolt
9 Aluminium bracket, adjustable in three directions
10 Facade post, extruded aluminium section, 50 x 140 mm
11 Sandblasted laminated safety glass, made from 2 No. 6 mm float glass, pane size 2500 x 600 mm
12 Glazing cap, aluminium with cedar wood cover strip
13 Facade post, extruded aluminium section, 50 x 100 mm
14 Loadbearing structure made from sheet steel, welded, with fire-resistant coating
15 Roof edge trim, sheet aluminium, bent to suit and insulated
16 Cladding, aluminium profile, 20 x 40 x 500 x 5 mm
17 Cladding cut to match curve of glass
18 Aluminium trim, 30 x 250 x 330 x 10 mm
19 Cedar wood soffit
20 Insulating glass, 2 No. 16 mm laminated safety glass
21 Fair-face concrete plinth

Office building

Würzburg, D, 1995

Architects:
Webler + Geissler, Stuttgart
Structural engineers:
Rudi Wolff, Stuttgart

Arch+ 113, 1993
Architectural Review 11/1996
A+U 05/1997
Bauwelt 27/1996
Byggekunst 08/1996
GLAS 06/1996
Techniques + architecture 434, 1997

- Non-segmented glass double facade with controllable ventilation flaps top and bottom
- Insulating glass used for both leaves to improve the thermal performance
- Vertical glass separating elements at the corners with integral axial-flow fans to distribute preheated supply air throughout the entire facade cavity
- Aluminium louvre blinds with different colours inside/outside
- Reflective outer face improves the sunshading function
- Inner face with absorbent coating permits use of solar energy to preheat cold supply air

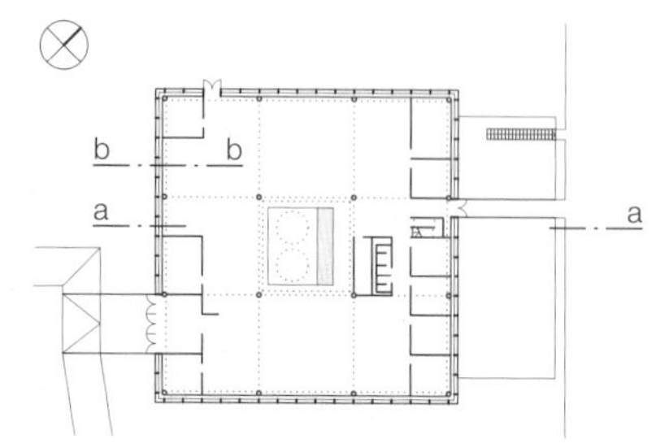

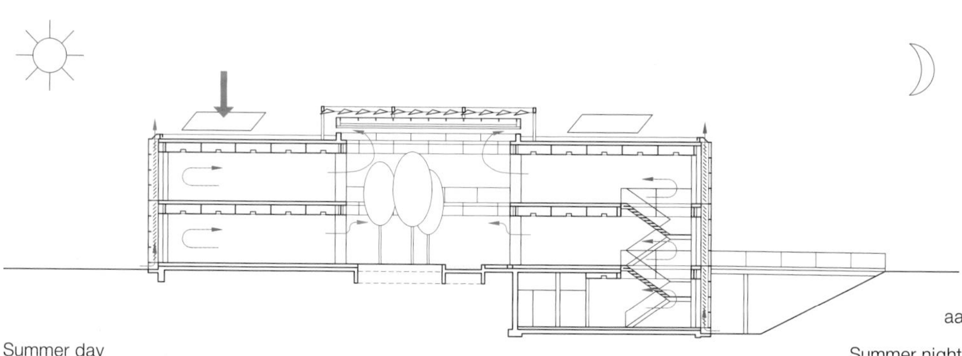

Summer day

Summer night

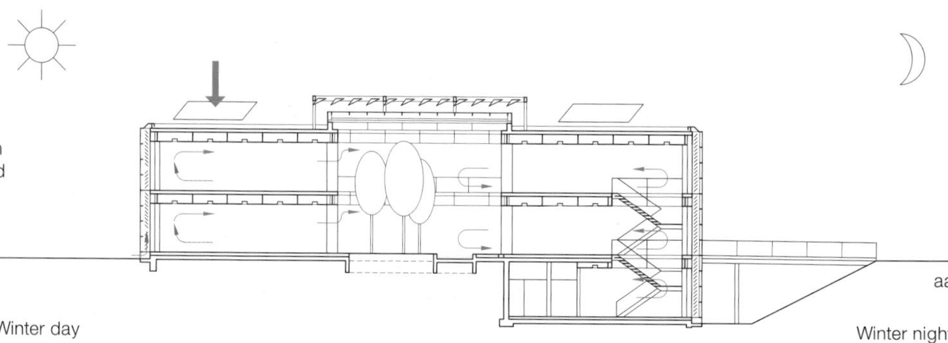

Winter day

Winter night

1 Sheet aluminium with anti-drumming coating
2 Upper ventilation flap with brush seals at the sides
3 Aluminium ventilation louvres, with insect screen
4 Outer glazing:
8 mm toughened safety glass (outside)
22 mm cavity filled with inert gas
6 mm float glass (inside), with low-e coating
5 Blind made from perforated lightweight metal louvres, upper section coated white both sides, lower section with dark coating on one side, both sections can be controlled independently
6 Aluminium open grid flooring
7 Inner glazing:
6 mm toughened safety glass (outside)
16 mm cavity filled with inert gas
6 mm float glass (inside), with low-e coating
8 Aluminium post-and-rail framing, with thermal break
9 Lower ventilation flap

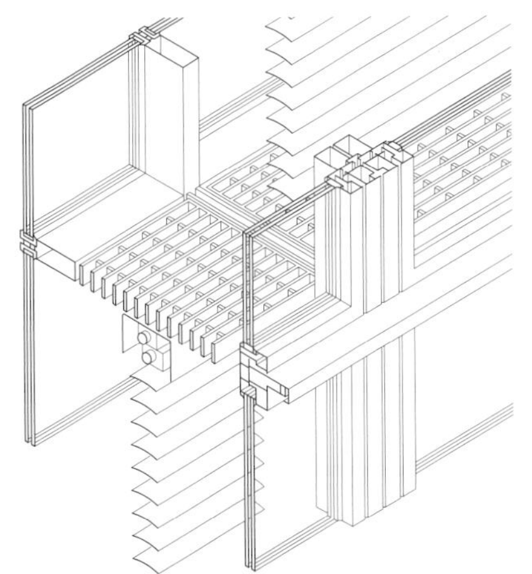

Plan, scale 1:1500
Sections, scale 1:500
Vertical section, scale 1:20
Isometric view (not to scale)

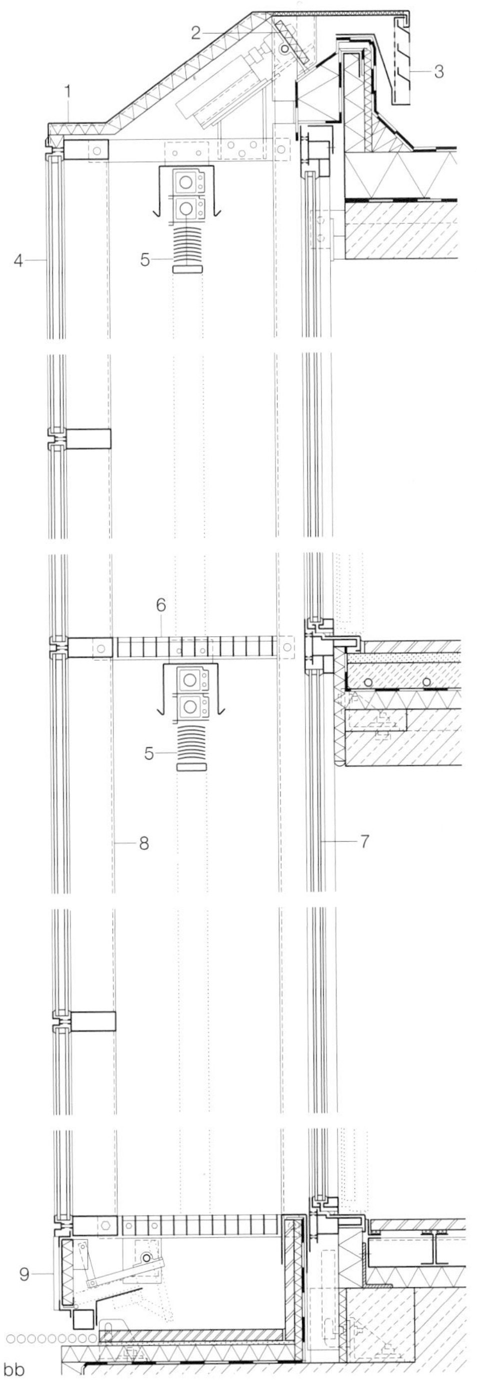

Gallery for Architecture and Labour

Gelsenkirchen, D, 1995

Architects:
Pfeiffer Ellermann und Partner, Lüdinghausen
Assistants:
Andrzej Bleszynski, Axel Rüdiger
Structural engineers:
Spiess Schäfer Keck, Dortmund
Interior climate concept:
Kahlert, Haltern

Bauwelt 27/1996
Byggekunst 08/1998
GLAS 06/1996

- Non-segmented second skin positioned 600 mm in front of lightweight concrete wall
- Window openings similar to double windows, closed off from the surrounding facade cavity
- Facade cavity has hinged openings to the outside at the top and bottom of the facade
- Distance between two glazing planes = approx. 900 mm
- Controllable ventilation openings between non-segmented glass double facade and double-window elements to enable use of preheated external air in winter

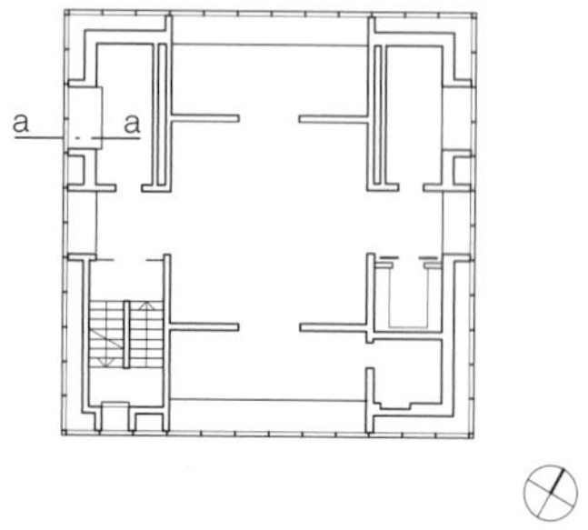

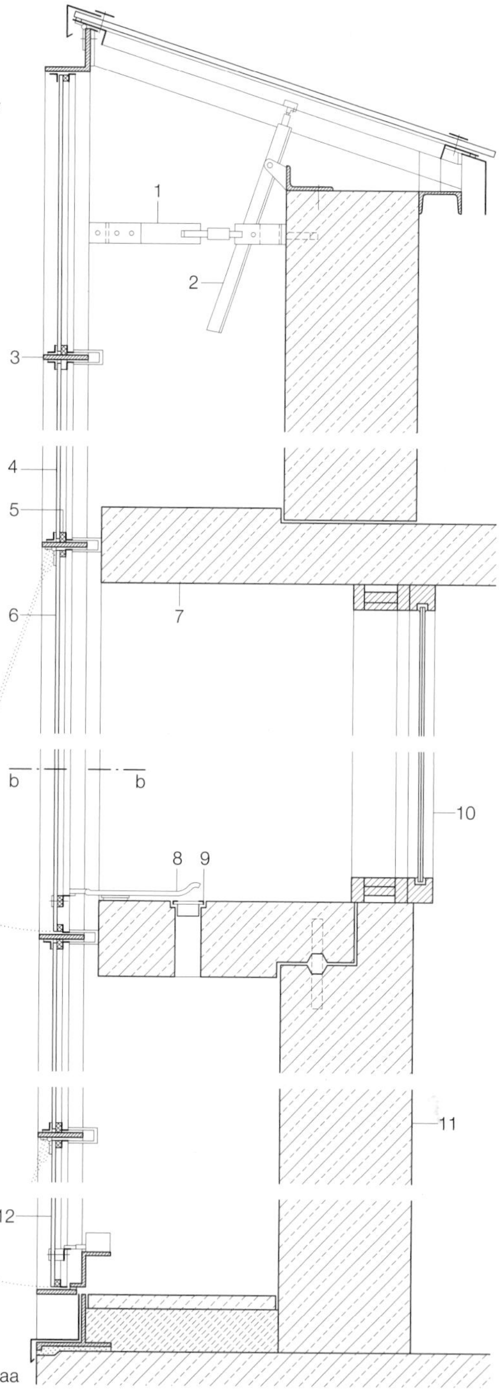

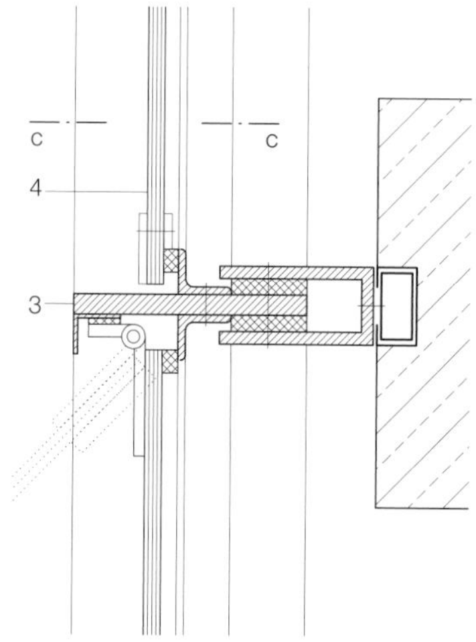

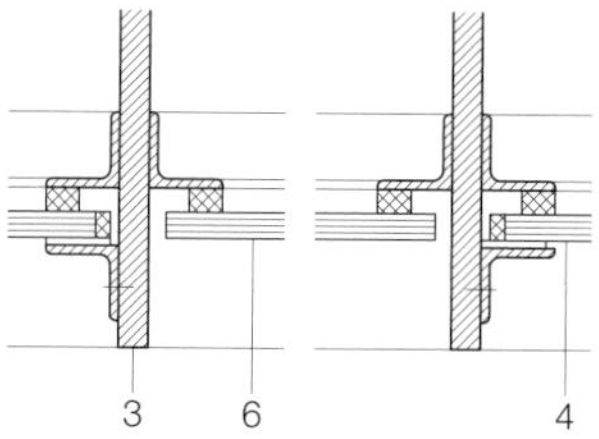

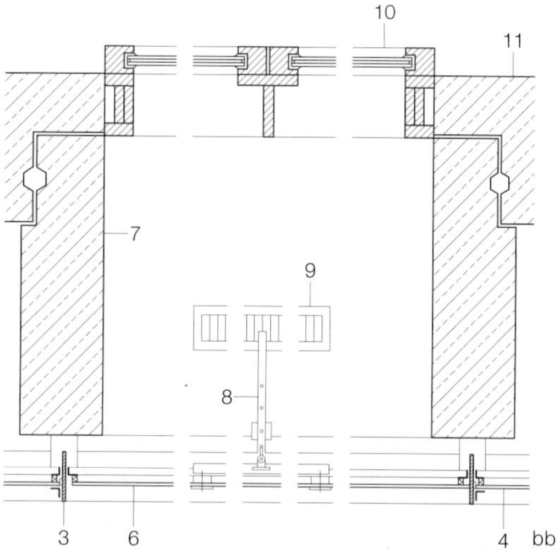

Plan of upper floor, scale 1:400
Horizontal section • Vertical section
scale 1:20
Details, scale 1:5

1 Adjustable facade fixing
2 Electrically operated window opener
3 Steel frame made from 60 mm steel flats, frame size 1020 x 1020 mm
4 Fixed light, 8 mm toughened safety glass, frameless, individual fixings
5 Peripheral gasket, 10 x 6 mm
6 Toughened safety glass, 8 mm, frameless top-hung opening light, individual hinges
7 Reinforced concrete "frame", 200 mm
8 Window stay
9 Individually adjustable ventilation openings
10 Wooden window, untreated fir, with insulating glass
11 Fair-face lightweight concrete, 365 mm
12 As No. 6 but electrically operated

Petrol and service station

Lechwiesen, D, 1997

Architects:
Herzog + Partner, Munich
with Arthur Schankula, Roland Schneider

db 04/1998
Herzog, Thomas: Architektur + Technologie. Munich/London/New York 2001
Herzog, Thomas et al.: Timber Construction Manual, Basel/Boston/Bonn, 2004, p. 233

- Multi-purpose, double-leaf glass facade provides sound insulation and acts as air duct in winter
- In winter the stale air from the interior is drawn downwards through the facade cavity and fed to the heat recovery system
- This heats up the inner glass surfaces during cold weather to improve the level of comfort
- Height-adjustable parapet at top of double-leaf glass facade enables hot interior air to escape in summer
- Interior layout and roof pitch optimised to suit flow requirements

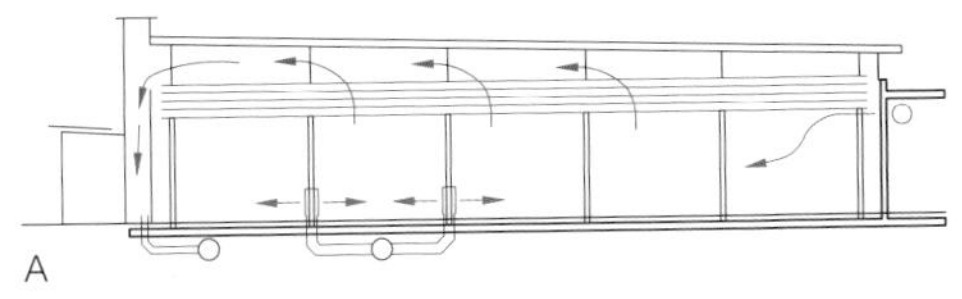

A

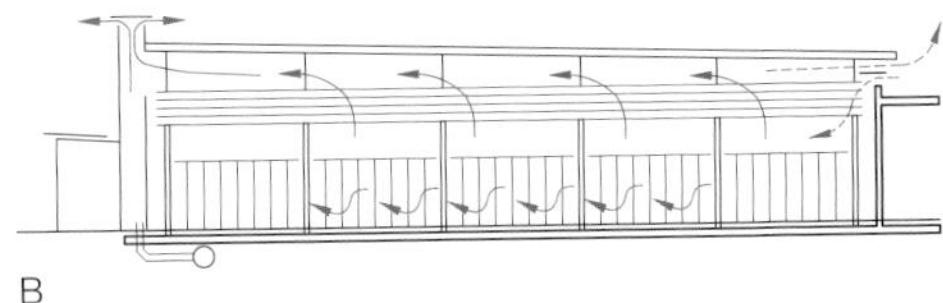

B

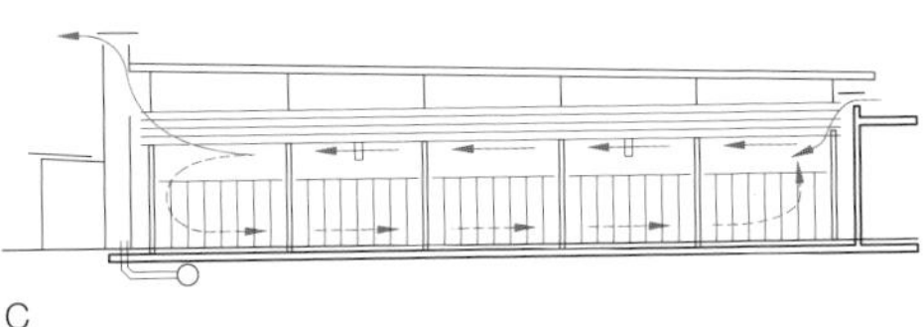

C

System sections (not to scale)
Vertical section through south elevation
scale 1:20

A Winter: mechanical ventilation, ventilation flaps closed
B Summer: natural ventilation, ventilation system in operation
C Summer: natural ventilation with mechanical backup provided by solar-powered axial-flow fans

1 Capping, 2 mm stainless steel sheet
2 Steel angle frame, 100 x 50 x 6 mm
3 Electrically operated rack-and-pinion lifting mechanism
4 Frame, 65 x 115 mm
5 Steel channel frame
6 Cover strip, 50 x 25 mm channel
7 Heat-absorbing glass, 28 mm
8 Steel channel, 50 x 38 mm
9 Frame of steel flats, 120 x 15 mm and 147 x 15 mm, welded
10 Tension/compression stay, 30 dia. x 2 mm
11 Wind bracing, 16 dia. x 1.8 mm
12 Timber beam grid with stays
13 Toughened safety glass, 8 mm
14 Transom, glued maple
15 Steel open grid flooring, 20 x 3 x 16 mm, over supply-air inlet

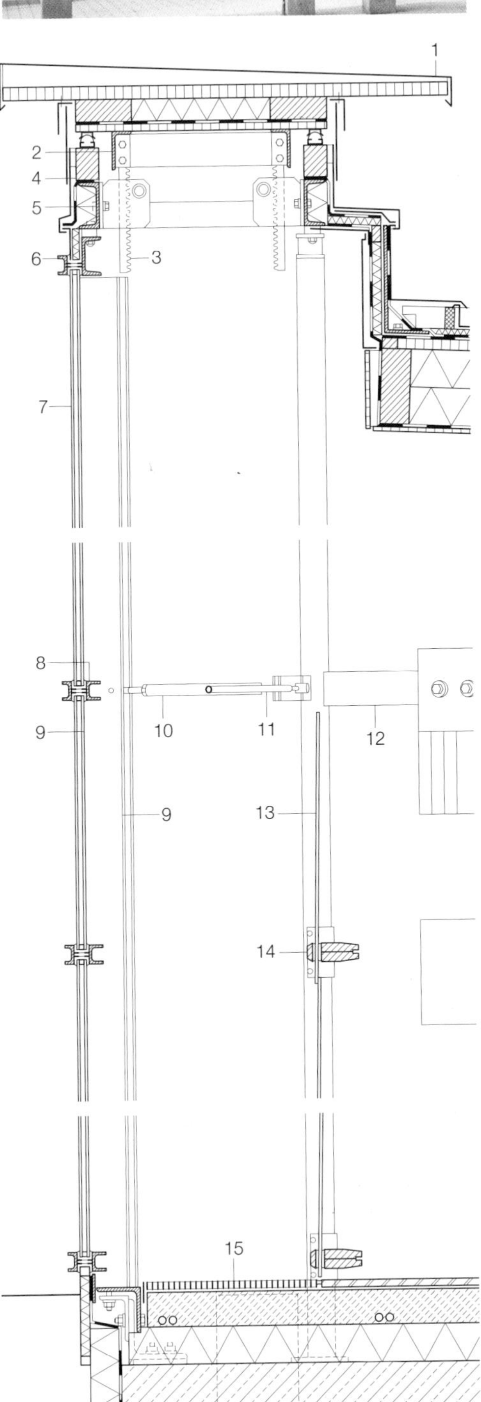

Office block facade refurbishment

Stuttgart, D, 1996

Architects:
Behnisch Sabatke Behnisch, Stuttgart
Project architect:
Carmen Lenz

Bauwelt 43–44/1996
GLAS Sonderheft 02/1997
Knaack, Ulrich: Konstruktiver Glasbau. Cologne, 1998
Schittich, Christian et al.: Glass Construction Manual, Munich/Basel, 1998

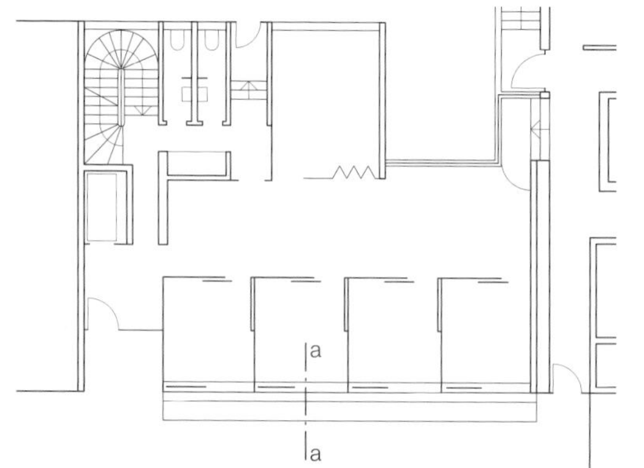

- Complete refurbishment of a prefabricated reinforced concrete structure dating from 1969
- Non-segmented glass double facade with outer glass envelope of glass louvres controllable storey by storey
- Maximum flow of air through facade cavity when glass louvres are open provides good protection against overheating
- Cross-ventilation thanks to ventilation openings in the corridor walls plus night-time cooling in summer

Plan of ground floor, scale 1:250
Vertical section, scale 1:20

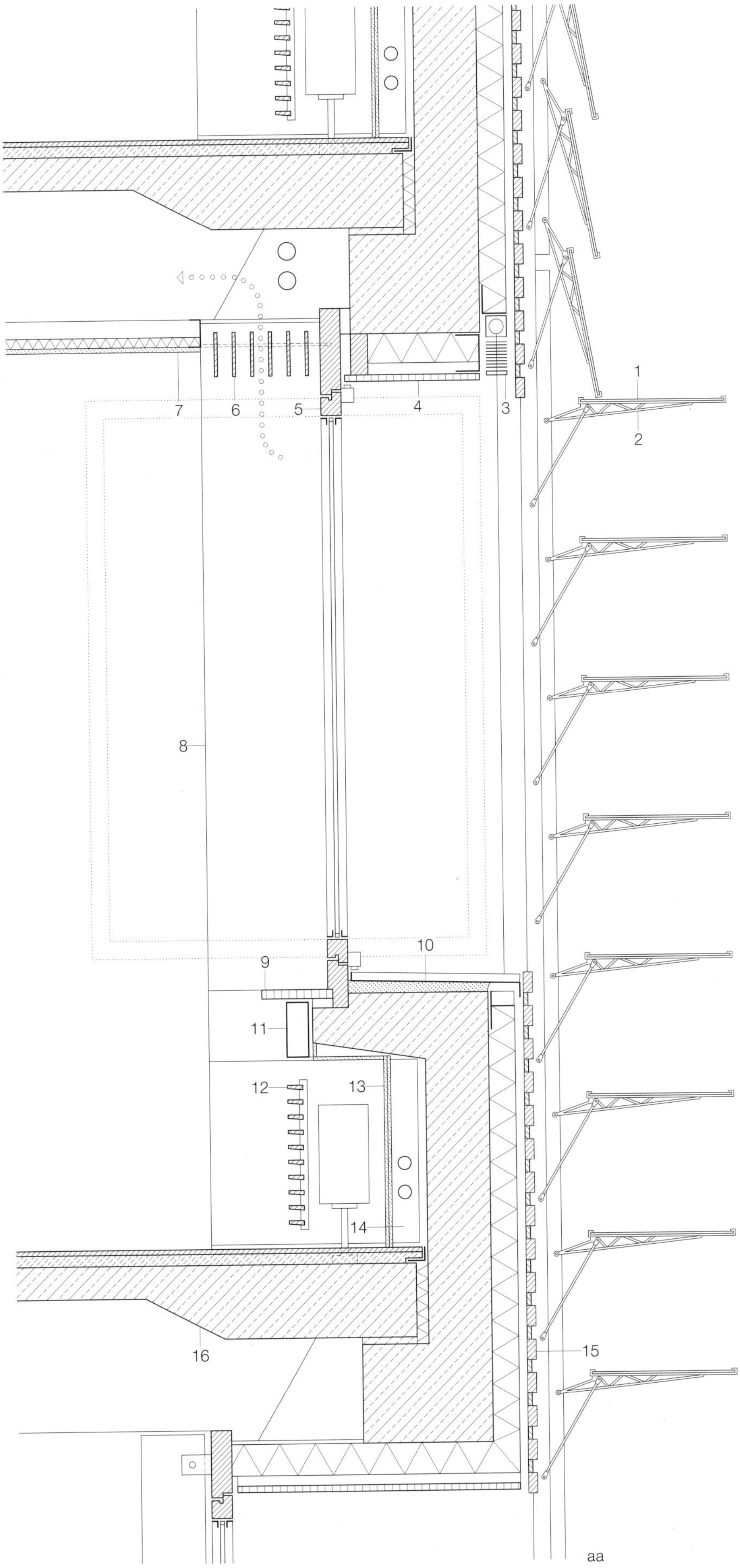

aa

1 Toughened safety glass, 6 mm
2 Glass clamp fitting connected to supporting framework
3 Aluminium louvre blind
4 Timber soffit
5 Vertical pivot window in wooden frame
6 Aluminium grille bars, 10 x 140 mm
7 Suspended plasterboard ceiling
8 Glass divider, 14 mm toughened safety glass
9 Veneer plywood window board, 200 mm
10 Window sill, sheet aluminium
11 Conduit for cables
12 Wooden grille
13 Plasterboard
14 Services duct
15 Facade construction: timber batten cladding, 20 x 60 mm, on framework
30 mm ventilated cavity
80 mm closed-cell insulation
precast concrete spandrel panel
16 Precast concrete floor

Office building

Kronberg im Taunus, D, 2000

Architects:
schneider+schumacher, Frankfurt am Main

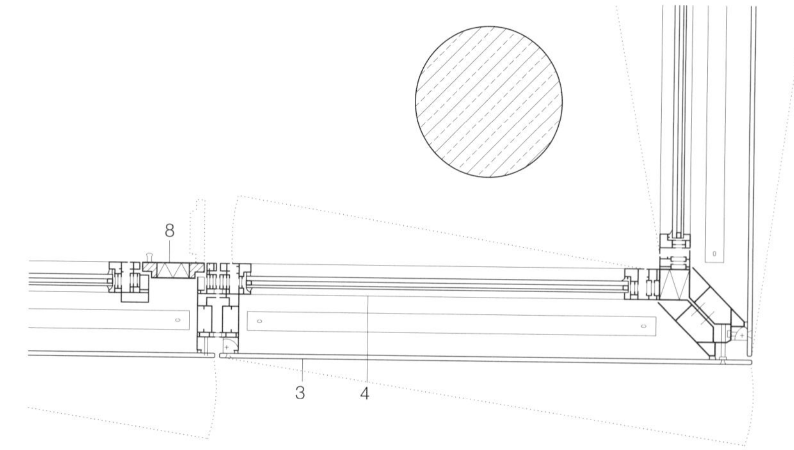

Architektur Aktuell 246–247, 2000
db 03/2000
DBZ 01/2001
Fassade/Façade 04/2001
GLAS 05/2000
Schittich, Christian (ed.): Gebäudehüllen. Munich/Basel 2001

- Double-window facade with electrically operated, outward-opening, single-glazed frameless side-hung windows
- Inner facade with fixed insulating glass, openable for cleaning and maintenance purposes only
- Ventilation flaps made from thermally insulated aluminium panels with thermal break
- Offices on the periphery are ventilated naturally
- Night-time cooling of offices in summer

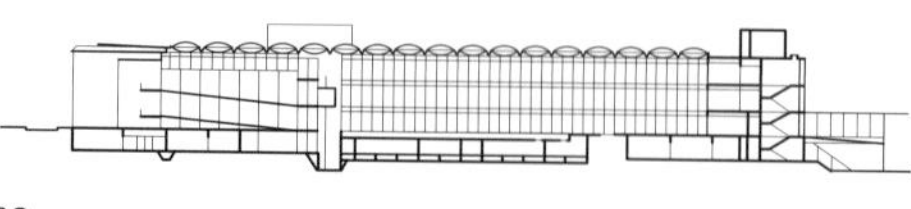

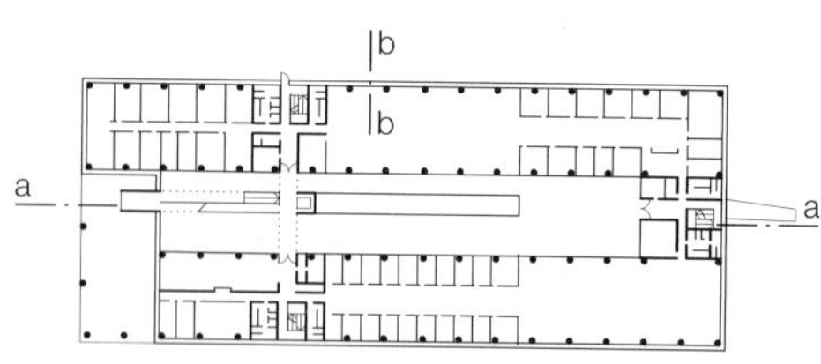

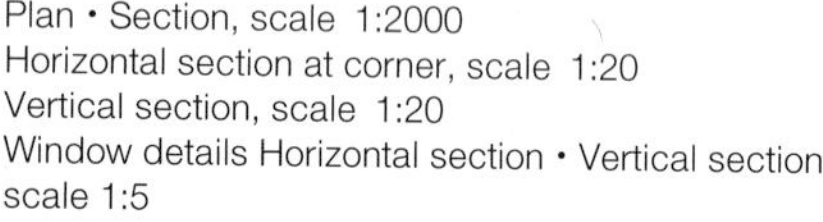

Plan • Section, scale 1:2000
Horizontal section at corner, scale 1:20
Vertical section, scale 1:20
Window details Horizontal section • Vertical section scale 1:5

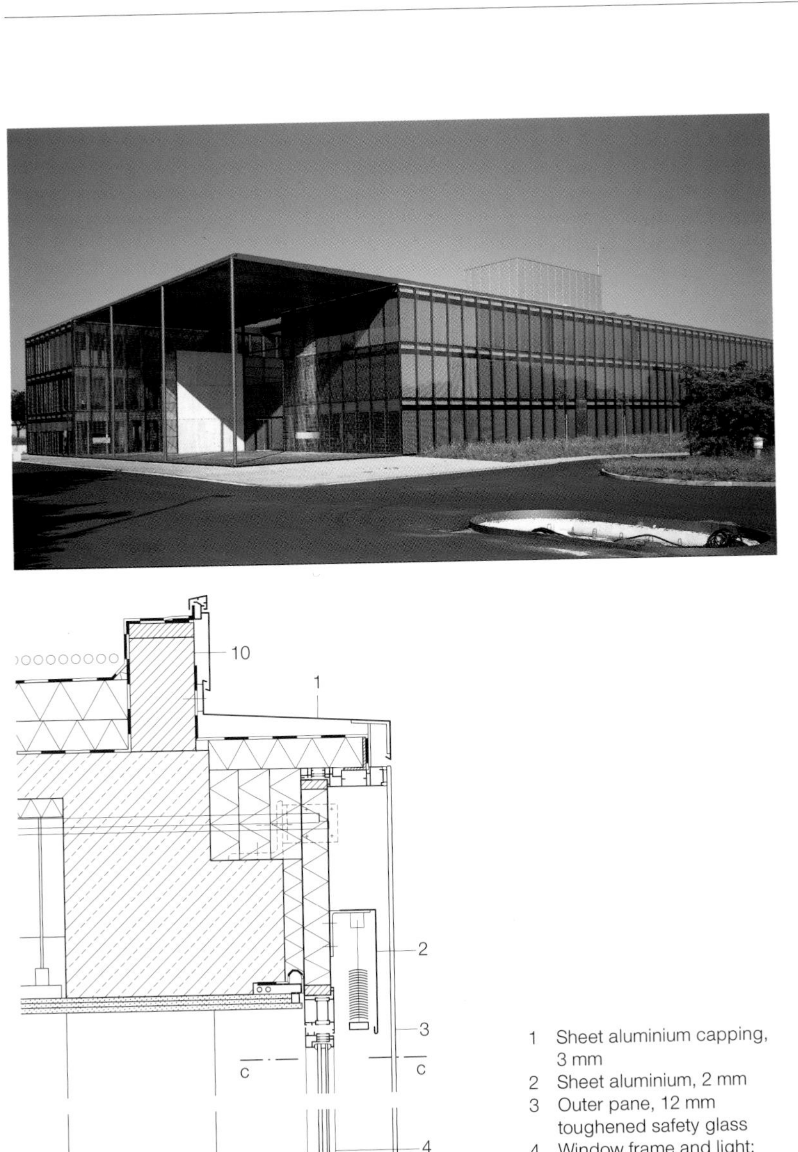

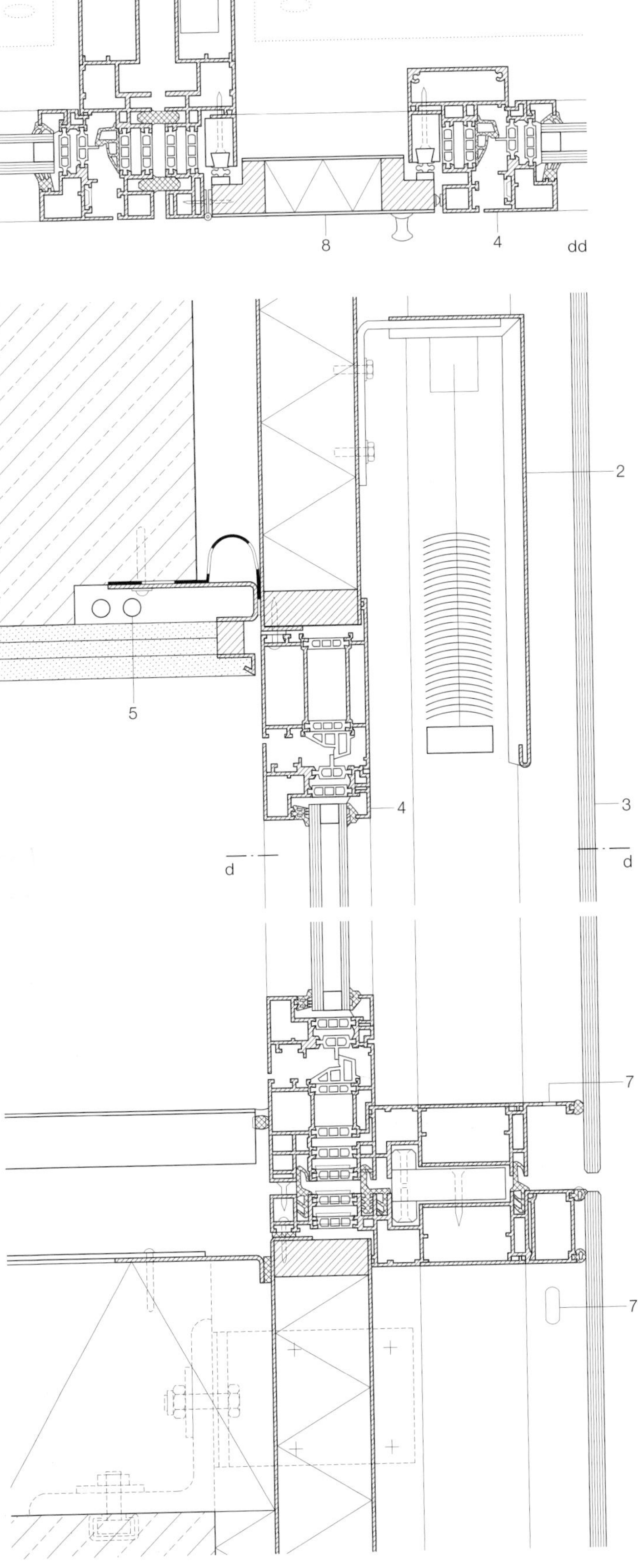

1 Sheet aluminium capping, 3 mm
2 Sheet aluminium, 2 mm
3 Outer pane, 12 mm toughened safety glass
4 Window frame and light: extruded aluminium section, with thermal break glazing: 6 mm toughened safety glass + 14 mm cavity + 8 mm laminated safety glass
5 Water pipes for heating/cooling ceiling, 20 mm dia.
6 Grating over air inlet to open-plan area
7 Slits for equalising differential air pressures
8 Ventilation flap: sheet aluminium facing both sides, with magnetic seal
9 Electric drive
10 Lightweight clay masonry

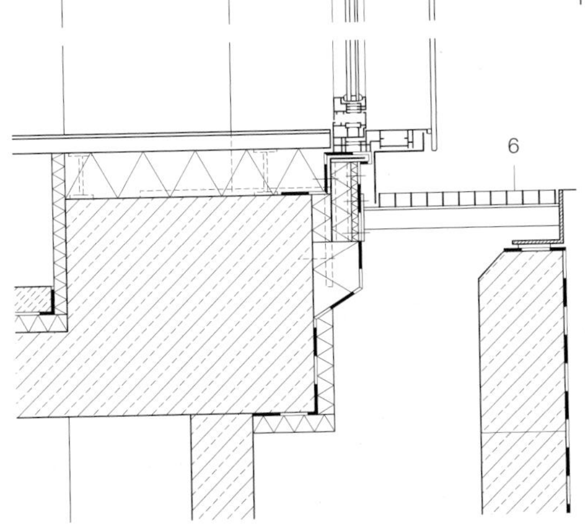

Stadttor development

Düsseldorf, D, 1997

Architects:
Draft design and planning permission:
Overdiek Petzinka und Partner, Düsseldorf
Working drawings and realisation:
Petzinka Pink und Partner, Düsseldorf
Structural engineers:
Ove Arup und Partner, Düsseldorf
Facade consultant:
Erich Mosbacher, Friedrichshafen

Oesterle, Eberhard et al.:
Doppelschalige Fassaden. Munich, 1999

- Corridor facade with a 1.4 m wide, accessible facade cavity
- Safety barrier also acts as structural member to withstand wind loads
- Adjustable ventilation openings
- Diagonally offset openings prevent mixing of incoming and outgoing air

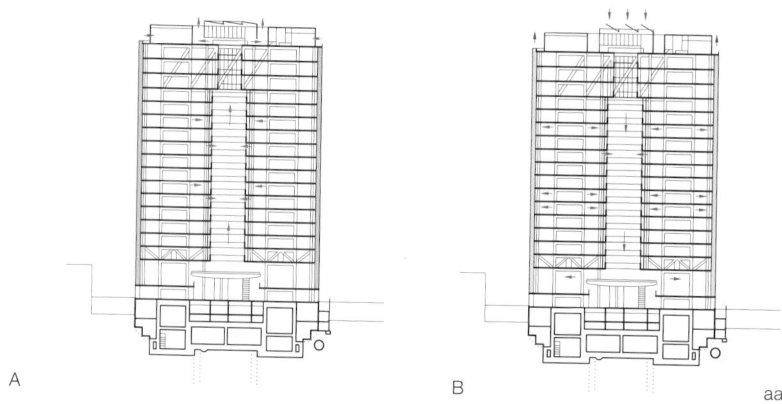

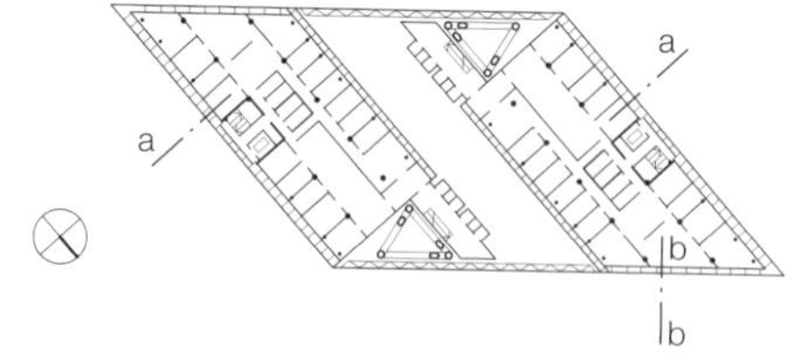

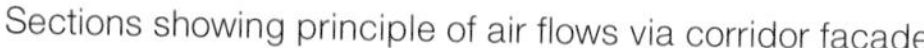

Sections showing principle of air flows via corridor facade

A Winter
ventilation elements:
closed at night (thermal insulation function) and open during the day (for preheating external air); preheated air fed into atrium

B Summer
ventilation elements:
open; temperature of cooling ceiling adjusted by means of cool air from underground

Plan of standard floor • Section
scale 1:2000
Vertical section, scale 1:20
Horizontal section through safety barrier/facade bracing, scale 1:5

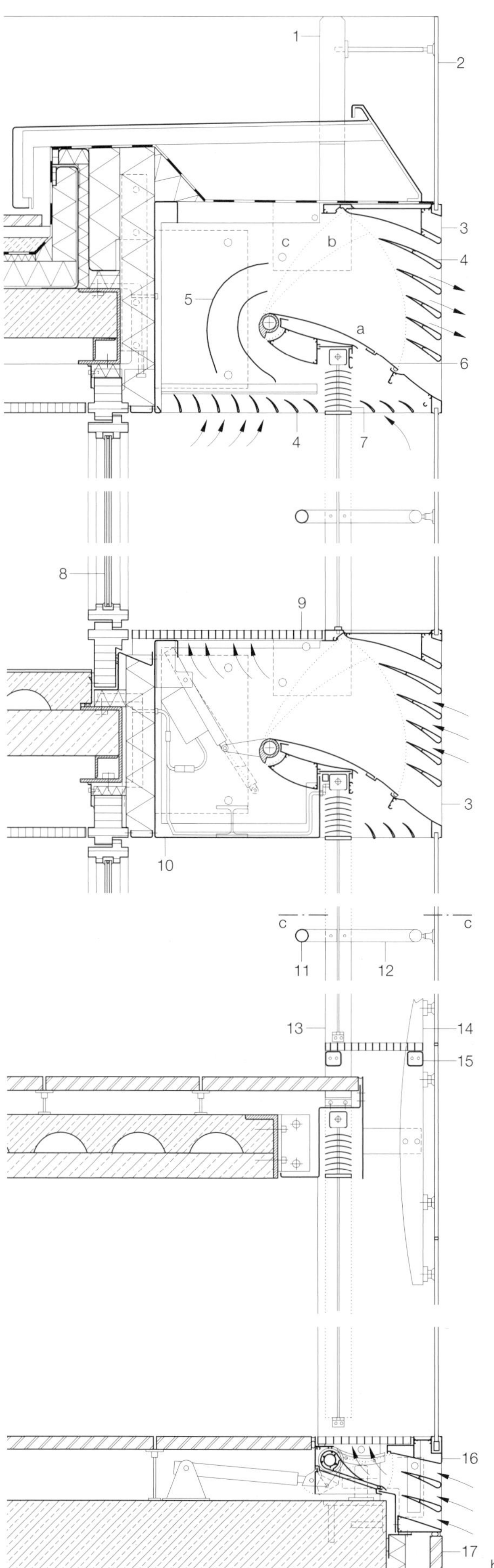

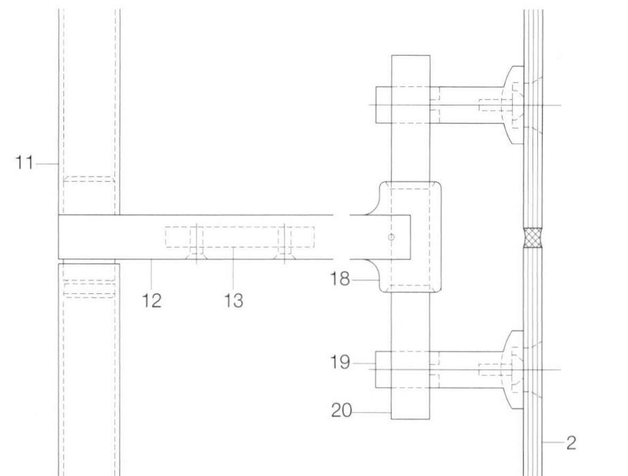

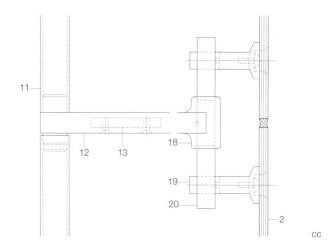

1 Safety barrier upright, steel flat, 80 x 8 mm
2 Glazing, 12 mm toughened safety glass
3 Alternate ventilation inlets/outlets, 3 mm stainless steel sheet
4 Ventilation louvres, aluminium
5 Air flow guide plate, extruded aluminium
6 Ventilation control plate, aluminium
   a open
   b rain position
   c closed
7 Sunshade, electrically operated louvre blind
8 Wooden window (beech) with insulating glass
9 Open grid flooring alternating with stainless steel non-slip plates
10 Stainless steel sheet, 3 mm
11 Handrail, 40 dia. x 4 mm aluminium tube
12 Cast aluminium arm, 28 x 40 mm
13 Safety barrier upright, steel flat, 100 x 12 mm
14 Steel fin, 15 mm steel flat, with stainless steel glass fixings
15 Steel square hollow section, 50 x 50 x 3 mm
16 Special section with opening for condensation water
17 Wall construction:
   40 mm natural stone
   40 mm natural stone
   60 mm thermal insulation
18 Tensioning sleeve
19 Stainless steel glass fixing
20 Solid stainless steel bar, 24 mm dia.

Trade fair headquarters

Hannover, D, 1999

Architects:
Herzog + Partner, Munich
with Roland Schneider

📖 Detail 03/2000
Herzog, Thomas (ed.):
Nachhaltige Höhe – Sustainable Height.
Munich/London/New York 2000

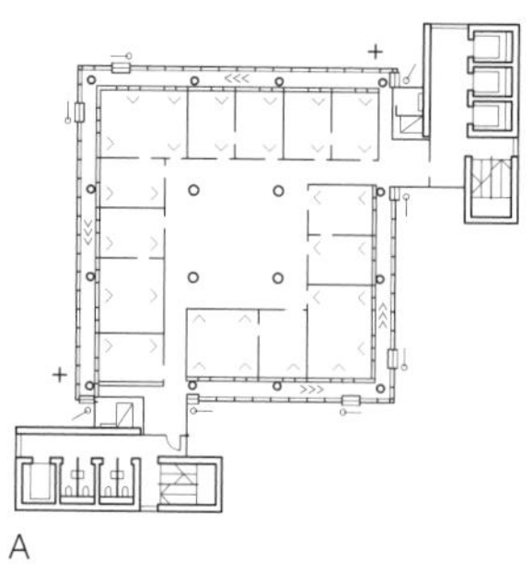

A

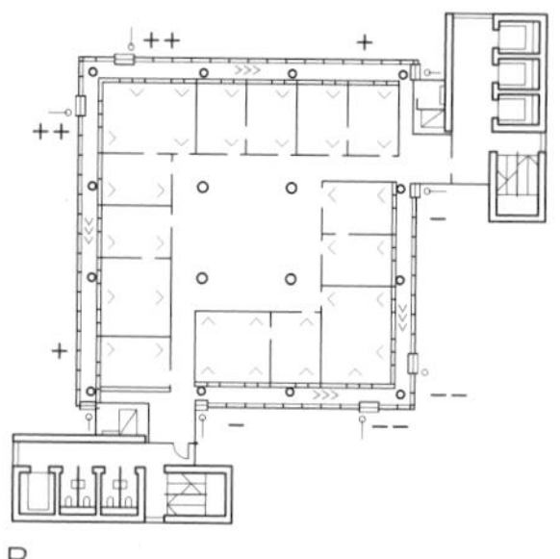

B

Principle of natural air flows via corridor facade (wind blowing from the north)
A Winter
B Summer

+ Wind pressure
– Wind suction
–o Facade opening, temperature sensor controls air flow in corridor facade
>>> Supply air
▭ Exhaust-air duct

- Structure, facade and building services interwoven to form a complete system in terms of energy usage
- High level of comfort and good-quality working environment, low energy consumption
- Ventilation tower to exploit the stack effect (thermal buoyancy) for natural ventilation of the entire building
- Controllable ventilation elements in the outer facade permit adjustments to suit different pressure conditions
- Possibility of individual natural ventilation via sliding doors opening on to the facade cavity

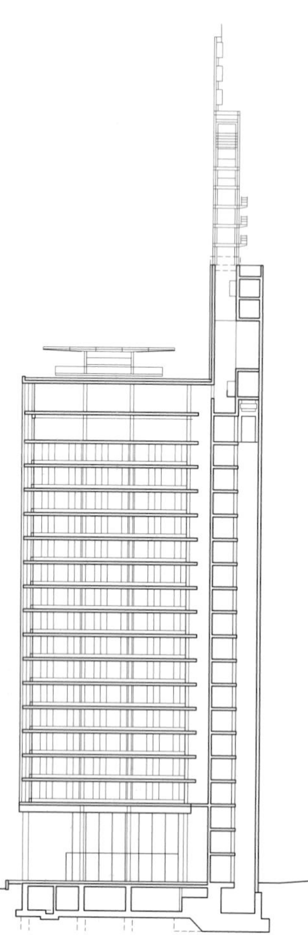

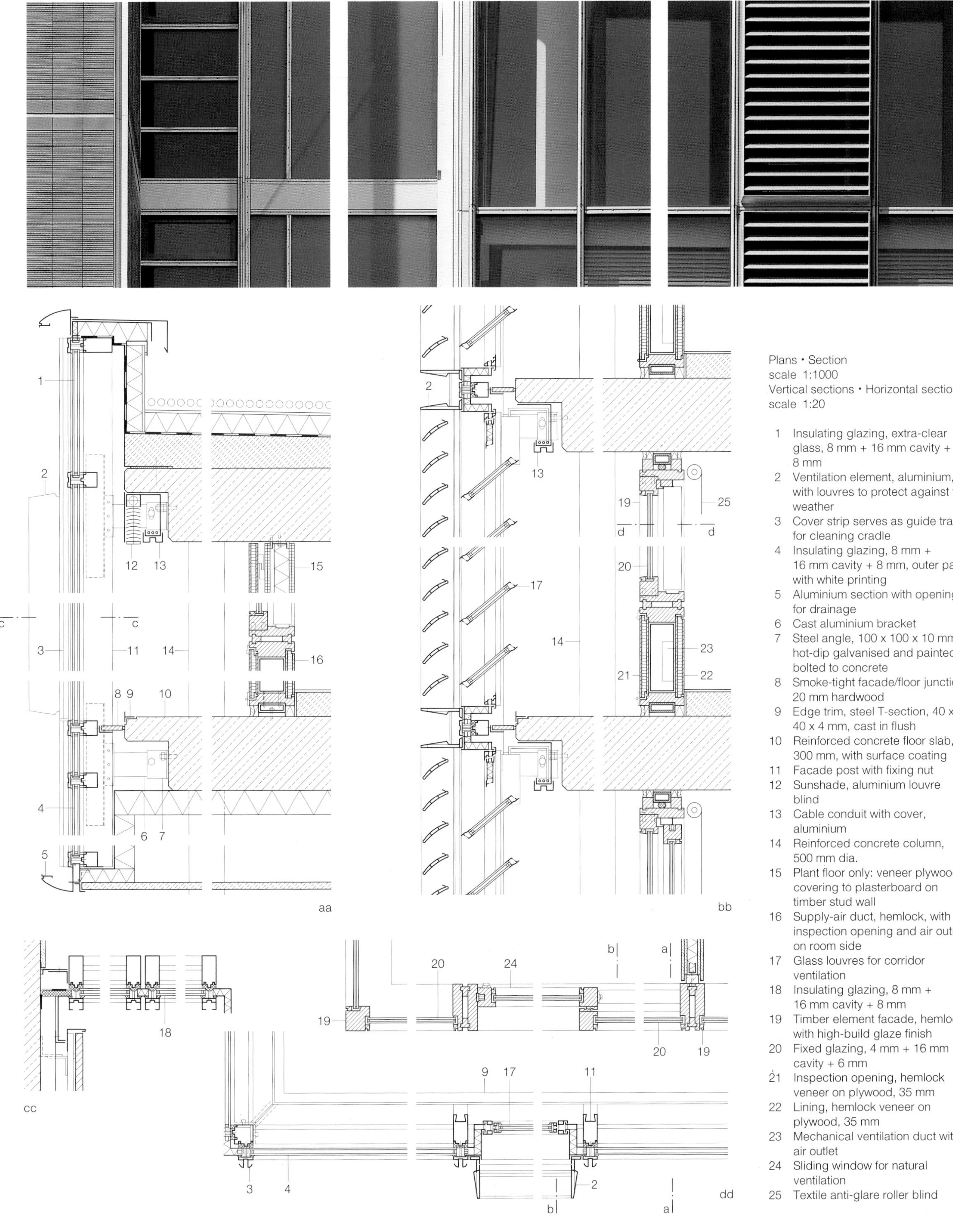

Plans • Section
scale 1:1000
Vertical sections • Horizontal sections
scale 1:20

1 Insulating glazing, extra-clear glass, 8 mm + 16 mm cavity + 8 mm
2 Ventilation element, aluminium, with louvres to protect against the weather
3 Cover strip serves as guide track for cleaning cradle
4 Insulating glazing, 8 mm + 16 mm cavity + 8 mm, outer pane with white printing
5 Aluminium section with opening for drainage
6 Cast aluminium bracket
7 Steel angle, 100 x 100 x 10 mm, hot-dip galvanised and painted, bolted to concrete
8 Smoke-tight facade/floor junction, 20 mm hardwood
9 Edge trim, steel T-section, 40 x 40 x 4 mm, cast in flush
10 Reinforced concrete floor slab, 300 mm, with surface coating
11 Facade post with fixing nut
12 Sunshade, aluminium louvre blind
13 Cable conduit with cover, aluminium
14 Reinforced concrete column, 500 mm dia.
15 Plant floor only: veneer plywood covering to plasterboard on timber stud wall
16 Supply-air duct, hemlock, with inspection opening and air outlet on room side
17 Glass louvres for corridor ventilation
18 Insulating glazing, 8 mm + 16 mm cavity + 8 mm
19 Timber element facade, hemlock, with high-build glaze finish
20 Fixed glazing, 4 mm + 16 mm cavity + 6 mm
21 Inspection opening, hemlock veneer on plywood, 35 mm
22 Lining, hemlock veneer on plywood, 35 mm
23 Mechanical ventilation duct with air outlet
24 Sliding window for natural ventilation
25 Textile anti-glare roller blind

RWE headquarters

Essen, D, 1997

Architects:
Ingenhoven Overdiek Kahlen and Partner, Düsseldorf

db 04/1997
Fassadentechnik 05/1997, 06/1997, 01/1998
Schittich, Christian et al.: Glass Construction Manual, Munich/Basel, 1998
Briegleb, Till (ed.): Hochhaus RWE AG Essen. Basel/Berlin/Boston 2000

1 Safety barrier, toughened safety glass, 12 mm extra-clear glass, with tubular aluminium handrail, 100 mm dia.
2 Grating over drainage channel
3 Sheet metal fascia panel
4 Post to two-storey rooftop terrace glazing, 50 x 280 mm aluminium tube, stove-enamelled
5 Metal grating
6 Sheet metal gutter, 4 mm, heated, gutter outlets positioned on facade grid lines above suspended ceiling
7 Facade cavity ventilated through perforated sheet aluminium in alternate bays (neighbouring bay closed), 4 mm, anodised, natural colour
8 Sunshade, aluminium louvres
9 Textile anti-glare roller blind

- Height of building = 127 m, diameter = 32 m
- Reinforced concrete structure
- Storey-height glazing to optimise daylight utilisation
- 8.40 m high facade at ground floor level, extra-clear glass with countersunk screw fixings, insulating glazing with toughened safety glass outside and laminated safety glass inside, supply air routed via facade posts made from aluminium tubes
- Centrally controlled sunshades (in facade cavity) and anti-glare blinds (in rooms)
- Standard floor with prefabricated double-window facade for natural ventilation, 1970 x 3591 mm
- Alternate fixed and sliding (manually operated) glazed elements
- Multi-functional ventilation element at floor level with air inlets and outlets offset laterally

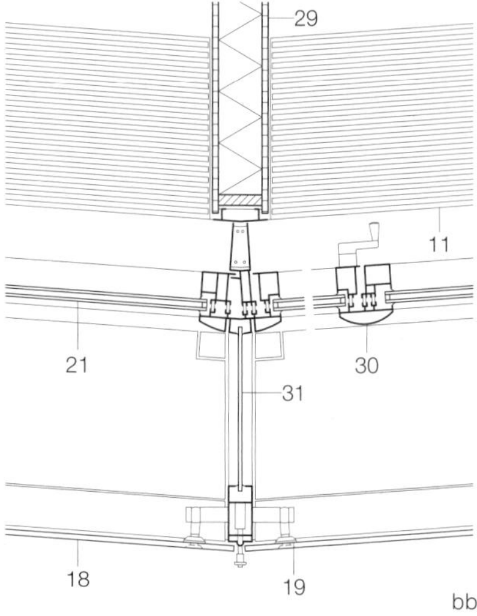

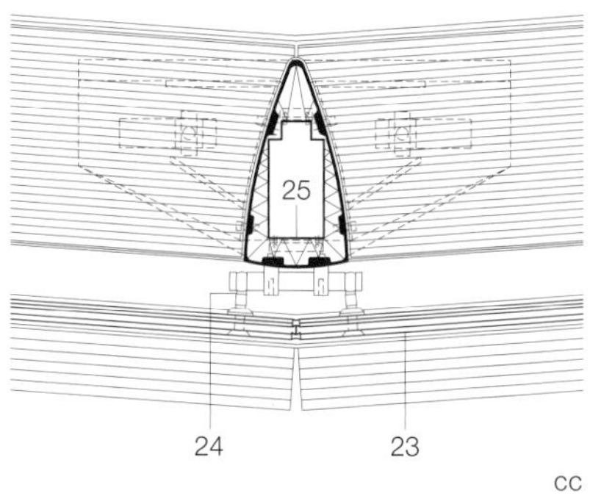

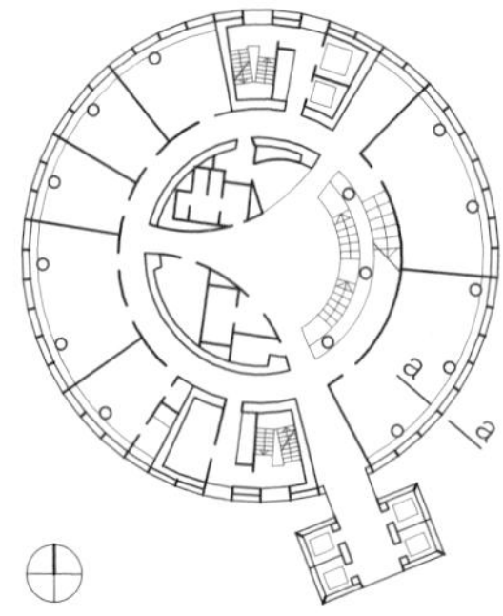

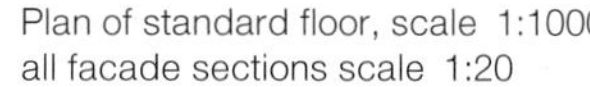

Plan of standard floor, scale 1:1000
all facade sections scale 1:20

Horizontal section showing junction with partition
Horizontal section through ground floor
Vertical section through entrance lobby at ground floor and upper facade junction

10 Multi-functional ceiling element, stove-enamelled sheet metal, partly perforated
11 Underfloor convector
12 Hinged, upward-opening plain sheet aluminium (perforated in neighbouring bay), 4 mm, anodised, natural colour
13 Walkway for cleaning and inspection
14 Butt joint for erection purposes
15 Fixing for travelling cradle
16 Horizontal ventilation slit with aluminium air flow guide louvres, anodised, natural colour
17 EPDM gasket
18 Outer facade, 10 mm toughened safety glass, extra-clear glass
19 Stainless steel countersunk screw fixing
20 Facade post, aluminium section, 50 x 120 mm
21 Inner facade, storey-height heat-absorbing glass, extra-clear glass in aluminium frame
22 Silicone joint on backer rod
23 Insulating glazing, 10 mm toughened safety glass + 14 mm cavity +12 mm laminated safety glass
24 Stainless steel countersunk screw fixing
25 Aluminium facade post
26 Metal grating
27 Adjustable base to post
28 Aluminium glazing bar
29 Prefabricated office partition, 175 mm, perforated beech panels, matt finish
30 Sliding door element in alternate bays, with crank handle for opening
31 Facade divider, toughened extra-clear safety glass

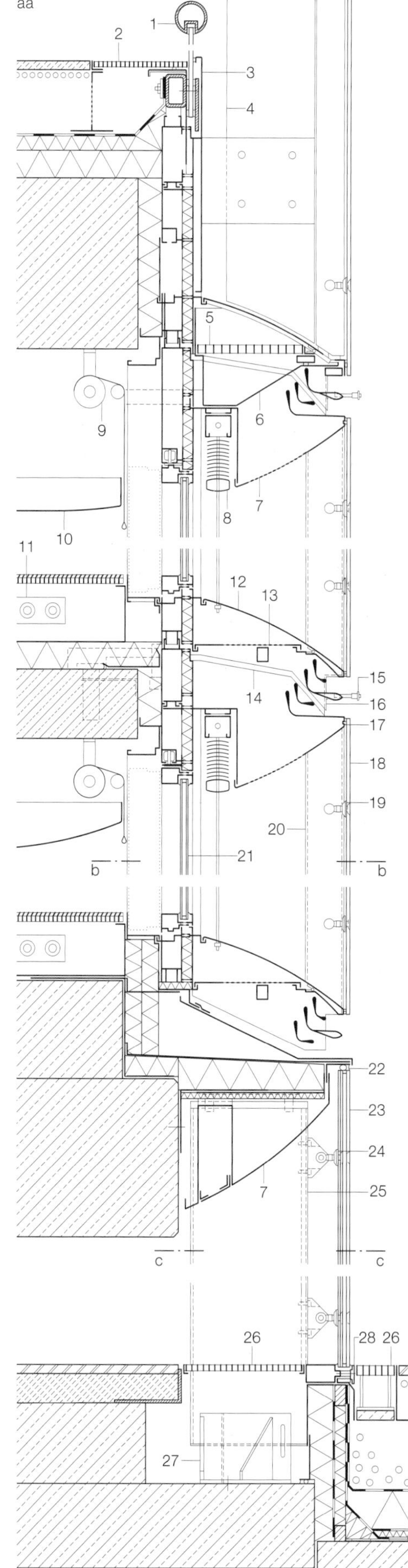

B 2.2 Manipulators

This section describes the components of the building by which the nature and magnitude of internal and external influencing factors – plus their interaction – can be influenced. On the one hand, where the external wall is a plain, unbroken surface, it acts as a divider between the interior and exterior climates. But on the other hand, its colour, its materials, its dimensions and its proportions can turn it into a means of intermediate storage for flows of energy between inside and outside, in both directions.

When openings are also included, permeable for light, heat, air and views, these are the criteria that determine the quality of the interior conditions.

The interplay between the external climate (weather, day/night rhythm, seasons, etc.) and the internal variables (heat sources, constant or changing moisture levels, etc.) therefore results in conditions within the building that are generally very different to the extreme conditions of the external climate and come rather closer to the level of comfort desirable in the interior.

Depending on demands and requirements, the characteristics of the openings, which are the preferred means of enabling an exchange of air, light, heat and moisture, can be intentionally designed to be variable. Increasing or decreasing the permeability hence becomes a control measure. As a result, we are given the opportunity to manipulate the internal climate through the operation of appropriate, variable components.

Doors and windows are the simplest and best-known forms of such "manipulators" [1]. Their changing degree of openness/closure and the materials of their construction have had a fundamental influence on the interior climate and the appearance of the facade since time immemorial.

So it stands to reason that effects like the greenhouse effect (a temperature rise in the interior due to solar energy entering via transparent areas in the building envelope, using the natural solar radiation in temperature ranges much higher than the temperature of the air outside) are just as achievable as they are avoidable. In the case of an undesirable heat gain, this is can be avoided by operating the sunshading devices accordingly. At the same time, this temporary protection against the heat – as with means of darkening the interior – is one way the user can have a direct influence on the thermal and lighting conditions in the building, with the option of regulating the changes as required at any time.
Manipulators have become increasingly important in conjunction with the conscious use of environmental energy sources, especially solar energy. Depending on internal requirements and the external climatic conditions, manual, regulatory corrections to the building envelope can adjust the internal climate without any appreciable inflow of energy from the outside. This is similar to the way in which we are accustomed to using clothing.

In addition, operated correctly, a drastic reduction in the other means employed to influence the interior climate is the logical and desirable consequence. As a rule, the "other means" are technical systems for heating, cooling, ventilating, lighting, and so on.

The ongoing development of these systems in the building envelope, which are already available, is – chiefly for architects – an urgent and worthwhile task, owing to the interplay involved with the entire building–energy balance. The role of the architect has always been defined as the person responsible for the overall composition and hence also for the overall optimisation of a structure, including the proper integration of the relevant subsystems.

Light-permeable components (windows)
In the past, besides glass, other materials were also used as an infill for light-permeable window areas: alabaster, marble, animal horn and skins, canvas, paper, and so on.
It was the Romans who, by using glass, first turned the window opening into a technologically developed part of the building. However, up until the eleventh or twelfth century a glazed window opening was the exception.

The first translucent or transparent windows were generally immovable items. Although side-hung opening lights have been known from ancient times, they are regarded as an invention of the Middle Ages. Sliding windows in which the opening light moves horizontally and parallel to the plane of the window first appeared in the thirteenth century.

Non-light-permeable components
The simplest form of non-light-permeable infill to a window opening is the shutter. Throughout history, shutters of wood, stone and iron have been used as an infill to the window opening, sometimes to provide additional protection (fig. B 2.2.3). The different types are distinguished by the ways they are fitted and used [2]:

- loose shutters: panel-type screening arrangements clamped in position as required
- folding shutters: fitted above or below the window by means of hinges, first used in the twelfth century
- hinged shutters: fixed to hinges at the side of the window, used since ancient times
- horizontal sliding shutters: fitted internally or externally in a frame, mostly for smaller window openings, first used in classical Greek architecture
- vertical sliding shutters: fitted above or below the window and usually let into the facade cladding, found primarily in eastern Switzerland in the fifteenth to eighteenth centuries [3].

B 2.2.1 Apartment block, Square Mozart, Paris (F), 1954, Jean Prouvé

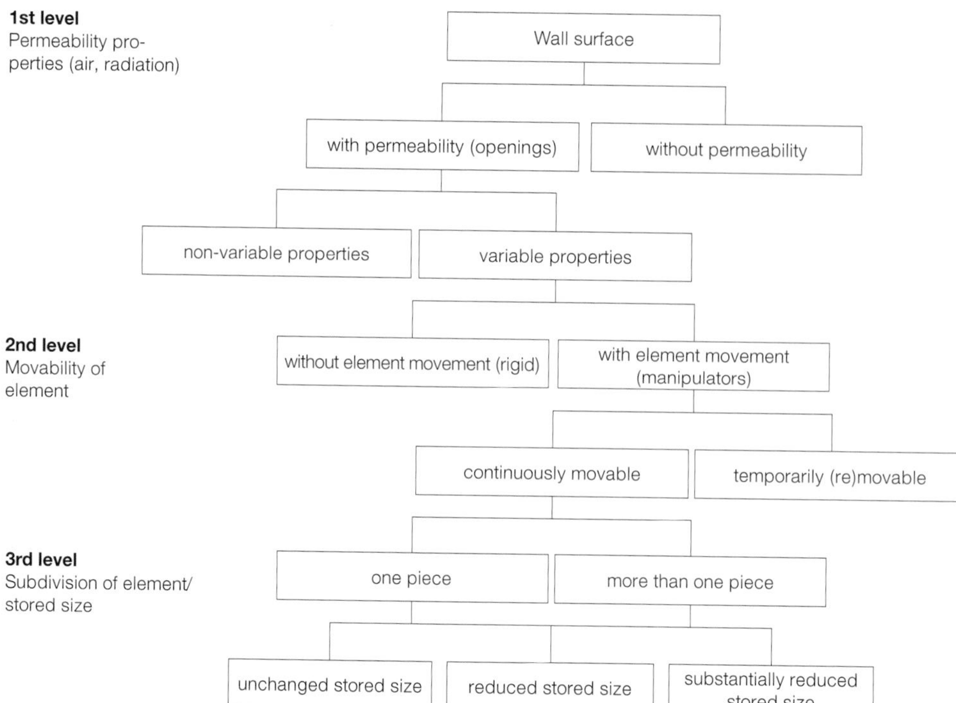

B 2.2.2

B 2.2.2 Classification of the term "manipulator"
B 2.2.3 Stone shutters, Torcello (I)
B 2.2.4 Facade openings with side-hung shutters and perforated tympana for breaking up the incoming daylight and providing finely controlled ventilation, Montagnana (I)
B 2.2.5 Light-permeable elements, traditional house, Takayama (J)
B 2.2.6 Combination of various manipulators at the "Palazzo Pitti", Florence (I)
B 2.2.7 Classification of common manipulators
The note above each diagram refers to the change in size of the movable elements when they are stored, i.e., are not providing a screening function.

B 2.2.3

B 2.2.4

B 2.2.5

It is only since the fifteenth century that shutters have been employed to supplement glazing. And only from the eighteenth century onwards do we find shutters being used almost exclusively in conjunction with transparent window infill materials [4].

Besides horizontal sliding and hinged shutters (the latter often referred to as folding shutters in colloquial speech), the following types are also common:

- roller shutters and louvre blinds: made from narrow transverse slats attached in a row to cords or chains, known since the eighteenth century
- louvre shutters: hinged shutters with infills of angled (sometimes movable) horizontal slats for regulating the amount of light and air admitted, known since the early eighteenth century, mainly in France

Like the general technological developments (which led to changes in the performance profiles of buildings), the functions of the window and the elements positioned in front of the openings in the building envelope are diverse and complex.

There has been an increase in the variety of movement mechanisms for manipulators in recent years. In this context we are also seeing a tendency among manufacturers to return to a wider range of movement mechanisms for windows as well. These represent alternatives to the tilt-and-turn windows ubiquitous throughout Germany, and which are not without their problems when looked at in terms of heating energy consumption criteria.

Classification

We shall attempt to classify the great diversity of well-known varieties in this section. This might also serve as inspiration for new functional, geometrical and technical combinations.

When classifying types of manipulators, we need to consider them on three different levels:

- permeability of the elements
- movability of the elements
- subdivision and "storage" of the elements (change in volume/size)

Permeability
We can distinguish areas according to their permeability to air, light, heat and moisture, or their lack (or virtual lack) of permeability.
The properties with respect to permeability can be:

- non-variable
- variable

The nature and extent of the permeability govern the function of an area. If the functional performance profile of an area is to assume different states, the permeability of the area must be variable.

B 2.2.6

Movability
Elements with variable properties can be subdivided into:

- stationary elements
- non-stationary elements

The stationary elements include, for example, thermotropic coatings and gasochromic or electrochromic glazing.

The non-stationary elements can be further defined:

- moveable (in the sense of: can be moved)
- moving (in the sense of: designed to move)

There are therefore two further criteria for defining non-stationary elements [5]:

- (temporarily) (re)movable, e.g., storm windows
- (continuously or potentially continuously) moving

The term "manipulator" applies to areas of a wall or facade with variable properties and non-stationary elements. Generally, manipulators are (continuously or potentially continuously) moving components.

Subdivision
The change in size of non-stationary elements (stored size) is crucial for the construction, function and appearance of the facade. We distinguish the possible changes to the dimensions as follows:

- no change
- reduced
- substantially reduced

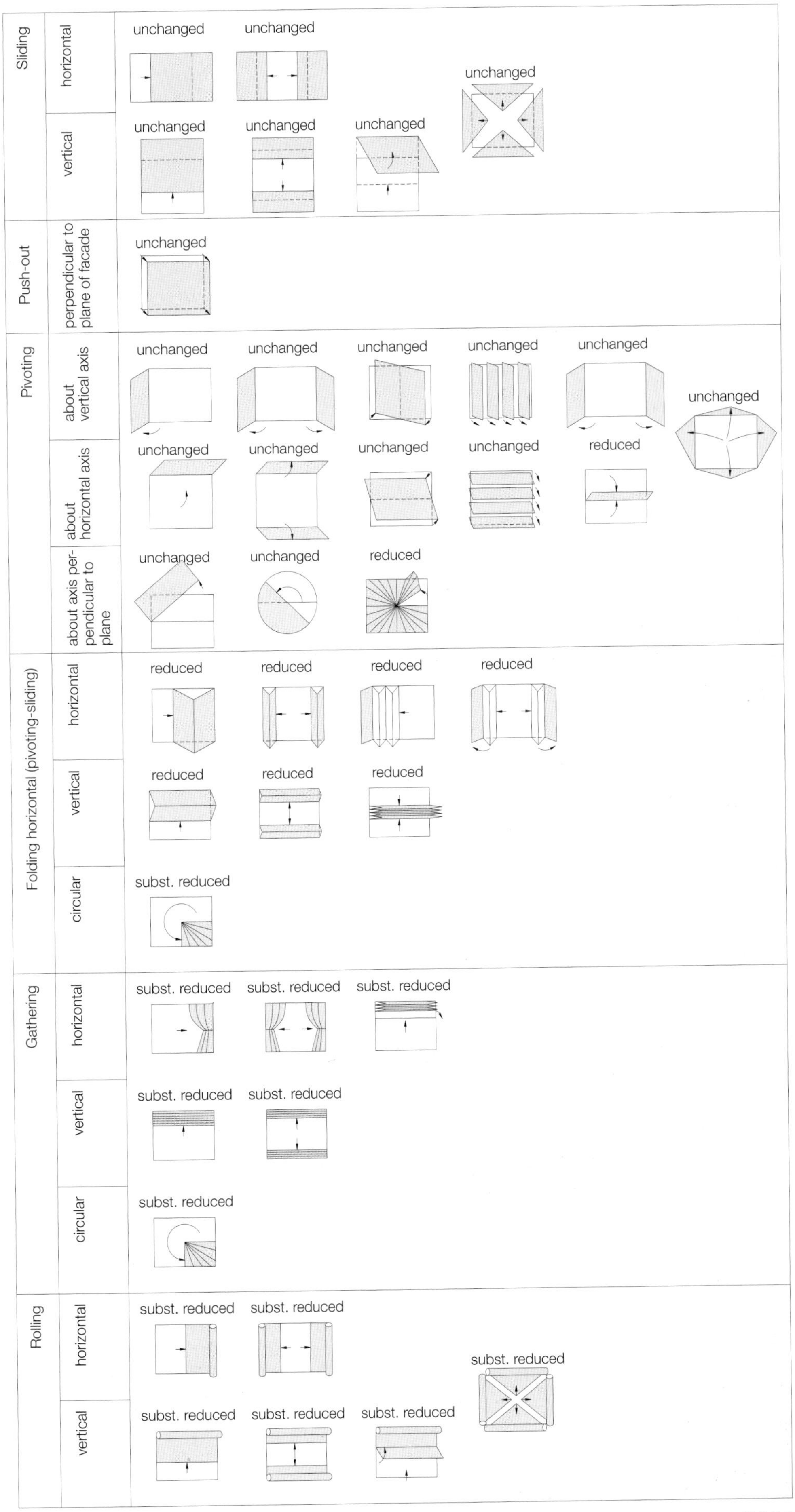

Sliding	horizontal	unchanged	unchanged				unchanged
	vertical	unchanged	unchanged	unchanged			
Push-out	perpendicular to plane of facade	unchanged					
Pivoting	about vertical axis	unchanged	unchanged	unchanged	unchanged	unchanged	unchanged
	about horizontal axis	unchanged	unchanged	unchanged	unchanged	reduced	
	about axis perpendicular to plane	unchanged	unchanged	reduced			
Folding horizontal (pivoting-sliding)	horizontal	reduced	reduced	reduced	reduced		
	vertical	reduced	reduced	reduced			
	circular	subst. reduced					
Gathering	horizontal	subst. reduced	subst. reduced	subst. reduced			
	vertical	subst. reduced	subst. reduced				
	circular	subst. reduced					
Rolling	horizontal	subst. reduced	subst. reduced				subst. reduced
	vertical	subst. reduced	subst. reduced	subst. reduced			

B 2.2.7

B 2.2.8

Normally, the manipulator consists of one or more parts, and each part can be further subdivided into two or more parts. This fact, combined with the type of movement, results in the different states, and hence the performance profiles of the areas with variable properties. The stored size has a direct influence on the operation. Besides the functional properties, the stored size is the other factor responsible for the constructional and architectural aspects.

Further distinguishing features

We can make further distinctions on a fourth level, e.g.:

- position in relation to the climate boundary: outside and clear of the opening, outside and integrated into the plane of the window, inside
- position in relation to the opening: above, central, below, to the side, on one or more sides

These arrangements have a direct effect on functional issues. For example, an anti-glare device mounted above the opening can prevent daylight reaching the far corners of a room. A sunshading device fitted internally can lead to undesirable heat gains.

Type and direction of movement

The fundamental types of movement for elements in the facade were classified in chapter A 2.2 "Edges, openings" with respect to the movement mechanisms for windows. The types of movement employed for manipulators are often a combination of various movement principles. Fig. B 2.2.7 illustrates the variety of movement options for manipulators, along with the direction of movement [6]. This overview is based on the types of movement met with in practice, but does not claim to be exhaustive.

If a system consists of a combination of various movable elements, the movement mechanisms employed is highly significant. The elements can then be moved independently only when they do not hinder each other [7]. The various contributory aspects place high demands on a facade system with respect to the integration of the elements to be accommodated. Efficient regulation of the interior conditions is then possible when the components for controlling lighting, acoustic and thermal variables can be operated independently of each other – as is indeed the case with their historical predecessors.

Control of manipulators

Manipulators may be operated manually or mechanically. Manual operation is carried out by the occupants of the building to suit their requirements. Depending on the movement mechanisms and the effort needed, it is possible to operate several manipulators simultaneously.

When a mechanical drive is provided, manipulators can be controlled automatically, and this type of operation must be taken into account in developing the energy concept of the building. In such situations, the occupants can usually carry out individual adjustments within a certain range.
As a result of the combination of various elements, regulating the permeability of the building envelope may also contribute to optimising both the comfort of occupants and the consumption of energy.

States of manipulators

Apart from the closed and the open state, manipulators are able to assume intermediate states. Depending on the type of movement, it is therefore possible to effect a fine control of the permeability property. To appreciate this characteristic we need only to think of hinged shutters and louvre blinds; both of these can be used to regulate the amount of incoming light. In the case of hinged shutters, regulating the relationship with the outside world is possible only to a limited extent, whereas with louvre blinds the amount of incoming light can be finely controlled and hence the associated view out also controlled by the angle of the louvres.
Furthermore, it is important to remember that the window can also be used for ventilation. A sliding movement allows the linear gap between the shutters to be altered at will and easily adjusted to ensure the right degree of ventilation. With a pivoting movement on the other hand, the possibilities in this respect are more limited (please refer to chapter A 2.2 "Edges, openings").

Applications

Virtually all the materials employed in the construction industry are used in the multitude of manipulator forms. The infill materials may be solid (panels, sheets, fabrics with and without frames, etc.) or semi-solid (vertical/horizontal louvres, adjustable or rigid, perforated sheet metal, etc.).

In addition, it is possible to achieve a wealth of combinations of individual components and different positions with respect to the climate boundary:

- horizontal sliding shutters
- vertical sliding shutters
- external side-hung shutters
- internal side-hung shutters
- top- or bottom-hung shutters
- horizontal shutters folding to top or bottom
- vertical shutters folding to side(s) (with or without ventilation openings)
- horizontal shutters folding to middle
- horizontal louvres
- push-out window
- side-hung opening lights
- horizontal sliding windows
- vertical sliding windows
- folding windows
- gathered awnings
- roller awnings

Through their movement, the use of manipulators leads to changes in the appearance. The function of varying the permeability thus has a crucial effect on the architecture of the facade.

B 2.2.8 St Mark's Square, Venice (I)
B 2.2.9–16 Building envelopes varied functionally and aesthetically by means of manipulators

B 2.2.9

B 2.2.10

Notes

[1] The term "manipulator" for movable elements in or on the building envelope stems from a dissertation – "Verfahren zur Beurteilung kinetischer Manipulatoren im Bereich der Gebäudehülle als Maßnahme zur Regulierung des Gebäudeklimas" – by Waldemar Jaensch, supervised by Thomas Herzog, Kassel, 1981, p. 28.
The term "manipulator" is derived from the Latin word *manus* (= hand) and the Latin suffix *-ator* (used for forming agent nouns) via the French *manipuler* (= to operate, handle, arrange). A manipulator is a device used for handling with dexterity, managing, working or treating by manual or mechanical means, especially to gain an advantage (*The New Shorter Oxford Dictionary*, Oxford, 1993).
[2] *Reallexikon zur Deutschen Kunstgeschichte*, vols 7 and 8, Munich, 1981.
[3] Herzog, Thomas; Natterer, Julius (ed.): *Gebäudehüllen aus Glas und Holz*, Lausanne, 1984.
[4] Gerner, Manfred; Gärtner, Dieter: *Historische Fenster*, Stuttgart, 1996, p. 68.
[5] Krippner, Roland: Entwicklung beweglicher Manipulatoren im Bereich der Außenwände mit wärmedämmenden und weiteren Funktionen; in: ISOTEG final report, Munich Technical University, Chair of Building Technology, 2001 (unpublished manuscript).
[6] An extension of the illustrations in: ibid.
[7] This chapter includes passages from an ongoing dissertation by Daniel Westenberger, currently being prepared at Munich Technical University, Chair of Building Technology. The work deals with the use of vertical sliding mechanisms for windows and other movable components at openings in the facade, with special emphasis on the resulting combination options.

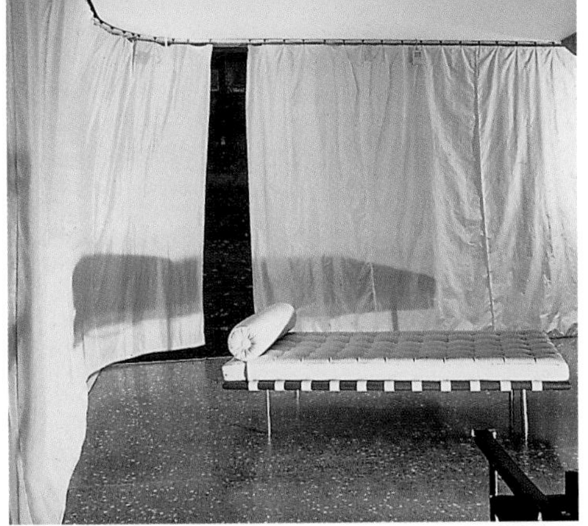
B 2.2.11

B 2.2.12

B 2.2.13

B 2.2.14

B 2.2.15

B 2.2.16

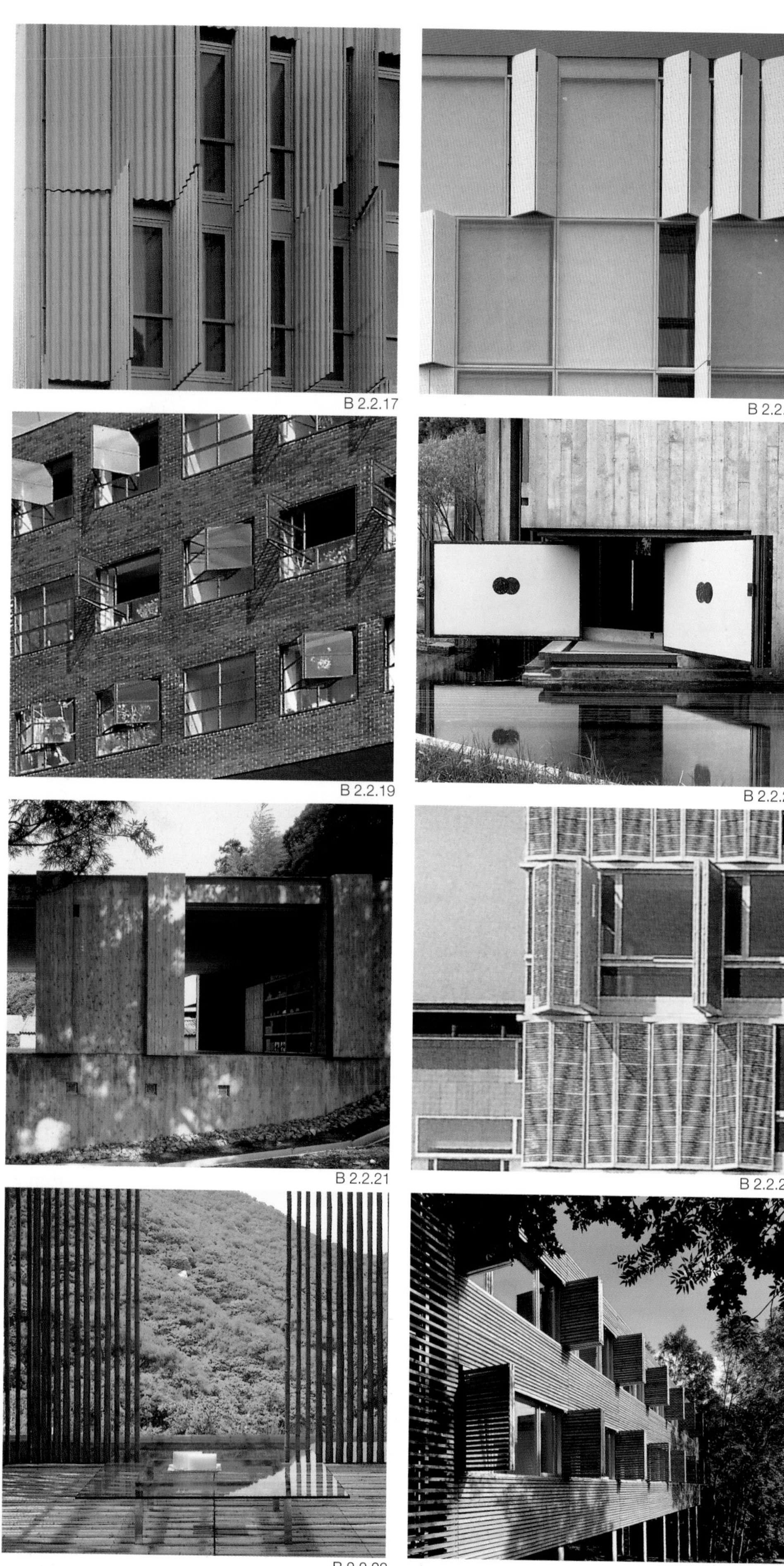

B 2.2.17

B 2.2.18

B 2.2.19

B 2.2.20

B 2.2.21

B 2.2.22

B 2.2.23

B 2.2.24

B 2.2.17–32 Building envelopes varied functionally and aesthetically by means of manipulators

B 2.2.25

B 2.2.26

B 2.2.27

B 2.2.28

B 2.2.29

B 2.2.30

B 2.2.31

B 2.2.32

Institut du Monde Arabe

Paris, F, 1987

Architect:
Jean Nouvel, Paris
Assistants:
Gilbert Lezenes, Piere Soria,
Architecture Studio

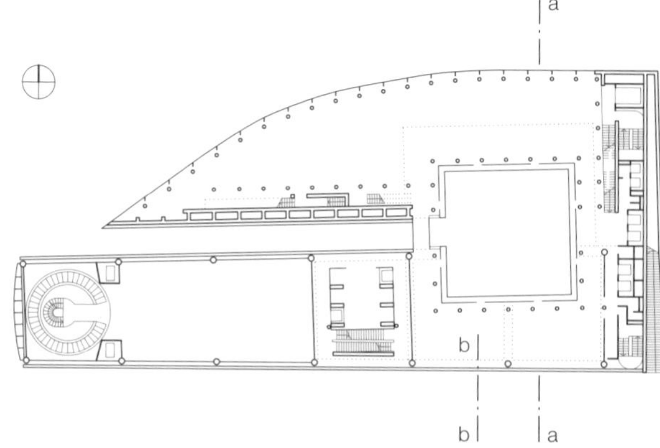

📖 l'ARCA 15, 1988
l'architecture d'aujourd'hui 12/1998
Architectural Review 1088, 1987
and 1113, 1989
El Croquis 65–66, 1994: Jean Nouvel
Progressive architecture 09/1995

- Multiple openings operating on the principle of a camera shutter control the amount of daylight entering the interior
- Mechanisms and controls are left exposed
- Geometric layout of repetitive ornamentation is a reference to traditional motifs of Islamic architecture (*mushrabeyeh* = ornate window gratings)
- Mechanism is delicate and requires considerable maintenance

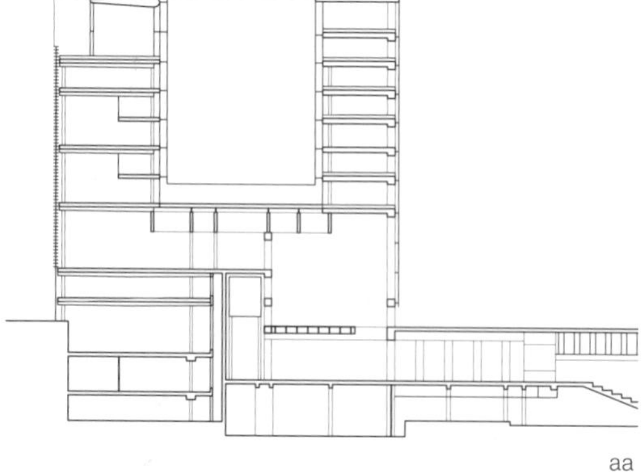

Plan of 4th floor • Section
scale 1:1000
Vertical section through shutter
Horizontal section through shutter
scale 1:5

1 EPDM joint
2 Seals interrupted to allow ventilation of void
3 Perforated infill
4 Toughened safety glass, 6 mm
5 Vent hole
6 Polyurethane thermal break
7 "Camera shutter" mechanism
8 Insulating glazing, 4 mm +12 mm cavity + 4 mm
9 Toughened safety glass, 8 mm
10 Facade support member

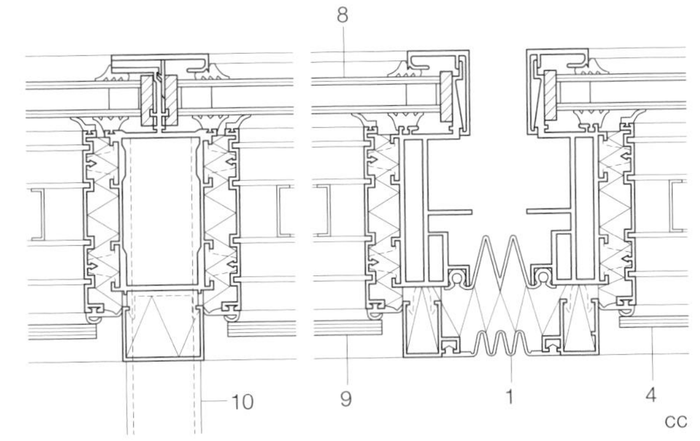

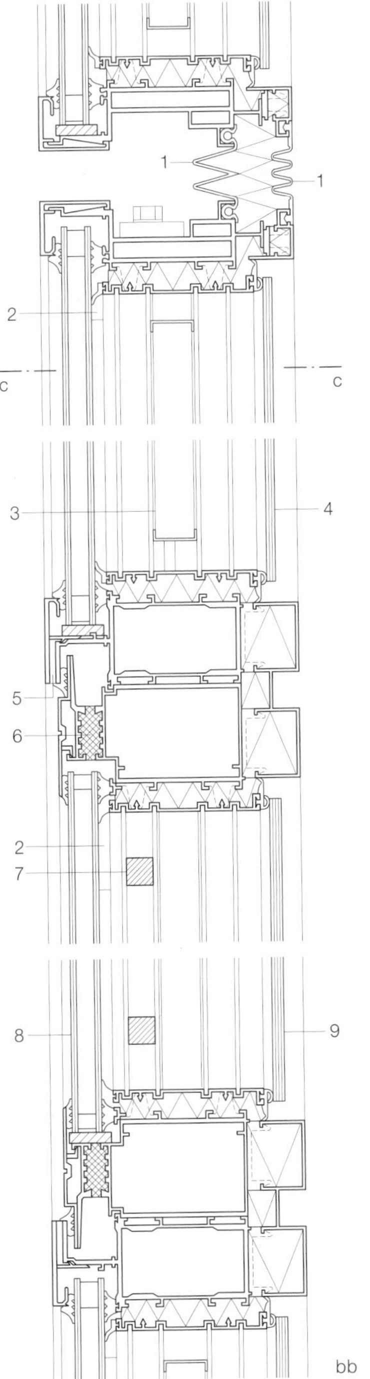

Factory building for Dial-Norm AG

Kirchberg, CH, 1971

Architect:
Fritz Haller, Solothurn
Fassadenplanung:
Hans Diehl, Neuenhof Baden

- Opening light in the form of a rotating semicircle
- "MAXI" steel building system (Fritz Haller)
- Minimum proportion of joints per unit surface area thanks to use of large facade panels
- Use of prefabricated panels resulted in short construction period
- Facade construction without direct metal connection between inside and outside

Plan • Section
scale 1:500
Horizontal section • Vertical section
scale 1:5

A Fixed light
B Butt joint between elements
C Opening light

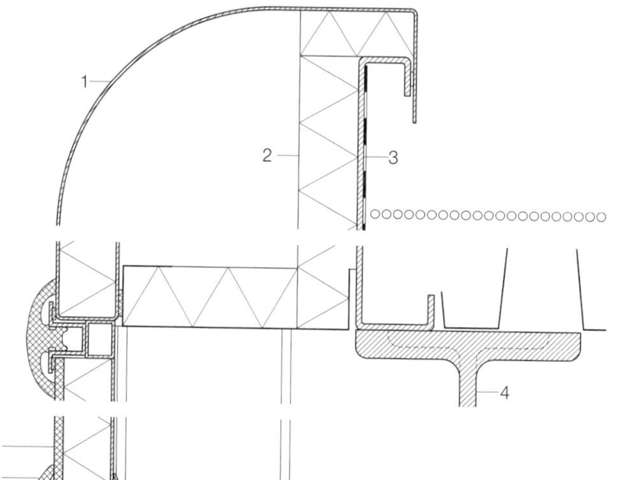

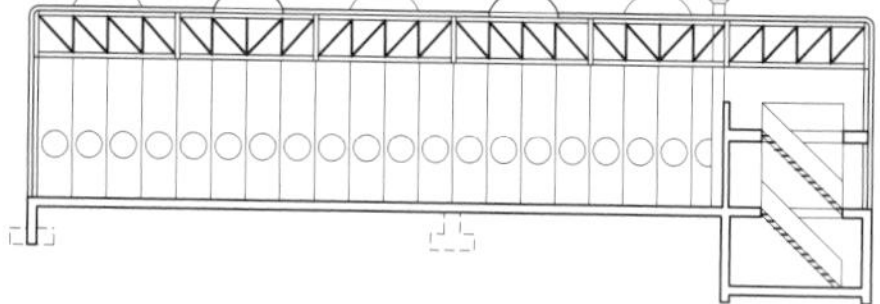

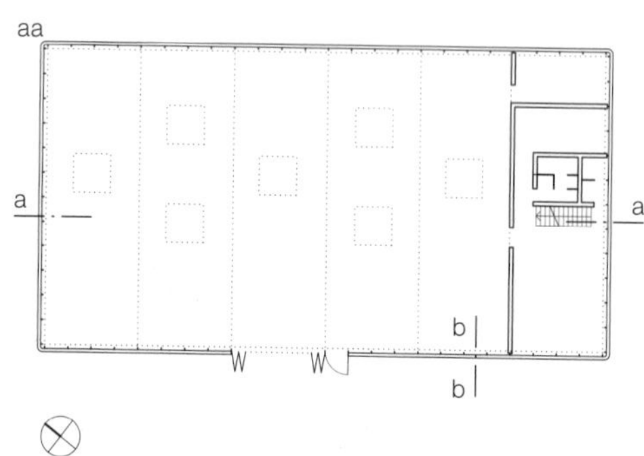

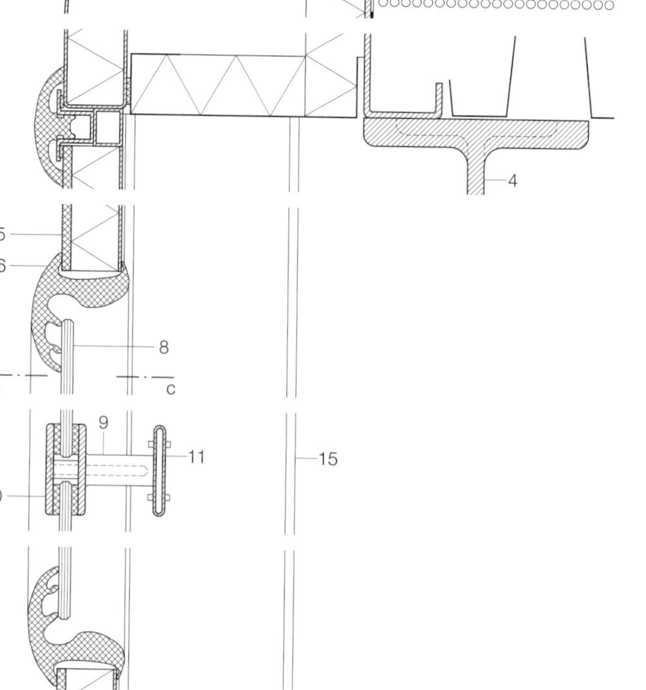

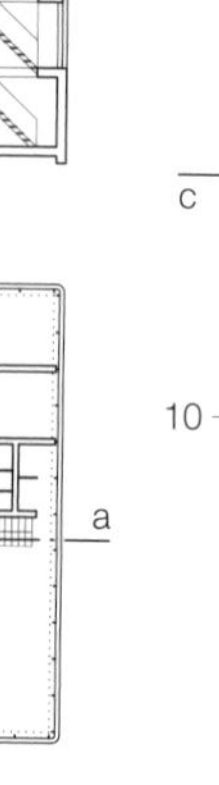

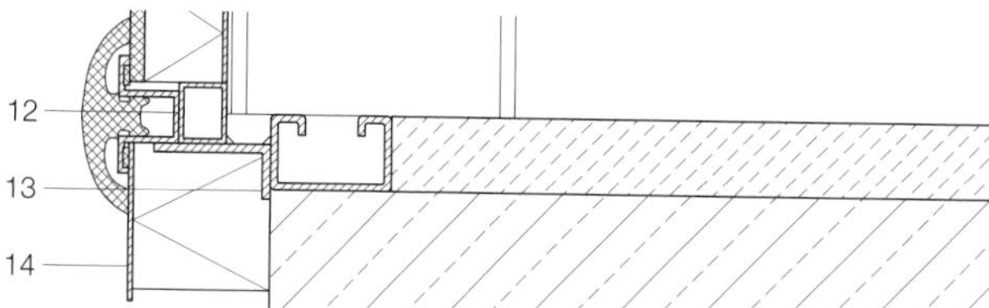

1 Sheet aluminium, 2 mm, r = 150 mm
2 Thermal insulation, 40 mm
3 Sheet steel, bent to suit, 3 mm
4 "MAXI" structural steelwork, IPE 400 or IPE 220 (on short side)
5 Facade construction: structural sandwich panel, 1 mm stove-enamelled sheet aluminium facings, synthetic core
PU foam thermal insulation
3 mm stove-enamelled sheet aluminium
6 EPDM split gasket, outer part
7 EPDM split gasket, inner part
8 Reflective solar-control glass, toughened, 8 mm
9 Central glass holder, special chromium-plated steel section
10 Chromium-plated steel disc, 60 mm dia.
11 Chromium-plated steel handle
12 Steel rectangular hollow section, 25 x 20 x 2 mm
13 Steel angle, 50 x 20 x 3 mm
14 Sheet aluminium
15 Steel stanchion, IPE 120
16 Aluminium clip
17 EPDM capping strip

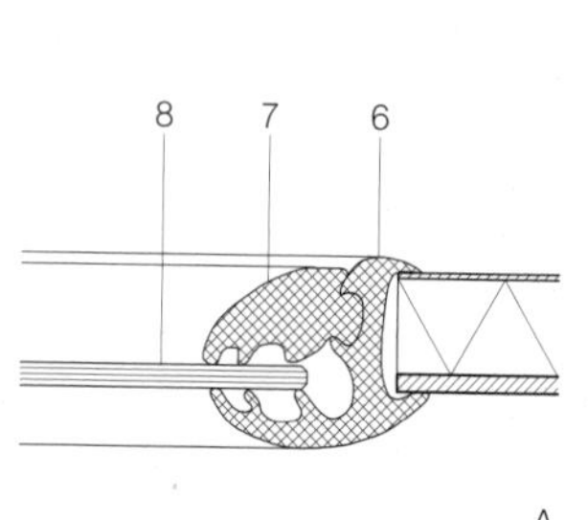

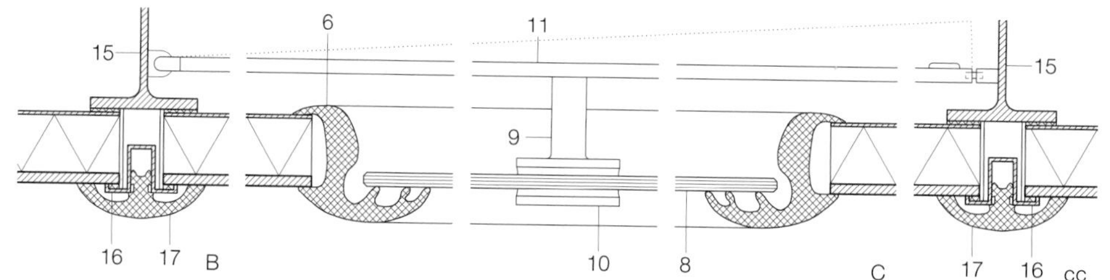

Nakagin Capsule Tower

Tokyo, J, 1972

Architects:
Kisho Kurokowa & Associates, Tokyo

📖 l'architecture d'aujourd'hui 06/2000
Kurokawa, Kisho: From Metabolism to symbiosis. London/New York 1992
Detail/jpn 33, 1972

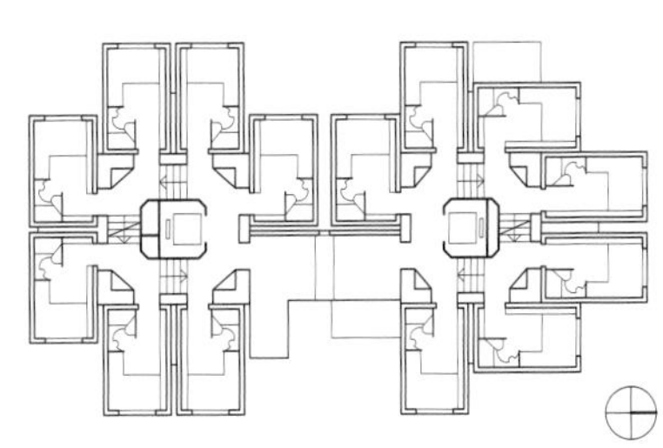

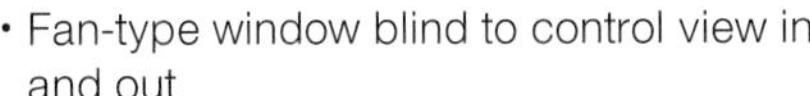

- Fan-type window blind to control view in and out
- Industrially prefabricated room modules (2.3 x 3.8 x 2.1 m) suspended from two concrete access towers
- Window diameter = 1.30 m

Plan, scale 1:500
Interior view • Details •
Vertical section, scale 1:5

1 Circular base plate, 140 mm dia., glued to glass
2 2-piece inner guide ring, polished brass, screwed to No. 4
3 Circular inner cover plate 120 dia. x 5 mm, screwed to No. 11
4 Divider, 1.2 mm sheet metal
5 Frame for No. 6, 2 mm sheet aluminium, bent to suit
6 Plastic-coated paper
7 Aluminium retainer with clip for No. 5
8 Outer guide rail
9 Fixing for No. 8, attached to window reveal
10 Threaded sleeve, 20 mm dia.
11 Fixed light, 6 mm toughened safety glass, 1300 mm dia.
12 Rubber gasket
13 Sheet aluminium, screwed to aluminium angle, 40 x 40 x 4 mm

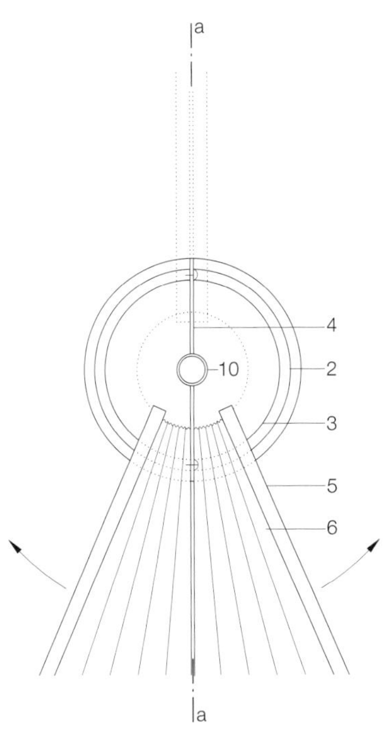

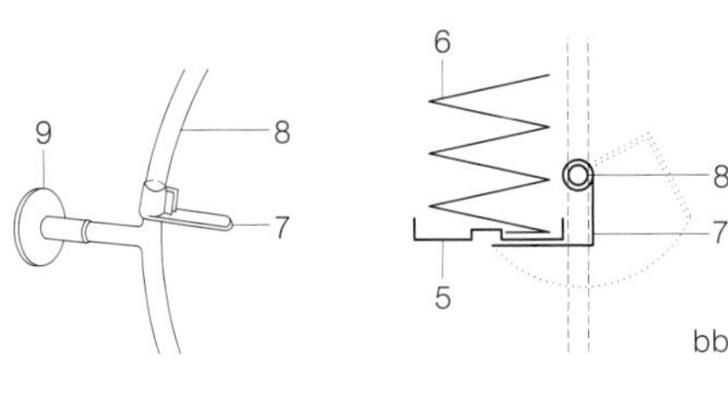

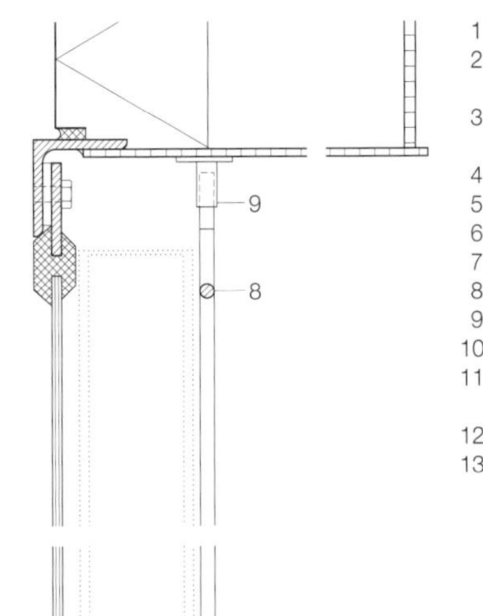

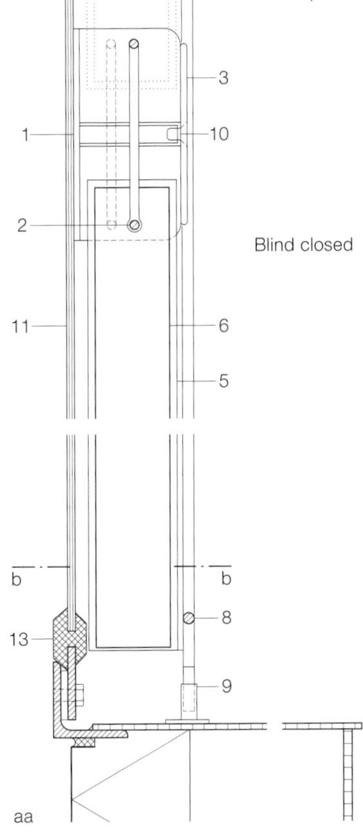

Mixed residential and commercial development

Munich, D, 1996

Architect:
Von Seidlein, Munich
Peter C. von Seidlein, Horst Fischer, Egon Konrad, Stephan Röhrl
Responsible for facade:Stephan Röhrl

Detail 03/1998
Von Seidlein, Peter C.: Zehn Bauten 1957–97. exhibition catalogue, Architekturgalerie Munich, 1997

- Horizontal louvre blinds fitted externally
- Large sliding elements provide a link between apartments and surroundings on south elevation
- Large vertical sliding windows in pitched roof (not shown here)
- Metal facade fixed to timber framing in front of reinforced concrete structure to avoid thermal bridges

Section • Plan of 1st floor
scale 1:750
Horizontal section • Vertical section
scale 1:20

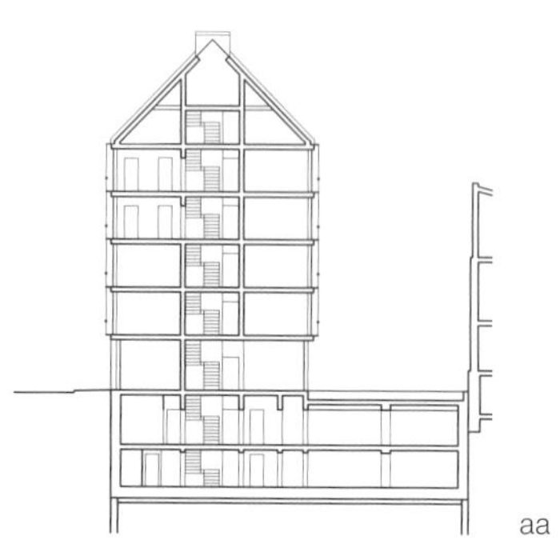

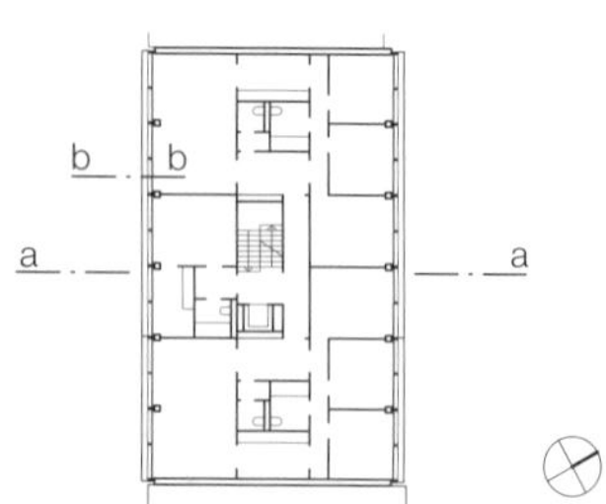

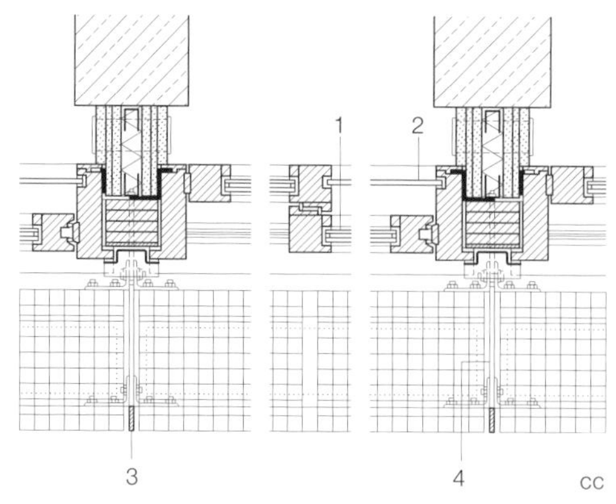

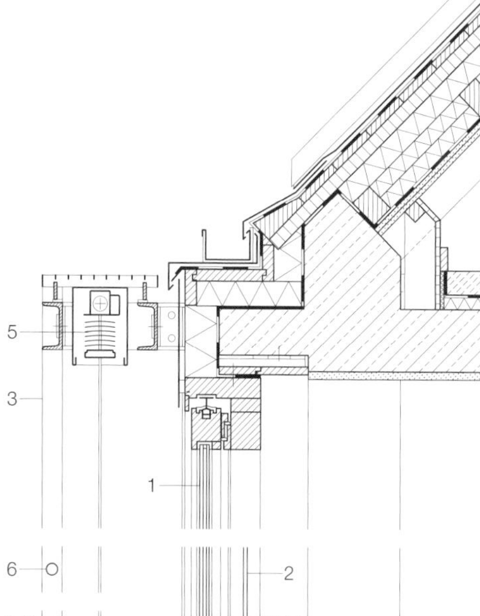

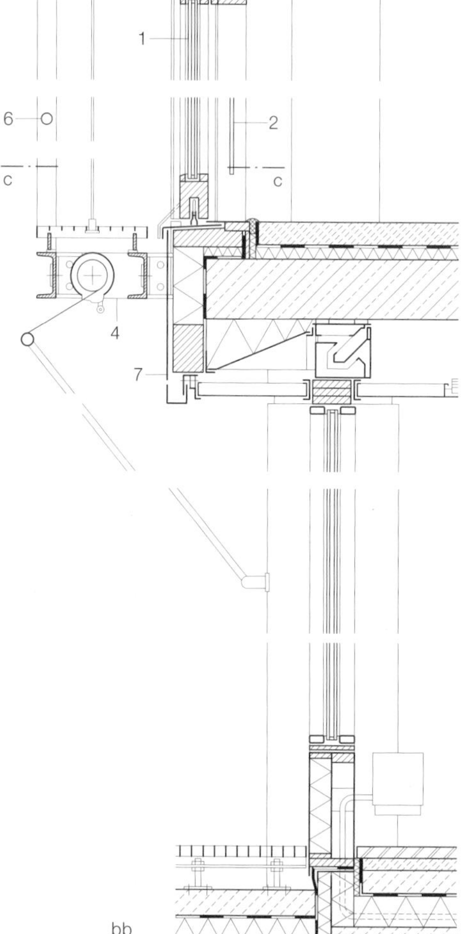

1 Horizontal sliding window: laminated niangon wood insulating glass, 10 mm laminated safety glass + 15 mm cavity + 4 mm float glass
2 Internal glazed spandrel panel, 10 mm toughened safety glass
3 Steel flat, 10 x 55 mm
4 Steel flat bracket, 10 x 120 mm, connected to reinforced concrete frame at floor level via 100 x 100 mm gluelam post
5 Aluminium louvre blind, plain louvres, cord guides with electric drive, 2 mm sheet aluminium housing
6 Handrail, steel tube, 31 dia. x 2.25 mm
7 Sheet aluminium, 3 mm

All steel parts zinc-sprayed and powder-coated.

Development centre

Ingolstadt, D, 1999

Architects:
Fink + Jocher, Munich
Structural engineer:Schittig, Ingolstadt

l'architecture d'aujourd'hui 07/2000
Bauwelt 08/1999
Detail 03/1999
Intelligente Architektur 11–12/2000
World architecture 07–08/2000

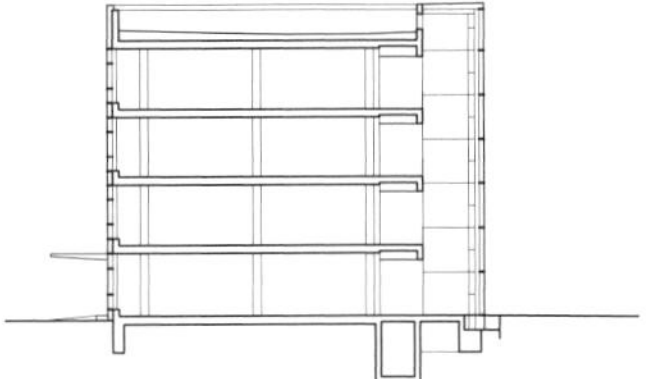

- Louvre blind fitted in cavity between panes (south facade)
- South orientation of building forms one aspect of the energy concept
- Facade continuous over four storeys

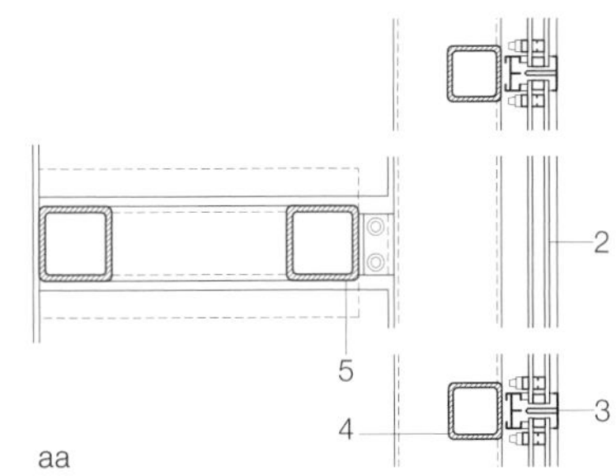

1 Sheet aluminium, 2 mm, bent to suit, rigid foam thermal insulation
2 Insulating glazing, 6 mm + 22 mm cavity + 5 mm, light-redirecting aluminium louvres fitted in cavity, b = 16 mm, stove-enamelled, white outside, silver-grey inside
3 Aluminium glazing bar
4 Post-and-rail construction, steel hollow sections, 90 x 90 mm and 180 x 100 mm, micaceous iron oxide finish
5 Vierendeel column, steel square hollow sections, 120 x 120 mm
6 Aluminium open grid flooring
7 Steel open grid flooring in angle frame
8 Supply-air flap:
2 mm sheet aluminium
40 mm rigid foam thermal insulation
2 mm sheet aluminium

Section, scale 1:750
Horizontal section • Vertical section, scale 1:20
Details, scale 1:5

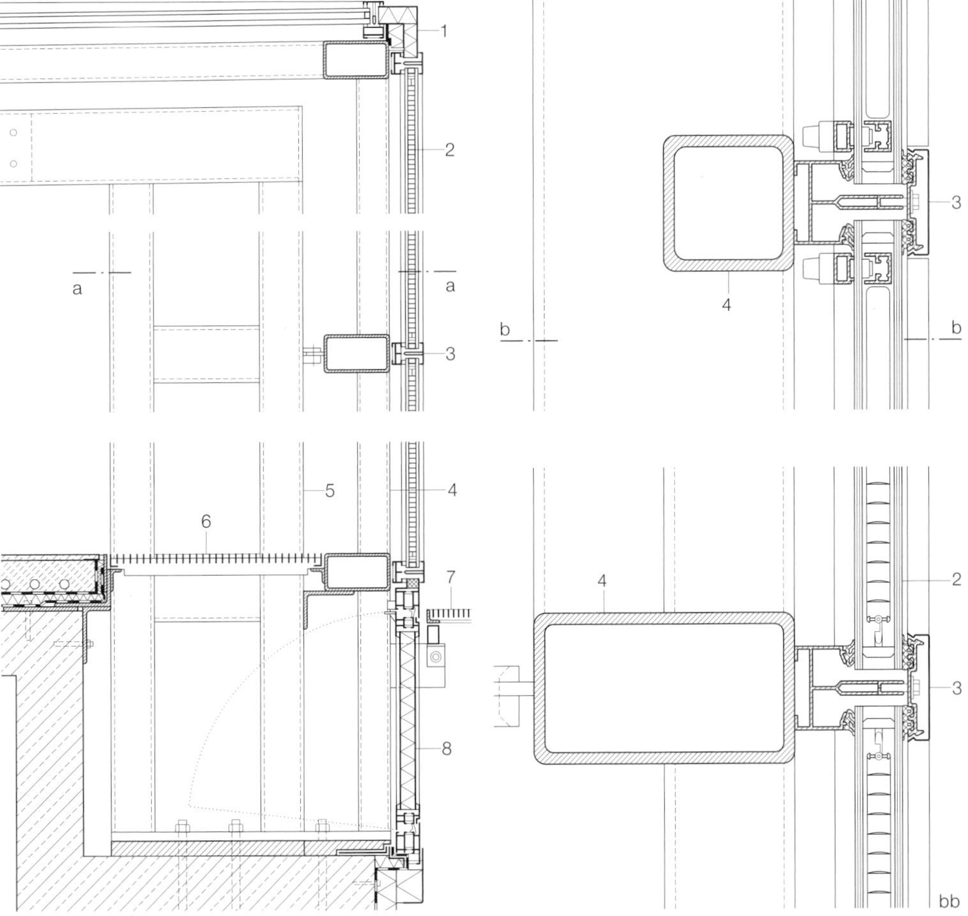

Museum of Paper

Shizuoka, J, 2002

Architect:
Shigeru Ban, Tokyo
Opening facade elements:
Bunka Shutter, Shinjuku-ku, Tokyo

Detail 07–08/2003
domus 03/2003

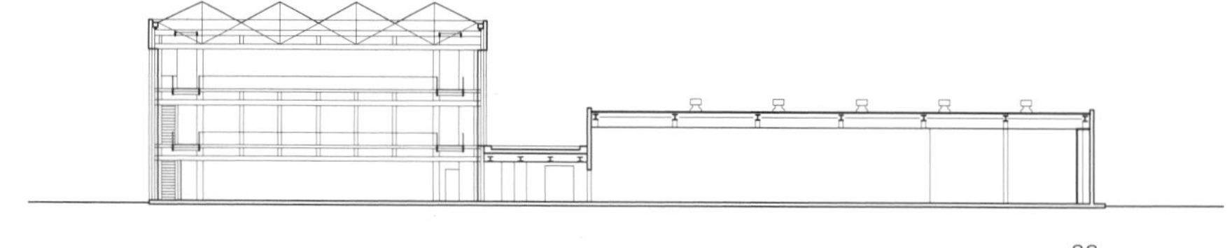

- Large roller shutter doors 10 m high on end elevations (east and west)
- Components on south side of museum can be folded upwards and outwards (max. 90°) to provide shade – *shitomido*, an element of traditional Japanese architecture
- Storey-height facade segments on south side of gallery can be slid outwards on cantilevering guide rails to create a canopy over the terrace in front of the building
- Translucent GFRP multiple-web sheets used in different ways

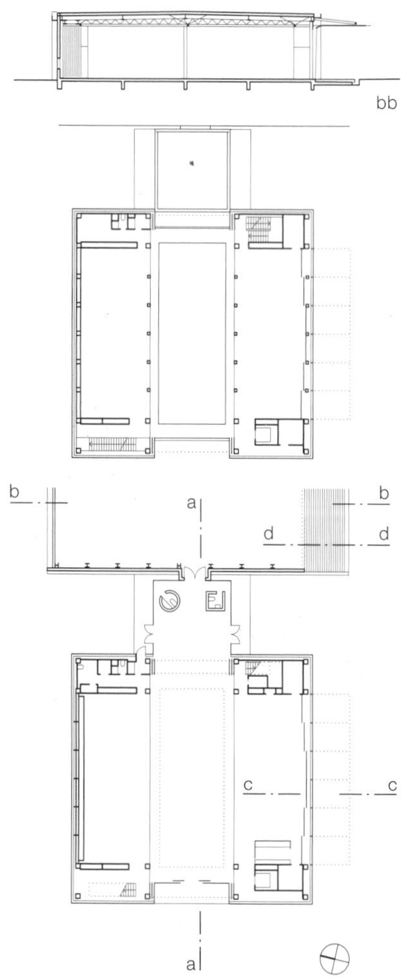

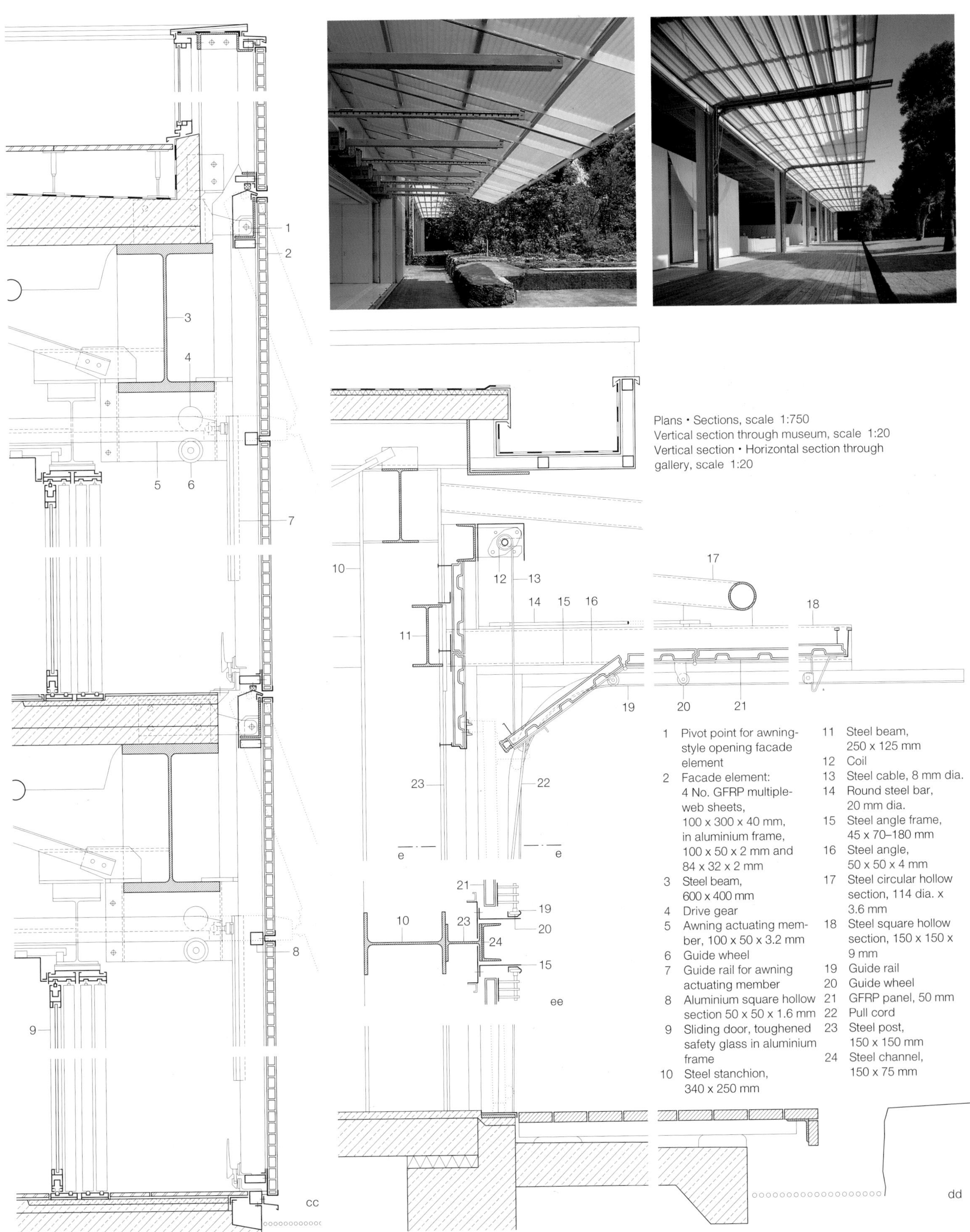

Plans • Sections, scale 1:750
Vertical section through museum, scale 1:20
Vertical section • Horizontal section through gallery, scale 1:20

1 Pivot point for awning-style opening facade element
2 Facade element: 4 No. GFRP multiple-web sheets, 100 x 300 x 40 mm, in aluminium frame, 100 x 50 x 2 mm and 84 x 32 x 2 mm
3 Steel beam, 600 x 400 mm
4 Drive gear
5 Awning actuating member, 100 x 50 x 3.2 mm
6 Guide wheel
7 Guide rail for awning actuating member
8 Aluminium square hollow section 50 x 50 x 1.6 mm
9 Sliding door, toughened safety glass in aluminium frame
10 Steel stanchion, 340 x 250 mm
11 Steel beam, 250 x 125 mm
12 Coil
13 Steel cable, 8 mm dia.
14 Round steel bar, 20 mm dia.
15 Steel angle frame, 45 x 70–180 mm
16 Steel angle, 50 x 50 x 4 mm
17 Steel circular hollow section, 114 dia. x 3.6 mm
18 Steel square hollow section, 150 x 150 x 9 mm
19 Guide rail
20 Guide wheel
21 GFRP panel, 50 mm
22 Pull cord
23 Steel post, 150 x 150 mm
24 Steel channel, 150 x 75 mm

Home for the elderly

Neuenbürg, D, 1995

Architects:
Mahler Günster Fuchs, Stuttgart
Structural engineer:
Wolfgang Beck, Dennach

Architectural Review 06/1997
Bauwelt 05/1997
Schunk, Eberhard et al.: Roof Construction Manual, Munich/Basel, 2002
Herzog, Thomas et al.: Timber Construction Manual, Munich/Basel, 2004

- Sliding wooden shutters
- Four identical separate buildings
- Reinforced concrete walls with thermally insulated, ventilated timber cladding
- Timber left untreated
- Solar collectors mounted on pitched roof beneath acrylic corrugated sheets
- Transparent roof covering allows view of timber roof structure underneath

Vertical sections • Horizontal sections, scale 1:5
A Large sliding shutter
B Small sliding shutter

1 Facade construction adjacent floors:
100 x 21 mm weatherboarding, divided up by vertical battens
22 mm ventilated cavity
water-repellent airtight membrane
80 mm thermal insulation
reinforced concrete
2 Steel T-section, 95 x 80 x 5 mm, screwed to vertical battens
3 Aluminium track
4 Cast-in track
5 Sliding element, 3-ply core plywood, 25 mm
6 Plastic rollers
7 Safety barrier
8 Steel angle, 95 x 40 x 5 mm, screwed to vertical battens

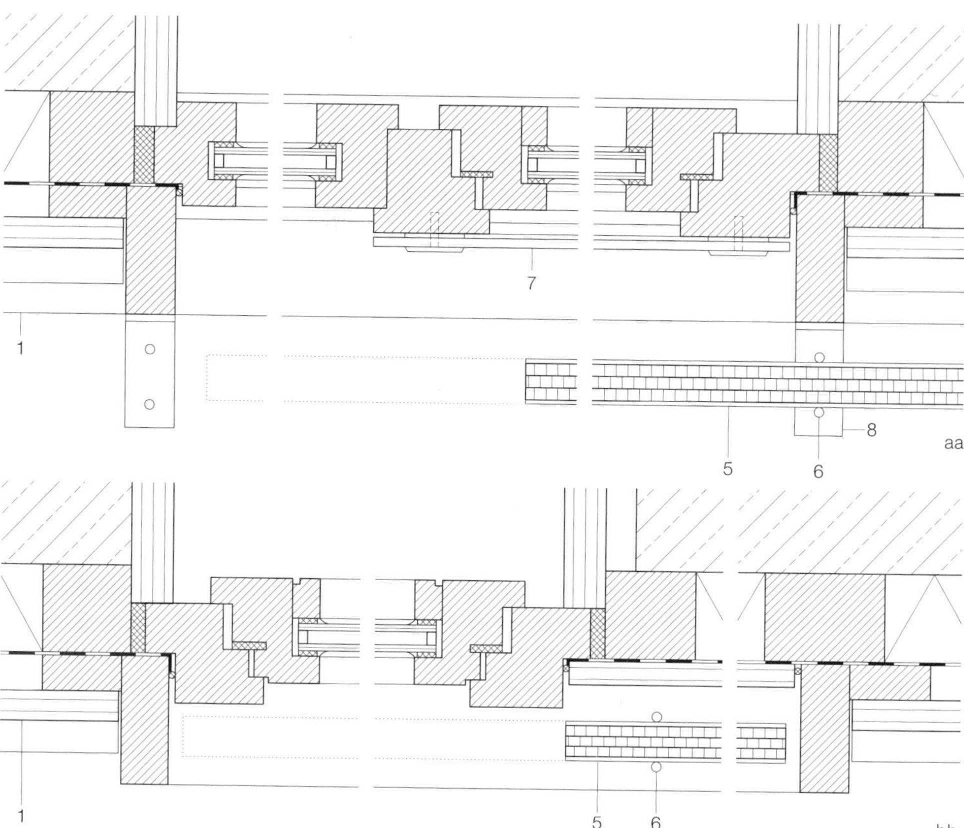

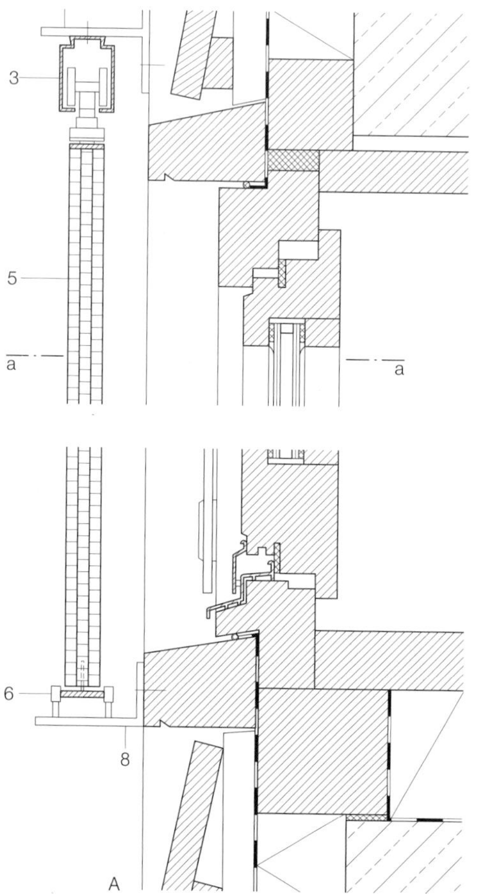

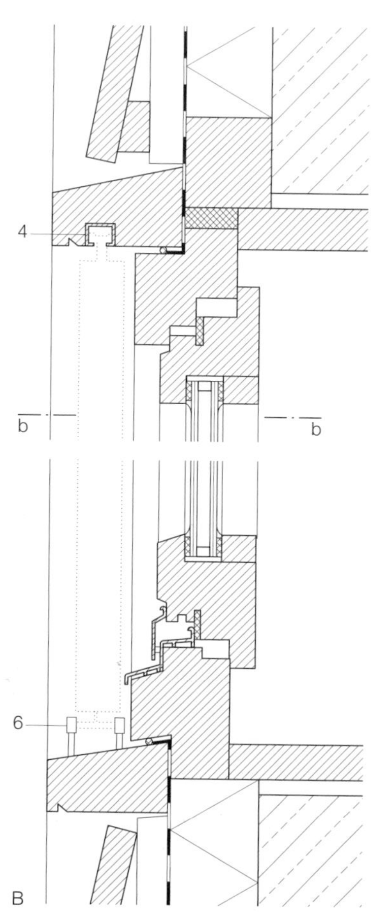

Office building

Unterschleißheim, D, 2002

Architects:
Baader + Schmid, Munich
Andrea Baader, Hanja Schmid
Assistant:
Maurice Mayne

📖 Baudokumentation. Hameln 2003

- Elements consisting of a membrane fitted to a frame form a second leaf and function as sunshade and anti-glare screen
- Horizontal, pivoting louvres, with membrane fitted to both sides
- Fixed louvres in front of spandrel panels fitted with open-pore membrane on one side only to allow occupants a view of their surroundings

Plan, scale 1:1000
Horizontal section • Vertical section
scale 1:20

1. Parapet panel with closed-pore membrane fitted to both sides
2. Fixed louvre comprising aluminium frame with membrane fitted to one side only, translucent open-pore material in front of spandrel panels, otherwise with closed pores for sunshading and glare protection
3. Movable louvre comprising aluminium frame with membrane fitted to both sides, PTFE-coated glass fibre fabric, 13% light transmittance, electric drive incorporated in No. 9, central and individual control
4. Sheet aluminium parapet capping, bent to suit
5. Hot-dip galvanised open grid flooring, 30 x 11 mm
6. Steel flat, 200 mm
7. Wall finish: insulated aluminium panel, 120 mm thermal insulation
8. Fixed light, insulating glass
9. Aluminium post, 120 x 55 mm
10. Opening light, insulating glass
11. Convector with outlet for displacement ventilation
12. Thermal insulation, 100 mm
13. Steel rectangular hollow section, 130 x 50 mm
14. Steel square hollow section, 120 x 120 mm

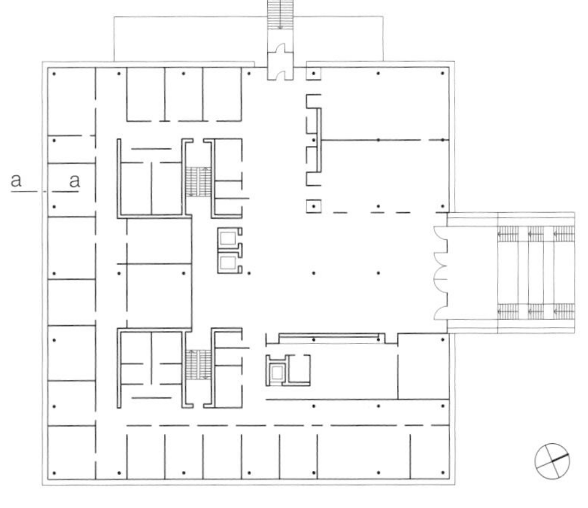

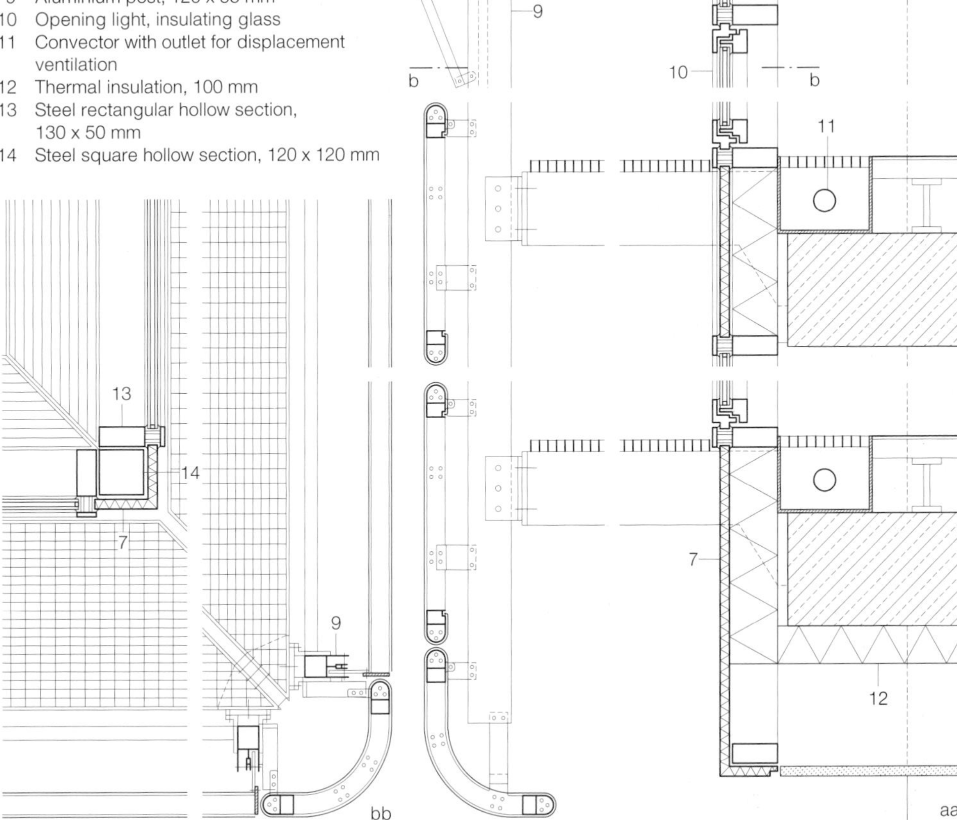

Housing complex

Hannover, D, 1999

Architects:
Fink + Jocher, Munich
Structural engineers:
Bergmann + Partner, Hannover

A+U 10/2001
db 07/2000
Pfeifer, Günter et al.: Masonry Construction Manual, Munich/Basel, 2001

- Wooden folding shutters
- Shutters fold within opening in masonry
- Room-height French window
- Low-energy standard
- Top-hung/sliding windows to staircases

Plan, scale 1:2000
Horizontal section • Vertical section
scale 1:20

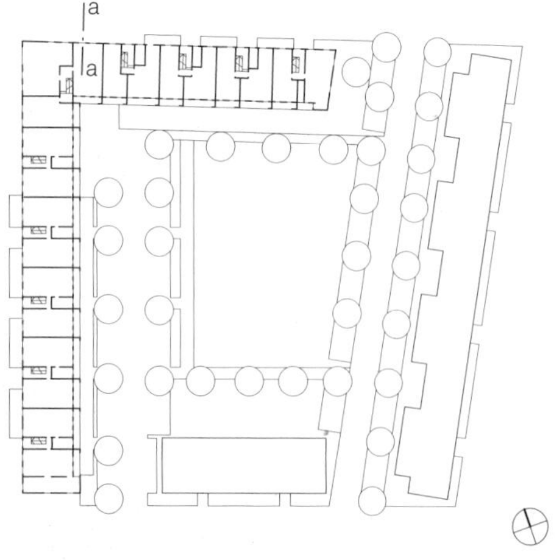

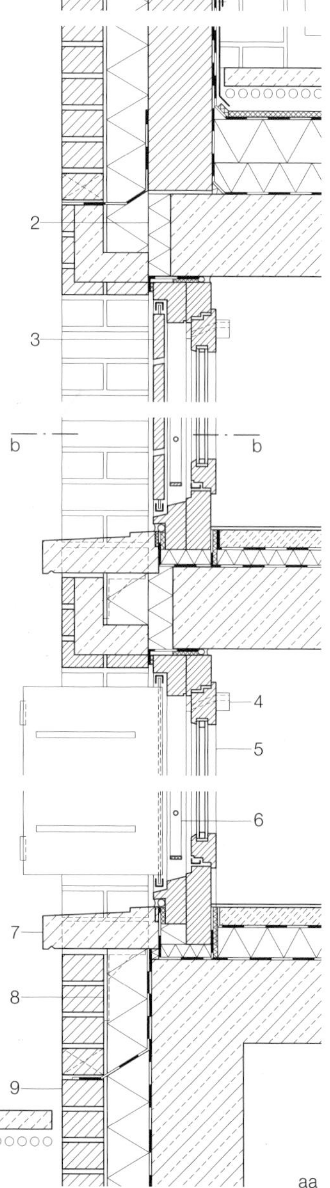

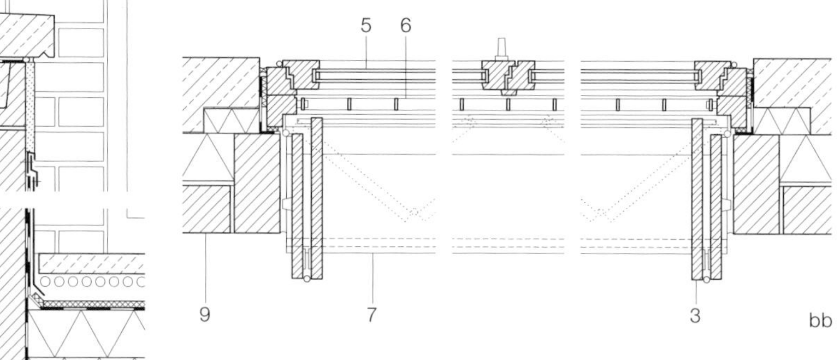

1 Parapet: peat-burned facing bricks in stretcher bond, 10 mm ventilated cavity
120 mm mineral fibre thermal insulation
175 mm aerated concrete
2 Thermal insulation, 60 mm rigid foam
3 4-part folding shutters of 3-ply core plywood, with 15 mm edge beading in weatherproof glue, guide tracks top and bottom, light grey paint finish, fixed to double thickness pieces at sides by means of galvanised strap hinges
4 Vent
5 Wooden window, 2 No. opening lights, with insulating glass
6 Safety barrier of galvanised steel flats, 35 x 8 mm, with micaceous iron oxide finish
7 Window sill, precast concrete, 50 mm overhang with rainwater drip
8 Steel angle as support for window sill
9 Plinth: peat-burned facing bricks in stretcher bond, NF 115 mm
10 mm ventilated cavity
120 mm mineral fibre thermal insulation
180 mm reinforced concrete

Private house

Amsterdam, NL, 2000

Architects:
Heren 5, Amsterdam
Project team:
Ed. Bijman, Jan Klomp,
Bas Liesker, Dirk van Gestel
Steel facade:
Limelight, Breda

Architectural Review 06/2001
Werk Bauen + Wohnen 01–02/1999
Schittich, Christian (ed.): Gebäudehüllen.
Munich/Basel 2001

- Shutters pivot and fold about a horizontal axis
- Weathering steel on north and south elevations as a reference to historical industrial structures

Plans of ground and upper floors, scale 1:400
Vertical section • Horizontal section through south elevation, scale 1:20

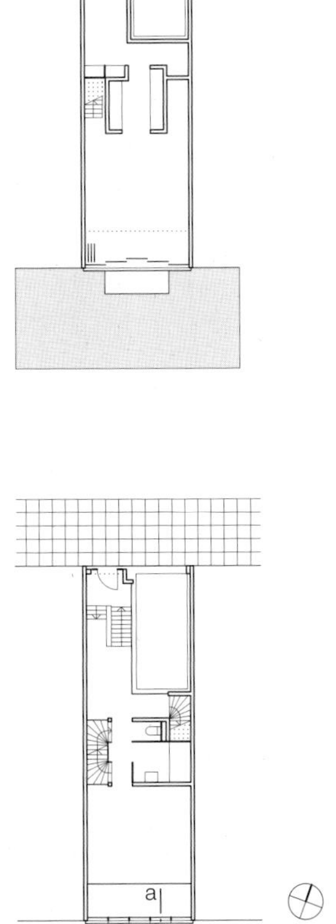

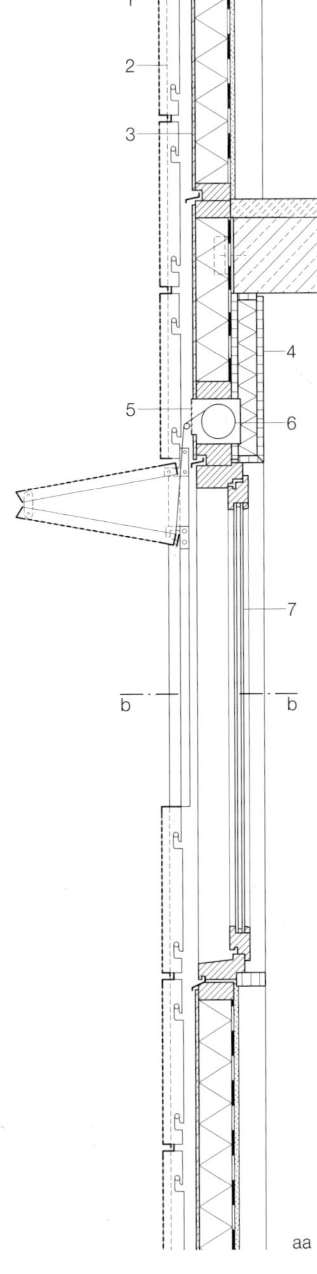

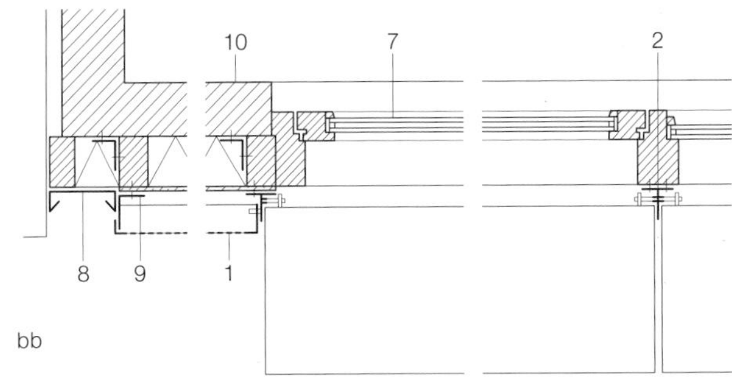

1 Pre-weathered perforated sheet steel, bent to suit, 485 x 30 mm
2 Steel T-section, 70 x 70 x 8 mm
3 Prefabricated facade element: 5 mm fibre-cement sheet 90 mm insulation vapour check 12.5 mm plasterboard
4 Interior panel: 18 mm veneer plywood 50 mm insulation
5 Aluminium grille, 100 x 5 mm
6 Drive for shutters
7 Insulating glass
8 Galvanised steel channel
9 Galvanised steel angle, 50 x 70 x 5 mm
10 Calcium silicate masonry, 115 mm

Housing complex

Innsbruck, A, 2000

Architects:
Baumschlager & Eberle, Lochau

Architectural Record 02/2002
Architectural Review 06/2001
Bauwelt 16/2001
Casabella 698, 2002
Detail 03/2002
Techniques + architecture 454, 2001

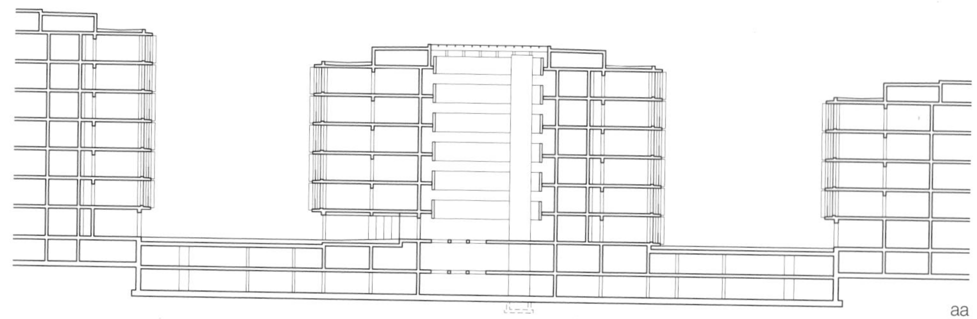

- Folding shutters mounted on framing
- Patination of copper to prevent glare (neighbouring airport)
- Six compact tower blocks (favourable A/V ratio) with identical layouts
- Height stepped to match slope of site (light admittance)
- Good relationship with landscape despite high density
- Unusually high standard of fitting-out for publicly assisted housing thanks to simplification and standardisation
- Passive energy systems with controlled ventilation to apartments
- Winner of the Energy Globe Award 2001 and the Mies van der Rohe Award 2001

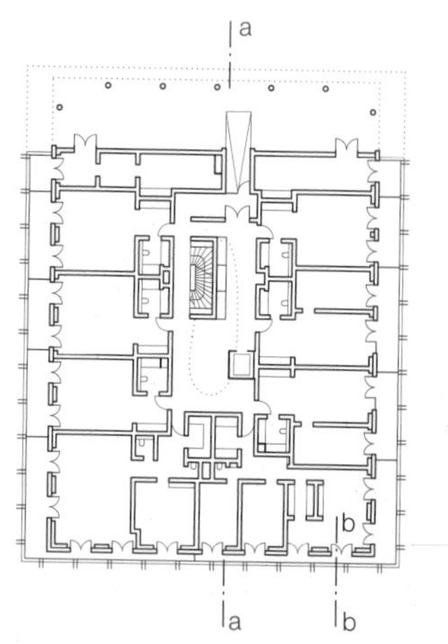
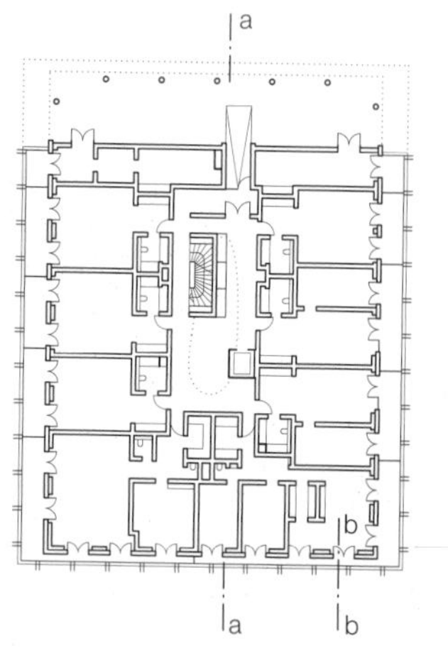

Section • Plan
scale 1:750
Vertical section • Horizontal section
scale 1:20

1 Wall construction:
18 mm pine boarding, with red-brown glaze finish
80 mm rockwool thermal insulation
200 mm rockwool thermal insulation
vapour barrier
180 mm reinforced concrete
15 mm plaster
2 4-part folding shutter, 0.6 mm pre-weathered sheet copper, glued and riveted to stainless steel tubular frame, 30 x 20 x 2 mm
3 Clip to lock shutters in position
4 Stainless steel handrail
5 Spandrel panel, 12 mm laminated safety glass, with matt PVB interlayer
6 Balcony partition, obscured 8 mm toughened safety glass
7 Glazed door with triple glazing
8 V 100 veneered chipboard
9 Precast concrete balcony floor, 6000 mm long, with thermal break

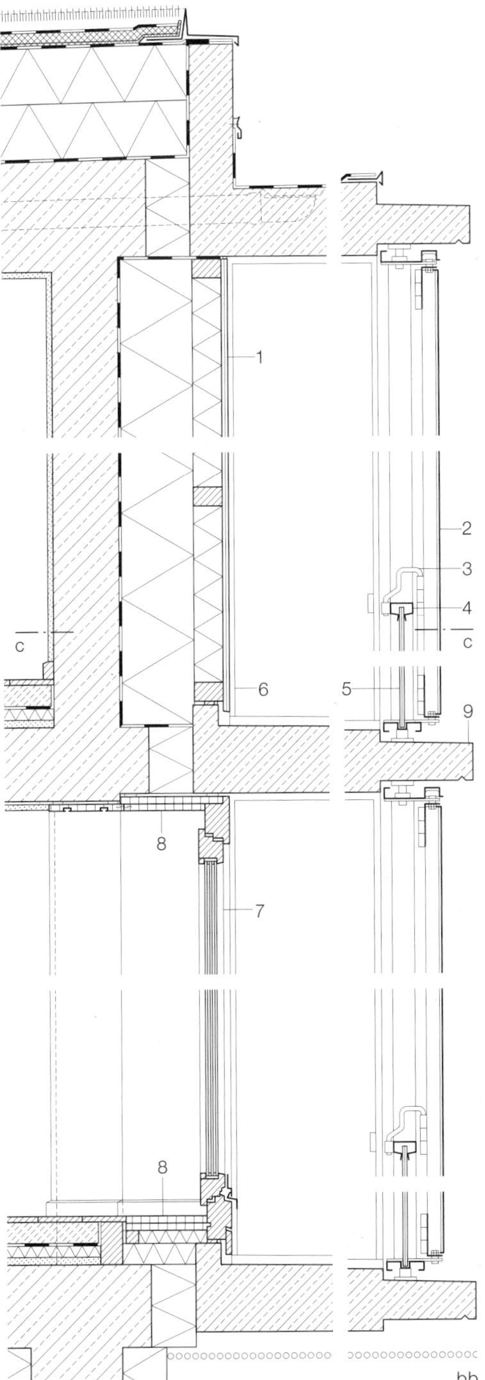

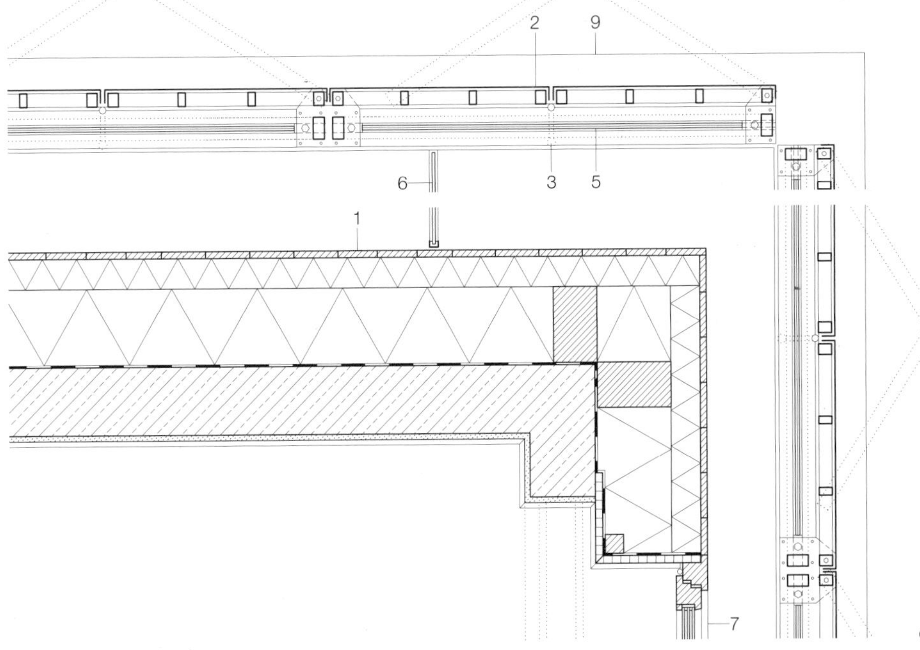

Private house

Munich, D, 1996

Architects:
b17, Munich
Martin Kühleis, Tobias de la Ossa,
Klaus Stierhof

l'architecture d'aujourd'hui 01/1999
Detail 07/1998

- West gable facade fitted with two movable sunshading elements: upper element pivots, lower element slides and can be folded upwards too when in the right-hand position, where it becomes a (sunshading) pergola
- Highly insulated timber frame construction comprising prefabricated wall and floor elements
- Low-energy house

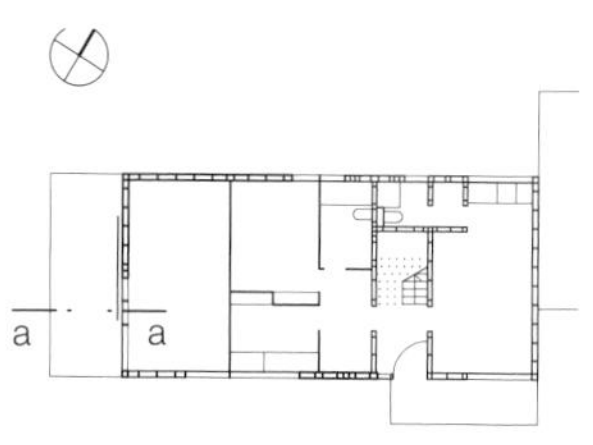

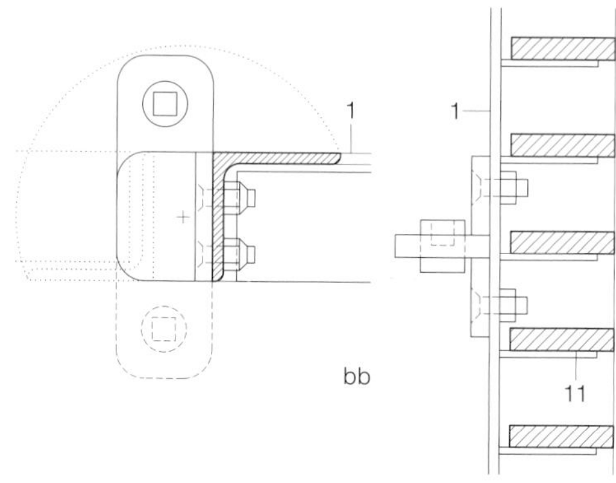

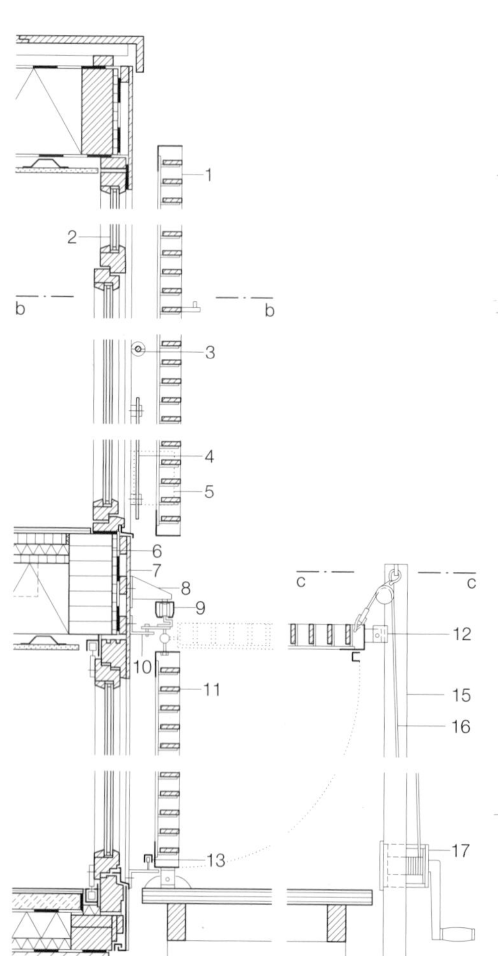

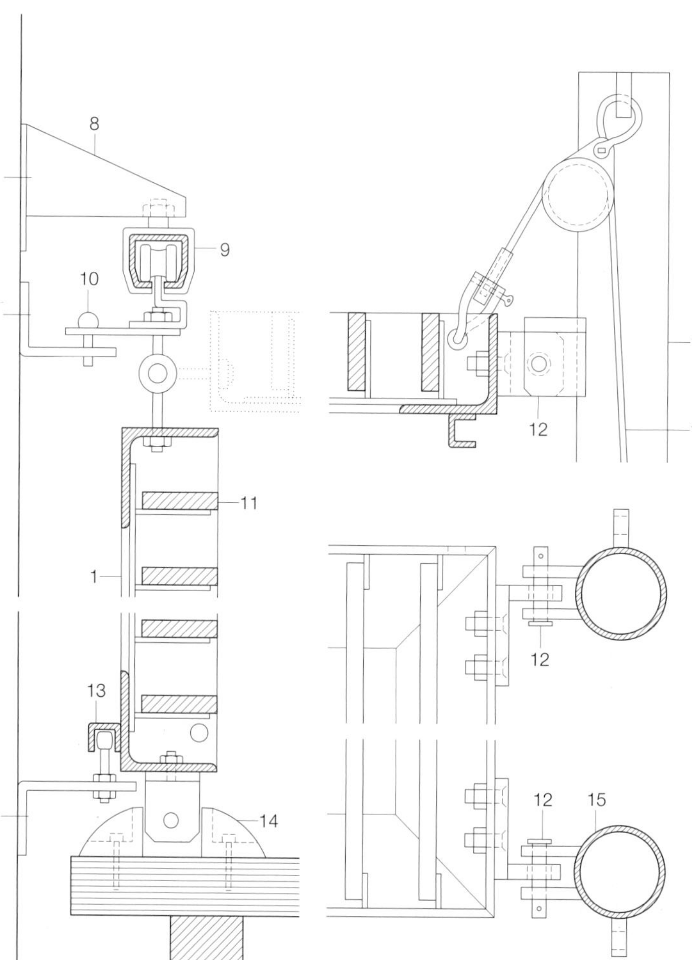

Plan, scale 1:500
Vertical section
scale 1:20
Detail of hinged shutter to upper floor, horizontal and vertical
Detail of hinged shutter to ground floor, connections top and bottom
Detail of hinged shutter to ground floor in sunshading position
vertical and horizontal
scale 1:5

1 Steel angle frame, 70 x 70 x 4 mm, with 3 mm sheet metal lugs welded on
2 Insulating glazing, toughened safety glass + cavity + laminated safety glass
3 Handle to fix shutter
4 Safety barrier, 10 mm toughened safety glass
5 Hinge on 1 mm rubber
6 Glulam beam, 265 x 120 mm
7 Fibre-cement sheet on preformed sealing strips
8 Steel flat bracket
9 Sliding shutter track
10 Retaining pin
11 Larch wood slats, 12 x 60 mm
12 Retaining pin adjuster
13 Guide track with wheel
14 Adjuster
15 Steel circular hollow section, 60.3 dia. x 4 mm
16 Operating cord
17 No-kickback aluminium winch

Office building

Berlin, D, 1999

Architects:
Sauerbruch Hutton, Berlin
Facade consultants:
Emmer Pfenniger + Partner, Munichstein

A+U 09/2002
Architectural Review 12/2000
Intelligente Architektur 21, 2000
Schittich, Christian (ed.):
Building Skins. Munich/Basel 2001

- Sliding/folding shutters made from perforated sheet metal, coloured paint finish externally
- Non-segmented prefabricated west facade (exhaust-air facade)
- Narrow on plan
- "Wind roof" (aerodynamic wing, venturi effect) for promoting the stack effect in the exhaust-air facade

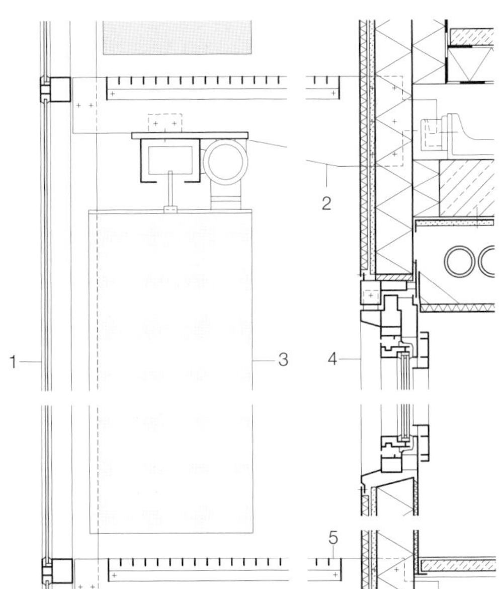

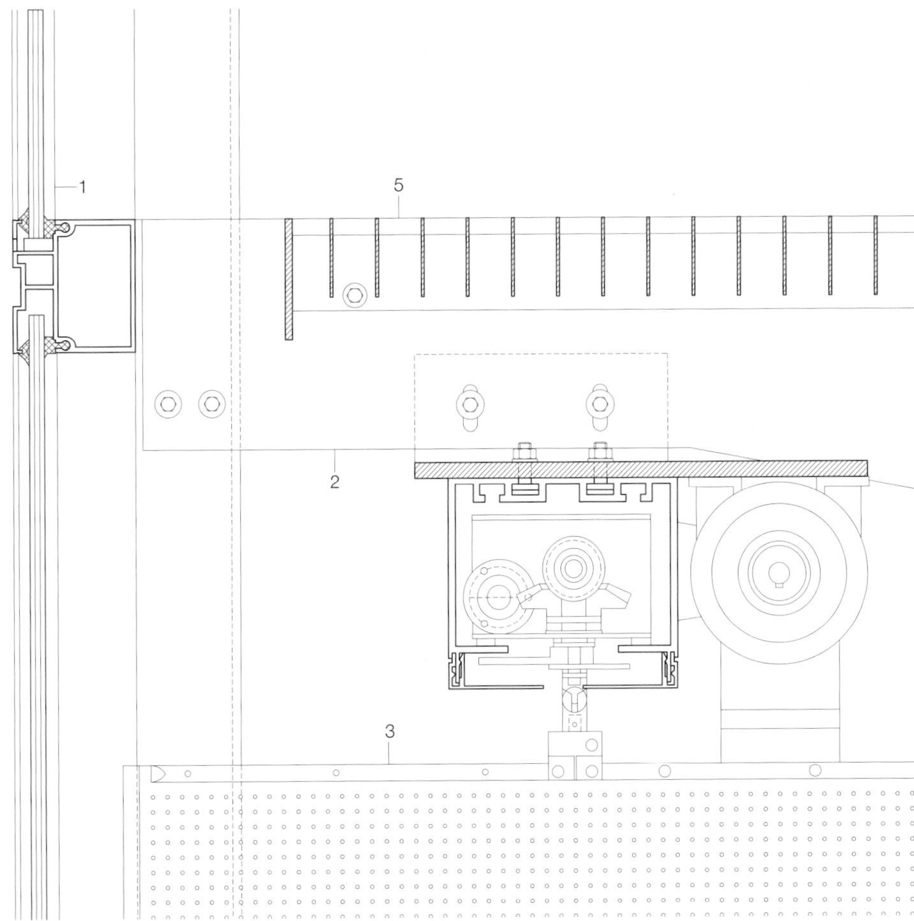

Part vertical section
scale 1:20
Detail
scale 1:5

1 External leaf, west elevation: extruded aluminium sections 10 mm toughened safety glass infill panels, 1800 x 3300 mm
2 Steel cantilever arm
3 Sunshading shutter, 600 x 2900 mm perforated sheet aluminium, 1.5 mm, shutter pivots and slides to side
4 Internal leaf, west elevation: suspended elements made from extruded aluminium profiles, 1800 x 3250 mm insulating glazing, 6 mm + 14 mm cavity + 8 mm
Spandrel panel:
2 mm perforated sheet aluminium
20 mm fleece-laminated mineral insulation
18 mm fire-retardant board on steel framing with integral 100 mm thermal insulation
5 Open grid flooring

Office building

Wiesbaden, D, 2001

Architects:
Herzog + Partner, Munich
Lighting design together with Lichtlabor Bartenbach, Aldrans
Structural design of outer leaf:
Ludwig + Weiler, Augsburg

Detail 07/2001
Dialogue Taiwan 68, 2003
THE PLAN 003/2003
Nikkei Architecture 04/2003

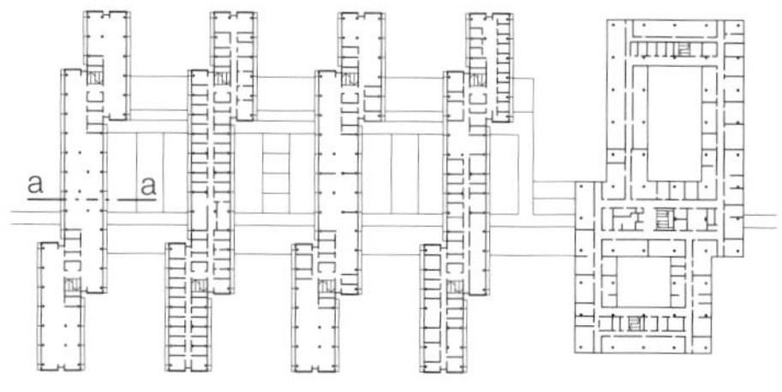

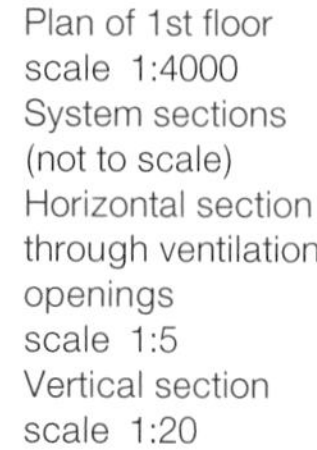

Plan of 1st floor
scale 1:4000
System sections
(not to scale)
Horizontal section through ventilation openings
scale 1:5
Vertical section
scale 1:20

- South side: combination of two shading elements pivoted about the horizontal axis: upper element with light-redirecting louvres for controlling daylight admittance, lower element acts like an awning to allow views of the outside
- South side: additional (diffuse) daylight capture even with an overcast sky by means of shading elements with light-redirecting profiles
- North side with stationary light-redirecting elements to capture overhead daylight like the south facade
- Opaque ventilation flaps with integral air inlets: controlled natural ventilation in combination with uncontrolled ventilation
- Building services for the offices integrated into the facade

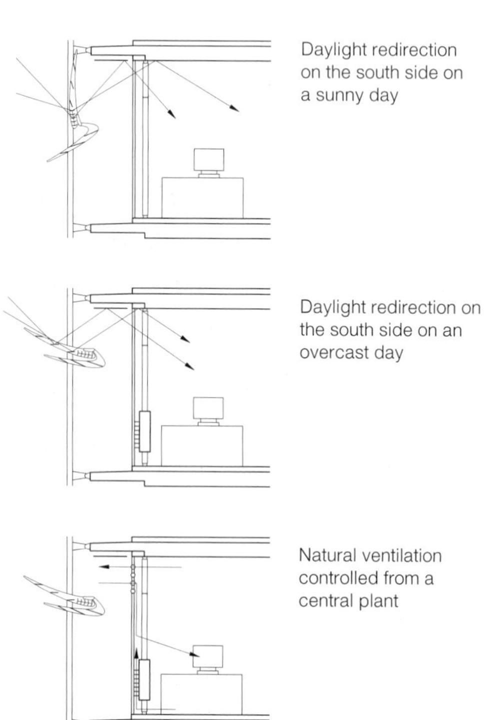

Daylight redirection on the south side on a sunny day

Daylight redirection on the south side on an overcast day

Natural ventilation controlled from a central plant

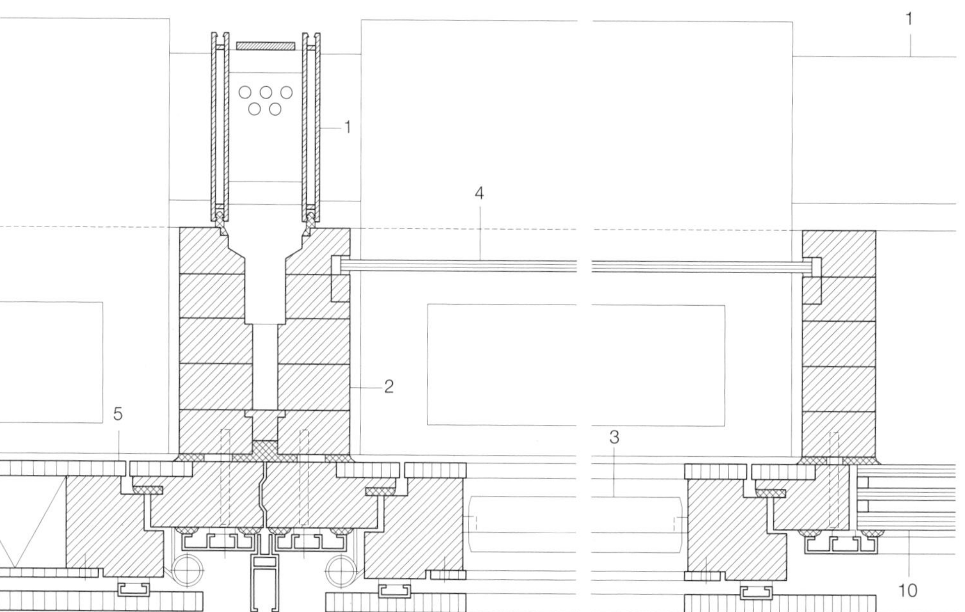

1 Aluminium cable duct
2 Frame of glued hemlock laminations, 50 x 15 mm
3 Plastic vent
4 Baffle plate behind vent, toughened safety glass
5 Ventilation flap:
15 mm plywood with makoré veneer, removable
9 mm air cavity
6 mm plywood with makoré veneer frame of glued spruce laminations, 60 mm, or
PU rigid foam insulation
10 mm plywood with makoré veneer 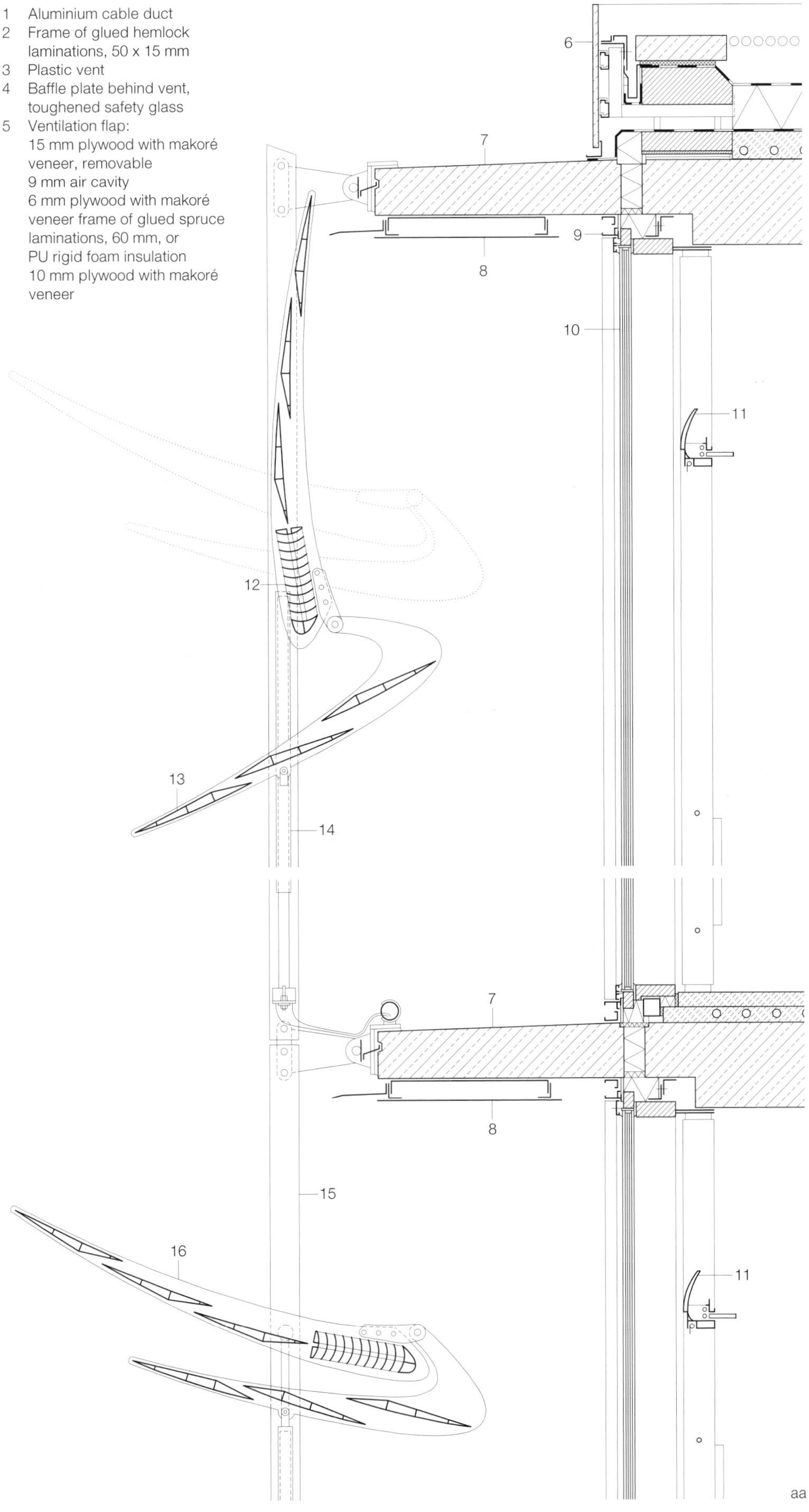
6 Fibre-cement fascia panel, 12 mm
7 Precast concrete cantilever element, 160 mm, with polyurethane coating
8 Aluminium daylight reflector
9 Extruded aluminium section with EPDM seal
10 Triple glazing with powder-coated aluminium glazing bars
11 Lighting unit with aluminium reflector, light-scattering glass diffuser and integral glare protection
12 Extruded section for redirecting daylight, highly reflective
13 Extruded section for shading and indirect light redirection, highly reflective
14 Drive motor
15 Steel flat, 100 x 12 mm, glass-bead-blasted finish
16 Aluminium arm, powder-coated

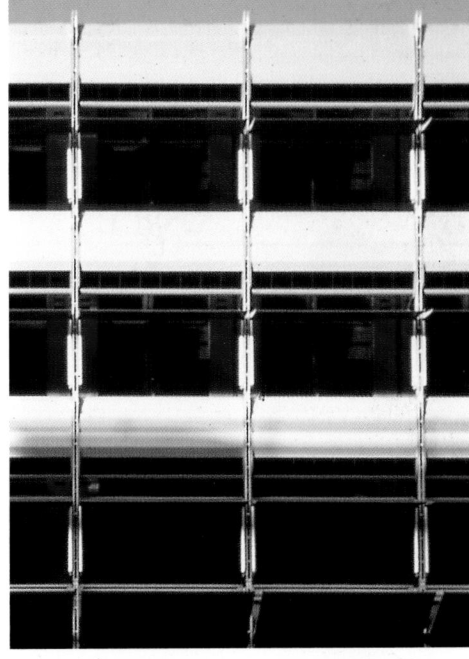

aa

University building

Brixen, I, 2004

Architects:
Kohlmayer Oberst, Stuttgart
Sunshading element developed with the help of Fraunhofer Institute for Solar Energy Systems (ISE), Freiburg (ISE), Freiburg

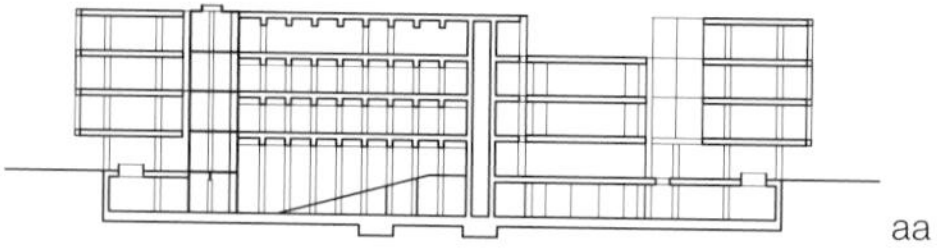

- External highly reflective roller sunshade made from special shaped stainless steel strips
- Full shading for solar altitude angles > 20°
- View of the surrounding landscape possible thanks to special geometry
- Prefabricated facade with projections and recesses, and push-out elements for ventilation positioned in recesses.

Sunshading element, enlarged, scale 2,5:1
Section • Plans of ground and 2nd floors, scale 1:1500
Sorizontal section • Vertical section, scale 1:20
Detail, scale 1:5

1 Sheet aluminium capping, 3 mm, bent to suit
2 Fascia panel:
3 mm sheet aluminium
waterproofing
80 mm extruded rigid foam insulation
3 Insulating glazing, laminated safety glass (8 + 6 mm) + 16 mm cavity + 10 mm toughened safety glass
4 Push-out light for ventilation, 3200 x 250 mm
5 Sheet aluminium, 3 mm, 2 parts, upper cover with slits
6 Sunshade: 6 mm wide stainless steel strips, spacing = 150 mm, with stainless steel louvres riveted to these, integral electric drive motor
7 Steel fin as connection for partition
8 Sheet aluminium, 4 mm, suitable for maintenance loads
9 Lighting unit, sheet steel, 350 x 180 x 1280 mm, with cold light reflectors
10 Insulating glazing, 10 mm toughened safety glass + 16 mm cavity + laminated safety glass (6 + 8 mm)
11 Mineral wool insulation, 100 mm
12 Steel flat, 20 mm
13 Waterproofing
14 Mineral wool insulation, 80 mm
15 Sheet aluminium, 3 mm, bent to suit
16 Stainless steel guide tube, 50 dia. x 2 mm, with guides at both ends
17 Rail, detaches for installing and removing sunshade

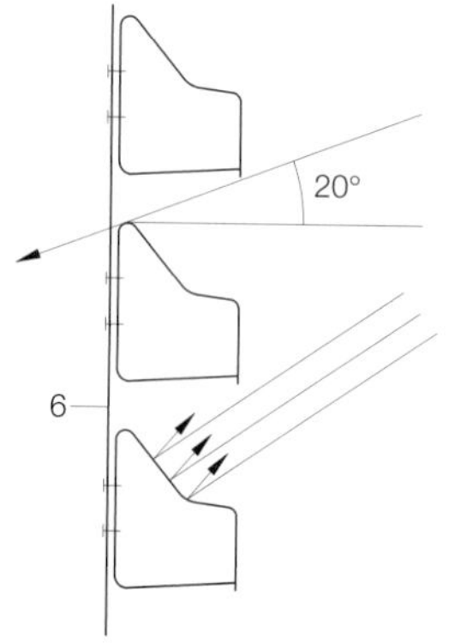

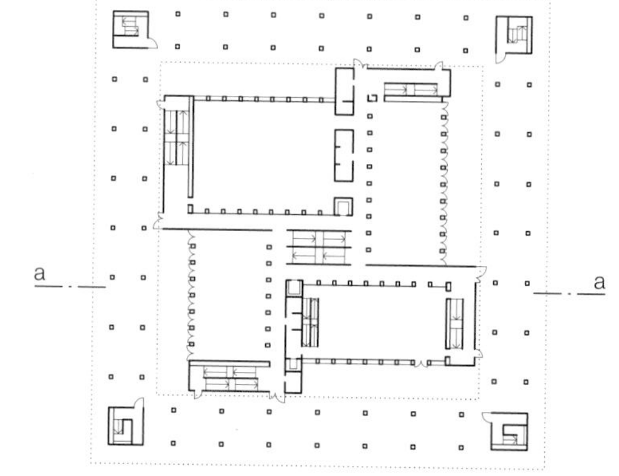

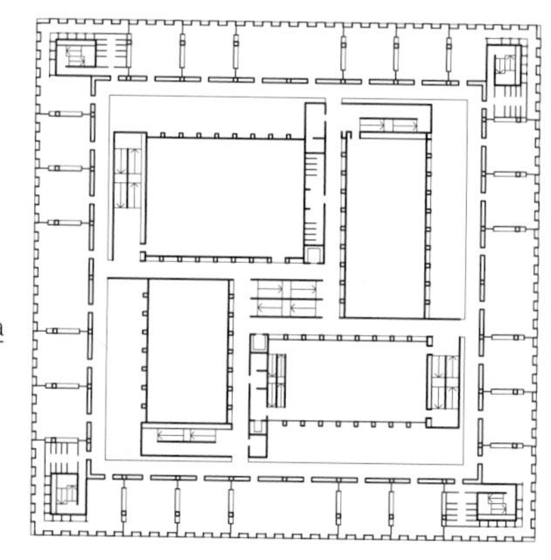

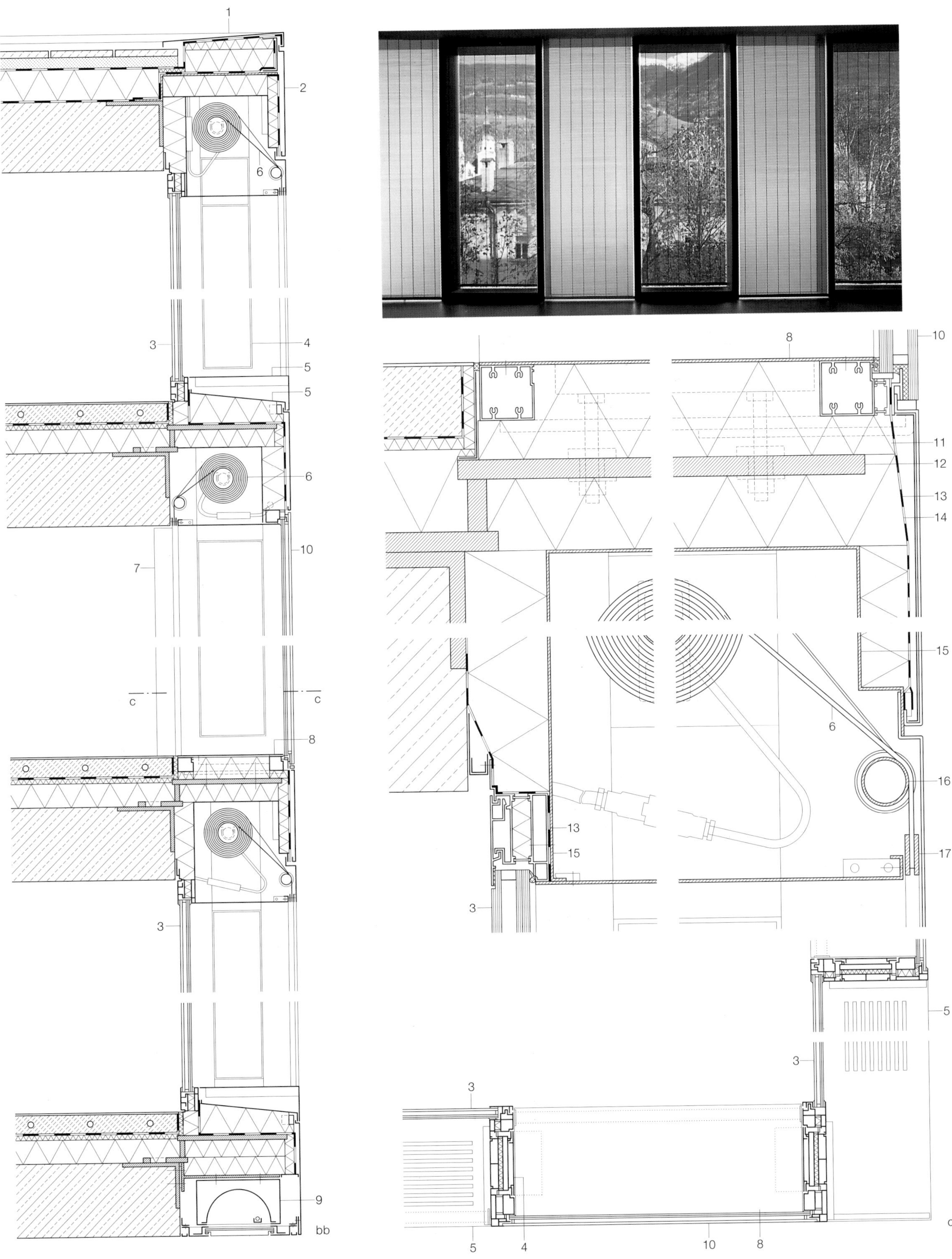
1
2
6
3
4
5
5
6
10
7
c
c
8
3
9
bb
8
10
11
12
13
14
15
6
16
13
15
17
3
5
3
3
cc
5
4
10
8

B 2.3 Solar energy

The building envelope represents the most important building subsystem in terms of the energy balance of a building. When integrating solar energy systems as the bridge between architecture and solar technology, the building envelope is the primary – also the visually effective – point of reference. One fundamental feature of the non-passive use of solar energy systems in buildings is their visible integration into roofs and walls. Such systems have to fulfil protective functions and must be coordinated with the construction, but also affect the appearance of a building.

In terms of thermal energy systems, since the early 1990s solar energy facades have become, more and more, those facades in which the wall, as a shield and buffer against the climate, is also required to function as an active supplier of heat. What we shall look at here is every form of building-related solar energy utilisation connected with the facade, from glass curtain walls to photovoltaic panels.

Direct and indirect use

Solar energy occurs in various forms, of which radiation represents the main useful source for buildings. We distinguish here between direct, i.e., "passive" use and indirect, i.e., "active" use. Direct use is the name given to the utilisation of specific constructional measures to collect, store and distribute the incident (i.e., irradiant) solar energy, as far as possible without using any technical equipment. These specific building features, especially those of the building envelope, for regulating the interior climate and the energy balance embrace fundamental principles of solar heating and cooling plus utilisation of daylight. Indirect use requires additional technical measures to collect, distribute and, if required, store the solar energy. Collector technology is especially important here, to complement the heating and cooling effects, plus photovoltaic systems for generating electricity. Numerous different systems can be assigned to these two forms of solar energy application. This results in a wide range of tools for building-related solar energy utilisation.

Climatic parameters and classification principles

Available solar radiation

The quantity of solar radiation available fluctuates very considerably over the course of a year, indeed during the course of a day, and is very heavily influenced by the prevailing local weather conditions. Whereas the irradiant energy of two consecutive days can vary by a factor of 10, the value on a bright summer's day can be 50 times as high as that on a dull winter's day.
In addition, in Central Europe the available solar radiation, whether taken over the course of a day or a year, hardly coincides with the actual heating requirement. Short-term changes can be offset by heat storage media. On the other hand, the seasonal fluctuations represent a serious problem. In Germany about three-quarters of the annual irradiant solar energy occurs over the six months in the middle of the year. This is energy which currently can be stored only by means of elaborate, expensive underground storage systems. These limitations to the availability place technical and economic constraints on the use of solar energy.

Incident energy (orientation and inclination)

Two important parameters determine the sensible use of solar energy in buildings: the exposure of the collecting surfaces, i.e., compass orientation and angle of inclination, and the freedom from (unwanted) shading.
The total solar radiation (global radiation) comprises the direct radiation from the sun plus the diffuse – indirect – radiation reflected from the sky and the surroundings (open sky radiation). In Central Europe more than 50% of the annual global radiation is diffuse radiation. Within Germany we notice small differences in the incident energy depending on geographical location (annual average up to 300 kWh/m²a, max. 25%).

From the shaded entrance hall to the energy facade

Primary, directly effective principles such as compact structures, orientation towards the south, graded interior layouts and passive sunshading can be traced back to ancient Greek architecture. The facade has therefore been used – intentionally or unintentionally – as a source of heat for many centuries. In doing so, the (window) openings in the wall acted as the first "collectors". Steps towards the optimisation of the external wall as a climate regulator led to the breakdown into and differentiation between different zones. Open intermediate or transitional zones such as shaded entrance halls and arcades provided the first protection from the weather and the sun, thus enabling these areas to be included in the useful floor space of a building exposed to a Central European climate. In order to use solar radiation more effectively, but also to reduce the loss of heat from heated rooms to the outside, it is necessary to create a "thermal break". These generally transparent, multi-leaf constructions (e.g., double windows, oriels, bay/bow windows, glazed loggias, lean-to conservatories) are intentional constructional (i.e., passive) strategies for utilising solar energy. The importance of such zones of intermediate temperature for heating purposes increased hand in hand with the possibility of producing ever-larger panes of glass. In addition, research into more efficient systems or completely new utilisation concepts began to speed up in the second half of the twentieth century.

B 2.3.1 Residential complex, Munich (D), 1982, Thomas Herzog and Bernhard Schilling [1]

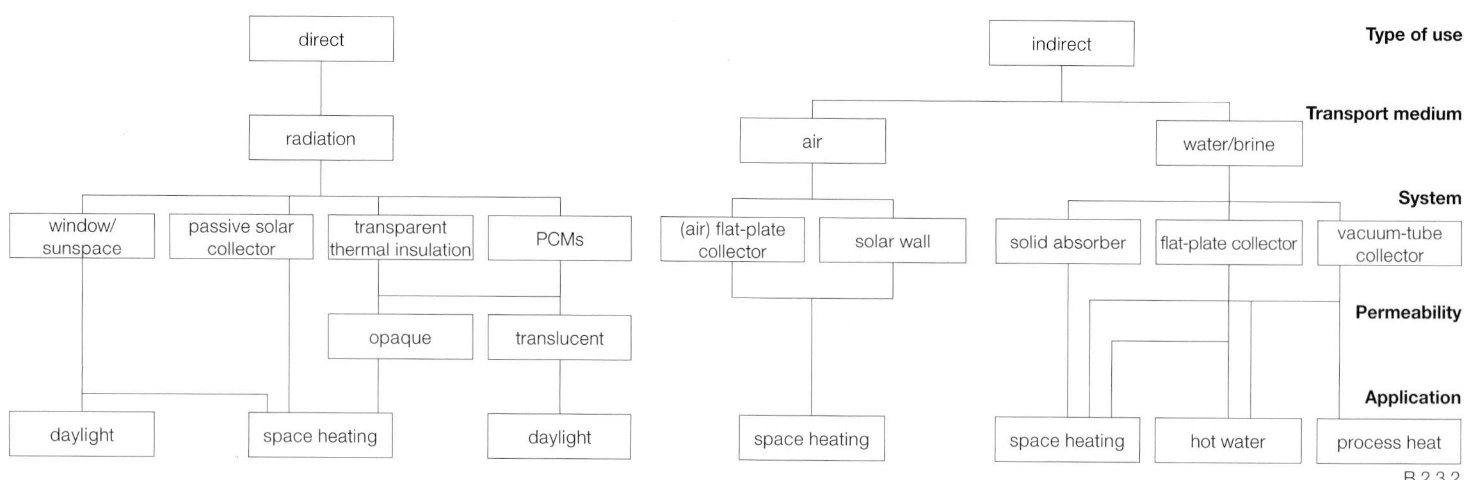

B 2.3.2

This direct form of solar energy usage was also accompanied by the development of technical systems for the indirect use of solar energy. Collectors for hot-water heating and photovoltaic panels for generating electricity have in the meantime become virtually everyday features of the building envelope. The range of options for designers has been given an enormous boost by the expansion of fundamental passive strategies and technical systems for utilising solar energy.

The multi-leaf building envelope

The superimposition of the different functional requirements to be met by the building envelope in terms of the general constructional properties gives rise to various (three-dimensional) zones between the heated interior and the outside world. Such a gradation of functional areas helps to decrease heat losses and increase solar gains. Furthermore, these zones of intermediate temperature allow additional utilisation or recovery of building heat losses and, if necessary, can preheat any incoming external air; they can also accommodate systems for protecting against overheating and glare.

We distinguish between three thermally effective basic types, which occur in practice in many different forms; these are the "airlock", the air collector and the thermal buffer. This construction principle extends from the formation of narrow air layers, cavities or spaces in front of the external wall right up to areas of additional useful floor space which are used only temporarily, such as:

- entrances, lobbies
- glazed loggias, balconies
- conservatories, lean-to greenhouses
- functional areas that act as thermal buffers and "airlocks" in addition to their primary function [2]

Direct "passive" systems

The best-known way of using solar energy directly is via window openings, which, in conjunction with the immediate rooms they serve, act as simple collection and storage systems. The proportion of usable solar energy depends on the climatic conditions and local circumstances, but also – crucially – on the compass orientation, inclination and size of the openings. In addition, the construction of the internal walls, ceilings and floors affects the degree of utilisation and has a decisive effect on the interior climate. Large expanses of glass without additional sunshading measures lead to overheating during the summer, a fact that must be given serious consideration in most cases, particularly with facades facing east or west. In other words, we must always strive to achieve an adequate balance between incident energy, size of opening, heating requirements, shading and thermal mass [3].

Sunspaces

Sunspaces are usually unheated rooms and represent simple "air collectors". These zones of intermediate temperature are found in numerous forms, as storey-height structures, or extending over more than one storey, or even enclosing the entire building. In Central Europe these unheated areas provide usable floor space for up to two-thirds of the year. The magnitude of the incident energy gains is likewise influenced by the exposure, area of glass and any shading caused by the building itself or neighbouring buildings and/or vegetation. As with windows, to avoid overheating in summer it is usually prudent to include some form of sunshading, but above all, effective ventilation options. In order to be able to exploit some of the "excess" solar heat, specific building and technical measures are necessary. The wall surfaces between the glazed area and the adjoining rooms can serve here as thermal masses which pass on the stored heat to the rooms after a time lag. Various types of passive solar collectors work on this principle.

Passive solar collectors

One of the first passive solar collector concepts [4] was developed by Felix Trombe and Jaques Michel. They used a combination of a south-facing glazed surface, a solid wall painted matt black and an air layer to create a thermal mass system. To improve the degree of efficiency, the collector zone was linked to the adjoining room by means of ventilation flaps at the top and bottom of the wall. If the temperature of the absorber, which can reach 70°C in direct sunlight, is higher than the room temperature, the air starts to circulate. Due to the thermal buoyancy effect, the rising hot air can be used relatively easily to feed heat directly into the interior. Sunshading measures are absolutely essential to avoid overheating in summer. The yield of such a passive solar collector is also heavily dependent on the specific capacity of the materials used. As the specific heat capacity of water is (in terms of volume) about 2-4 times higher than that of a solid wall material, trials with water tanks positioned/stacked in the facade were carried out in the 1970s and 1980s. There have been attempts to "regulate" the release of heat more efficiently by including a layer of insulation on the room side so that the heat is released by way of convection through ventilation flaps.

Transparent thermal insulation

This represents yet another form of direct solar energy utilisation [5]. The combination of suitable thermal insulation and direct solar energy gains enables the heating energy consumption to be lowered even further. The concept is based on the use of radiation-permeable thermal insulation [6], a principle that not only decreases the transmission heat losses but can also increase the magnitude of the solar gains. We distinguish between two basic approaches here:

- solid wall systems (opaque)
- direct gain systems (translucent)

Solid wall systems

Compared to conventional insulating materials, a transparent thermal insulation system placed in front of a solid wall provides thermal insulation

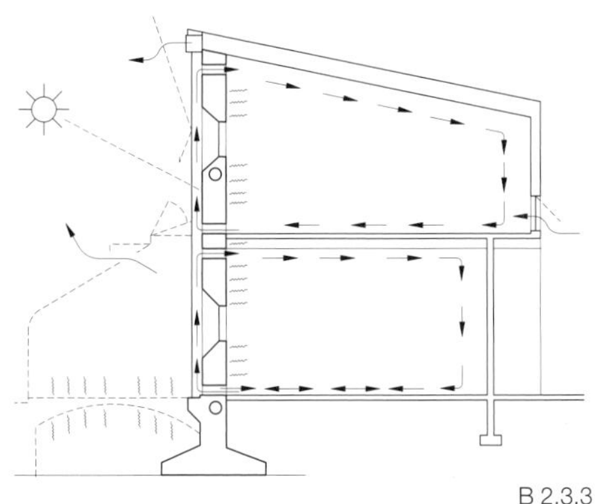

B 2.3.3

B 2.3.2 Classification of thermal energy systems
B 2.3.3 Sketch of the Trombe wall principle
B 2.3.4 Part of the cloisters at San Giorgio Maggiore, Venice (I), 1575, Andrea Palladio
B 2.3.5 Glazed balconies, Barcelona (E), c. 1900
B 2.3.6 "The growing house", prototype development, Berlin (D), 1932, Martin Wagner
B 2.3.7 Private house, New Mexico (USA), 1972, Steve Bear

B 2.3.4

B 2.3.5

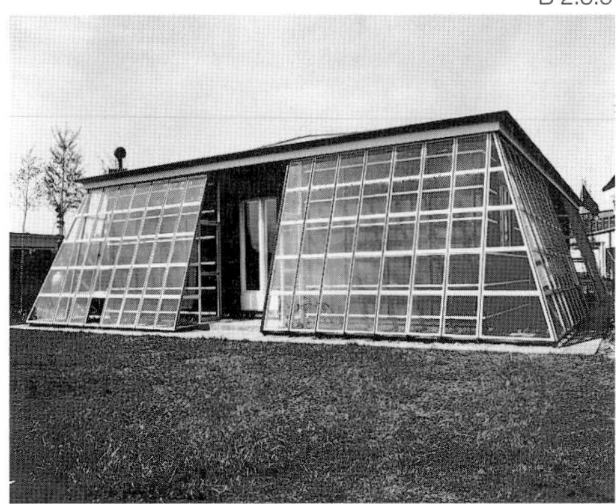

B 2.3.6

B 2.3.7

and also brings additional solar gains. This system is based on the principle of raising the temperature of an absorbent layer, i.e., a layer of translucent thermal insulation – generally with a structure perpendicular to the absorbent surface – is positioned in front of a solid wall with a high thermal mass. The high total thermal resistance of the transparent thermal insulation means that a large proportion of the absorbed solar energy is stored in the wall. A glass outer leaf protects against the weather. Solar radiation penetrates the layer of transparent thermal insulation and is absorbed by the dark wall surface, where up to 95% of the radiation is converted into heat. Whereas the construction of the transparent thermal insulation virtually rules out any heat losses, the solid wall absorbs the heat, stores it and then releases it to the adjoining rooms after a delay of – depending on material and thickness – about 6–8 hours. This is an efficient short-term measure for overcoming the discrepancy between available solar radiation and heating requirement. Even if there is increased reflection at the panes of glass in the summer, the areas of transparent thermal insulation must be protected against overheating by means of sunshading devices. When the transparent thermal insulation element sizes lie between 5% and 15% of the usable floor space, passive measures such as overhanging eaves, balconies and plantings are usually sufficient. However, manipulators are generally required for systems covering a large area.

It is difficult to compare the different systems in terms of the basic materials used and the different forms of construction. The primary parameters are UV resistance plus mechanical and thermal stability. The typical transparent thermal insulation materials include polymethyl methacrylate (PMMA), polycarbonate (PC) and glass. Very recently, cardboard honeycomb structures and sawn "wooden slats" have been used.

Direct gain systems

These are special glazing systems with the transparent thermal insulation material placed between the inner and outer panes of glass. This form of construction provides good thermal insulation and still allows the passage of daylight, but does severely impede the view through the glazing. The exploitation of the solar irradiation takes place via the thermal masses within the room. That means that here again it may be necessary to incorporate measures to prevent overheating in the summer. Besides the plastics already mentioned, glass and silica aerogels are also suitable materials.

Phase change materials

The first trials with PCMs took place way back in the 1940s in conjunction with the development and construction of passive solar collectors. The task of temporarily storing excess heat and releasing it into the interior later calls for materials with high admittance (i.e., energy storage capacity). The thermally effective mass can be increased by using materials with a high specific heat capacity in the primary construction. Owing to the usually lower admittance of building materials and hence their lower efficiencies, the use of heat storage effects without a change of phase but with noticeable heat requires a higher specific weight or a large surface area. In this respect, PCMs are extremely promising new materials because of their ability to store relatively large quantities of heat in a relatively small temperature range. In one concept dating from the 1970s, glass blocks are filled with Glauber salt (melting point 32°C) [7]. In recent years, experiments with PCMs have been carried out in an attempt to increase the thermal mass of lightweight structures. Furthermore, PCMs can also be used for direct gain systems – in a way similar to the use of transparent thermal insulation. Placed in "containers" made from transparent plastic materials, PCMs guarantee a high thermal mass, passage of daylight and limited transparency.

Solid absorbers

The solid absorber represents a sort of hybrid system. By the mid-1990s in Germany the area covered by such systems was equal to that of solar collectors.
Solid absorbers are planar, solid concrete external wall components with internal circulation

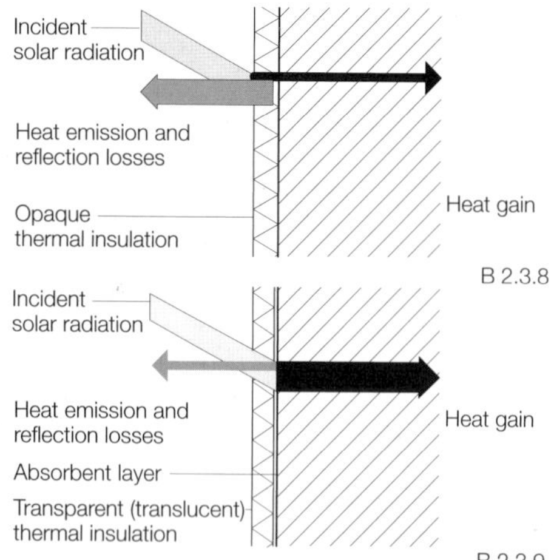

B 2.3.8

B 2.3.9

Inclination of surface (values for April to September)

	0°	20°	40°	60°	90°
Orientation					
East	>95 %	93 %	86 %	72 %	46 %
South-east	>95 %	>95 %	93 %	81 %	50 %
South	>95 %	**100 %**	95 %	82 %	49 %
South-west	>95 %	>95 %	93 %	81 %	50 %
West	>95 %	93 %	86 %	72 %	46 %

B 2.3.10

Inclination of surface (values for October to March)

	0°	20°	40°	60°	90°
Orientation					
East	58 %	57 %	53 %	45 %	32 %
South-east	58 %	75 %	83 %	83 %	69 %
South	58 %	82 %	96 %	**100 %**	88 %
South-west	58 %	75 %	83 %	83 %	69 %
West	58%	57%	53%	45%	32%

B 2.3.11

B 2.3.12

B 2.3.13

B 2.3.8 The principle of opaque insulation
B 2.3.9 The principle of transparent (translucent) insulation
B 2.3.10 Energy gains for different collector orientations and inclinations (April to September)
B 2.3.11 Energy gains for different collector orientations and inclinations (October to March)
B 2.3.12 Vacuum-tube collector facade, "City of Tomorrow", Malmö (S), 2001, Månsson + Dahlbäck
B 2.3.13 Semi-detached house, Pullach (D), 1989, Thomas Herzog, Michael Volz and Michael Streib

pipes acting as a heat exchanger exposed to the ambient heat. These elements are installed mainly above ground and absorb heat from air, rain, partly also from snow, and the moisture in the air through their surfaces. When combined with thermal mass on or in the ground (ground floor slab, foundations, etc.), the solid absorber heating system – in conjunction with a heat pump – can also be operated without additional heating [8].

In principle, solid absorbers can be used with all building components or systems in contact with the outside air. However, apart from new experimental trials [9], this approach has currently been shelved owing to the (high) power requirement of the heat pump.

Indirect "active" systems

Solar collectors

A solar collector is a technical system for absorbing radiation, converting it into heat and transferring this to a flowing transport medium (water, air). The element in which the energy conversion and the heat transfer take place is called the absorber. Collectors are employed mainly for hot-water heating and space heating. However, there are also special systems available for generating process heat (e.g., for industrial applications) and for cooling. The collector is the heart of a thermal energy system and together with the traditional building services components (pipework, heat exchanger, pumps, storage media) forms a total system. Depending on the type of use, different system configurations are possible. In terms of conventional collectors, we distinguish between the flat-plate and vacuum-tube varieties.

Air collector systems
This is a special form of collector in which air can be used directly as the transport medium, that is, without a heat exchanger, for space heating or drying. There is no risk of problems due to frost or corrosion, and the sealing requirements for these components are also less stringent. However, the specific heat capacity of air is only a quarter that of water. Relatively large quantities of air and correspondingly large duct cross-sections plus powerful fans are therefore required.

Collector systems
Flat-plate collectors
This is the most common type of collector. In contrast to solid absorbers, the flat-plate collector is fitted with a metal absorber – usually copper – and covered with a pane of transparent, hail-proof safety glass. These days, instead of giving the absorber a coat of matt black paint, special selective coatings are used which absorb virtually all of the solar radiation (up to 95%) and convert this into heat, while exhibiting much lower heat radiation losses (degree of emission ≤12%).

Vacuum-tube collectors
In this type of collector the air between the absorber and the enclosure is evacuated to reduce the convection and conduction heat losses substantially. A collector module contains up to 30 vacuum tubes alongside each other. These tubes merge in a thermally insulated manifold before being connected to the solar energy circuit. We distinguish between two principles:

- the direct connection with a coaxial pipe for the separate flow and return of the heat transport medium
- the indirect, "dry" connection with a "heat-pipe" in which the transport medium and the solar energy circuit are separate

In new products the absorber is also made from a glass tube, which with ever slimmer cross-sections results in an almost transparent appearance. One of the advantages of the high degree of modularity is that it is possible to replace individual tubes while the system is still in operation. The heat losses of vacuum-tube collectors are much lower than those of flat-plate collectors, which is especially advantageous in the case of high operating temperatures (for process heat).

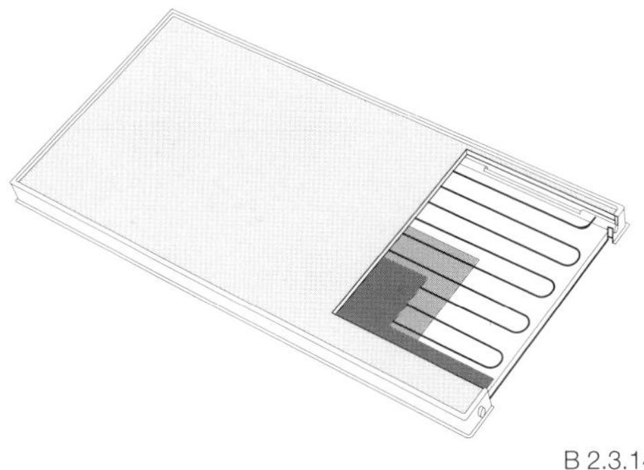

B 2.3.14

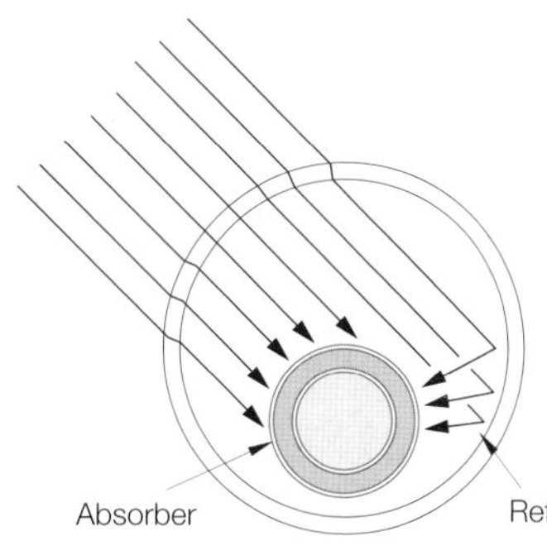

B 2.3.15

B 2.3.16

B 2.3.14 Flat-plate collector
B 2.3.15 Vacuum-tube collector just a few millimetres thick, with glass absorber pipe (production started in 2003)
B 2.3.16 The first application for the vacuum-tube collector shown in fig. B 2.3.15, Centre for Environmental Communication, Osnabrück (D), 2003, Herzog + Partner

Applications

Hot-water heating

The geographical and climatic conditions that prevail in Central Europe mean that solar collectors are primarily used for hot-water heating. The operating temperature lies between 30°C and 60°C. Typical flat-plate collectors achieve favourable degrees of efficiency in this temperature range. As the energy demand for hot water remains more or less constant over the whole year, the high level of solar radiation available in the summer can be used to best effect.
The size of a collector system must be matched to the true energy requirement (number of persons, consumption figures, equipment, etc.) and the level of coverage required. With an optimum south orientation, the hot-water supply for a four-person household can be met by collectors in the facade covering an area of just 6.0–7.5 m² (and a 300 l tank). This means that during the summer months normal hot-water requirements are essentially met, and taken as an average over the year the level of coverage achievable is about 50–60%.

Space heating

Looking at the whole year, in Central Europe the available solar radiation and the space heating requirement do not coincide (and of course are mutually dependent). Whereas about 60% of the annual space heating requirement occurs in the core heating period from November to February, the irradiant energy over the same period for an inclined south-facing surface is only 12–15%. This inescapable fact places greater demands on the options for using a system for solar space heating.

In order to be able to transfer usable heat to the heating circuit, the operating temperature of the space heating must lie between < 30°C (low-temperature heating) and 90°C, depending on the method of heat transfer. Flat-plate collectors with a selective coating and vacuum-tube collectors are suitable for such purposes. A coverage of about 20–25% of the annual heating energy requirement of a detached house calls for a collector area equal to about one-quarter of the heated floor space. In a very well-insulated house this corresponds to collectors covering an area of about 12 m² (vacuum-tube collectors) to 18 m² (flat-plate collectors).

Photovoltaics

A photovoltaic system is a technical installation that converts solar radiation directly into electricity. The heart of such a system is the group of solar cells assembled to form a panel. Such systems generate d.c. electricity which, for the majority of household appliances, must be converted into a 230 V/50 Hz a.c. supply by means of an inverter. Photovoltaic installations are generally operated as mains-coupled systems, i.e., are connected to the public electricity grid, which serves as a storage medium. Stand-alone (autonomous) systems in which the excess electricity is stored in batteries (e.g., rechargeable batteries) are less common.
In terms of the available solar radiation, the degree of exposure and the inclination of the panel surface determine the annual yield of a photovoltaic system. In contrast to thermal collectors, even incident energy < 200 W/m² can still contribute to generating electricity. The maximum annual quantity of radiation in Central Europe for a south-facing rigid array is achieved with an inclination of 30° to the horizontal. However, the yield from a vertical facade panel is very much lower. The output of a photovoltaic system is mostly specified in terms of Wp or kWp, where p stands for "peak". This specifies the peak output that can be transferred to the electric circuit connected to the system. The value is generally based on 1000 W/m² incident energy with a cell temperature of 25°C. Taken as an average over the year (summer/winter, day/night), this figure is roughly one-tenth of the peak output. Where possible, neighbouring buildings or other parts of the building itself should not cast a shadow on the photovoltaic surfaces because even small shadows (e.g. aerials, frame profiles, etc.) can lead to a considerable drop in output. As all the units in a system connected in series are reduced to the smallest output in the system, small areas in the shade can put large panels out of action. Parallel wiring can limit the decrease in output, but with the disadvantages of lower voltages and higher currents.

In principle, when integrating photovoltaic panels into the building envelope we distinguish between rigid and movable systems. The alternatives to the rigidly mounted unit are the one- or two-axis tracking systems. The axis of rotation can be horizontal or vertical, depending on the orientation and the actual installation. Theoretically, photovoltaic panels with two-axis tracking can exploit twice as much solar radiation over the year as a properly aligned rigid array. However, since the yield from a one-axis tracking system is only marginally inferior to that of a two-axis system (owing to the system's energy requirements), it is worth considering whether the more elaborate engineering and the additional integration work is really worthwhile. On the whole, it is necessary to carry out a costs-benefits analysis for a tracking system because, taken as an average over the year, about 50% of the radiation occurs in the form of diffuse radiation. Concentrating the radiation on solar cells with holograms achieves a higher yield with (semi-)transparent panels.

Solar cells

The basic material for typical solar cells is the semiconductor material silicon. Cells made from monocrystalline and polycrystalline silicon are produced in thicknesses of 200–300 μm. In addition, there are also solar cells based on thin-film technology, with copper indium diselenide (CIS) and amorphous silicon being common forms. Solar cells have a relatively low degree of efficiency, which depends on the material of the cell. The theoretical maximum degree of efficiency of conventional (silicon) cells is about 25%. For simplicity, we can distinguish the typical solar cells currently available as follows:

- monocrystalline silicon cells with a very pure, completely consistent crystal lattice structure: difficult and expensive to produce; degree of efficiency of industrially manufactured cells = 15–20%
- polycrystalline silicon cells, characterised by the lower purity of the material and the only partly consistent crystal lattice structure: simpler to produce and hence less expensive; degree of efficiency = 12–17%

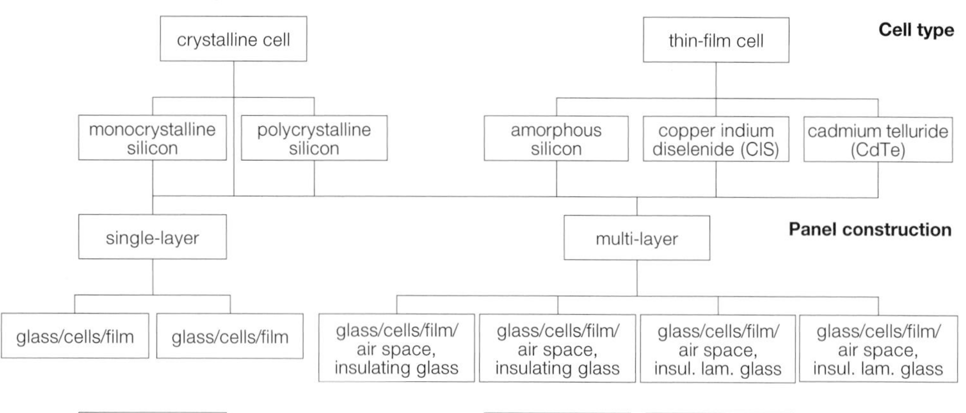

B 2.3.17

B 2.3.17 Classification of photovoltaic systems
B 2.3.18 Photovoltaic cells:
a monocrystalline silicon cells
b amorphous silicon cells, semi-transparent type
c polycrystalline silicon cells
d CIS thin-film cells

Thin-film technology holds great technical and architectural potential. These cells save materials because layers only a few micrometres thick (1–6 µm) are sufficient for light absorption. In addition, there is the possibility of greater automation during production, which promises enormous cost-savings. Thin-film cells exhibit a number of advantages in terms of less dependence on incident energy and temperature, and they can also tolerate more shading. Diffuse and poor light is used (somewhat) better and the decline in output as the temperature rises is less marked. Furthermore, the long, narrow bands of cells prevent individual cells being completely in shadow. We distinguish between:

- amorphous silicon cells: thin-film cells in which the silicon is applied wafer-thin to a backing material by way of vapour deposition; method of manufacture saves costs and materials, degree of efficiency = 5–7%, particularly suitable for coating large areas
- CIS thin-film cells: new cell technology, consisting mainly of copper, indium and selenium; low material requirement; can also be applied to large areas with virtually any shape by means of vapour deposition, degree of efficiency = approx. 10%

The development of multiple layer cells in which two (tandem cells) or three (triple cells) layers are applied one over the other leads to an improvement in the degree of efficiency. To optimise the output of the cells, for example, each layer of the triple cell is designed for a different spectral range (short-, medium-, long-wave radiation).
Another advantage of thin-film technology is its relatively free formability. As these cells, unlike the crystalline types, are not limited to the standard wafer sizes, the panels can be designed with different geometrical shapes and sizes, and also fitted to curved or flexible backing materials. This type of cell is ideal for integrating into areas of the building where adequate ventilation is not always guaranteed or (partial) shading can occur.

The appearance of the panels is dominated by homogeneous areas in which texture is provided by extremely narrow and transparent dividing lines. These are due to the method of manufacture, i.e., the electrical isolation and wiring of the layers.

Varying the width or including additional horizontal dividing lines enables this effect to be used as an intentional architectural feature. While the colour range of the crystalline cells can be expanded by reflective layers, the darker colours, from black to reddish brown or dark green (still) dominate thin-film technology.

Photovoltaic panels
As a rule, about 30–40 cells are fitted together to form a large prefabricated unit with an area of 0.5–1.0 m². These photovoltaic panels have a multi-layer construction, i.e., the cells are either embedded in a synthetic resin between two panes of glass, or placed between glass and a plastic laminate. The rear face is either opaque, translucent (obscured glass/light-scattering film) or transparent (clear glass/transparent film) depending on requirements. Amorphous silicon cells can also be applied to flexible backing materials like plastic films. In addition, "sawn", semi-transparent cells are available and thin-film cells with printing (but not covering the whole cell) as well. Many manufacturers offer panels with various standard dimensions and designs.

Integration of solar energy systems
When we come to consider the constructional aspects of the integration of solar energy systems, we notice first that the installation conditions – especially the types of fixing and the sealing around the edges – are being constantly refined and improved by the manufacturers. Besides simplifying assembly and speeding up installation, new types of frame section minimise the height and the visible width. In the meantime, solar energy systems can be integrated into the building envelope relatively "flexibly" thanks to numerous options. Further, more and more complete solutions are appearing in which thermal energy and photovoltaic systems can be combined within one

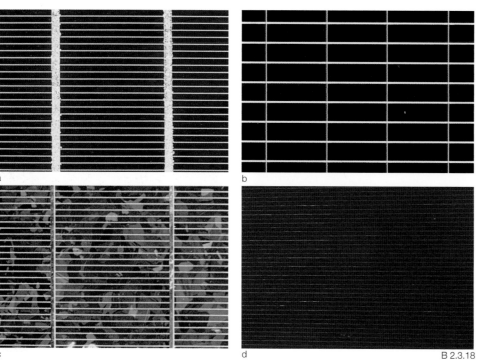

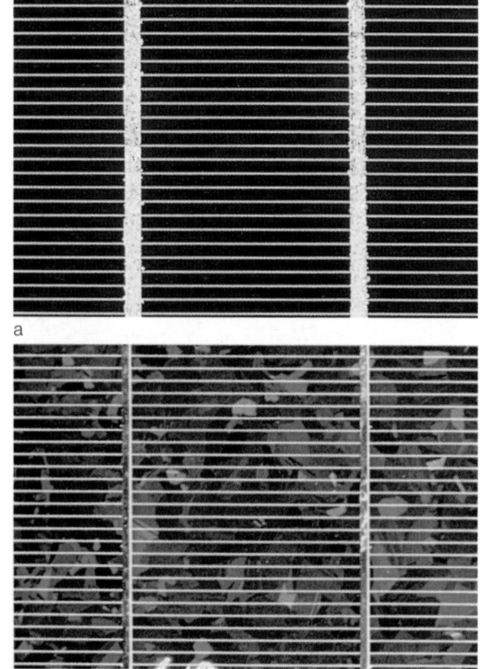
a
b
c
d
B 2.3.18

B 2.3.19 Energy gains for different orientations and inclinations of photovoltaic panels (100% = 1055 kWh/m²a)
B 2.3.20 Hot-rolled strip slitting plant, Duisburg (D), 1962/2002, Cerny und Gunia
B 2.3.21 "Technology and Business Start-up Centre", Herten (D), 1992, Kramm + Strigl

Inclination of surface	0 °	30 °	60 °	90 °
Orientation				
East	93 %	90 %	78 %	<60 %
South-east	93 %	96 %	88 %	66 %
South	93 %	100%	91 %	68 %
South-west	93 %	96 %	88 %	66 %
West	93 %	90 %	78 %	<60 %

B 2.3.19

B 2.3.20

type of construction and with other building envelope elements. Consequently, a multitude of tried-and-tested systems are available for all the more usual types of facade [10].

It is essential that collectors and photovoltaic panels be integrated into the building services. This may require service ducts and additional equipment, depending on the type of use. Owing to the relatively slim assemblies and flexible cables with small cross-sections, photovoltaic systems are ideal for integrating into facades. Water collectors, on the other hand, require much larger pipes and attention must be paid to sealing the units, and such systems are usually filled with an anti-freeze agent.

In terms of the criteria governing the architectural language, solar energy systems can employ a wide range of design options for their integration. These days, manufacturers try to comply with all the architect's wishes. The range of colours for absorber surfaces and the numerous frame sections available affect the appearance of the systems just as much as the junction pieces that connect panels to the roof covering or the facade. Architects are frequently told that the wide range of colours is a special bonus offered by photovoltaics. However, the use of additional colours, likewise additional shapes, in the building envelope is a particularly sensitive area with respect to the appearance and needs to be handled very carefully.

But the architectural integration of solar energy systems into the building envelope has a far greater significance. It represents the insertion of a component into the walls and roof such that this component takes on the same functional and constructional tasks. These requirements and characteristics of the structure must be correlated with the architectural criteria and the rules governing the energy system in an overall concept. The appropriateness of the integration is influenced by the design, the materials and the surface finishes, also the size, proportion and subdivision of the components. These must always take into account the constructional system as a whole [11].

Notes

[1] The photovoltaic panels and vacuum-tube collectors were used for the first time in 1982 in this residential development in Munich (D) by Thomas Herzog and Bernhard Schilling together with the Fraunhofer Institute for Solar Energy Systems in Freiburg.
[2] Herzog, Thomas et al.: *Gebäudehüllen aus Glas und Holz. Maßnahmen zur energiebewussten Erweiterung von Wohnhäusern*, Lausanne, 1986, pp. 8 & 15.
[3] Koblin, Wolfram et al.: *Handbuch Passive Nutzung der Sonnenenergie*, series 04 "Bau- und Wohnforschung" published by the Federal Ministry for Regional Planning, Building and Urban Development, issue 04.097, Bonn, 1984, pp. 93–99.
[4] ibid., pp. 118 & 135ff.
[5] Schild, Kai; Weyers, Michael: Transparente Wärmedämmsysteme (TWDS); in: Schild, Kai; Weyers, Michael: *Handbuch Fassadendämmsysteme*, Stuttgart, 2003, pp. 151–68.
[6] Although frequently called transparent thermal insulation, the adjective "transparent" is misleading because these materials, although permeable (i.e., transparent) to radiation, are very limited in terms of being able to look through them. Since, in the building industry, we make a clear distinction between "translucent" and "transparent", we should really speak of translucent thermal insulation.
[7] Hebgen, Heinrich: *Bauen mit der Sonne.* Vorschläge und Anregungen, pub. by RWE-Anwendungstechnik, Essen/Heidelberg, 1982, pp. 81 & 88.
[8] Von Primus, Illo-Frank (ed.): *Massivabsorber.* Die Wärmequelle für die Wärmepumpe, Düsseldorf, 1995, pp. 34ff.
[9] Krippner, Roland: Holzleichtbeton; in: DBZ 12/2002, p. 76.
[10] Krippner, Roland: Die Gebäudehülle als Wärmeerzeuger und Stromgenerator; in: Schittich, Christian (ed.): *Building Skins*, Munich/Basel, 2001, pp. 55–58.
[11] Krippner, Roland: Architektonische Aspekte solarer Energietechnik; in: *9th Symposium on Thermal Solar Energy*, proceedings, Regensburg, 1999, p. 237.

B 2.3.21

Housing complex

Batschuns, A, 1998

Architect:
Walter Unterrainer, Feldkirch

db 10/2000
Detail 03/1999

- Active solar technology integrated into the building envelope
- Group of four two-storey and two three-storey housing units according to low-energy design standards
- Compact structures, high degree of insulation and airtightness to save on heating systems
- Heating requirement met by controlled ventilation and heat pump
- Water collectors in the facade and on the flat roof (with 750 l tank per housing unit) for hot-water heating

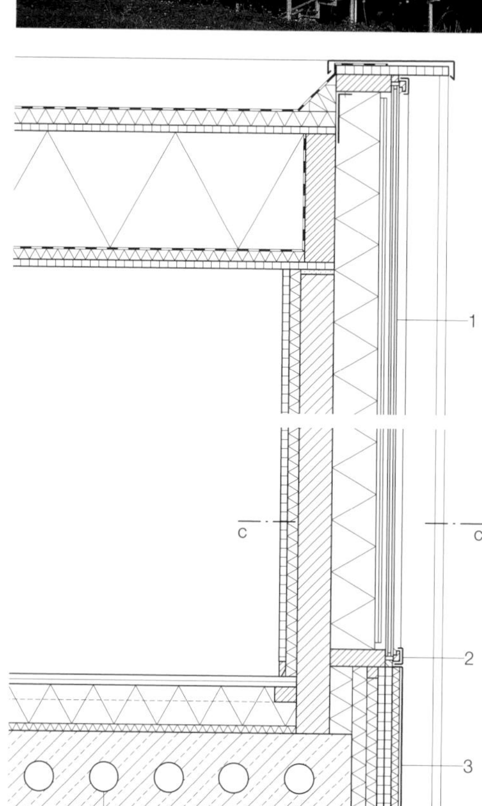

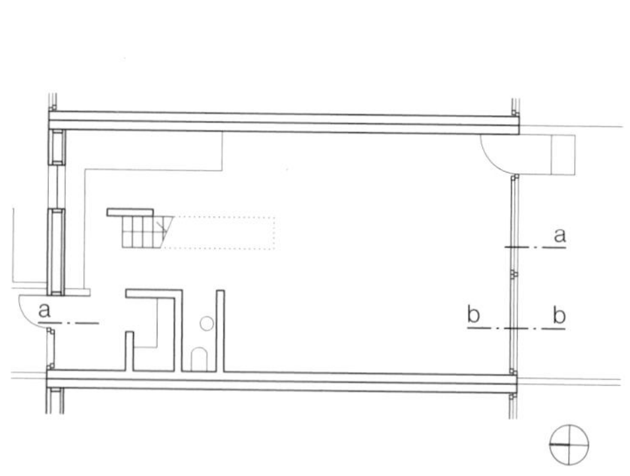

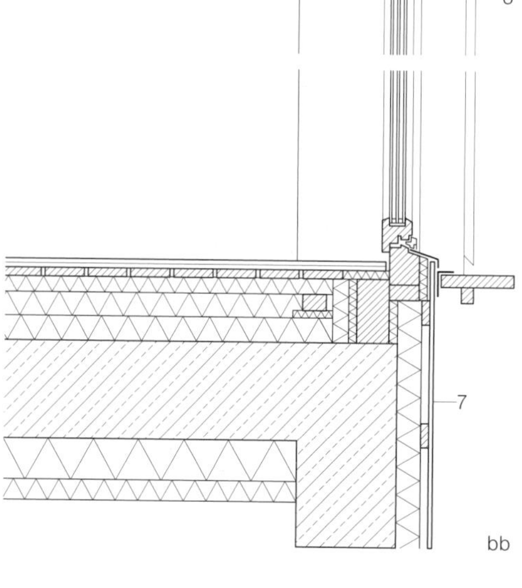

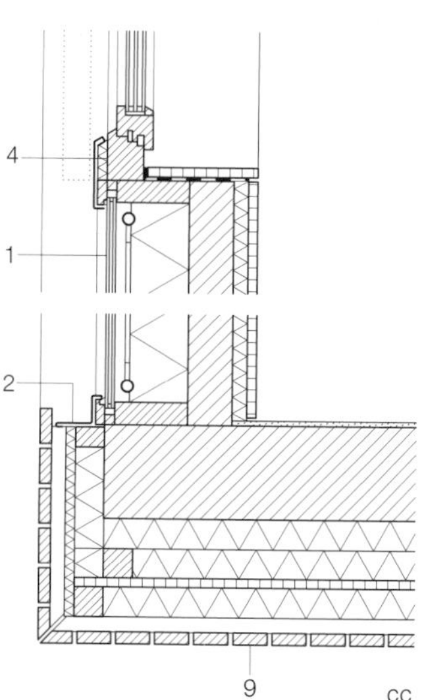

Section scale 1:250
Vertical section • Horizontal section
scale 1:20

1 South facade:
insulating glass
hot water collector/absorber
120 mm mineral wool insulation
90 mm clay masonry
30 mm flax insulation
19 mm 3-ply core plywood
2 Aluminium clamping batten
3 Panel construction:
sheet aluminium, bent to suit
20 mm foam insulation
2 No. 19 mm 3-ply core plywood
40 + 30 mm thermal insulation
4 Wooden window, larch with aluminium cover
5 Timber batten, 40 x 140 mm
6 Triple glazing
7 Plinth construction:
fibre-cement sheet on framework
60 mm peripheral insulation
250 mm reinforced concrete
8 Reinforced concrete floor, 240 mm, with 80 mm dia. ventilation pipes
9 Facade construction:
24 mm vertical larch boards
30 x 50 mm battens
60 mm foam insulation
18 mm 3-ply core plywood
2 No. 60 mm foam insulation
180 mm porous clay masonry
8 mm plaster
10 Aluminium louvre blind

House of the future

Wildhaus, CH 1999

Architects:
Architheke, Brugg
Beat Klaus

Der Architekt 11/2002
bauen mit holz 10/2000
mikado 01/2000

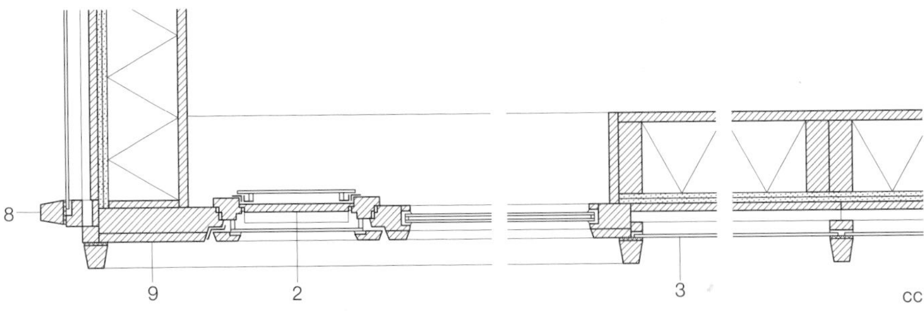

- "Lucido" facade, a development from Giuseppe Fent and Hermann Blumer
- External wall system achieves maximum efficiency in passive solar energy utilisation
- Special routing in the timber boards results in good insulation properties but also good protection against overheating in summer without the need for additional shading measures
- Prototype building exclusively in timber

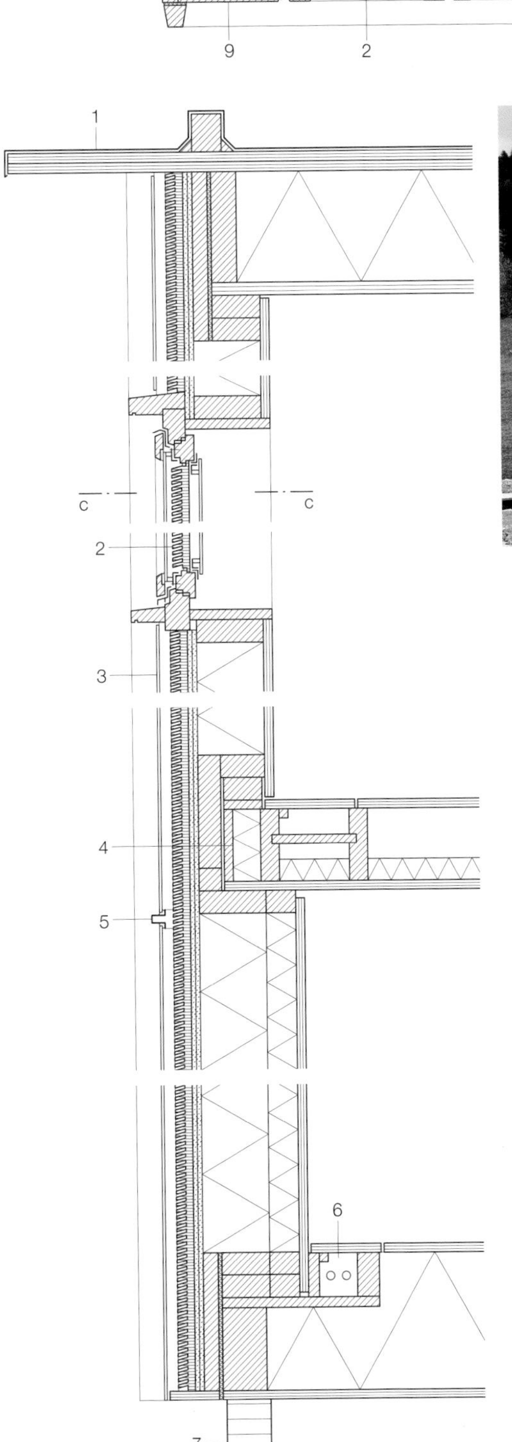

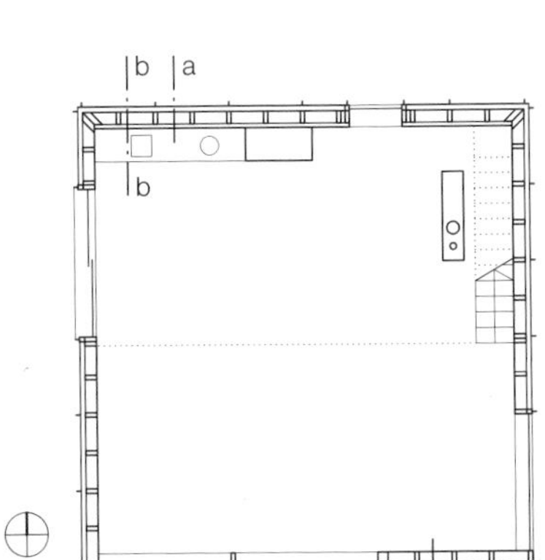

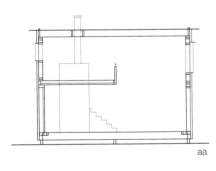

Section • Plan of ground floor
scale 1:200
Vertical section • Horizontal section
scale 1:20

1 Sheet zinc on 3-ply core plywood, larch outer ply
2 Ventilation flap, 2 No. 6 mm toughened safety glass, satin finish, with timber end-grain "nuggets" in between
3 Facade construction:
6 mm low-iron solar-control toughened safety glass
30 mm ventilated cavity
absorber, timber end-grain "nuggets"
plasterboard
cellulose insulation
3-ply core plywood, larch
4 Thermal insulation, 80 mm
5 Sheet metal, bent to suit
6 Floor duct, 117 x 150 mm
7 Gluelam plinth
8 Larch cover strip
9 Larch fascia board

Factory building

Eimbeckhausen, D 1992

Architect:
Thomas Herzog, Munich
with Bernd Steigerwald, Holger Gestering

Arch+ 126, 1995
Architectural Review 01/1994Flagge, Ingeborg (ed.) et al.: Thomas Herzog Architektur + Technologie, Munich/London/New York, 2001
Götz, Gutdeutsch: Building in Wood, Basel/Berlin/Boston, 1996, p. 56
Herzog, Thomas et al.: Timber Construction Manual, Basel/Berlin/Boston, 2004, p. 359

- Factory building based on ecological aspects, i.e., functionally differentiated building concept, timber loadbearing structure and timber facades
- Natural lighting and ventilation to production areas, transparent thermal insulation panels also used to redirect incoming daylight
- Photovoltaic canopy with frameless semi-transparent ASI panels (4 kWp) for the power supply to the electric fork-lift trucks
- Extensive planting on roof of building to protect against overheating in summer, reduce noise emissions and delay rainwater run-off

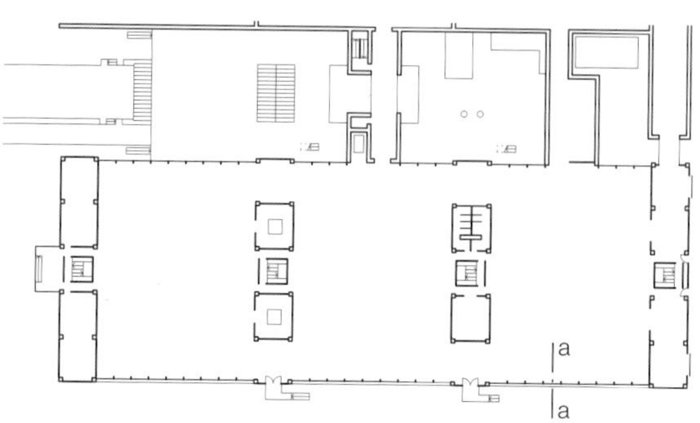

Solar energy

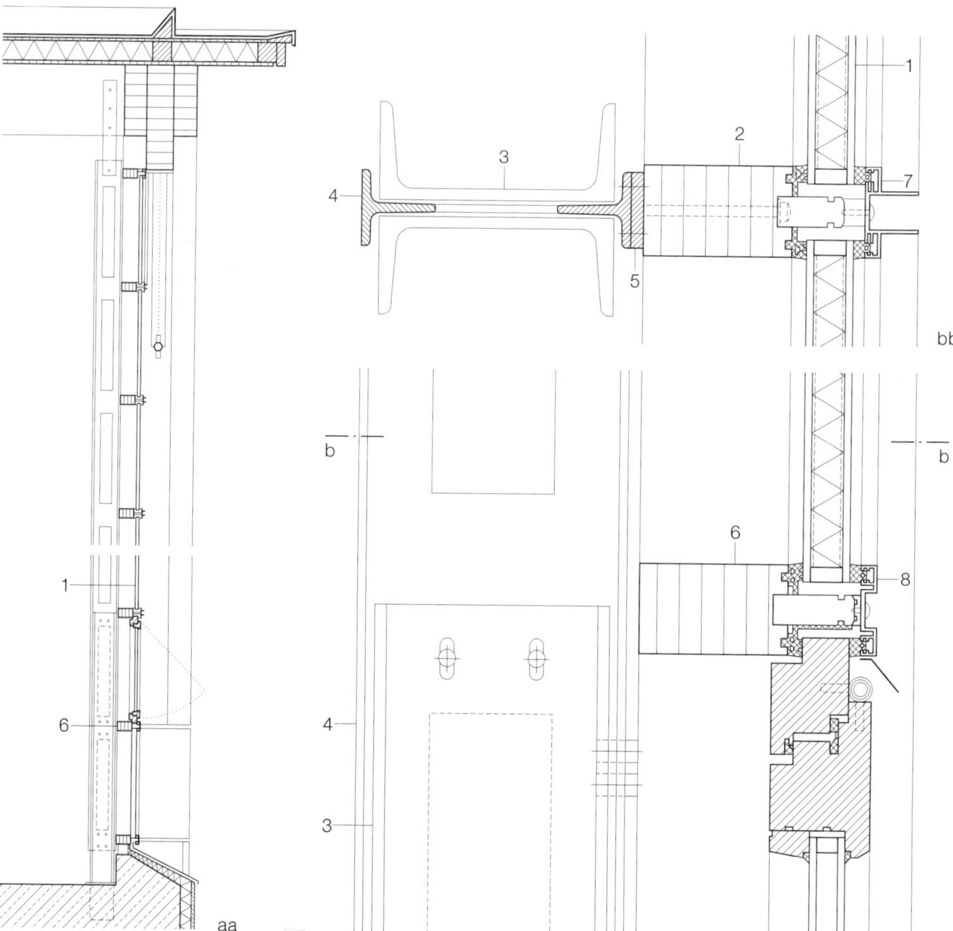

Plan scale 1:1500
Vertical section scale 1:50
Details scale 1:5

1 Transparent thermal insulation element:
5 mm float glass
glass fibre fleece
24 mm capillary infill
glass fibre fleece
5 mm float glass
2 Glulam post, 60 x 100 mm
3 2 No. steel channels
4 2 No. steel T-sections, glazing bar for resisting wind forces, welded to steel flat
5 Steel flat, 50 x 40 x 10 mm
6 Glulam rail, 60 x 100 mm
7 Extruded aluminium section, vertical
8 Extruded aluminium section, horizontal

Solar house

Ebnat-Kappel, CH, 2000

Architect:
Dietrich Schwarz, Domat/Ems
Structural engineers:
Conzett Bronzini Gartmann, Chur

Erneuerbare Energien 05/2001
Detail 06/2002

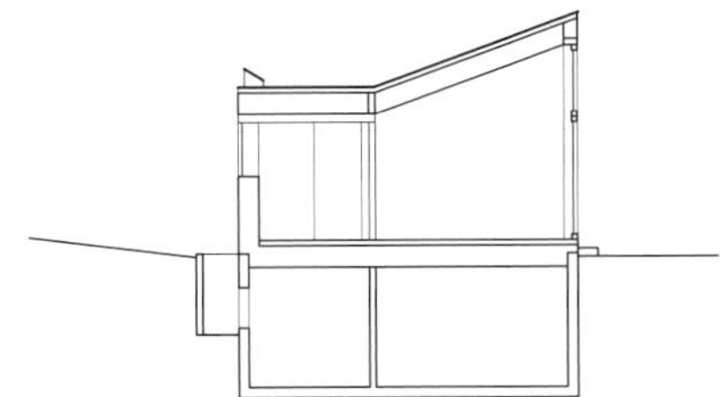

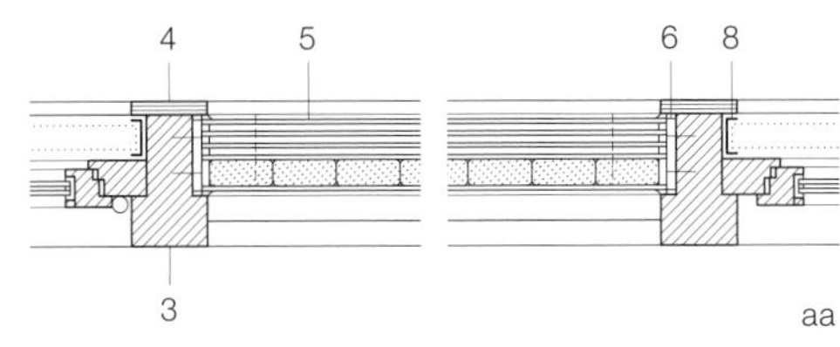

- Light-permeable passive solar collector wall with infill of dyed plastic containers filled with paraffin, developed by the architect and used by him for the first time
- Outer leaf of prismatic glass prevents overheating in summer
- Winner of Swiss Solar Prize 2001 as "best-integrated solar system"

Section scale 1:250
Horizontal section • Vertical section
scale 1:20

1 3-ply core plywood, untreated larch
2 Gluelam beam, oiled spruce, 2 No. 200 x 80 mm
3 Gluelam column, oiled spruce
4 Wooden glazing cap
5 Passive solar collector wall:
6 mm toughened safety glass with low-e coating
12 mm argon filling
6 mm PMMA prisms
12 mm argon filling
6 mm toughened safety glass with low-e coating
100 x 100 x 40 mm self-supporting paraffin containers
6 mm toughened safety glass with silk-screen printing
6 GFRP bracket, structural sealant connection
7 Stainless steel cover strip
8 Guide track for sunshade
9 Oak shims, 2 mm
10 Oak strip, 20 mm, to compensate for unevenness
12 Reinforced concrete, 200 mm

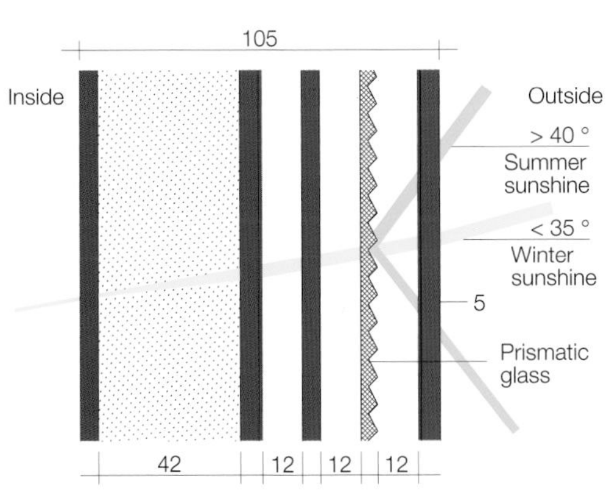

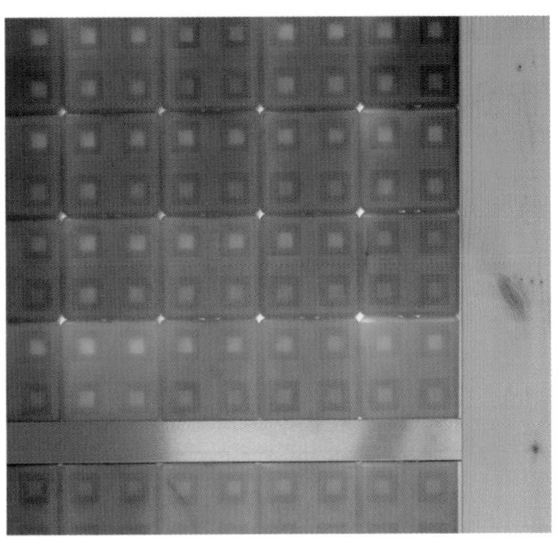

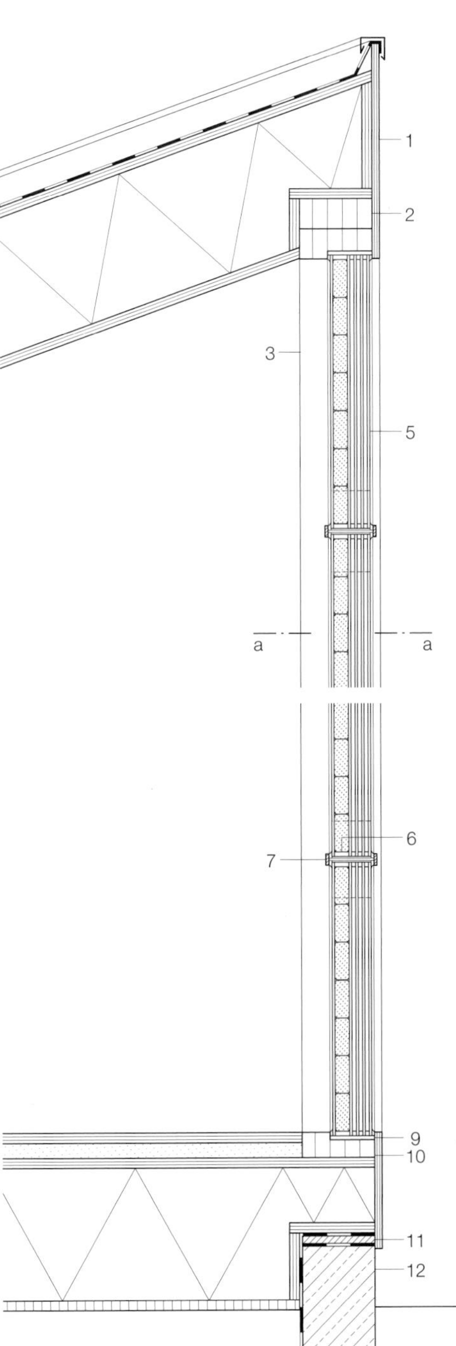

House and studio

Gleißenberg, D, 2001

Architect:
Florian Nagler, Munich

archicrée 309, 2003
db 01/2003
Bavaria Chamber of Architects (pub.): Bavaria Architecture Yearbook, Munich, 2002
Schunk, Eberhard et al.: Roof Construction Manual – Pitched Roofs, Basel/Boston/ Berlin, 2003, pp. 342-43

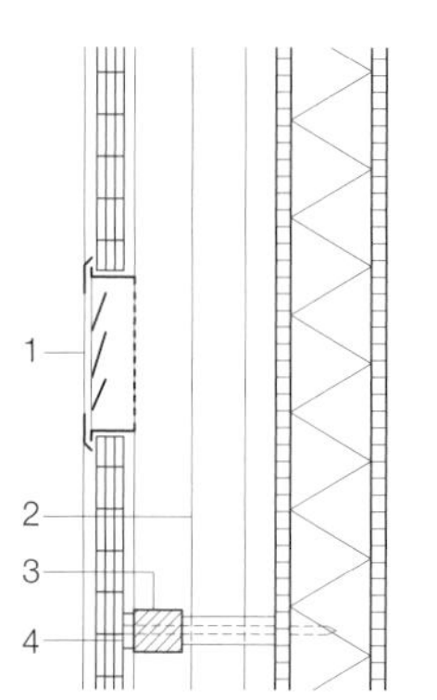

- Passive solar collector wall (multiple-web sheet/simple timber panel construction)
- A set-back basement positioned across the slope supports the two-storey building
- Transparent weatherproof outer skin made from plastic as a low-cost and weather-resistant material
- Gable ends translucent, eaves sides function as temperature buffer and weather protection to timber wall behind
- Roof covering of red cedar shingles

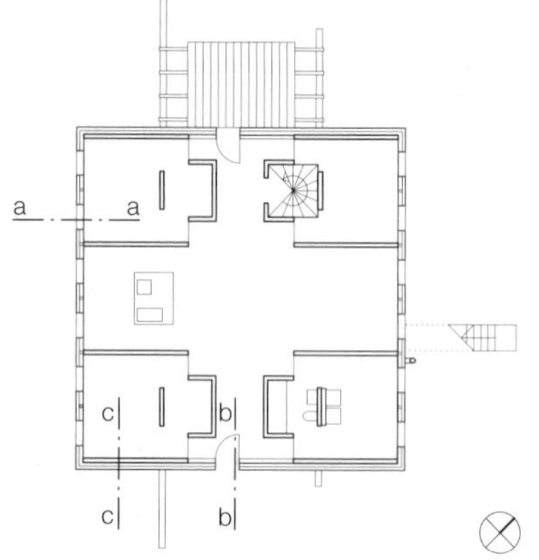

Plan scale 1:400
Ventilation opening, vertical
scale 1:20
Vertical sections • Horizontal section
scale 1:20

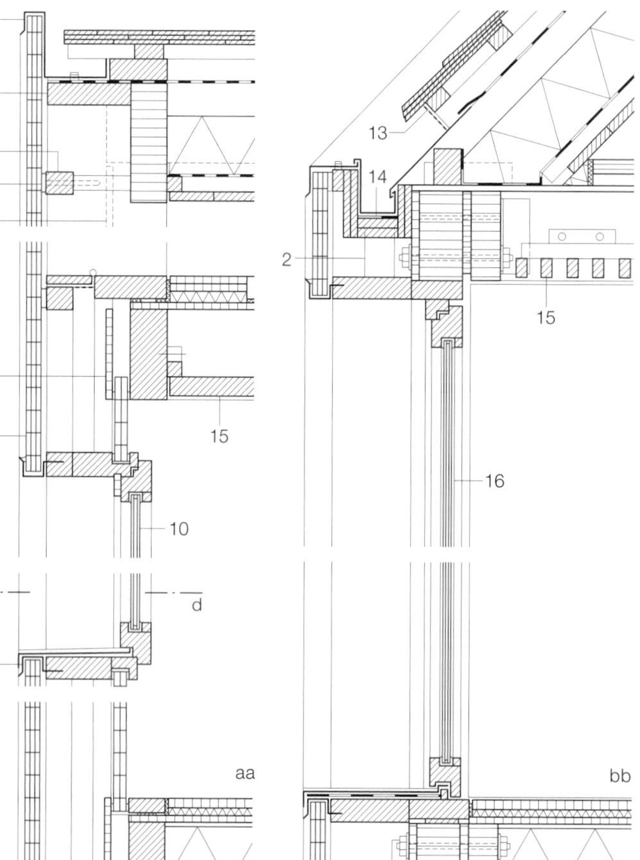

1 Weather and insect screen, screw fixings, bottom edge perforated to enable drainage of condensation water
2 Rainwater downpipe, galvanised steel, 40 dia. x 2 mm
3 Timber batten, 60 x 80 mm, screw fixings
4 Flat suction tie, screw fixings
5 Sheet aluminium, bent to suit, butt joints covered with aluminium foil
6 Verge gutter board
7 Oriented strand board to clamp polycarbonate sheets in place at corner of building
8 Oriented strand board, 18 mm
9 Wall construction:
PC multiple-web sheet
220 mm ventilated cavity
polycarbonate sheet
10 Wooden window, insulating glass
11 Sheet aluminium, bent to suit, as non-rigid clamp to allow for longitudinal expansion of polycarbonate sheets
12 Timber board, 60 x 240 mm, with ventilation inlets
13 Insect screen
14 Box gutter, sheet titanium-zinc on separating layer
15 Timber battens, 30 x 50 mm
16 Wooden door, insulating glass
17 Wall construction:
PC multiple-web sheet
220 mm ventilated cavity
22 mm oriented strand board
120 mm thermal insulation
22 mm oriented strand board
18 Galvanised steel bracket
19 Timber facade post
20 Junction with partition

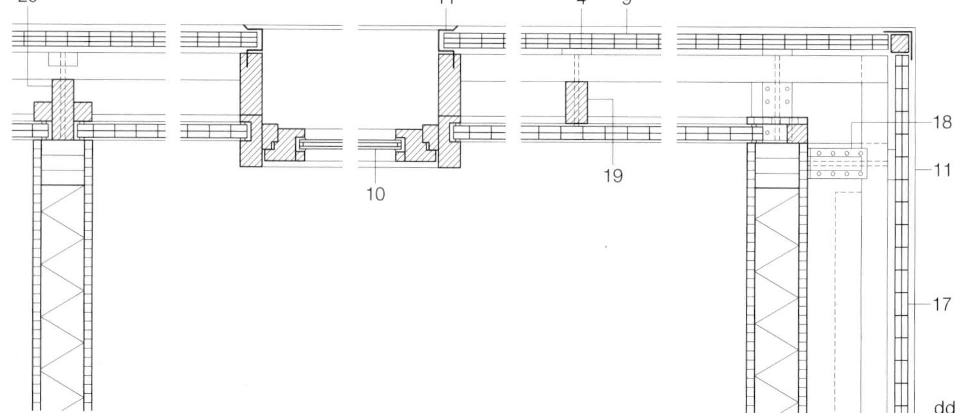

Pharma Service Center

Binzen, D, 2003

Architects:
Pfeifer Roser Kuhn, Freiburg
Project manager:
Wolfgang Stocker
Facade engineer:
Silke Gauthier, Radebeul

Der Architekt 11/2002
DBZ 01/2003

- Production, logistics and office building
- Large thermal mass in external walls and ground floor slab, plus building zoning, regulates the large heat emissions from the production process
- Wall functions as air collector
- Controlled cavity ventilation helps to cool concrete wall in summer by means of natural thermal buoyancy; in winter the air heated by solar gains reduces the heat losses

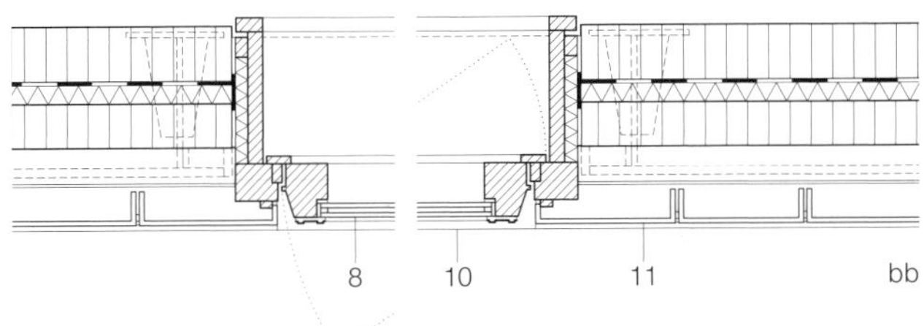

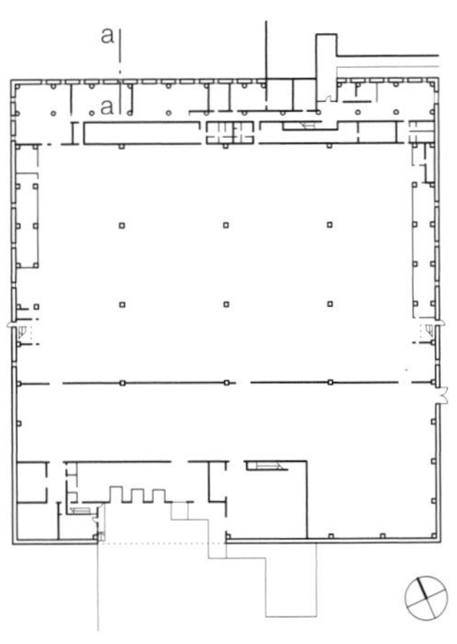

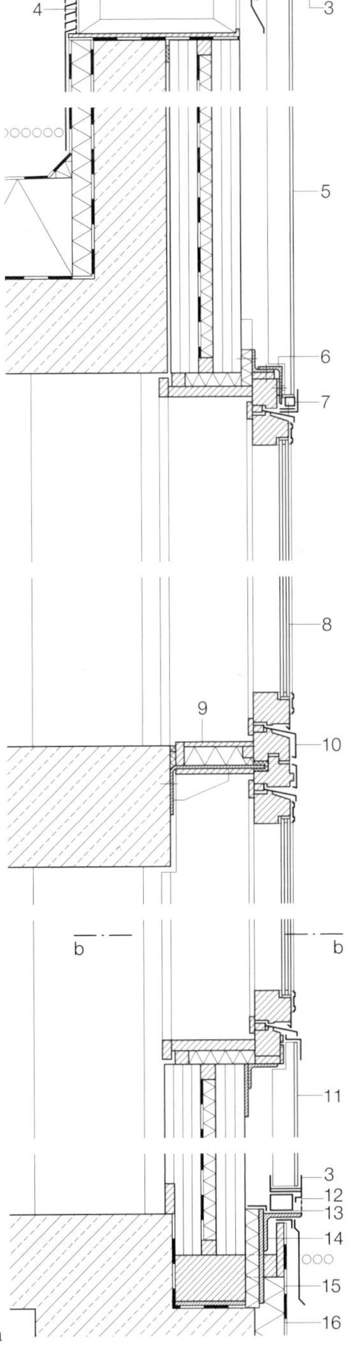

Plan scale 1:2000
Vertical section • Horizontal section
scale 1:20

1 Sheet aluminium capping, 1.5 mm
2 Loadbearing steel angle
3 Aluminium channel
4 Ventilation device with screen to protect against weather
5 Parapet construction:
patterned glass
135 mm ventilated cavity
double-leaf edge-glued timber element
mineral wool with roofing felt (open to diffusion), 80 + 40 + 80 mm, in between
separating layer
200 mm reinforced concrete
vapour barrier
60 mm PUR rigid foam
synthetic roofing felt
6 Galvanised steel angle
7 Aluminium tube, 32 x 25 mm
8 Wooden window, larch, with insulating glass
9 Floor panel:
13 mm larch
50 mm sound insulation
continuous steel bracket
13 mm plywood
10 Aluminium flashing
11 Wall construction:
patterned glass
150 mm ventilated cavity
double-leaf edge-glued timber element mineral wool with roofing felt (open to diffusion), 80 + 40 + 100 mm, in between
vertical acoustic profiling on inside
12 Insect screen
13 Aluminium tube
14 Loadbearing steel angle
15 Peripheral insulation, 80 mm
16 Waterproofing

Business start-up centre

Hamm, D, 1998

Architects:
Hegger Hegger Schleiff, Kassel
General contractor:
Hering Bau, Burbach
Building services:
Gerhard Hausladen, Munich
Rempe + Polzer, Giessen

DBZ 10/1998
Hausladen, Gerhard (ed.): Innovative Gebäude-, Technik- und Energiekonzepte, Munich, 2001
Pfeifer, Günter et al.: Masonry Construction Manual, Basel/Boston/Berlin, 2001, pp. 331–33

- Business start-up centre on site of disused colliery
- Complex consists of a four-storey office block plus business units in single-storey wings
- Heavyweight masonry office block with upper floors in timber-concrete composite construction (with edge-glued timber elements)
- Business units heated via ground duct (exploits cooling/heating effect of soil) or via four-storey-high collector facade (120 m²)
- Winner of architecture prize "Architecture and Solar Thermal Energy System 2000"

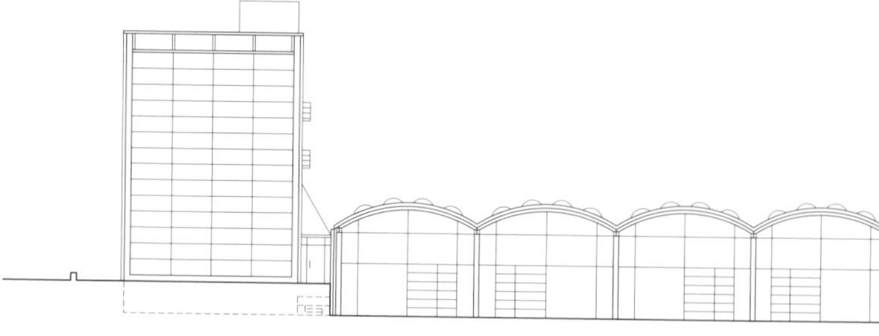

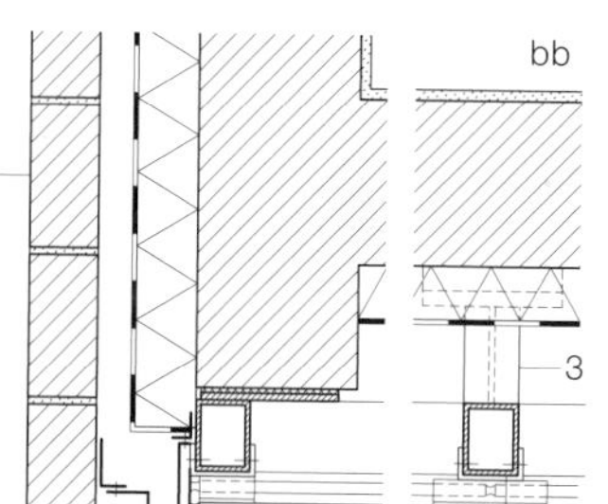

1 Parapet coping, timber plank with sheet zinc covering
2 Steel frame construction 100 x 80 x 4 mm steel support for collectors/ ventilation grille
110 mm ventilated cavity
airtight membrane
80 mm thermal insulation
240 mm calcium silicate masonry
15 mm plaster
at parapet only:
waterproofing
80 mm thermal insulation
20 mm render
3 Steel section, IPE 120, with end plate, EPDM pad serves as thermal break and compensates for tolerances
4 Wall construction:
217 x 100 x 66 mm recycled bricks
50 mm ventilated cavity
airtight membrane
90 mm thermal insulation
240 mm calcium silicate masonry
15 mm plaster

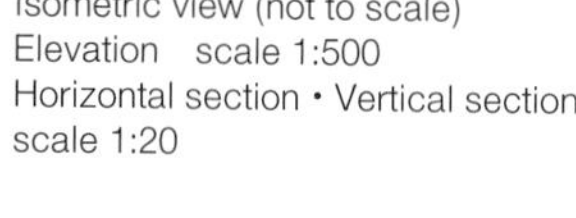
Isometric view (not to scale)
Elevation scale 1:500
Horizontal section • Vertical section
scale 1:20

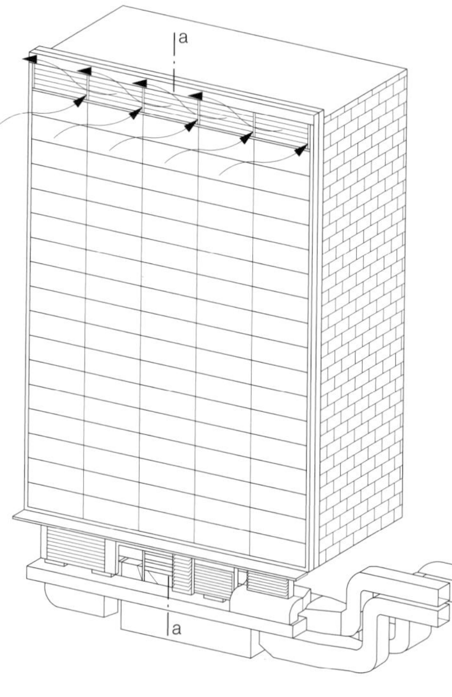

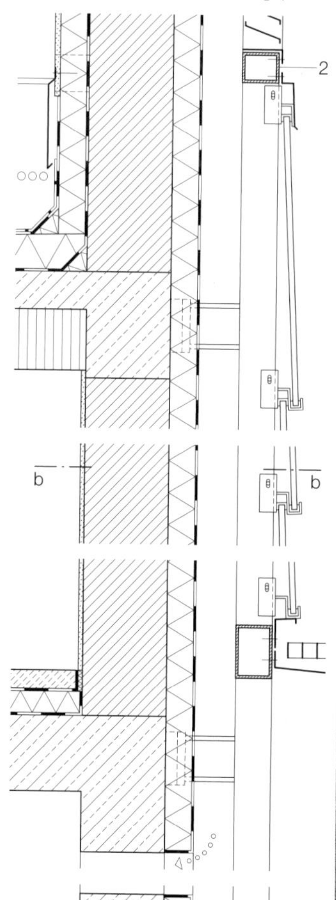

Private house

Herisau, CH, 1998

Architect:
Peter Dransfeld, Ermatingen

Detail 03/1999

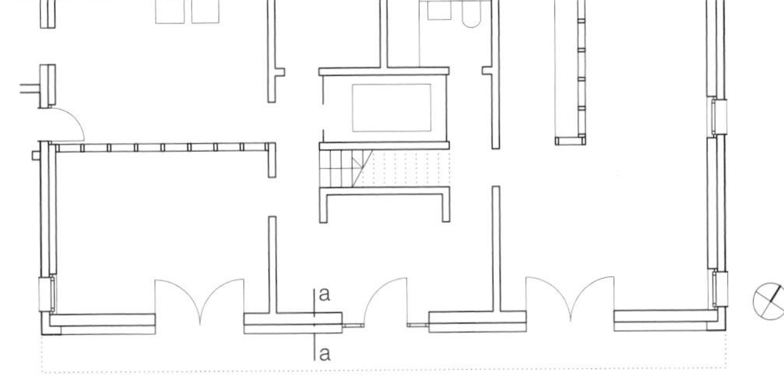

- Energy concept based on a compact, highly insulated structure with transparent thermal insulation in front of the south-facing masonry wall
- Central wood-burning night storage stove for meeting the heating requirement
- Vacuum-tube collectors behind the building
- Shading louvres protect top section of transparent thermal insulation against overheating, integral plastic prisms protect lower section

Plan of ground floor
scale 1:200
Section through south facade •
Horizontal section through south-east corner
scale 1:20

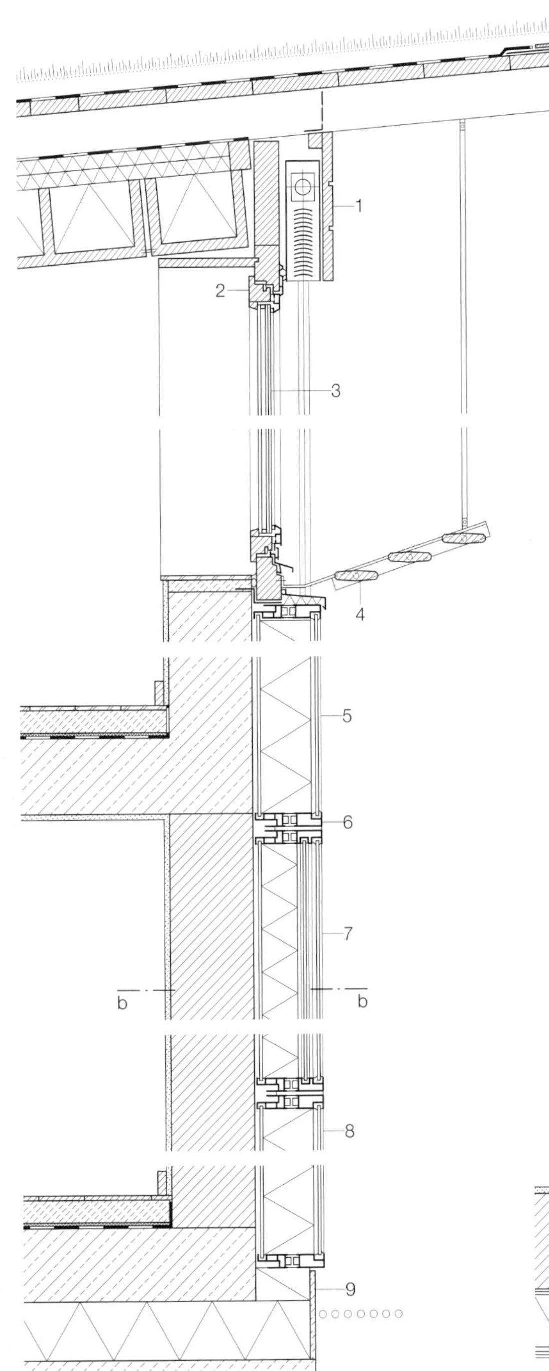

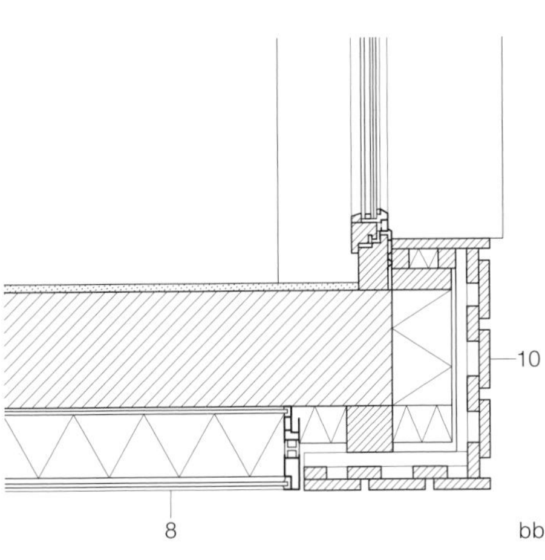

1 3-ply core plywood, spruce, with horizontal grooves to accommodate stresses
2 Wooden window frame with aluminium facing
3 Triple glazing
4 Sunshading louvres, solid wood, for screening the upper section of transparent thermal insulation
5 Transparent thermal insulation in aluminium frame:
5 mm low-iron solar-control glass
12 mm cavity
140 mm plastic tube insulation
5 mm glass
cavity
250 mm reinforced concrete, outer face painted black
15 mm plaster
6 Extruded aluminium section, powder-coated, with thermal break
7 Transparent thermal insulation in aluminium frame:
5 mm low-iron solar-control glass
plastic prism sheet in cavity for reflecting radiation in summer
100 mm plastic tube insulation
5 mm glass
cavity
250 mm calcium silicate masonry, outer face painted black
15 mm plaster
8 Transparent thermal insulation as No. 5 but without shading element
9 Fibre-cement sheet
10 Corner construction:
vertical rough-sawn timber boards, spruce, with 3-coat high-build glaze finish, red
ventilated cavity
140 mm thermal insulation

Technical college

Bitterfeld, D, 2000

Architects:
scholl, Stuttgart
Building services:
ARGE HLSE, Leipzig/Bitterfeld
Facade consultants:
PBI, Wiesbaden

AIT 05/2001
Bauwelt 26/2001
Beton Prisma 81, 2002
Intelligente Architektur 30, 2001
L'ARCA 178, 2003

- New complex to complement existing arts centre (1953) and swimming pool
- Low-energy design
- Opaque fair-face concrete surfaces
- Multi-storey collector wall on south side, 70 m long, integrated into fair-face concrete
- Further targets: use of ecologically neutral materials, seepage of rainwater on the plot

Plan of ground floor
scale approx. 1:3000
Section scale 1:500
Vertical section scale 1:20
Vertical section • Horizontal section
scale 1:5

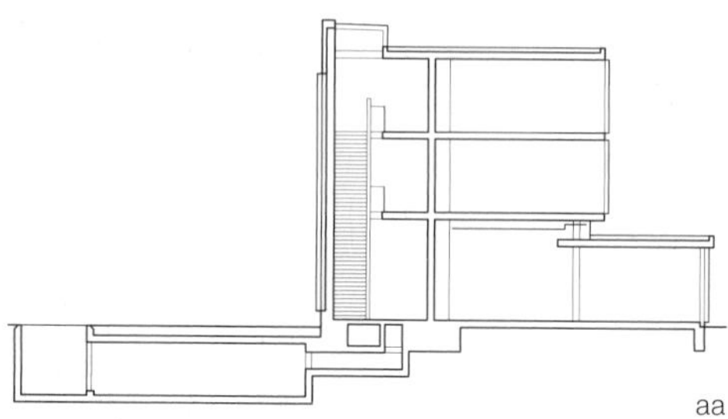

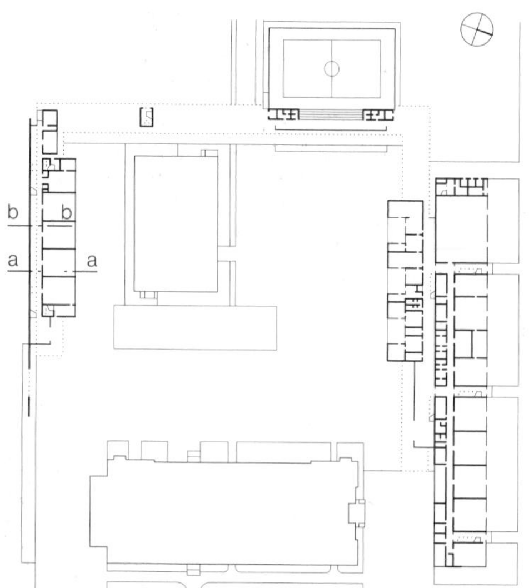

1 Precast concrete parapet, 170 mm, fair-face finish
2 Parapet construction:
80 mm mineral fibre thermal insulation
350 mm reinforced concrete, fair-face finish to inside
3 Collector wall:
4 mm solar-control toughened safety glass
water collectors, copper absorber with selective coating
multi-ply pine board backing
vertical timber framing
in 80 mm ventilated cavity
horizontal timber framing
between 120 mm thermal insulation
350 mm reinforced concrete, fair-face finish to inside
4 Horizontal glazing cap, anodised aluminium
5 Fresh-air inlet, aluminium grille on steel brackets
6 Gap for drainage
7 Sheet aluminium, bent to suit
8 Splice plate
9 Sheet metal side cladding
10 Insect screen
11 Water run-off membrane
12 Butt joint between collector elements
13 Permanently elastic seal

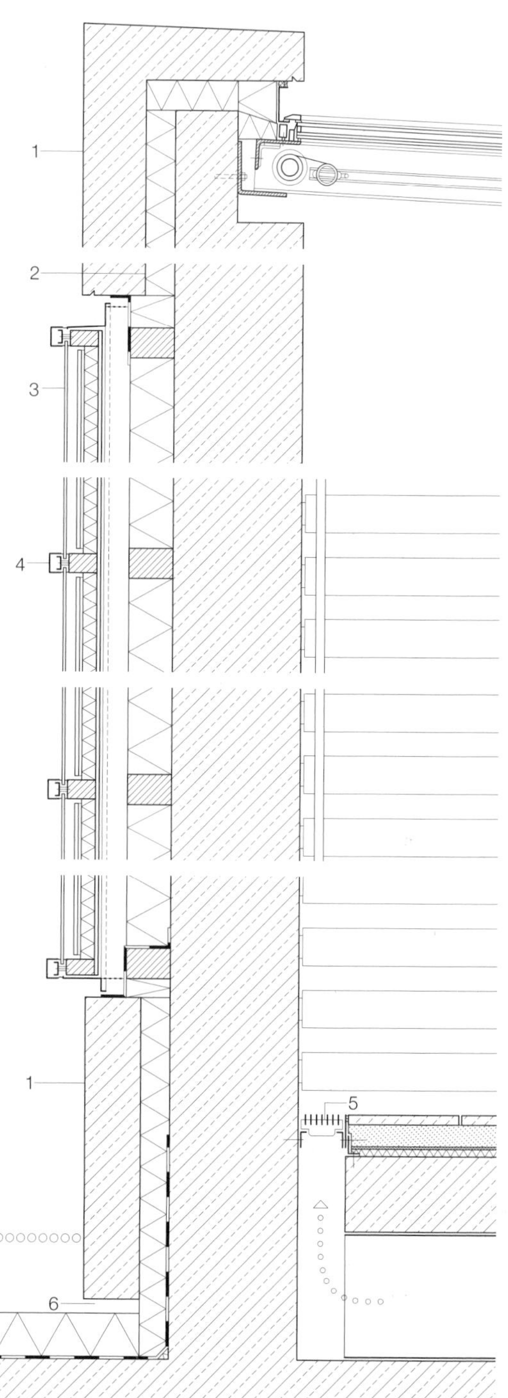

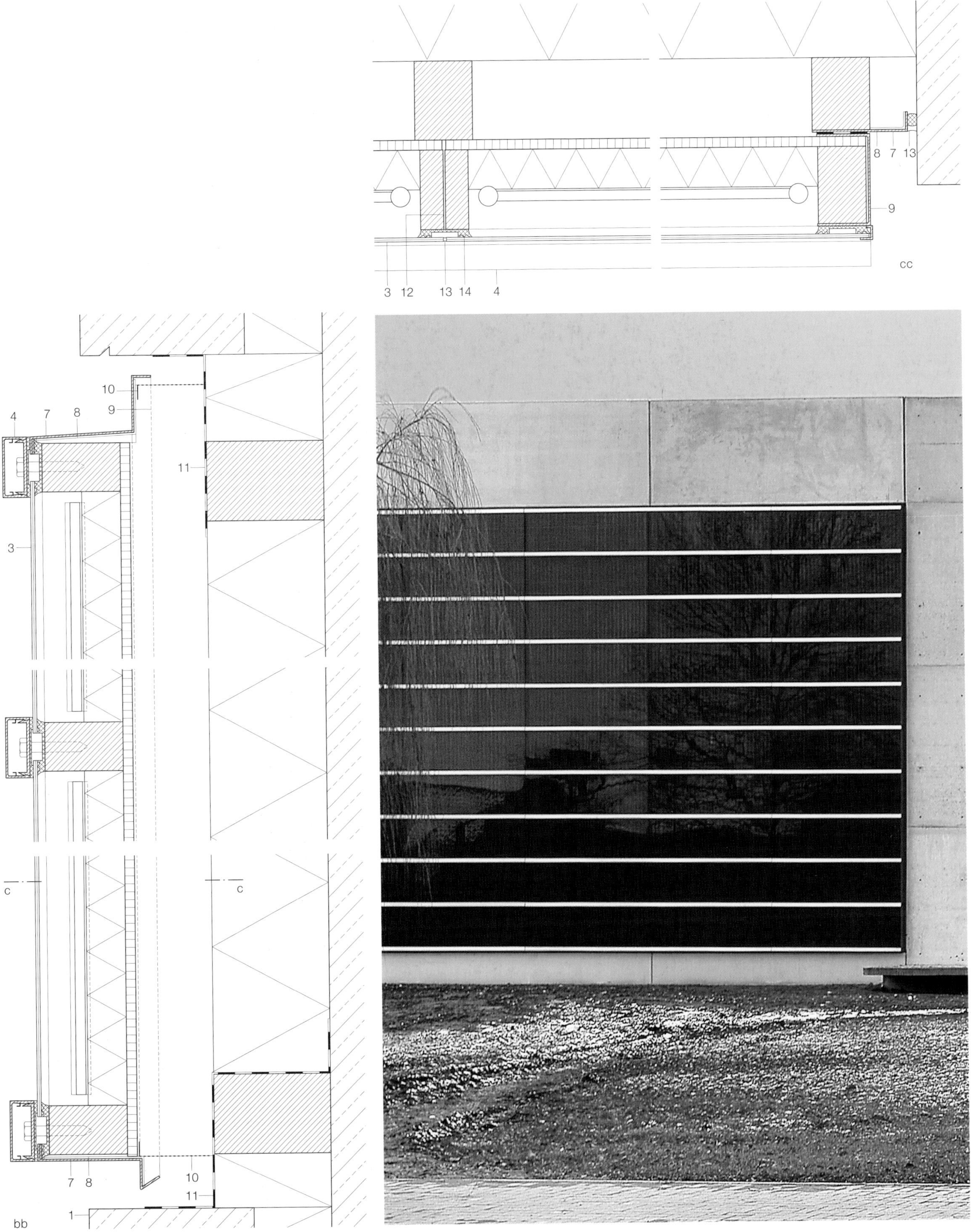
8 7 13
9
cc
3 12 13 14 4
10
4 7 8 9
11
3
c c
7 8 10
11
1
bb

Office building

Freiburg, D, 2001

Architects:
Harter + Kanzler, Freiburg

- South-west facade virtually free of shadows over a height of 60 m
- Frameless toughened safety glass/film standard panels (1900 x 700 mm) with monocrystalline solar cells
- Colour of film matches cells and results in homogeneous appearance
- Panels clipped to supporting construction at six points
- Air cavity, approx. 200 mm, ensures good ventilation, reinforced by the stack effect

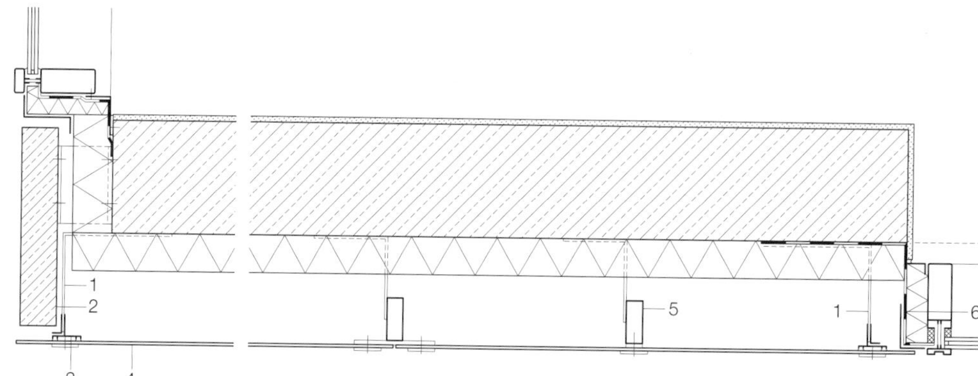

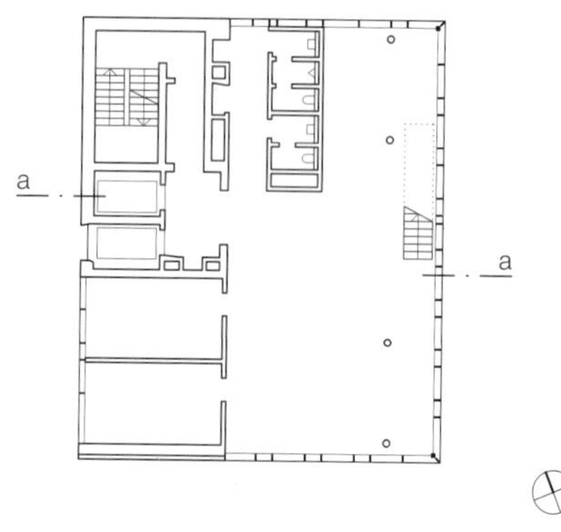

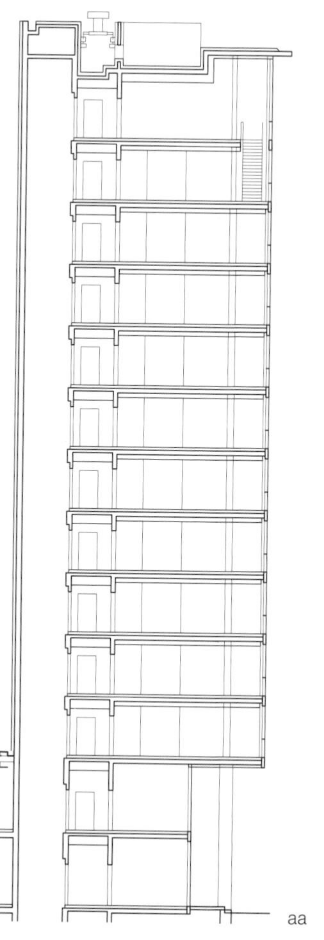

Plan of 17th floor • Section
scale 1:400
Horizontal section, scale 1:20

1. Special bracket, 260 x 300 mm
2. Precast concrete panel, 100 x 600 mm
3. Aluminium section, black finish, 25 x 50 mm
4. Wall construction:
   9 mm frameless solar panel
   186 mm ventilated cavity
   100 mm fleece-laminated thermal insulation, black
   300 mm reinforced concrete
   15 mm plaster
5. Supporting construction:
   220 x 200 mm bracket
   110 x 40 mm aluminium tube with clips
6. Waterproofing

Control tower, plant building and offices

Militärflughafen Sion, CH, 1997

Architects:
Claudine Lorenz + Florian Musso, Sion/Munich
Facade consultants:
Bitz & Savoye, Sion

amc 102, 1999
Fassade/Façade 03/1999
Werk, Bauen + Wohnen 05/1999

- Glass-glass photovoltaic panels
- Pond in front of building increases radiation yield due to reflection
- When used in conjunction with the roof panels, more electricity is generated than is needed
- Energy production (electricity) = 460 MJ/m^2/a
- Plan layout enables window-less photovoltaic facade
- Winner of Swiss Solar Prize 1997

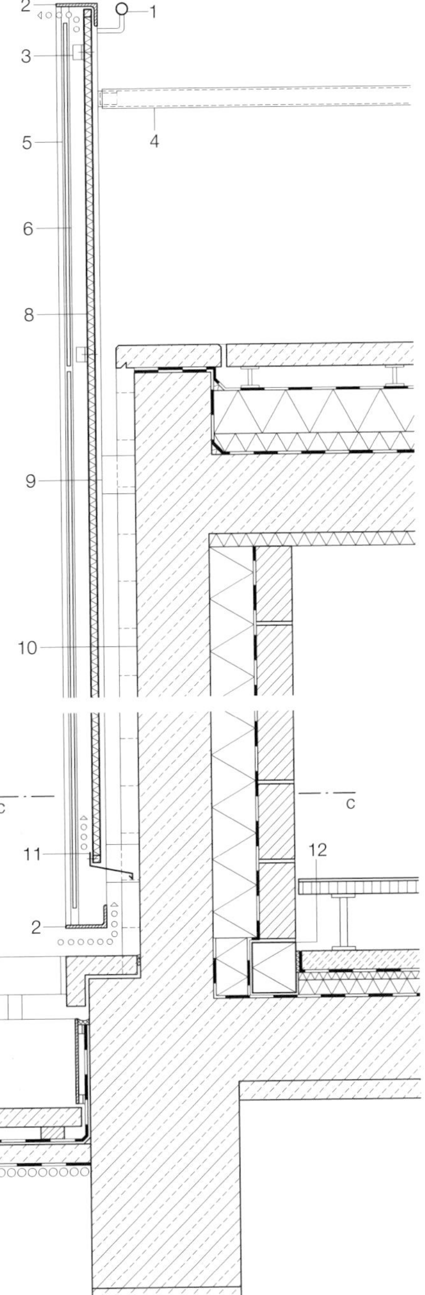

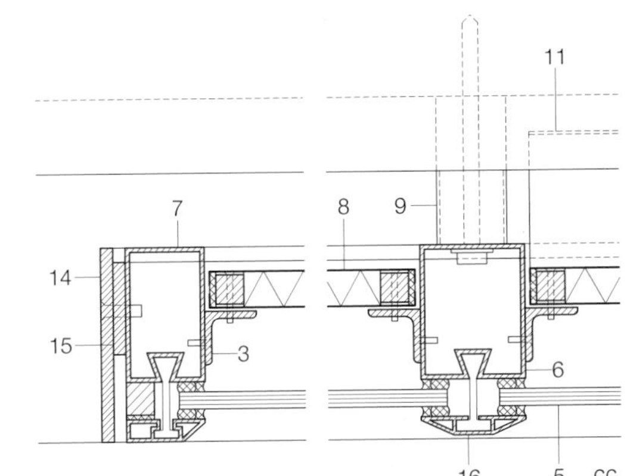

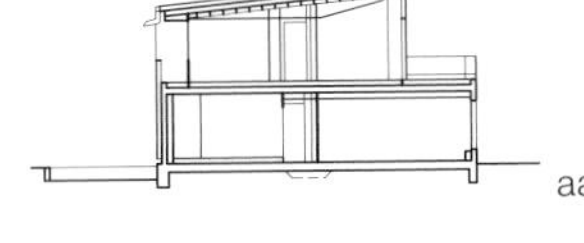

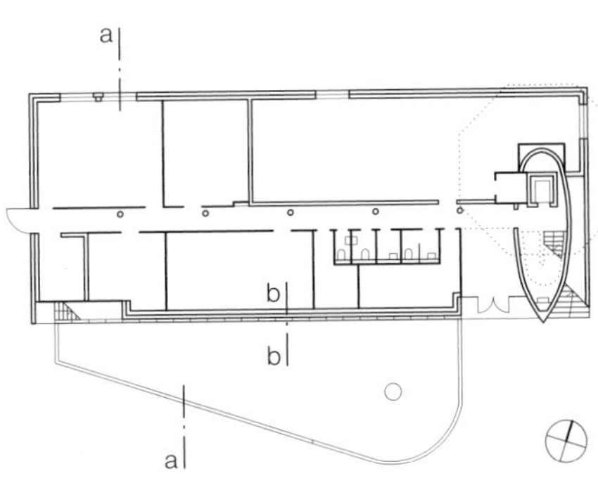

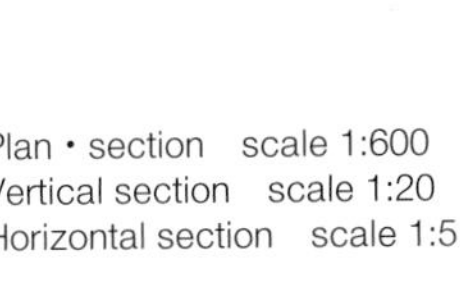

Plan • section scale 1:600
Vertical section scale 1:20
Horizontal section scale 1:5

1 Handrail, steel tube, 33.7 mm dia.
2 Frame, 2 No. steel angles, 109 x 60 mm, steel flats, 112 x 8 mm at sides, welded to No. 14
3 Steel angle, 30 x 30 mm
4 Adjacent to stairs: steel tube, 50 mm dia., as bracing to facade
5 Photovoltaic panel, 10 mm, 72 No., 1355 x 970 mm
6 Central "mullion", steel rectangular hollow section, 60 x 75 mm, welded to No. 2
7 Side "mullion", steel rectangular hollow section, 45 x 75 mm, welded to No. 2
8 Insulated panel to avoid overheating: 1.5 mm stainless steel extruded polystyrene
9 Steel rectangular hollow section, 100 x 40 mm, welded to No. 6, bolted to concrete wall
10 Wall construction: 200 mm reinforced concrete 120 mm thermal insulation waterproofing 100 mm masonry
11 Stainless steel sheet, 1 mm, for redirecting air flow
12 Insulated make-up piece to avoid thermal bridge
13 Bored pile, 13 m long
14 Steel flat, 112 x 8 mm, welded to No. 2
15 Steel flat, 50 x 7 mm
16 Aluminium glazing cap, 45 x 15 mm, anodised, with bevelled edges to avoid casting a shadow

Library

Mataró, E, 1995

Architect:
Miquel Brullet i Tenas, Barcelona

Detail 03/1999
Werk, Bauen + Wohnen 09/1998
Herzog, Thomas (ed.): Solar Energy in Architecture and Urban Planning, Munich/London/New York, 1996

- South facade in the form of a glass double facade
- Polycrystalline solar cells fitted to outside in the form of glass-glass panels (thermally toughened glass, size = 2 m², glued to frame)
- Cavity (150 mm) ensures effective ventilation to photovoltaic panels in summer and preheats incoming air in winter
- The semi-transparent solar cells fitted clear of the facade generate electricity and act as sunshades and still allow transmission of daylight
- At the time of completion, one of the largest buildings in Europe with a photovoltaic system integrated into the building

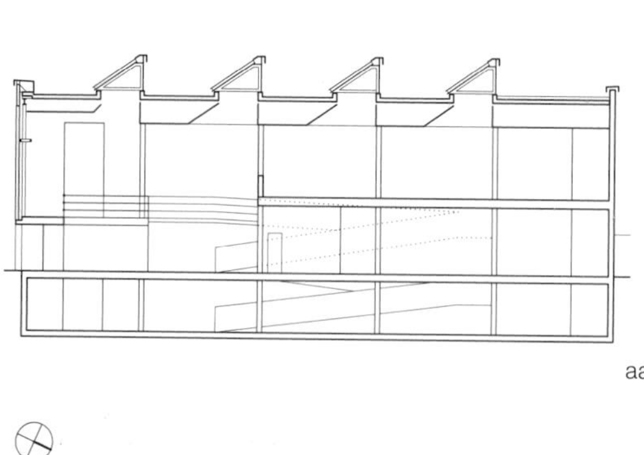

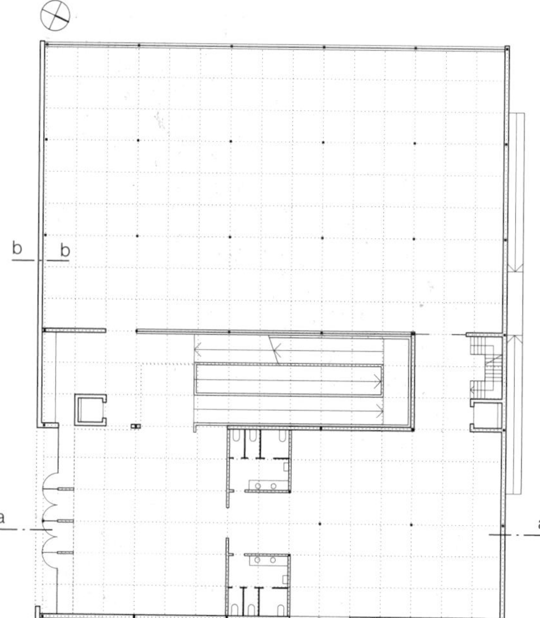

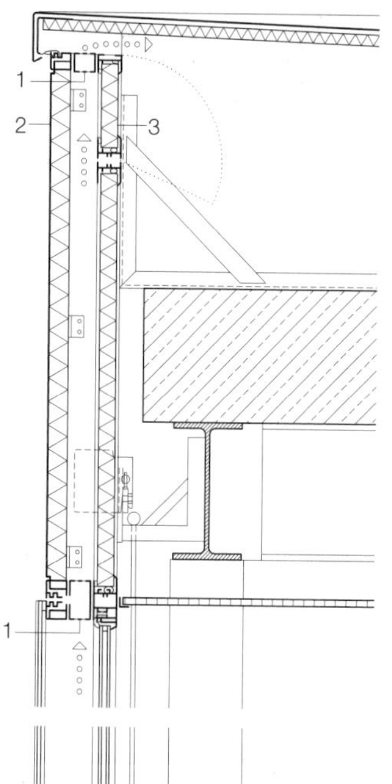

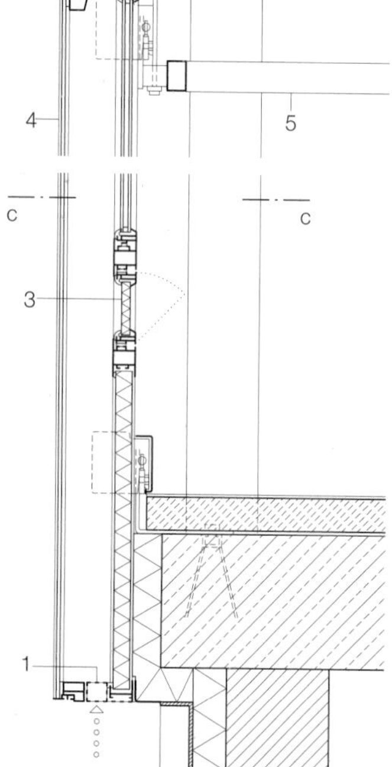

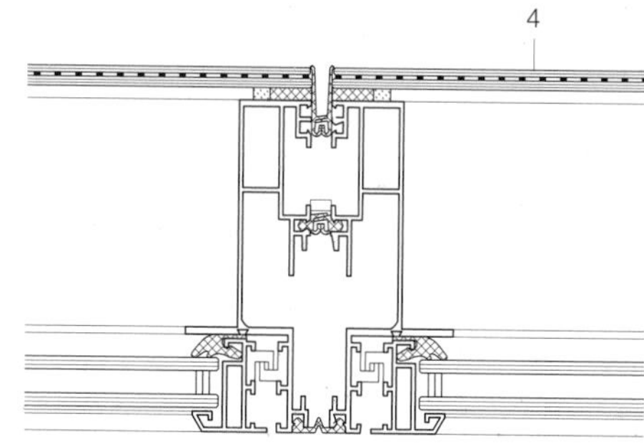

Section scale 1:500
Vertical section scale 1:20
Horizontal section scale 1:5

1 Ventilation opening with filter
2 Plain facade element:
40 mm insulated metal panel
60 mm ventilated cavity
40 mm insulated metal panel
3 Exhaust-air flap
4 Photovoltaic panel, south facade (6495 x 1050 mm):
laminated safety glass with integral solar cells, fixed to frame with adhesive
150 mm cavity
insulating glass
5 Horizontal facade support member

Training academy

Herne, D, 1999

Architects:
Jourda et Perraudin, Paris
Hegger Hegger Schleiff, Kassel
Structural engineers:
Ove Arup und Partner, Düsseldorf
Schlaich Bergermann und Partner, Stuttgart

Architectural Record 12/1999
Architectural Review 10/1999
Detail 03/1999
Hagemann, Ingo B.: Gebäudeintegrierte Photovoltaik, Cologne, 2002
Herzog, Thomas et al.: Timber Construction Manual, Basel/Berlin/Boston, 2004, p. 336

- Glass building acts as "microclimatic envelope" for passive use of solar energy
- Approximately half of the surface area of roof and facades is fitted with photovoltaic glass panels, total output max. 1 MWp
- Monocrystalline photovoltaic cells replace 30% of the glazing in the facade
- Shading for internal building components
- Different photovoltaic panels permit a modular d.c.–a.c. conversion concept for efficient energy conversion

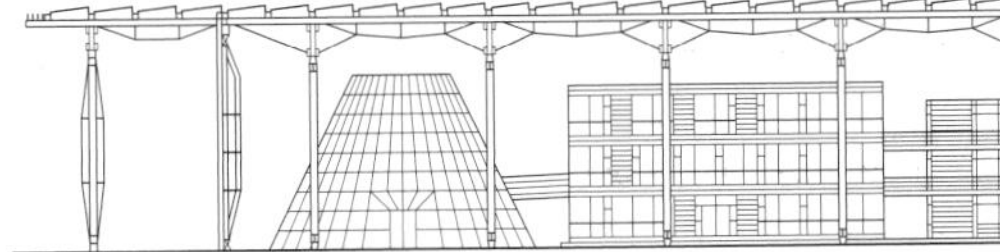

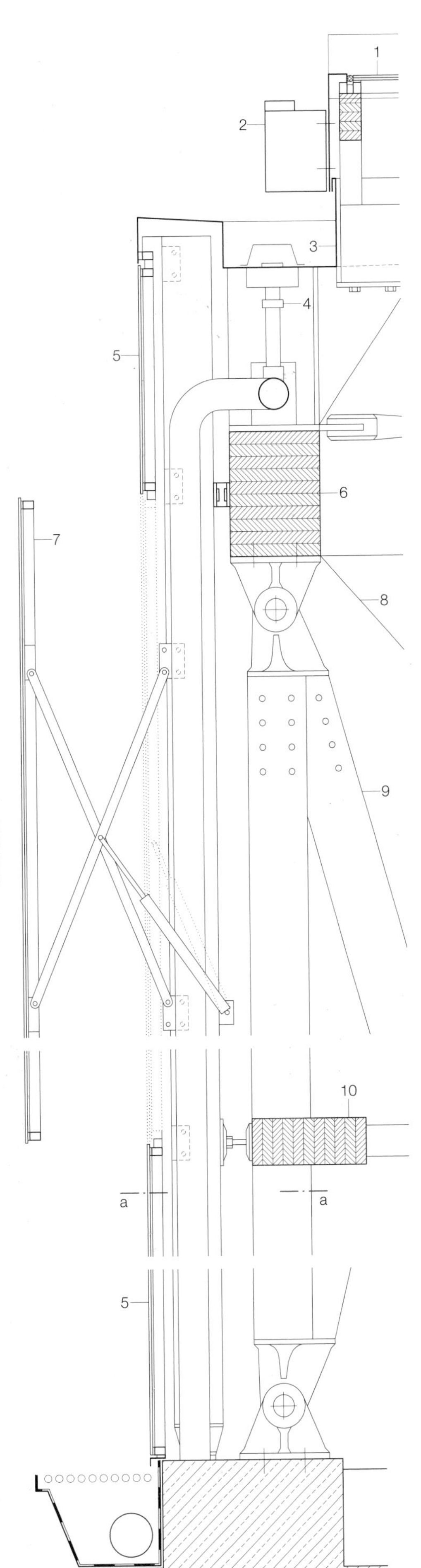

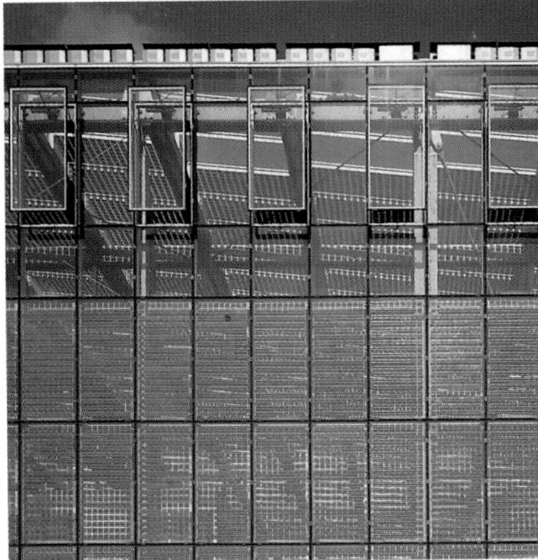

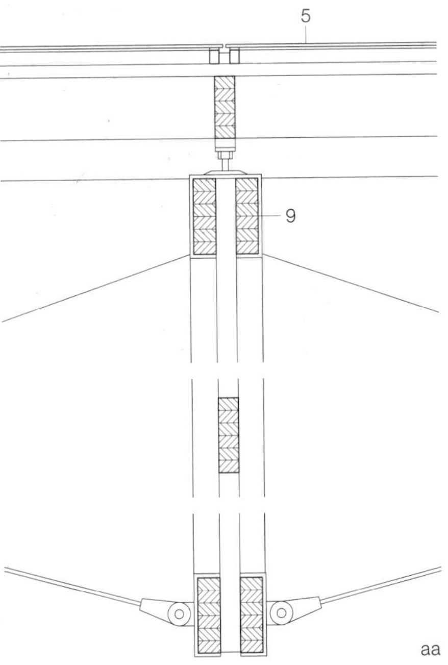

Elevation scale 1:1000
Vertical section • Horizontal section
scale 1:20

1. Roof glazing, laminated safety glass:
   6 mm heat-treated extra-clear glass
   photovoltaic cells in casting resin, 2 mm
   8 mm heat-treated glass
2. Inverter
3. Galvanised steel gutter
4. Rainwater quick-drain system
5. Facade construction:
   structural sealant single glazing
   160 x 60 mm gluelam facade posts
   individual photovoltaic panels in certain areas
6. Gluelam edge beam, 300 x 400 mm
7. Opening light
8. Timber roof girder
9. Timber frame to resist wind forces
10. Gluelam facade rail

Solar factory

Freiburg, D, 1999

Architects:
rolf + hotz, Freiburg
Project architect:
Karin Sinnwell

AIT 09/1999
Detail 03/1999

- Carbon dioxide-neutral office building with production area
- 17° slope to glass facade
- Photovoltaic panels act as permanent sunshades
- Atrium ventilated naturally via ground ducts
- Winner of German "Facade Prize/Special Photovoltaics Prize 2002"

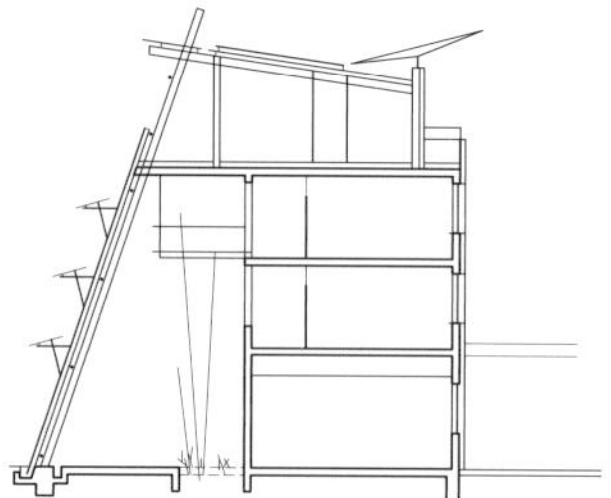

Section scale 1:500
Vertical section • Horizontal section
scale 1:20

1. Open-grid flooring
2. Photovoltaic panel
3. Steel square hollow section, < 50 x 50 mm
4. Steel angle, 2 No. 55 x 6 mm
5. Handrail, stainless steel tube, 33 mm dia
6. Insulating glazing, with integral solar cells at ground floor level
7. Steel section, IPE 100
8. Steel circular hollow section, 101.6 mm dia.
9. Facade support member, steel rectangular hollow section 50 x 280 mm
10. Drainage channel
11. Cable duct
12. Splice in steel beam, IPE 270, with thermal break
13. Rainwater drip

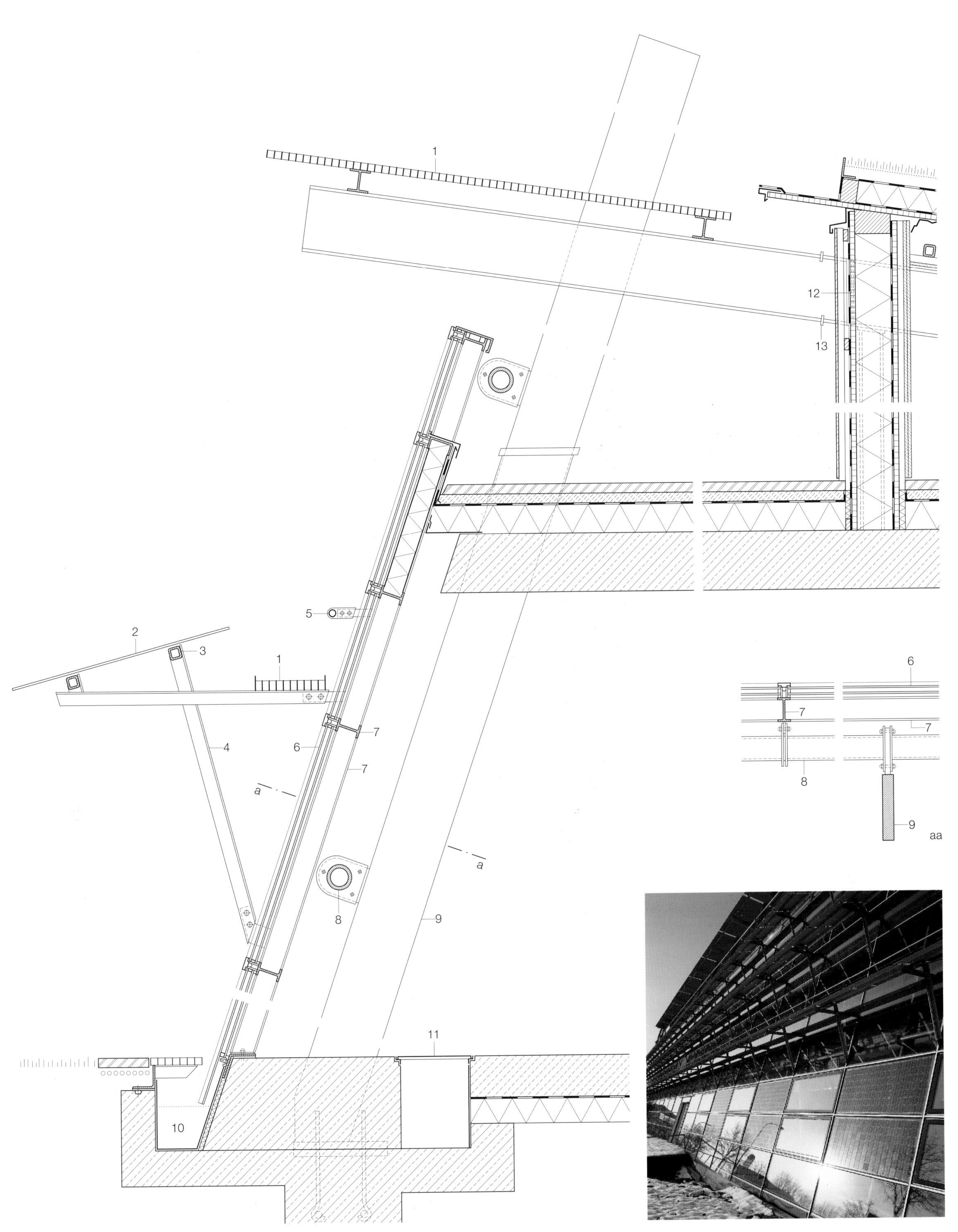
1
12
13
5
2
3
1
6
4
6
7
7
a
a
8
9
7
7
8
9
aa
11
10

Plant building for solar residential development

Emmerthal, D, 2000

Architects:
Niederwöhrmeier + Wiese, Darmstadt
Structural engineers:
Bollinger + Grohmann, Frankfurt am Main

db 10/2000
Fassade/Façade 04/2001
Hagemann, Ingo B.: Gebäudeintegrierte Photovoltaik. Cologne, 2002

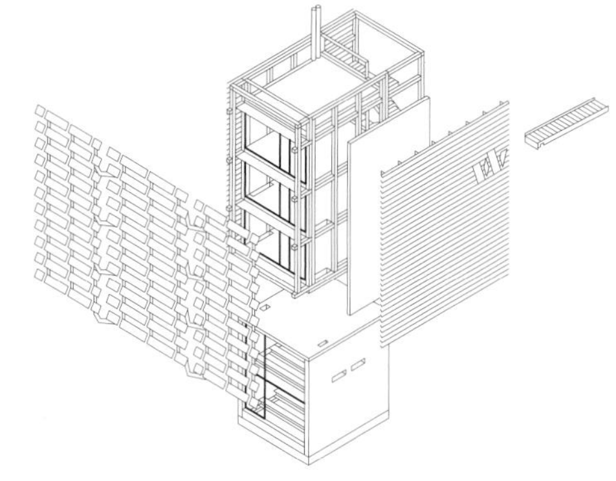

- Energy supply is a combination of heat pump and photovoltaic system
- Photovoltaic panels, single glass pane-film construction with different photovoltaic cells, one-axis tracking in front of tower facade, "solar paddles" rotate through 180° with two-axis tracking
- Analyses indicate solar yields up to 38% higher than with facade integration
- The energy tower was a registered project for EXPO 2000

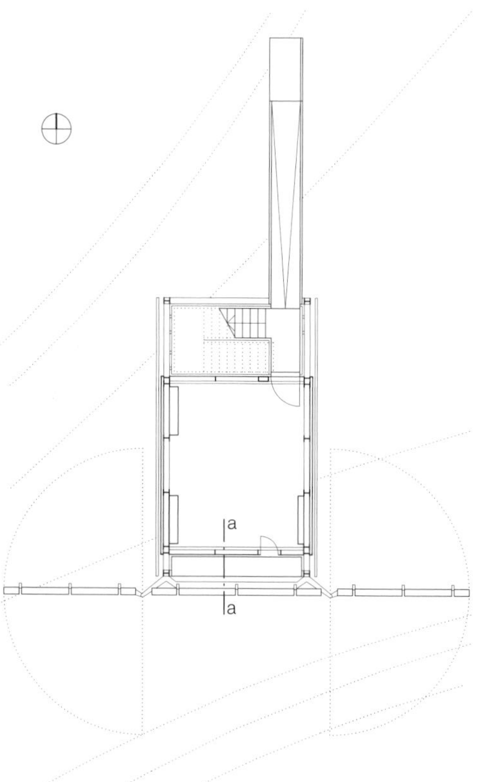

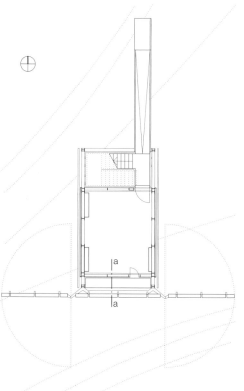

1 Aluminium louvres, 100 mm, fixed (providing protection against UV radiation and weather)
2 Wall construction:, roofing felt
27 mm LVL board, colourless impregnation
50 x 80 mm timber, black stain finish/50 mm insulation
steel sections, HEB 220, hot-dip galvanised
3 Installation grid, 30 x 30 mm, hot-dip galvanised, with steel angle frame, 150 x 100 x 10 mm
4 Sheet aluminium, 2 mm, coated
5 Steel sections, HEB 220, hot-dip galvanised
6 Steel beam, HEB 220, as primary loadbearing support for photovoltaic system
7 Steel flats, 2 No. 150 x 15 mm, hot-dip galvanised, as vertical support bracket, connected to primary support via steel flat, 100 x 10 mm, hot-dip galvanised
8 Photovoltaic panel, 1730 x 480 mm, supported at six points, heat-treated glass, PVB interlayer
9 Supporting frame to photovoltaic panel:
steel/EPDM clamp fixing at two points, 6 mm steel ribs, torsion tube, 42.4 dia. x 2.6 mm, hydraulic altitude angle tracking, steel flat bracket, 50 x 10 mm, with plastic bearing for tracking
10 Platform construction:
open grid flooring, 30 x 30 x 3 mm, hot-dip galvanised
steel angle frame, 40 x 40 x 5 mm
steel circular hollow section spacer, 20 dia. x 4 mm
steel channel supporting construction ⊔⊔ 140 mm
11 Post-and-rail construction: steel rectangular hollow sections, 60 x 50 x 4 mm
10 mm laminated safety glass, 0.38 mm film
aluminium glazing bars
12 Solar paddle construction:
diagonal bracket, steel flat, 50 x 10 mm, with plastic bearing for altitude angle tracking
100 x 60 mm steel section
strut, 60.3 mm dia. steel circular hollow section connection to 168.3 mm dia. torsion tube, with 2 No. steel channels, 100 x 50 x 6 mm, with screw fixings
13 Azimuth angle tracking for "solar paddles", steel torsion tube, 168.3 mm dia., with electric drive

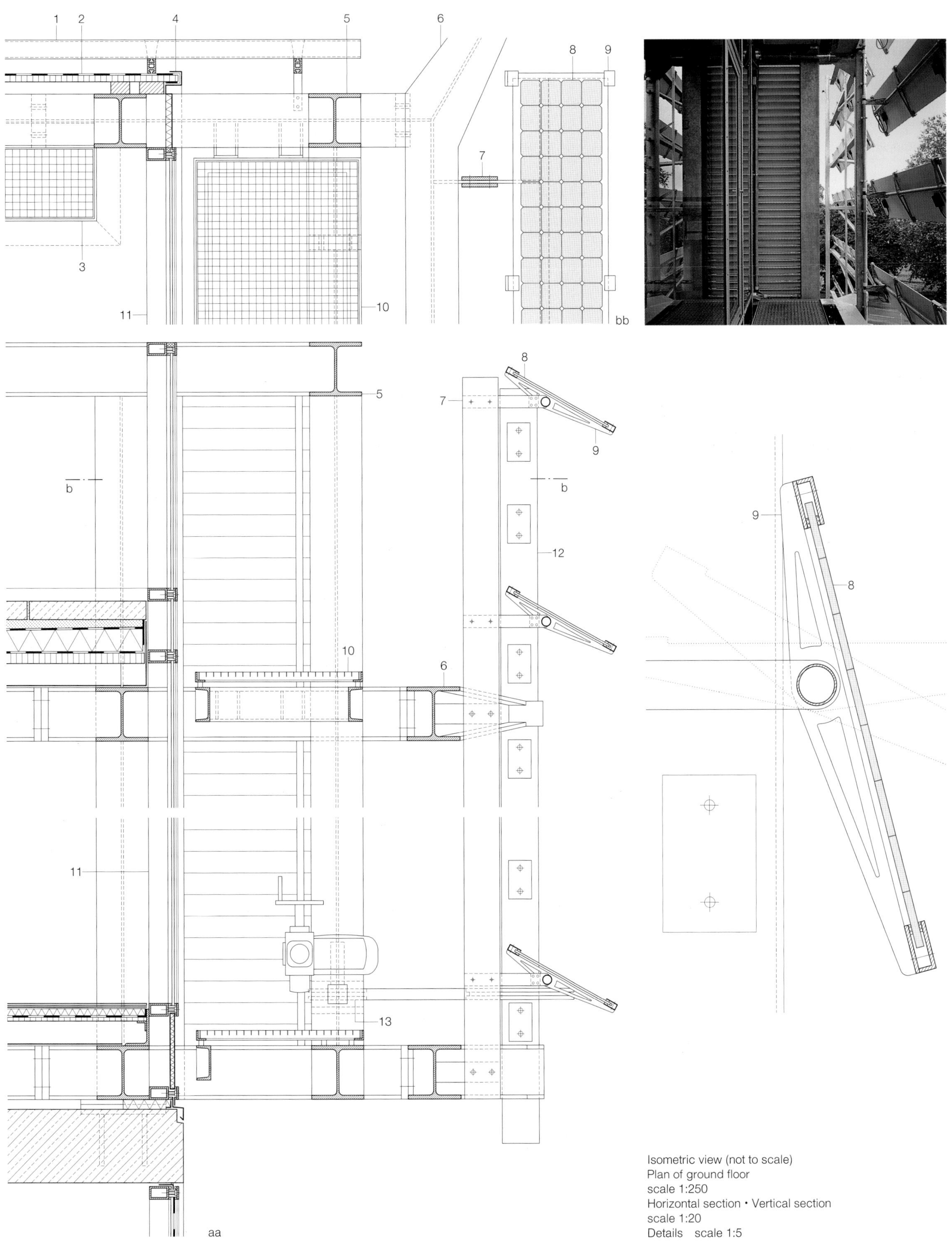

Isometric view (not to scale)
Plan of ground floor
scale 1:250
Horizontal section • Vertical section
scale 1:20
Details scale 1:5

Statutory instruments, directives, standards

The following compilation represents a selection of statutory instruments, directives and standards applicable in Europe and of special interest to designers and contractors involved with facades. The intention is to help ensure that the planning is optimised and related to the project. However, the list is neither exhaustive nor final.
The latest edition of a standard always applies. The new European standards will replace national standards in the foreseeable future. We must make a clear distinction between product, application and testing standards. When comparing materials it is essential to make sure that the numerical results being compared are based on the same methods of analysis and testing. In the case of glass this fact is particularly critical for the radiometric data and the U-values. The German "Ü" symbol (conformity verification for building authority standards and directives) specifies that products subject to building authority requirements or standards included in construction legislation comply with the directives. In Germany the corresponding standards are listed in the "Construction Products List". In the future, the CE symbol will replace the "Ü" symbol. The intention of technical standards, statutory instruments and directives is to create a "framework" for all design and construction work, and to outline the requirements. They are the result of experience and should be subject to continuous revision. Standards and directives that form part of construction legislation must be adhered to. Deviations are possible when appropriate proof of serviceability can be provided. However, the same level of safety must be achieved.
Sets of technical rules (codes of practice, etc.) provide the user with advice on methods of design and construction which experience has shown to be advantageous. It is therefore also possible to achieve the same success using other methods and materials when the requirements are satisfied. This paves the way for new approaches to design and construction.
Voluntary agreements regarding the strict adherence to standards not called for in construction legislation, plus additional properties and requirements must be stipulated in the contract. A clause in the contract stating that all standards are to be adhered to is meaningless and cannot form part of future contracts. To avoid inconsistencies, it is essential to stipulate which standards are relevant and which elements in the standards shall apply in the event of different levels of requirements.

Part A Fundamentals

1 Internal and external conditions

DIN 1341 Heat transfer; concepts, dimensionless parameters, Oct 1986
DIN 18073 Roller shutters, solar shading and black-out equipment in building construction; concepts and requirements, Nov 1990
DIN 18351 Contract procedures for building works – pt C: General technical specifications for building works; facade works, Dec 2002
DIN EN 13363 pt 1 Solar protection devices combined with glazing – Calculation of solar and light transmittance – Simplified method, Oct 2003
DIN EN ISO 12569 Thermal performance of buildings – Determination of air change in buildings – Tracer gas dilution method, Mar 2001

2.1 The structural principles of surfaces

DIN 3750 Seals; terms, Aug 1957
DIN 18351 Contract procedures for building works – pt C: General technical specifications for building works; facade works, Dec 2002
DIN 18516 pt 1 Cladding for external walls, ventilated at rear – pt 1: requirements, principles of testing, Dec 1999
DIN 18540 Sealing of exterior wall joints in building using joint sealants, Feb 1995
DIN 18545 pt 1 Glazing with sealants; rebates; requirements, Feb 1992
DIN EN 12365 pt 1 Building hardware – Gaskets and weatherstripping for doors, windows, shutters and curtain walling – pt 1: Performance requirements and classification, Dec 2003
VDI 2221 Systematic approach to the development and design of technical systems and products, May 1993
VDI 2222 Bl. 1 Methodic development of solution principles, Jun 1997

2.2 Edges, openings

ASR 7/1 Visual contact with the outside, Apr 1976
DIN 107 Building construction; identification of right and left side, Apr 1974
DIN 1946 pt 6 Ventilation and air conditioning – pt 6: Ventilation for residential buildings; requirements, performance, acceptance (VDI ventilation code of practice), Oct 1998
DIN 33417 Description of position, orientation and direction of movement of objects, Aug 1987
DIN EN 12464 pt 1 Light and lighting – Lighting of work places – pt 1: Indoor work places, Mar 2003
DIN EN 12519 Windows and pedestrian doors – Terminology, Nov 1996
DIN EN 13829 Thermal performance of buildings – Determination of air permeability of buildings – Fan pressurisation method, Feb 2001
DIN EN ISO 7730 (draft standard) Ergonomics of the thermal environment – Analytical determination and interpretation of thermal comfort using calculation of the PMV and PPD indices and local thermal comfort, Oct 2003
EnEV Energiesparverordnung (Energy Economy Act), Nov 2001

2.3 Dimensional coordination

DIN 18000 Modular coordination in building, May 1984
DIN 18201 Tolerances in building – terminology, principles, application, testing, Apr 1997
DIN 18202 Dimensional tolerances in building construction – buildings, Apr 1997
DIN 30798 Modular systems; modular coordination
pt 1 Terminology, Sept 1982
pt 2 Principles, Sept 1982
pt 3 Principles for the application, Sept 1982
pt 4 Representation in drawings, Apr 1985

3 Planning advice for the performance of the facade

DIN 4102 Fire behaviour of building materials and building components, Mar 1991
DIN 4108 Thermal protection and energy economy in buildings, Jul 2001
DIN 4108 pt 4 Thermal insulation and energy economy in buildings – pt 4: characteristic values relating to thermal insulation and protection against moisture, Feb 2002
DIN 4109 Sound insulation in buildings; requirements and testing, Nov 1989
DIN V 4701 pt 10, (pre-standard) Energy efficiency of heating and ventilation systems in buildings – pt 10: Heating, domestic hot water, ventilation, Feb 2001
DIN V 4701 pt 12 (pre-standard) Energetic evaluation of heating and ventilation systems in existing buildings – pt 12: Heat generation and domestic hot water generation, Feb 2004
DIN 5034 Daylight in interiors
DIN 18073 Roller shutters, solar shading and black-out equipment in building construction; concepts and requirements, Nov 1990
DIN 5036 pt 3 Radiometric and photometric properties of materials; methods of measurement for photometric and spectral radiometric characteristics, Nov 1990
DIN 52345 Testing of glass; determination of dew point temperature of insulating glass units; laboratory test, Dec 1987
DIN 52619 pt 3 Testing of thermal insulation; determination of the thermal resistance and the thermal transmission coefficient of windows; determination on frames, Feb 1985
DIN EN 673 Glass in building – Determination of thermal transmittance (U-value) – Calculation method, Jun 2003
DIN EN 12865 Hygrothermal performance of building components and building elements – Determination of the resistance of external wall systems to driving rain under pulsating air pressure, Jul 2001
DIN EN 13125 Shutters and blinds – Additional thermal resistance – Allocation of a class of air permeability to a product, Oct 2001
DIN EN 13363 pt 1 Solar protection devices combined with glazing – Calculation of solar and light transmittance – pt 1: Simplified method, Oct 2003
DIN EN 13947 (draft standard) Thermal performance of curtain walling – Calculation of thermal transmittance – Simplified method, Jan 2001
DIN EN ISO 10211 pt 1 Thermal bridges in building construction – Heat flows and surface temperatures – pt 1: General calculation methods, Jun 2001
VDI 2719 Sound isolation of windows and their auxiliary equipment, Aug 1987

Part B Case studies in detail

1.1 Stone

DIN 18516 pt 3 Cladding for external walls, ventilated at rear – pt 3: Natural stone; requirements, design, Dec 1999
DIN 18332 Contract procedures for building works – pt C: General technical specifications for building works; ashlar works, Dec 2002
DIN EN 771 pt 6 Specification for masonry units – pt 6: Natural stone masonry units, Jan 2001
DIN EN 1341 Slabs of natural stone for external paving – Requirements and test methods, Mar 2000
DIN EN 1469 (draft standard) Natural stone products – Slabs for cladding – Requirements, Jan 2003
DIN EN 12059 (draft standard) Natural stone – Finished products, massive stone work – Specifications, Jan 1996
DIN EN 12326 pt 1 (draft standard) Slate and stone products for discontinuous roofing and cladding – pt 1: Product specifications, Nov 2003

1.2 Clay

DIN 105 Clay bricks
DIN 398 Granulated slag aggregate concrete blocks; solid, perforated, hollow blocks, Jun 1976
DIN 1053 Masonry
DIN 18515 pt 1 Cladding for external walls – pt 1: tiles fixed with mortar; principles of design and application, Aug 1998
DIN 18516 pt 1 Cladding for external walls, ventilated at rear – pt 1: requirements, principles of testing, Dec 1999
ENV 1996-1-1 Eurocode 6: Design of masonry structures, pt 1-1: General rules for buildings – Rules for reinforced and unreinforced masonry

1.3 Concrete

DIN 4226 pt 1 Aggregates for concrete; aggregates of dense structure (heavy aggregates); terminology, designation and requirements, Apr 1983
DIN 18151 (preliminary standard) Lightweight concrete hollow blocks, Oct 2003
DIN 18152 (preliminary standard) Lightweight concrete solid bricks and blocks, Oct 2003
DIN 18153 (preliminary standard) Normal-weight concrete masonry units, Oct 2003
DIN 18333 Contract procedures for building works – pt C: General technical specifications for building works; Cast stone works, Dec 2000
DIN 18500 Cast stones; terminology, requirements, testing, inspection, Apr 1991
DIN 18515 pt 1 Cladding for external walls – pt 1: tiles fixed with mortar; principles of design and application, Aug 1998
DIN 18516 pt 5 Cladding for external walls, ventilated at rear – pt 5: Manufactured stone; requirements, design, Dec 1999

DIN EN 197 pt 1 Cement – pt 1: Composition, specifications and conformity criteria for common cements, Feb 2001
DIN EN 206 pt 1 Concrete – pt 1: Specification, performance, production and conformity, Jul 2001
DIN EN 12878 (draft standard) Pigments for colouring of building materials based on cement and/or lime – Specifications and methods of test, Dec 2003

1.4 Timber
DIN 18334 Contract procedures for building works – pt C: General technical specifications for building works; carpentry and timber construction works, Dec 2002
DIN 68364 Characteristic values for wood species; strength, elasticity, resistance, May 2003
DIN 68800 Protection of timber
pt 2 preventive constructional measures in buildings, May 1996
pt 3 Protection of timber; preventive chemical protection, Apr 1990
pt 4 Wood preservation; measures for the eradication of fungi and insects, Nov 1992
pt 5 Protection of timber used in buildings; preventive chemical protection for wood based materials, May 1978

1.5 Metal
DIN 18203 pt 2 Tolerances in building; prefabricated steel components, May 1986
DIN 18335 Contract procedures for building works – pt C: General technical specifications for building works; steel construction works, Dec 2002
DIN 18339 Contract procedures for building works – pt C: General technical specifications for building works; sheet metal works, Dec 2002
DIN 18360 Contract procedures for building works – pt C: General technical specifications for building works; metal construction works, Dec 2002
DIN 18364 Contract procedures for building works – pt C: General technical specifications for building works; corrosion protection of steel and aluminium structures, Dec 2002
DIN 18516 pt 1 Cladding for external walls, ventilated at rear – pt 1: requirements, principles of testing, Dec 1999
DIN 18807 pt 1 Trapezoidal sheeting in building; trapezoidal steel sheeting – pt 1: General requirements and determination of loadbearing capacity by calculation, Jun 1987
DIN EN ISO 12944 Paints and varnishes – Corrosion protection of steel structures by protective paint systems pt 1–7, July 1998
VDI 3137 Forming – definitions, terms, characteristic quantities, Jan 1976

1.6 Glass
DIN EN 1051 pt 1 Glass in building – Glass blocks and glass pavers
pt 1 Definitions and description, Apr 2003
pt 2 Evaluation of conformity, May 2003
DIN 1249 pt 10 Glass for use in building construction; chemical and physical properties, Aug 1990
DIN 1249 pt 11 Glass in building; glass edges; concept, characteristics of edge types and finishes, Sept 1986
DIN 1259 Glass
pt 1 Terminology for glass types and groups, Sept 2001
pt 2 Terminology of glass products, Sept 2001
DIN 1286 Insulating glass units
pt 1 air filled, Mar 1994
pt 2 gas filled, May 1989
DIN 4242 Glass block walls; construction and dimensioning, Jan 1979
DIN 12116 Testing of glass – Resistance to attack by a boiling aqueous solution of hydrochloric acid – Method of test and classification, Mar 2003
DIN 18545 pt 1 Glazing with sealants; rebates; requirements, Feb 1992
DIN 52290 Security glazing
DIN EN 356 Glass in building – Security glazing – Testing and classification of resistance against manual attack, Feb 2000
DIN EN 572 pt 1 Glass in building – Basic soda lime silicate glass products – pt 1: Definitions and general physical and mechanical properties, Jan 2004
DIN EN 1063 Glass in building – Security glazing – Testing and classification of resistance against bullet attack, Jan 2000
DIN EN 1279 Glass in Building – Insulating glass units
pt 1 Generalities, dimensional tolerances and rules/guidance for the product/type description, Sept 1995
pt 2 Long term test method and requirements for moisture penetration, Jun 2003
pt 3 Long term test method and requirements for gas leakage rate and for gas concentration, May 2003
pt 4 Methods of test for the physical attributes of the edge seals, Oct 2002
pt 5 Evaluation of conformity, Oct 2001
pt 6 Factory production control and audit tests, Oct 2002
DIN EN 1863 pt 1 Glass in building – Heat strengthened soda lime silicate glass – pt 1: Definition and description, Mar 2000

1.7 Plastics
DIN 53350 Testing of plastics films and coated textile fabrics, manufactured using plastics; determination of stiffness in bending, method according to Ohlsen, Jan 1980
DIN 53362 (draft standard) Testing of plastics films and textile fabrics (excluding nonwovens), coated or not coated with plastics – Determination of stiffness in bending – Method according to Cantilever, Oct 2003
DIN 53363 (draft standard) Testing of plastic films – Tear test using trapezoidal test specimen with incision, Oct 2003
DIN 53370 (draft standard) Testing of plastic films – Determination of the thickness by mechanical feeling, Apr 2004
DIN 53386 Testing of plastics and elastomers, exposure to natural weathering, Jun 1982
DIN EN ISO 305 Plastics – Determination of thermal stability of polyvinyl chloride, related chlorine-containing homopolymers and copolymers and their compounds – Discoloration method, Oct 1999
DIN EN ISO 527 pt 1 Plastics – Determination of tensile properties – pt 1: General principles
DIN EN ISO 2578 Plastics – Determination of time–temperature limits after prolonged exposure to heat, Oct 1998

2.1 The glass double facade
Sound insulation
DIN EN ISO 717 pt 1 Acoustics – Rating of sound insulation in buildings and of building elements – pt 1: Airborne sound insulation, Jun 1997
VDI 2058 Bl. 2 Assessment of noise with regard to the risk of hearing damage, Jun 1988
VDI 2058 Bl. 3 Assessment of noise in the working area with regard to specific operations, Feb 1999
VDI 2719 Sound isolation of windows and their auxiliary equipment, Aug 1987
Aerophysics
DIN 1946 pt 2 Ventilation and air conditioning; technical health requirements (VDI ventilation rules), Jan 1994
DIN 33403 pt 3 Climate at the workplace and its environments – pt 3: Assessment of the climate in the warm and hot working areas based on selected climate indices, Apr 2001
VDI 2083 Bl. 5 Cleanroom technology – Thermal comfort, Feb 1996

2.2 Manipulators
AGI F 20 Sonnen- und Blendschutzsysteme: Leitfaden zur Auswahl, Jun 1995
DIN 18055 Windows; air permeability of joints, water tightness and mechanical strain; requirements and testing, Oct 1981
DIN 18056 Window walls; design and construction, Jun 1966
DIN 18357 Contract procedures for building works – pt C: General technical specifications for building works; mounting of window and door fittings, Dec 2002
DIN EN 12207 Windows and doors – Air permeability – Classification, Jun 2000
DIN EN 12208 Windows and doors – Watertightness – Classification, Jun 2000
DIN EN 12210 Windows and doors – Resistance to wind load – Classification, Aug 2003
DIN EN 12216 Shutters, external blinds, internal blinds – Terminology, glossary and definitions, Nov 2002
DIN EN 12400 Windows and pedestrian doors – Mechanical durability – Requirements and classification, Jan 2003
DIN EN 13115 Windows – Classification of mechanical properties – Racking, torsion and operating forces, Nov 2001
DIN EN 13120 Internal blinds – Performance requirements including safety, Mar 1998
DIN EN 13125 Shutters and blinds – Additional thermal resistance – Allocation of a class of air permeability to a product, Oct 2001
DIN EN 13126 pt 1 Building hardware, fittings for windows and door height windows – Requirements and test methods – pt 1: Requirements common to all types of fittings, Apr 1998
DIN EN 13561 External blinds – Performance requirements including safety, Jul 1999
DIN EN 13659 (draft standard) Shutters – Requirements and classification, Oct 1999
DIN EN 14501 Blinds and shutters – Thermal and visual comfort – Assessment of performances, Aug 2002
FensterTürRL Richtlinie über Fenster und Fenstertüren (FenTür), Nov 2002
GUV-R 1/494 Richtlinien für kraftbetätigte Fenster, Türen und Tore, Jul 1990
Hadamar Technical Directive, Institut des Glaserhandwerks für Verglasungstechnik und Fensterbau; pub. 2: Windlast und Glasdicke; pub. 12: Fensterwände, Bemessung und Ausführung, Erläuterung zur DIN 18056; pub. 20: Montage von Fenstern
VDI 2719 Sound isolation of windows and their auxiliary equipment, Aug 1987

2.3 Solar energy
DIN 4757 pt 2 Solar heating plants operating on organic heat transfer media; Requirements relating to safe design and construction, Nov 1980
DIN 18015 pt 1 Electrical installations in residential buildings
pt 1 Planning principles, Sept 2002
pt 2 Nature and extent of minimum equipment, Aug 1996
pt 3 Wiring and disposition of electrical equipment, Apr 1999
DIN 18516 pt 4 Cladding for external walls at rear: tempered safety glass; requirements, design, testing, Feb 1990
DIN EN 410 Glass in building – Determination of luminous and solar characteristics of glazing, Dec 1998
DIN EN 674 Determination of thermal transmittance (U-value) – Guarded hot-plate method, Jan 1999
DIN EN 12975 pt 1 Thermal solar systems and components – Collectors – pt 1: General requirements, Mar 2001
DIN EN ISO 10077 Thermal performance of windows, doors and shutters: Calculation of thermal transmittance
pt 1 (draft standard) General, Dec 2003
pt 2 Numerical method for frames, Dec 2003
DIN EN 61277 Terrestrial photovoltaic (PV) power generating systems – General and guide, Feb 1999
DIN IEC 60364 pt 1 (draft standard) Erection of low voltage installations – pt 100: Fundamental principles, assessment of general characteristics, definitions, Aug 2003
DIN VDE 0100-300 Erection of power installations with nominal voltages up to 1000 V – pt 3: Assessment of general characteristics of installations, Jun 1994

Picture credits

The authors and publishers would like to express their sincere gratitude to all those who have assisted in the production of this book, be it through providing photos or artwork or granting permission to reproduce their documents or providing other information. All the drawings in this book were specially commissioned. Photographs not specifically credited were taken by the architects or are works photographs or were supplied from the archives of the magazine DETAIL. Despite intensive endeavours we were unable to establish copyright ownership in just a few cases; however, copyright is assured. Please notify us accordingly in such instances.
The numbers refer to the figures.

Envelope, wall, facade

1 Cremers, Stefan; Karlsruhe
2 Herzog-Loibl, Verena; Munich
3 Cremers, Jan; Munich
4 Schittich, Christian; Munich
5 Merisio, Pepi; Bergamo, in: Merisio, Pepi; Barzanti, Roberto: Italien. Zürich 1975, p. 216
6 Bednorz, Achim; Cologne
7 Merisio, Pepi; Bergamo, in: Merisio, Pepi; Barzanti, Roberto: Italien. Zürich 1975, p. 218
9–11 Herzog-Loibl, Verena; Munich
13 Pictor International
14 Cremers, Jan; Munich
15 Kaltenbach, Frank; Munich
17 Herzog-Loibl, Verena; Munich
18 Ogawa, Shigeo/Shinkenchiku-sha; Tokyo

Part A

p. 16 in: Lampugnani, Vittorio Magnago : Architecture of the 20th century in drawings: Utopia and reality, Stuttgart, 1982

Internal and external conditions
A 1.1 Herzog, Thomas; Cremers, Jan; Munich
A 1.2 Cremers, Jan; Munich
A 1.3–5 Federal Ministry for Regional Planning, Building and Urban Development (pub.): Handbuch Passive Nutzung der Sonnenenergie, No. 04.097, 1984, pp. 78/52
A 1.6 DIN 4710
A 1.9 Kunzel and Gertis, 1969
A 1.10 Federal Ministry for Regional Planning, Building and Urban Development (pub.): Handbuch Passive Nutzung der Sonnenenergie, No. 04.097, 1984, p. 14
A 1.11 German Weather Service, Hamburg
A 1.12 Cremers, Jan; Munich
A 1.13–15 Kind-Barkauskas, Friedbert et al.: Concrete Construction Manual, Munich/Düsseldorf, 2001, p. 79
A 1.19–23 Cremers, Jan; Munich
A 1.24 European Wind Atlas
A 1.25–31 Cremers, Jan; Munich

The structural principles of surfaces
A 2.1.1 Bonfig, Peter; Munich
A 2.1.2–6 Bonfig, Peter; Munich
A 2.1.7 Herzog, Thomas; Nikolic Vladimir: Petrocarbona Außenwandsystem, Bexbach, 1972
A 2.1.8–20 Bonfig, Peter; Munich

Edges, openings
A 2.2.1 Leistner, Dieter/artur; Cologne
A 2.2.2 Westenberger, Daniel; Munich
A 2.2.3 Schittich, Christian (ed.): Solares Bauen, Munich/Basel, 2003, p. 63
A 2.2.4–5 Westenberger, Daniel; Munich
A 2.2.6 Zürcher, Christoph; Frank, Thomas: Bauphysik, vol. 2, Bau und Energie – Leitfaden für Planung und Praxis, Zürich/Stuttgart, 1998, p. 80
A 2.2.7–8 Westenberger, Daniel; Munich
A 2.2.9–10 Fassade / Façade 03/2002, p. 24f.; db 09/ 2003, p. 87f.

Dimensional coordination
A 2.3.1 Neuhart, Andrew; El Segundo
A 2.3.2 Yoshida, Tetsuro: Das japanische Wohnhaus, Berlin, 1954, p. 69
A 2.3.3 Durand, Jean-Nicolas-Louis: Précis des leçons II. Paris 1819
A 2.3.4 Kunstverein Solothurn (pub.): Fritz Haller. Bauen und Forschen. Solothurn 1988, p. 3.1.4
A 2.3.7 Bussat, Pierre: Die Modulordung im Hochbau. Stuttgart 1963, p. 31
A 2.3.9 DIN 18000.
A 2.3.13 Girsberger, Hans (ed.): ac panel. Asbestzement-Verbundplatten und -Elemente für Außenwände, Zürich ,1967, pp. 46-49

Planning advice for the performance of the facade
A 3.1 Kaltenbach, Frank; Munich
A 3.2–3 Schüco International
A 3.4–5 Hart, Franz et al.: Stahlbau Atlas, Brussels, 1982, p. 338f.
A 3.6 Schüco International
A 3.7–8 Pfeifer, Günter et al.: Masonry Construction Manual, Munich/Basel, 2001, pp. 186, 190
A 3.9–10 Herzog, Thomas et al.: Timber Construction Manual, Munich/Basel, 2003, p. 71
A 3.11 Detail 9/2002, p. 1070

Part B

p. 60 Wimmershoff, Heiner; Aachen

Stone
B 1.1.1 Bonjoch, Eloi; Barcelona
B 1.1.2–3 Herzog-Loibl, Verena; Munich
B 1.1.4 Schittich, Christian; Munich
B 1.1.5 Herzog-Loibl, Verena; Munich
B 1.1.6 Luciano Chiappini: Ferrara und seine Kunstdenkmäler, Bologna, 1979, p. 39
B 1.1.8 Merisio, Pepi; Bergamo, in: Merisio, Pepi; Barzanti, Roberto: Italien, Zürich, 1975, p. 247
B 1.1.9 Bonjoch, Eloi; Barcelona
B 1.1.10 Müller, Friedrich: Gesteinskunde, Ulm, 1994, pp. 196-97
B 1.1.11 Hugues, Theodor et al.: Naturwerkstein, Munich 2002, p. 72
B 1.1.12 Braun, Zooey/Architekton; Mainz
B 1.1.13 Heinz, Thomas A.; Illinois
B 1.1.14–16 Havixbeck Sandstone Museum
B 1.1.17 Stein, Alfred: Fassaden aus Natur- und Betonwerkstein, Munich, 2000, p. 58
B 1.1.18–22 Detail 06/1999, p. 1026
B 1.1.23 Herzog-Loibl, Verena; Munich
B 1.1.24 Müller, Friedrich: Gesteinskunde, Ulm, 1994, p. 171
B 1.1.25–26 Detail 06/1999, p. 1032
B 1.1.27–30 Gahl, Christian; Berlin
B 1.1.38 Gundelsheimer Marmorwerk, Treuchtlingen
B 1.1.39 Müller, Friedrich: Gesteinskunde, Ulm, 1994, pp. 196–97
B 1.1.40–41 Hugues, Theodor et al.: Naturwerkstein, Munich, 2002
S. 72 Fanconi, Doris; Zürich
S. 73 Peda, Gregor; Passau
S. 74 Ruault, Philippe; Nantes
S. 76 Lenzen, Thomas; Munich
S. 77 Müller, Stefan; Berlin
S. 78 Mühling, André; Munich
S. 79 top: Brigola, Victor S.; Stuttgart
bottom: Mühling, André; Munich
S. 80 Steiner, Rupert; Wien
S. 81 Kaltenbach, Frank; Munich

Clay
B 1.2.2 Enders, Ulrike; Hannover
B 1.2.3 Pfeifer, Günter: Masonry Construction Manual, Munich/Basel, 2001, p. 57
B 1.2.5 Hirmer Fotoarchiv; Munich
B 1.2.6 Budeit, Hans Joachim; von Kuenheim, Haug: Backstein, die schönsten Ziegelbauten zwischen Elbe und Oder, Munich, 2001, p. 33
B 1.2.7 Klinkott, Manfred; Karlsruhe
B 1.2.8 Chabat, Pierre (ed.): Victorian Brick and Terra-Cotta Architecture, New York, 1989, p. 18
B 1.2.9 Halfen GmbH & Co. KG
B 1.2.10 Enders, Ulrike; Hannover
B 1.2.11 Halfen GmbH & Co. KG
Pfeifer, Günter et al.: Masonry Construction Manual, Munich/Basel, 2001, p. 125
B 1.2.12 Berlin Art Library
B 1.2.13 Fischer-Daber, in: l'architecture d'aujourd'hui 205, 1979, p. 8
B 1.2.14 Chemollo, Allessandra, in: Acocella, Alfonso: An architecture of place, Rome, 1992, p. 96
B 1.2.15–17 Halfen GmbH & Co. KG
B 1.2.18–20 Avellaneda, Jaume; Barcelona
B 1.2.21–23 Acocella, Alfonso; Florence
B 1.2.24–29 Moeding Keramikfassaden GmbH, Marklkofen
B 1.2.30 Herzog-Loibl, Verena; Munich
B 1.2.31 Bonfig, Peter; Munich
B 1.2.32 Moeding Keramikfassaden GmbH, Marklkofen
B 1.2.33–34 Acocella, Alfonso; Florence
B 1.2.35 Lang, Werner; Munich
B 1.2.36 Decorated walls of modern architecture, Tokyo, 1983, p. 30
B 1.2.37–38 Acocella, Alfonso; Florence
B 1.2.39 Tectónica 15/2003, p. 21
B 1.2.40–41 Herzog-Loibl, Verena; Munich
B 1.2.42–43 Tectónica 15/2003, p. 18
B 1.2.44 Ciampi, Allessandro; Florenz, in: Acocella, Alfonso: Involucri in cotto, Florence, 2002, p. 96
B 1.2.45 Acocella, Alfonso: Involucri in cotto, Florence, 2002, p. 98
B 1.2.46 Ciampi, Allessandro; Florenz, in: Acocella, Alfonso: Involucri in cotto, Florence, 2002, p. 98f.
p. 90 Klomfar, Bruno; Vienna
p. 91 Wood, Charlotte; London
p. 92–93 Leistner, Dieter/artur; Cologne
p. 94 Mosch, Vincent; Berlin
p. 95 Cano, Enrico; Como
p. 96 Kinold, Klaus; Munich
p. 98 Hajd, Jozef; Budapest
p. 99 Szentivani, Janos; Pilisszentivan

Concrete
B 1.3.1 Herzog, Thomas; Munich
B 1.3.2 Kinold, Klaus; Munich
B 1.3.3 Verlag Bau + Technik, Düsseldorf
B 1.3.4 Brandenburg Technical University, Cottbus, Chair of Design (pub.): Architekt Bernhard Hermkes, Cottbus, 2003
B 1.3.5 Richters, Christian; Münster
B 1.3.6 MIT Press; Cambridge
B 1.3.7 Kinold, Klaus; Munich
B 1.3.8 Walti, Ruedi; Basel
B 1.3.9 Grimm, Friedrich; Richarz, Clemens: Hinterlüftete Fassaden, Stuttgart/Zürich, 1994, p. 161
B 1.3.11 DIN 18500 parts 1–3
B 1.3.12 Dyckerhoff Weiss Marketing und Vertriebsgesellschaft
B 1.3.13–16 Heeß, Stefan: Mehr als nur Fassade. Konstruktion von Betonfertigteil- und Betonwerkstein-Fassaden. Wiesbaden
B 1.3.17 Großformatige Fassaden. Fassaden mit Holzzement, pub. by Eternit AG. Berlin 2001, p. 12
B 1.3.18–20 Dyckerhoff Weiss Marketing und Vertriebsgesellschaft
p. 110 Weber, Jens; Munich
p. 111 Schwarz, Ullrich; Berlin
p. 112–113 Roth, Lukas; Cologne
p. 116–117 Richters, Christian; Münster
p. 120–121 Klomfar, Bruno; Vienna
p. 122 Malhão, Daniel; Lisbon
p. 123 Halbe, Roland/artur; Cologne

Timber
B 1.4.1 Shinkenchiku-sha, Tokyo
B 1.4.2 Sawyer, Peter: The Oxford Illustrated History of the Vikings. Oxford 1997, p. 191
B 1.4.3 Herzog, Thomas et al.: Timber Construction Manual, Munich/Basel, 2003, p. 26
B 1.4.4 Gellner, Edoardo; Cortina d´Ampezzo
B 1.4.5 Herzog-Loibl, Verena; Munich
B 1.4.6–7 Herzog, Thomas et al.: Timber Construction Manual, Munich/Basel, 2003, pp. 31–33
B 1.4.8 after Baus, Urslula; Siegele, Klin: Holzfassaden, Stuttgart/Munich 2001, p. 19
B 1.4.9–10 Herzog, Thomas et al.: Timber Construction Manual, Munich/Basel, 2003, pp. 34–46
B 1.4.11 Heyer, Hans-Joachim/Werkstatt für Photographie; Stuttgart University
B 1.4.12 Zeitler, Friedemann; Penzberg
B 1.4.13 Heyer, Hans-Joachim/Werkstatt für Photographie; Stuttgart University
B 1.4.15 Heyer, Hans-Joachim/Werkstatt für Photographie; Stuttgart University
B 1.4.17 Heyer, Hans-Joachim/Werkstatt für Photographie; Stuttgart University
B 1.4.18 Strandex Europe; Walmley
B 1.4.19 Cerliani, Christian; Zürich
B 1.4.20 Walti, Ruedi; Basel
B 1.4.21 Jonathan Levi; Boston
B 1.4.22–23 Richters, Christian; Münster
B 1.4.24 Hueber, Eduard; New York
B 1.4.25 Leistner, Dieter/artur; Cologne
B 1.4.26 Kaltenbach, Frank; Munich
B 1.4.27 Rieger, Annegret; Munich
B 1.4.28 Werner, Heike; Munich
B 1.4.29 Busam, Friedrich/architekturphoto, Düsseldorf
B 1.4.30 Führer, Reto; Felsberg
B 1.4.31 Richters, Christian; Münster
B 1.4.32–34 Widmann, Sampo; Munich
B 1.4.35–41 Informationsdienst Holz, Düsseldorf 1992
B 1.4.42 Herzog-Loibl, Verena; Cologne
B 1.4.43 Huthmacher, Werner/artur; Cologne
B 1.4.44 Kaltenbach, Frank; Cologne
B 1.4.45–47 Schweitzer, Roland; Paris
B 1.4.48–49 Theo Ott Holzschindeln GmbH; Ainring
p. 134 top: Freeman, Michael; London
p. 134 bottom: Widmann, Sampo; Munich
p. 136–137 Richters, Christian; Münster
p. 138 Havran, Jiri; Oslo
p. 139 Huttunen, Marko; Helsinki
p. 140–141 Helfenstein, Heinrich; Adliswil
p. 142–143 Bonfig, Peter; Munich
p. 144–145 Shinkenchiku-sha, Tokyo
p. 146 Richters, Christian; Münster
p. 147 Strauß, Dietmar; Besigheim
p. 150–151 Roth, Lukas; Cologne
p. 153 Malhão, Daniel; Lisbon

Metal
B 1.5.1 Reid, Jo & Peck, John; Newport
B 1.5.2 N. P. Goulandris Foundation, Museum of Cycladic Art, Athens
B 1.5.3 Munich City Museum
B 1.5.4 Gay, John; London, in: Murray, John (ed.): Cast Iron. London 1985, p. 28
B 1.5.5 Courtesy of the Estate of R. Buckminster Fuller; Santa Barbara
B 1.5.6 Sulzer-Kleinemeier, Erika; Gleisweiler
B 1.5.7 Miller, Ardean; New York, in: Airstream – The history of the land yacht. San Francisco, p. 69
B 1.5.9–10 Reid, Jo & Peck, John; Newport
B 1.5.11 Cremers, Jan; Munich
B 1.5.12 Herzog-Loibl, Verena; Munich
B 1.5.13 Cremers, Jan; Munich
B 1.5.14 Herzog-Loibl, Verena; Munich
B 1.5.15 Cremers, Jan; Munich
B 1.5.16 Gilbert, Dennis/view/artur; Cologne
B 1.5.17–20 Cremers, Jan; Munich
B 1.5.21 Hoesch Siegerlandwerke GmbH; Siegen
B 1.5.22 Alcan Singen GmbH; Singen
B 1.5.23 Cremers, Jan; Munich
B 1.5.24 Table: Cremers, Jan; Munich
Photos: Kaltenbach, Frank; Munich
B 1.5.25 Cook, Peter/view/artur; Cologne
B 1.5.26 Cremers, Jan; Munich
B 1.5.27 Heinrich Fiedler GmbH & Co. KG; Regensburg
B 1.5.28–32 Mevaco GmbH; Schlierbach
B 1.5.33–34 Alcan Singen GmbH; Singen
B 1.5.35 Werner, Heike; Munich
B 1.5.36–37 Heinrich Fiedler GmbH & Co. KG; Regensburg
B 1.5.38–39 Werner, Heike; Munich
B 1.5.40 Kaltenbach, Frank; Munich
B 1.5.41 Heinrich Fiedler GmbH & Co. KG; Regensburg
B 1.5.42 AIM; Nürtingen
B 1.5.44 Schittich, Christian; Munich
B 1.5.47 Werner, Heike; Munich
B 1.5.48 Schröter, V. Carl; Hamburg
B 1.5.49–50 Werner, Heike; Munich
B 1.5.51 Hauer und Boecker; Oelde
B 1.5.52 Werner, Heike; Munich
B 1.5.53–54 Gebr. Kufferath GmbH & Co. KG; Düren
p. 168 Lechner, Dieter; Munich
p. 170–171 Moosbrugger, Bernhard; Zürich
p. 172 Donat, John; London
p. 173 left: Lang, Werner; Munich
p. 173 right: Kirkwood, Ken; Desborough
p. 174 Huthmacher, Werner; Berlin
p. 176 Warchol, Paul; New York
p. 177 Richters, Christian; Münster
p. 178 Helfenstein, Heinrich; Zürich
p. 179 Ortmeyer,Klemens/architekturphoto; Düsseldorf
S. 180–181 Binet, Hélène; London

Glass
B 1.6.1 Gilbert, Dennis/view/artur; Cologne
B 1.6.2 Bednorz, Achim; Cologne
B 1.6.3 Daidalos 66/1997, p. 85
B 1.6.5 Fessy, Georges; Paris
B 1.6.6 Lang, Werner; Munich
B 1.6.7–9 Schittich, Christian et al.: Glass Construction Manual, Munich/Basel, 1998
B 1.6.11 Coyne, Roderick; London
B 1.6.12 Esch, Hans-Georg; Hennef
B 1.6.13 Fessy, Georges; Paris
B 1.6.14 Schittich, Christian; Munich
B 1.6.15 Schittich, Christian et al.: Glass Construction Manual, Munich/Basel, 1998, p. 90
B 1.6.16–17 Herzog, Thomas: Sonderthemen Baukonstruktion. Materialspezifische Technologie und Konstruktion – Gläser, Häute und Membranen, Munich, 1998, p. 11 (unpublished manuscript)
B 1.6.18–20 Schittich, Christian et al.: Glass Construction Manual, Munich/Basel, 1998
B 1.6 21 Bitter, Jan; Berlin
B 1.6 22 Herzog, Thomas: Sonderthemen Baukonstruktion. Materialspezifische Technologie und Konstruktion – Gläser, Häute und Membranen, Munich, 1998, p. 36 (unpublished manuscript)
B 1.6.23 Schittich, Christian et al.: Glass Construction Manual, Munich/Basel, 1998, p. 120
B 1.6.24–25 Kaltenbach, Frank (ed.): Transluzente Materialien, Munich, 2003
B 1.6.26–28 Schittich, Christian et al.: Glass Construction Manual, Munich/Basel, 1998
B 1.6.29 Sundberg, David; New York
p. 192 Young, Nigel; Surrey
p. 193 Hunter, Alastair
p. 194, 195 top: Holzherr, Florian; Munich
p. 195 bottom: Richters, Christian; Münster
p. 196 top: Gilbert, Dennis/ View; London
bottom: Linden, John; Woodland Hills
p. 197 Hempel, Jörg; Aachen
p. 198 top left: Kim Yong Kwan; Seoul
top right, bottom: Hursley, Timothy; Little Rock
p. 199 Kwan, Kim Yong; Seoul
p. 200 Denancé, Michel; Paris
p. 201 Schittich, Christian, Munich
p. 202, 203 top: Sakaguchi, Hiro; Tokio
p. 203 bottom: Wiegelmann. Andrea; Munich
p. 204, 205 van den Bossche, Jocelyne; London
p. 206 Ege, Hans; Waggis
p. 207 Linden, John; Woodland Hills
pp. 208, 209 Gilbert, Dennis/View; London

Plastics
B 1.7.1 Burt, Simon/APEX; Exminster
B 1.7.2 Hansen, Hans/Vitra; Hamburg
B 1.7.3 The MIT Museum, in: Hess, Alan: Googie. fifties coffee shop architecture. San Francisco 1986, p. 50
B 1.7.4–5 Centraal Museum; Utrecht
B 1.7.6 Buckminster Fuller Institute; Los Angeles
B 1.7.7 Otto, Frei; Warmbronn
B 1.7.8 Einzig, Richard/Arcaid; Kingston upon Thames
B 1.7.10 Kandzia, Christian; Stuttgart
B 1.7.12 Lang, Werner; Munich
B 1.7.13 Waki, Tohru/Shokokusha; Tokyo
B 1.7.14–16 Kaltenbach, Frank (ed.): Transluzente Materialien. Munich 2003
B 1.7.18–21 Detail 06/2000, pp. 1048–54
B 1.7.22 Kurth, Ingmar; Frankfurt
p. 218 de Calan, Jean; Paris
p. 219 Tiainen, Jussi; Helsinki
pp. 220, 221 Ruault, Philippe; Nantes
p. 222 Müller-Naumann, Stefan/artur; Cologne
p. 223 Janzer, Wolfram/artur; Cologne
p. 224 Bleda + Rosa; Moncada; Valencia
p. 225 Kaunat, Angelo; Graz
pp. 226, 227 top: Lang, Werner; Munich
pp. 228, 229 Knott, Herbie; London
pp. 230, 231 Skyspan (Europe) GmbH; Rimsting

The glass double facade
B 2.1.1 Braun, Zooey; Stuttgart/artur
B 2.1.2 Lang, Werner; Munich
B 2.1.5 Lang, Werner; Munich
B 2.1.7 Krase, Waltraud; Frankfurt
B 2.1.8 Schenkirz, Richard; Leonberg
B 2.1.11 Graf, Rudi; Munich
B 2.1.15 Bryant, Richard; Kingston upon Thames
B 2.1.18–19 Lang, Werner; Munich
B 2.1.22–23 Lang, Werner; Munich
B 2.1.26 Esch, Hans-Georg; Hennef
B 2.1.27 Schmidt, Jürgen; Cologne
p. 241 top: Bednorz, Achim; Cologne
bottom: Lang, Werner; Munich
p. 242 Malagamba, Duccio; Barcelona
pp. 243, 244, 245 bottom: Halbe, Roland/artur; Cologne
p. 246 Richters, Christian; Münster
p. 247 Müller-Naumann, Stefan; Munich
p. 248 Richter, Ralf; Düsseldorf
p. 249 top: Kandzia, Christian; Esslingen
centre: Richter, Ralf; Düsseldorf
bottom: Schodder, Martin; Stuttgart
pp. 250, 251 Hempel, Jörg; Aachen
p. 252 top: Leistner, Dieter/artur; Cologne
bottom: Riehle, Thomas/artur; Cologne
p. 253 Riehle, Thomas/artur; Cologne
pp. 254, 255 Leistner, Dieter/artur; Cologne
pp. 256, 257 Knauf, Holger; Düsseldorf

Manipulators
B 2.2.1 Hellwig, Jean-Marie/Prouvé-Archiv Peter Sulzer; Gleisweiler
B 2.2.2 Westenberger, Daniel; Munich
B 2.2.3–4 Herzog-Loibl, Verena; Munich
B 2.2.5 Zwerger, Klaus; Vienna
B 2.2.6 Herzog-Loibl, Verena; Munich
B 2.2.7 ISOTEG Final Report, Munich Technical University, Chair of Building Technology, Munich, 2001 (unpublished manuscript)
B 2.2.8 Lang, Werner; Munich
B 2.2.9 Spiluttini, Margherita; Vienna
B 2.2.10 Herzog-Loibl, Verena; Munich
B 2.2.11 Werlemann, Hans; Rotterdam
B 2.2.12 Heinrich, Michael; Munich
B 2.2.13 Gahl, Christian; Berlin
B 2.2.14 Halbe, Roland; Stuttgart
B 2.2.15 Hueber, Eduard; New York
B 2.2.16 Spiluttini, Margherita; Vienna
B 2.2.17 Richters, Christian; Münster
B 2.2.18 Korn, Moritz / artur; Cologne

B 2.2.19 Büttner, Dominic; Zürich
B 2.2.20 Kinold, Klaus; Munich
B 2.2.21 Shinkenchiku-sha; Tokyo
B 2.2.23 Asakawa, Satoshi; Tokyo
B 2.2.24 Beyer, Constantin; Weimar
B 2.2.25 Feiner, Ralph; Malaus
B 2.2.26 Wörndl, Hans-Peter; Vienna
B 2.2.27 Müller, Ritchie; Munich
B 2.2.28 Westenberger, Daniel; Munich
B 2.2.29 Gabriel, Andreas; Munich
B 2.2.30 Furer, René; Benglen
B 2.2.31 Lenzen, Thomas; Munich
B 2.2.32 Carter, Earl; St. Kilda
pp. 266, 267 right: Suzuki, Hisao; Barcelona
p. 267 top left: Fessy, Georges; Paris
p. 268 Beyeler, Therese; Bern
p. 269 Ohashi, Tomio; Tokyo
p. 270 Voth-Amslinger, Ingrid; Munich
p. 271 Heinrich, Michael; Munich
p. 272 Shinkenshiku-sha; Tokyo
p. 273 Hirai, Hiroyuki; Tokyo
p. 274 Richters, Christian; Münster
p. 275 top: Strauß, Dietmar; Beisigheim
p. 276 Roth, Lukas; Köln
p. 277 Hummel, Kees; Amsterdam
p. 278 Hueber, Eduard; New York
p. 279 Kaltenbach, Frank; Munich
p. 280 Brandl, Sonja; Munich
p. 281 top: Bitter, Jan; Berlin
bottom: Kisling, Annette; Berlin
p. 282 bottom: Ott, Thomas; Mühltal
p. 284, 285 Wett, Günter Richard; Innsbruck

Solar energy

B 2.3.1 Herzog-Loibl, Verena; Munich
B 2.3.4–5 Herzog-Loibl, Verena; Munich
B 2.3.6 Köster, Arthur/Stiftung Archiv der Akademie der Künste; Berlin
B 2.3.7 Krier, Robert
B 2.3.8–9 TWD. Eigenschaften und Funktionen. Info-Mappe 2 des Fachverbands TWD. Gundelfingen 2000, p. 5
B 2.3.10–11 Schittich, Christian (ed.): Building Skins, Munich/Basel, 2001, p. 51
B 2.3.12 Viessmannwerke; Allendorf
B 2.3.13 Leistner, Dieter/artur, Cologne
B 2.3.14–15 Viessmannwerke; Allendorf
B 2.3.16 Schott Glas; Mainz
B 2.3.19 Schittich, Christian (ed.): Building Skins, Munich/Basel, 2001, p. 54
B 2.3.20 Schneider, Astrid; Berlin
B 2.3.21 Helle, Jochen; Dortmund
p. 294 Spiluttini, Margherita; Vienna
p. 295 Küng, Toni; Herisau
p. 296 Leistner, Dieter/artur; Cologne
p. 297 top, centre: Comtesse, Frederic; Zürich
p. 298 Müller-Naumann, Stefan; Munich
p. 299 Walti, Ruedi; Basel
p. 300 Willebrand, Jens; Cologne
p. 310 Brändli, Nick; Zürich
pp. 302, 303 Halbe, Roland/artur; Cologne
p. 304 bottom: Brunner, Arnold; Freiburg
p. 305 Hofer, Robert; Sion
p. 306 Miralles, Jordi; Barcelona
p. 307 top: Richters, Christian; Münster
bottom: Entwicklungsgesellschaft Akademie Mont-Cenis mbH; Herne
p. 308 Kirsch, Guido; Freiburg
pp. 310, 311 Richters, Christian; Münster

Whole page plates

p. 8 Moldavian monastery, Sucevita (RO), 16th century
Photo: Cremers, Stefan; Karlsruhe
🕮 Vasiliu and Mendrea (ed.): Moldauklöster, 14.-16. Jahrhundert, Munich, 1998

p. 16 Schocken department store, Stuttgart (D), 1928
Architect: Erich Mendelsohn
Drawing: Erich Mendelsohn
🕮 Lampugnani, Vittorio Magnago: Architektur unseres Jahrhunderts in Zeichnungen. Utopie und Realität, Stuttgart, 1982

p. 26 Studio house, Munich (D), 1993
Architect: Thomas Herzog
Photo: Bonfig, Peter; Munich
🕮 DBZ 11/1994

p. 38 Private house, Paderborn (D), 1995
Photo: Thomas Herzog
Foto: Leistner, Dieter/artur; Cologne
🕮 A+U 06/1999

p. 46 Eames house, Pacific Palisades (USA), 1949
Architects: Charles and Ray Eames
Photo: Neuhart, Andrew; El Segundo
🕮 Steele, James: Eames House. London 1994
Neuhart, Marilyn: Eames House. Stuttgart 1994

p. 52 Swiss Re headquarters, London (GB), 2003
Architects: Foster and Partners
Photo: Kaltenbach, Frank; Munich
🕮 domus 865, 2003

p. 60 Wrapped Reichstag, Berlin (D), 1995
Artists: Christo & Jeanne-Claude
Photo: Wimmershoff, Heiner; Aachen
🕮 Baal-Teshura, Jacob: Christo & Jeanne-Claude. Cologne1995

p. 62 German Pavilion, Barcelona (E), 1929/1986
Architect: Ludwig Mies van der Rohe
Photo: Bonjoch, Eloi; Barcelona
🕮 l'architecture d'aujourd'hui 09/1998
de Solà-Morales, Ignasi, et al.: Mies van der Rohe. Barcelona Pavilion. GG Barcelona, 5th ed., 2000

p. 82 Apartment block, rue des Meaux, Paris (F), 1991
Architects: Renzo Piano Building Workshop
🕮 Buchanan, Peter: Renzo Piano Building Workshop. vol. 1, London 1993

p. 100 Art and Architecture Building, Yale University New Haven (USA) 1964
Architect: Paul Rudolph
Photo: Herzog, Thomas; Munich
🕮 Stoller, Ezra: The Yale Art and Architecture Building. Princeton 1999

p. 124 Komyo-Ji Pure Land Temple, Saijo (J) 2000
Architect: Tadao Ando
Photo: Shinkenchiku-sha; Tokyo
🕮 Casabella 689, 2001
THE PLAN 004/2003

p. 154 Distribution centre, Chippenham (GB), 1982
Architects: Nicholas Grimshaw & Partners
Photo: Reid, Jo & Peck, John; Newport
🕮 Colin, Amery: Architecture, Industry and Innovation. The Early Work of Grimshaw & Partners. London 1995

p. 182 Bauhaus, Dessau (D) 1926/1976
Architect: Walter Gropius
Photo: Gilbert, Dennis/view/artur; Cologne
🕮 Sharp, Dennis: Bauhaus Dessau. Walter Gropius. London 1993
Whitford, Frank (ed.): Das Bauhaus. Stuttgart 1993

p. 210 Eden Project, St. Austell (GB) 2001
Architects: Nicholas Grimshaw & Partners
Photo: Burt, Simon/APEX; Exminster
🕮 l'architecture d'aujourd'hui 07–08/2001
Architectural Record 01/2002

p. 232 Posttower, Bonn (D) 2003
Architects: Murphy/Jahn
Photo: Braun, Zooey; Stuttgart/artur
🕮 Architectural Review 08/2003
Architecture today 09/2003

p. 258 Square Mozart, Paris (F) 1954
Architect: Jean Prouvé
Photo: Hellwig, Jean-Marie/Prouvé-Archiv Peter Sulzer; Gleisweiler
🕮 Huber, Benedikt (ed.) et al.: Jean Prouvé. London 1971
Sulzer, Peter: Jean Prouvé, Oeuvre complète, vol. 3: 1944–1954. Basel/Berlin/Boston 2004

p. 286 Residential complex, Munich (D), 1982
Architects: Thomas Herzog and Bernhard Schilling
Photo: Herzog-Loibl, Verena; Munich
🕮 Werk, Bauen + Wohnen 05/1983
Flagge, Ingeborg (ed.) et al.: Thomas Herzog Architektur + Technologie. Munich/London/New York 2001

The authors and publishers would like to express their thanks to the following persons, manufacturers and companies who provided information and/or artwork for this book:

Barbara Finke, Berlin (D)
Böhmer Natursteinbau GmbH, Leutenbach (D)
Cordelia Denks, Munich (D)
Dach + Wand Wolf GmbH & Co.KG, Dornbirn (A)
Delzer Kybernetik GmbH, Lörrach (D)
F. Brüderlin Söhne GmbH, Schopfheim (D)
Götz GmbH, Würzburg (D)
Halfen GmbH & Co. KG, Langenfeld (D)
Hightex Group, Rimsting (D)
Jörg Eschwey, ESO Chile (RCH)
Josef Gartner GmbH, Gundelfingen (D)
Lavis Stahlbau GmbH, Offenbach (D)
Magnus Müller GmbH, Butzbach (D)
Metallbau A. Sauritschnig GmbH, St Veit/Glan (A)
MEW Manfroni Engineering Workshop, Bologna (I)
Moeding Keramikfassaden GmbH, Marklkofen (D)
nbk Keramik GmbH & Co., Emmerich (D)
NMP Naturstein Montage GmbH & Co. KG, Vienna (A)
Serge Lochu, Cosylva Paris-Ouest (F)
Stahlbau Wörsching GmbH & Co. KG, Starnberg (D)
Wortmann Projektbau GmbH, Wenden (D)

Index of names

Index